Advances in Experimental Medicine and Biology

Volume 996

More information about this series at http://www.springer.com/series/5584

Shamim I. Ahmad
Editor

Ultraviolet Light in Human Health, Diseases and Environment

Springer

Editor
Shamim I. Ahmad
Faculty of Science and Technology
Nottingham Trent University
Nottingham, UK

ISSN 0065-2598 ISSN 2214-8019 (electronic)
Advances in Experimental Medicine and Biology
ISBN 978-3-319-56016-8 ISBN 978-3-319-56017-5 (eBook)
https://doi.org/10.1007/978-3-319-56017-5

Library of Congress Control Number: 2017954385

Printed on acid-free paper

This Springer imprint is published by Springer Nature
The registered company is Springer International Publishing AG
The registered company address is: Gewerbestrasse 11, 6330 Cham, Switzerland

The editor dedicates this book to his late father, Abdul Nasir, and mother, Anjuman Ara, who played very important roles to bring him to this stage of academic achievements with their esteemed love, sound care, and sacrifice. Dedication also goes to his wife Riasat Jan for her patience and persistent encouragement to produce this book and also to the Ahmad family, Farhin, Mahrin, Tamsin, Alisha, and Arsalan, especially the latter two for providing him great pleasure with their innocent interruptions leading to his energy revitalization. Dedication also goes to skin cancer patients who innocently, due to lack of knowledge, may have overexposed themselves under the sun or man-made UV lamps.

Preface

Ultraviolet (UV) light is sometimes referred to as UV radiation. In this section, I will be using UV light which would mean the same. There are two major sources of this light – our sun and the man-made UV lamps. The UV light from the sun ranges from 180 to 400 nm. This light is classified in three main types: UVA (320–400 nm), UVB (290–320 nm), and UVC (180–290 nm). Recently UVA and UVB bands are subclassified into narrowband UVB (311–313 nm), UVA2 (320–340 nm), and UVA1 (340–400 nm). Also a variety of UV lamps of different wavelengths and different capacities and filters are available for various applications.

Solar UV light has been exerting major impacts on almost all forms of life including its useful roles in photosynthesis and the production of vitamin D. On the other hand, it plays damaging roles on human health including promoting various forms of skin diseases, the most devastating being skin cancers. The UVC light in laboratory conditions has been shown to be the most damaging to biological systems especially to DNA, leading to various forms of cancers. Fortunately most of it is absorbed by the stratospheric ozone layer. Some UVB however can reach us, and this appears to play roles in inducing skin cancers including melanoma, basal cell carcinoma, and squamous cell carcinoma. Out of these three types, melanoma is the most dangerous, and if not attended soon, its metastasis can occur fairly rapidly in the body. The gradual depletion of the atmospheric ozone layer during the past few years, increasing the incidence of solar UVC radiation on the Earth's surface, is one of the environmental concerns, because of the harmful effects of this radiation in all forms of life.

The indirect effect of UV light on human health is when it photoactivates biological or nonbiological compounds and the reaction products are reactive oxygen species (ROS). A large number of photolyzing chemicals have been identified to react with the UV light generating ROS. These chemically generated highly reactive species are hydrogen peroxide, superoxide, hydroxyl radical, and singlet oxygen. In vivo, ROS are formed as a natural by-product of the normal metabolism of oxygen and have important roles in cell signaling and homeostasis. ROS have also been implicated in a variety of inflammatory responses including cardiovascular disease, hearing impairment via cochlear damage, and mediation of apoptosis and ischemic injury. ROS also play important positive roles in driving certain biochemical reactions as well eliminating invading microbial pathogens.

For readers' easy reading, the contents of the chapter have been sectionalized below.

Part I: Chapter 1 includes the introduction, historical aspects, and sources of the UV light and their applications.

Part II: Chapters 2 and 3 cover the photoactive products of biological and nonbiological compounds, impacts of UVA and UVB on a biological system, and what we have learned from the UV studies on microorganisms. In Chap. 4 the author presents a detailed account of the impacts of UVA and UVB light on living organisms and human health and diseases.

Part III: This section covers a major area of UV studies combining Chaps. 5, 6, 7, 8, 9, 10, 11, 12 and 13 describing a number of human diseases induced by UV light. In Chap. 5 a detailed analysis of various forms of skin cancers is covered, and in Chap. 6, extreme UV sensitivity of children suffering from xeroderma pigmentosum and the precautionary measures required to reduce its effect are highlighted. Due to its prominent phenotype, coupled with the chronic and incurable nature, vitiligo has a significant negative impact on the quality of life of patients suffering from it. This has been critically highlighted in Chap. 7; also discussed is the supremacy of UV light in vitiligo as an established therapeutic option over and above several treatment modalities instituted over the years, with varying efficacy. Polymorphous light eruption, the commonest immuno-mediated photodermatosis, occurs after solar or artificial UV light exposure, and because the reaction mostly appears on exposed areas, the exclusive association of this light in inducing this kind of skin problem is emphasized. Apart from shedding light on the mechanism involved in its development, in Chap. 8, broad-spectrum sunscreens and antioxidants, PUVA, and narrowband UVB have been recommended to prevent the disease. A detailed description of UV damage mechanism to skin cells and the defense mechanism is depicted in Chaps. 9, 10 and 11. Generalized photobiology in dermatology and the roles of UVA1 in dermatological effects and diseases such as sunburn, immunosuppression, skin aging, carcinogenesis, and photoprotection are comprehensively described. Also highlighted is that when used under controlled conditions, UV radiation can be helpful in the diagnosis and treatment of many skin conditions.

The differences seen between the pigmented- and non-pigmented-type skins when exposed to UV light can be read in Chap. 12. Psychoneuroendocrine immunology is a new and novel field of study that investigates the link between bidirectional communications among the nervous system, the endocrine system, and the immune system and the correlations of this cross-talk with physical health. This unusual field of UV studies has been addressed in Chap. 13.

Part IV: This section provides the information we reap from the UV light. Vitamin D in recent years drew considerable attention from researchers and medics, playing important roles in a number of biochemical reactions; in Chap. 14, the author, a leading player in vitamin D synthesis, describes the historical aspect as well as the detailed mechanism of the synthesis of vitamin D when a human body is exposed to UVB light. As mentioned above, UV is also an inducer of skin cancer, thus raising the question of how much sun exposure should a human have to synthesize the required amount of vitamin D without risking cancer development. The biochemical aspect of vitamin D in health and diseases has also been adequately described at different angles in Chap. 15. Further importance of vitamin D in human health has been made clear in Chaps. 16, 17, 18, 19, 20, 21, and 22 where it has been shown that its deficiency can lead to rheumatoid arthritis (Chap. 16) and asthma and allergy (Chap. 17). Vitamin D levels and its metabolism also play roles in the development of atherosclerosis leading to cardiovascular disease (Chap. 17); also in this chapter, the authors discuss both normal and disordered vitamin D metabolism and major clinical trials regarding vitamin D levels and effects of its supplementation. Although the importance of vitamin D in type 2 diabetes remains a controversial issue, the authors in Chap. 19 have discussed this issue fairly critically. Some of the remaining diseases triggered by vitamin D deficiency have been pooled in Chap. 20 in the form of metabolic syndrome; these include obesity, dyslipidemia, and cardiovascular diseases including myocardial infarction, coronary artery disease, and stroke. Also highlighted is the role of vitamin D in skeletal growth and maintenance. Furthermore the ubiquitous expression of vitamin D receptor in body cells such as immune, vascular, and myocardial cells, pancreatic beta cells, neurons, and osteoblasts points to an involvement of vitamin D-mediated effects on metabolic syndrome. Chapter 21 focuses on how important genome stability is for human health and, when genomes are damaged by UV exposure, what overall effects can be seen on health. In Chap. 22 the author presents a comprehensive treatise of the current knowledge of vitamin D effects from a cardiovascular health perspective and roles of vitamin D in relation to cardiovascular diseases such as ischemic heart disease and stroke; the traditional cardiovascular risk factors such as hypertension, abnormal blood lipids, and obesity; and the emerging risk factors such as hyperparathyroidism, microalbuminuria, chronic obstructive pulmonary diseases, and nonalcoholic fatty liver disease due to vitamin D deficiency.

Part V: Chapters 23, 24, 25 and 26 have been dedicated to understand the roles of UV light in sterilization and their impacts on human health. Out of several different ways that food and common-use materials are sterilized, it is suggested that exposure to UV light is one cheap, clean, and efficient method to get rid of unwanted pathogenic and nonpathogenic contaminations. Some microbial species have evolved their mechanisms to produce biofilms to protect themselves from the killing effect of UV light. UV treatment of water, besides eliminating the pathogenic microbes, can have the side effects of photoactivation of certain organic compounds leading to the production of mutagenic/genotoxic by-products including certain nitrogen-containing aromatic compounds; these are formed by the photolytic products of nitrate with natural organic matter. Hence, while treating the water by UV light, precaution must be taken to filter out such by-products before water is supplied for consumption; this has been addressed in Chap. 24. Chapter 26 comprehensively addresses that the presence of bacteria, viruses, and other pathogens in municipal wastewater can adversely affect the environment, human health, and economic activity. One way to mitigate these effects is a terminal disinfection step using UV light. The advantage of this method of disinfection, when compared to traditional chlorine disinfection, is that no chlorinated by-products and no chemical residues are produced by the former method. Chapter 25 emphasizes that the UV disinfection method holds promise for reducing the level of contamination in operating rooms and thereby lowering the risk of infection to patients. In Chap. 23 the production of biofilm and its use in adaptation against UV radiation are presented, and also the application of UV light to monitor and destroy biofilms in man-made surfaces is addressed.

Part VI: In this section, various methods of UV phototherapy of different skin diseases, which have been in use for many years, are highlighted; for this it is either that UV light is employed on its own or in combination with certain photolyzing compounds. Chapter 27 addresses atopic dermatitis which is one of the most common chronic inflammatory skin diseases that are treated by a variety of methods including exposure of the affected areas to broadband UVB (290–320 nm), narrowband UVB (311–313 nm), UVA1 therapy (340–400 nm), UVA therapy plus 8-methoxypsoralen (PUVA), 308 nm excimer laser (EL), and full-spectrum light (FSL). Currently, narrowband UVB phototherapy is the most employed treatment due to its availability, security, ease of administration, and efficacy. Chapter 28 emphasizes that phototherapy remains the only therapeutic option for certain patient groups where modification of the systemic immune reactions is contraindicated, such as HIV, internal malignancy, or pregnancy, and for this the UVB treatment is highly cost-effective. Chapter 32 addresses the safety and efficacy of phototherapy in the management of eczema treatment, and in Chap. 29 it has been claimed that the UV light can be used to decontaminate blood. This process was extensively used in the 1940s and 1950s to treat blood for diseases such as septicemia, pneumonia, tuberculosis, arthritis, asthma, and even poliomyelitis, but then for some reason, it could not prevail. The author suggests that ultraviolet blood irradiation is a valuable process to eliminate any pathological contamination and hence must be revitalized. Chapter 30 addresses the sunscreen commonly used to prevent or reduce the damaging effects of UV light on skin especially during holiday period at the seaside places as well to those working in the open field. A number of sunscreens with variable composition and strength (known as SPF or sun protection factor) are available in the market, and their effective use and possible abuse is included in the discussion. In Chap. 31, the author has presented a novel weed *Parthenium hysterophorus* found to grow ubiquitously in various continents with warmer climates; it has been found to induce contact dermatitis and possible mechanisms of photosensitization. The influence of UV light on the pattern of parthenium dermatitis is also discussed.

Parts VII and VIII: These sections address the use and abuse of tanning salons (Chap. 33) to tan human bodies and the importance of dosimeters when UV lamps are employed for skin tanning, for phototherapy, or for sterilization and decontamination (Chap. 34).

It is hoped that the materials presented in this book will present immense benefit and will stimulate both novice and expert researchers in the field with excellent overviews of the current status of research and pointers to future research achievements. Clinicians, medical general practitioners, technicians, and staff working in UV-related industries and especially those working in tanning salons should benefit from the information presented in the safe handling of UV light. Also the insight obtained should prove valuable for further understanding at the molecular level of damages caused by UV light and allows the development of new biomarkers, novel diagnostic tools, and highly therapeutic drugs and preventive measures.

Nottingham, UK Shamim I. Ahmad
3-2-2017

Acknowledgment

The editor cordially acknowledges various authors for their contribution of chapters with in-depth knowledge, comprehensive inclusion, and proficient presentations. Without their input, it would not have been possible to produce such an important book. The editor also acknowledges the hard work, friendly interaction, and patience of the Springer Publication staff, especially Marleen Moore, Melissa Morton, and Sara Germans for their efficient handling of this publication. Acknowledgment also goes to the staff of the IT services of Nottingham Trent University for providing technical help in solving computer problems especially with new packages regularly pushed in.

Contents

About the Editor

Shamim I. Ahmad, after obtaining his master's degree in botany from Patna University, Bihar, India, and his PhD in molecular genetics from Leicester University, England, joined Nottingham Polytechnic as grade 1 lecturer and was subsequently promoted to the senior lecturer post. Nottingham Polytechnic subsequently became Nottingham Trent University where after serving for about 35 years, he took early retirement yet still serving as a part-time senior lecturer. He is now spending much of his time producing/writing medical books. For more than three decades, he researched on different areas of molecular biology/genetics including thymineless death in bacteria, genetic control of nucleotide catabolism, development of an anti-AIDS drug, control of microbial infection of burns, phages of thermophilic bacteria, and microbial flora of Chernobyl after the accident at the nuclear power station. But his main interest which started about 30 years ago is DNA damage and repair specifically by near-ultraviolet light especially through the photolysis of biological compounds and production of reactive oxygen species and their implications on human health including skin cancer. He is also investigating near-ultraviolet photolysis of nonbiological compounds such as 8-methoxypsoralen and mitomycin C and their importance in psoriasis treatment and in Fanconi anemia. By collaborating with the University of Osaka, Japan, in his latest research publication, he and his colleagues were able to show that a number of naturally occurring enzymes were able to scavenge the reactive oxygen species.

In 2003 he received the prestigious "Asian Jewel Award" in Britain for "Excellence in Education." To fulfill his longtime ambition to produce medical books, he took early retirement in 2007 and since then has been able to publish, by Landes Bioscience/Springer Publication, *Molecular Mechanisms of Fanconi Anemia*; *Molecular Mechanisms of Xeroderma Pigmentosum*; *Molecular Mechanisms of Cockayne Syndrome*; *Molecular Mechanisms of Ataxia Telangiectasia*; *Diseases of DNA Repair*; *Neurodegenerative Diseases*; *Diabetes: An Old Disease, a New Insight*; *Obesity: A Practical Guide*; and *The Thyroid Disorder: Basic Science and Clinical Practice*. Also he is a coauthor of the book *Diabetes: A Comprehensive Treatise for Patients and Caregivers*. Recently Taylor &

Francis/CRC Press has published a book on *Reactive Oxygen Species in Biology and Human Health* and has now work going on in a book on *Aging: Exploring a Complex Phenomenon* to be published by the same publisher to come out by November 2017.

A Tribute to Aziz Sancar: The Nobel Prize Winner in Chemistry, 2015

It gives me immense pleasure to write about one of the most renowned scientists in the field of molecular biology. As this book covers a variety of chapters on ultraviolet (UV) light, including its major role in determining the molecular aspects of DNA damage, repair, and mutagenesis, and as Professor Sancar has extensively used the UV light in his investigations, I consider it appropriate to present a tribute to him here. Although studies in this area of molecular biology started with work on *Escherichia coli*, the knowledge gained on this bacterium has been successfully applied to higher eukaryotes, including man.

Dr. Aziz Sancar, after receiving his medical degree and practicing for about 2 years, in Turkey, in 1971 won a NATO fellowship to fund his PhD research and selected the USA as his destination. He initially started his studies at Johns Hopkins University, moving after a period to the University of Texas at Dallas to work under the supervision of Dr. Claud S. Rupert.

In this laboratory, Dr. Sancar started working on photolyase, an enzyme discovered in 1958 by Dr. Rupert. His isolation of an *E. coli* mutant deficient in photolyase synthesis and subsequent cloning of the photolyase gene earned him the PhD degree awarded in 1977. Dr. Sancar subsequently moved to Yale University and used his molecular biology experience to clone the *uvrA*, *uvrB*, and *uvrC* genes of *E. coli*, responsible for repairing DNA damaged by UV light. The first major breakthrough came when he developed a unique technique (maxicell) to identify any plasmid-encoded protein. The report of this ingenious method has provided his most cited research paper until now. He employed the maxicell technique to purify UVR ABC proteins and named them "ABC exinuclease." These proteins participate in the repair of DNA damaged by the UV light, a process called nucleotide excision repair. This discovery created excitement in the scientific community when in 1982 his results were presented in France at an International Meeting on Recombination and Repair.

Despite his significant achievements in research, in 1981 his 50 applications to find a faculty position in the USA were unsuccessful, until Dr. Mary Ellen Jones, chair of the Department of Biochemistry at the University of North Carolina (UNC) at Chapel Hill, offered him a position. He moved to UNC in 1982. Since then he has worked ceaselessly in the field of molecular biology, publishing no less than 358 research papers (ref. PubMed June 2016) and winning the most prestigious award in science – The Nobel Prize in Chemistry – in December 2015, sharing it with two of his academic friends, Dr. T. Lindahl of Cancer Research UK, Hertfordshire, England, and Dr. P. Modrich of Duke University Medical Center, Durham, NC, USA.

After making significant progress in the study of photoreactivation and nucleotide excision repair, Dr. Sancar moved to study another phenomenon of excision repair termed transcription-coupled repair (TCR). This was discovered in 1985–1986 by another renowned biological scientist, Professor Phil Hanawalt of Stanford University, California. Dr. Sancar elucidated the mechanism of TCR. He found that when RNA polymerase arrives at the damage in DNA, it is stalled and this is the rate-limiting step in NER. Dr. Sancar discovered the transcriptional repair coupling factor (TRCF) protein which recognizes the stalled RNA polymerase, displaces it from the damaged site, and facilitates the assembly of excision nucleases to accelerate the repair rate. This seminal study was published in *Science* in 1993 and played an important role in the selection of DNA repair enzyme as the "Molecule of the Year" by *Science* magazine in 1994.

Based on the wealth of knowledge he accumulated on the mechanism of the NER system in *E. coli*, Dr. Sancar subsequently moved to working on NER-induced DNA repair in humans. The motivation came from the work published by Dr. James Cleaver, who in a 1968 publication in *Nature* showed that skin biopsy cells from patients suffering from xeroderma pigmentosum were deficient in NER (details may be found in "Historical Aspects of Xeroderma Pigmentosum and Nucleotide Excision Repair" by James E. Cleaver in *Molecular Mechanisms of Xeroderma Pigmentosum*, Shamim I. Ahmad and Fumio Hanaoka (eds.), 2008, page 1–9, Landes Bioscience Publication). Dr. Sancar's studies showed that this system exists in humans as well as in *E. coli*, with certain important differences, in that there are seven genes, XPA to XPG, responsible for the removal of UV-induced photoproducts compared to three in *E. coli*. Furthermore, unlike in *E. coli* where there are only 3 NER proteins which remove DNA damage in 12–13-nucleotide-long oligomers, in humans there are sixteen proteins in six repair factors, and all these are necessary for making the dual incisions in DNA to remove the damage in 26–27-nucleotide-long oligomers. Other less important but remarkable differences were also observed.

Dr. Sancar subsequently moved to working on the cryptochrome (which is revolutionarily related to the bacterial photolyase involved in the circadian clock in humans, an innate timekeeping molecular mechanism that maintains daily rhythmicity in biochemical, physiological, and behavioral function independent of external input).

I (Shamim I. Ahmad), after spending a number of years studying the genetics and physiology of enzymes involved in the regulation of nucleotide catabolism and thymine metabolism in *E. coli* and receiving my PhD from Leicester University, England, in 1973, got a lectureship at Trent Polytechnic and was happy to accept this instead of a postdoctoral fellowship at NIH in Bethesda, USA. My active research restarted when I was invited as a post-doctoral fellow by Professor Abe Eisenstark (director of biological science) of the University of Missouri, Columbia. I went there in 1978 and was assigned to determine the mechanism of synergistic action of UVA light plus hydrogen peroxide. My 9 months' stay at this laboratory culminated in showing that the superoxide anion is produced in this process, and I also successfully isolated and partially analyzed an *E. coli* mutant hyper-resistant to UVC light. This was the point in my academic career when I started my deeper interest in DNA damage and repair and reactive oxygen species in biology.

Back at Trent Polytechnic, I managed to isolate another mutant of *E. coli* hyper-resistant to 8-methoxypsoralen and UVA (PUVA). A PAGE study identified a heavy spot of 55-kDa protein but I was unable to progress this time.

It was at this time that I wrote a letter to Professor Aziz Sancar sharing my results on PUVA-resistant *E. coli* and asking his help in identifying this 55-kDa protein. He agreed and I sent him the mutant strain. Then a letter arrived on September 14, 1987, which reads that Dr. Van Houten (an able scientist in his laboratory) tried to carry out the protein purification but the majority of this was removed from the cell-free extract by high-speed centrifugation, suggesting that the 55-kDa protein may be a membrane protein. From the small amount of protein in the soluble fraction, we managed to purify it but could not carry out the nicking or footprinting activity due to a high background of nuclease activity precluding any conclusion. Subsequent studies, however, identified this protein as malate dehydrogenase, suggesting the role of this enzyme (as well as a few others) in protecting against PUVA-induced DNA damage (Ahmad et al., J. Photochem Photobiol B, 2012, 116, 30–36)

In 2015, when I heard the news that Dr. Sancar has received the Nobel Prize, in the congratulation letter, I reminded him of the collaboration we had in 1987. I was delighted to receive a response from him saying that "I remember you very well." As I had signed a contract with Springer to produce a book on UV light, I decided to put a tribute to Dr. Sancar in it. I am thankful to him for sending me his biography and the text of the Nobel lecture with his permission to use them.

I wish to end this tribute with a warm congratulation to Dr. Sancar for receiving the honorable supreme "Nobel Prize Award" and, as he is continuing his research, a warm wish for him to receive further awards in the coming years.

Acknowledgment I am cordially indebted to my senior scientists and teachers, under whose kind umbrellas I have been trained and carried out my research in the field of microbial genetics; DNA damage, repair, and mutagenesis; and reactive oxygen species. They are the late Professor Robert Pritchard of Leicester University, Leicester, UK; Professor Abe Eisenstark of the University of Missouri, Columbia, MO, USA; and Professor Fumio Hanaoka of the University of Osaka (now at Tokyo), Japan. Also, thanks go to Drs. Phil Hanawalt, James Cleaver, and Kyoji Tanaka who have contributed valuable chapters to our books published by Landes Bioscience, Texas, USA, on molecular mechanisms of xeroderma pigmentosum, on Cockayne syndrome, and on Fanconi anemia (FA). Gratitude for their chapter contributions in those books also goes to Drs. G. Pagano, A. Sarasin, G. Spivak, K. Sugasawa, and J. A. Tainer and the co-editor of FA, S. Kirk.

Shamim I. Ahmad

Part I

Introduction, Historical Aspects and Sources of UV Light

History of UV Lamps, Types, and Their Applications

1

Shamim I. Ahmad, Luisa Christensen, and Elma Baron

Abstract

The use of ultraviolet (UV) light, for the treatment of skin conditions, dates back to the early 1900s. It is well known that sunlight can be of therapeutic value, but it can also lead to deleterious effects such as burning and carcinogenesis. Extensive research has expanded our understanding of UV radiation and its effects in human systems and has led to the development of man-made UV sources that are more precise, safer, and more effective for the treatment of wide variety of dermatologic conditions.

Keywords

UV light • UV radiation • UV lamps • Skin disease • Phototherapy

1.1 Introduction

Ultraviolet (UV) light is an electromagnetic radiation varying in its wavelength from 100–400 nm. At one side of this band is visible light which is longer than 400 nm starting with blue light and on the other is x-rays which is of shorter length than 100 nm. Two major sources of UV lights are: the sun and the man-made UV lamps. In earlier days the solar UV light, ranging from 180–400 nm, was divided in three basic types: UVA, 320–400 nm, UVB 290–320 nm and UVC 180–290 nm, but now is classified into narrowband UVB of 311–313 nm, man-made Excimer laser of 380 nm, UVA2 of 320–340 nm and UVA 340–400 nm. Ultraviolet light has played a major role in biological research and according to (PubMed, November 2016) 82,617 research papers have been published involving UV light.

S.I. Ahmad (✉)
Faculty of Science and Technology, Nottingham Trent University, Clifton Lane, Nottingham NG11 8NS, UK
e-mail: shamim.ahmad@ntu.ac.uk

L. Christensen • E. Baron
Department of Dermatology, UH Hospitals Cleveland Medical Center/Case Western Reserve University, Cleveland, OH 44106, USA

S. Ahmad (ed.), *Ultraviolet Light in Human Health, Diseases and Environment*, Advances in Experimental Medicine and Biology 996, https://doi.org/10.1007/978-3-319-56017-5_1

1.2 The Discovery of UV Light

UV light was discovered in 1801 by German Physicist, Johann Wilhelm Ritter. The discovery was based on his observation that there exist invisible rays of light beyond the violet end of the visible spectra, capable of darkening silver chloride-soaked paper more rapidly than the violet light itself, termed "Oxidizing rays". Soon after its discovery, its name was changed to "chemical rays" but finally these names were dropped in favor of ultraviolet rays. Then, in 1878 the effect of short-wavelength UV light on bacteria was discovered which was used in sterilization. The earliest record (PubMed, June 2016) of the role of UV light on a biological system is of 1903, when this light was used to treat Lupus in humans [14]. Then, in 1922 the importance of UV light on living organisms was shown on Drosophila; a detailed monograph can be seen published in 1928, implicating the importance and application of UV light in therapeutics [40, 46]. Although the exact type of UV light used is unknown, it is likely that short-wavelength UVC light was used in the experiments.

In earlier days most biological experiments were carried out using germicidal UV lamp emitting UVC band of light. Later, various filters were developed stopping unwanted bands of UV light to go through, which led to the production of various types of UV lamps including those emitting only UVB (TL01) and UVA also known as black light or Near UV Lamps or Philips TL12 lamp [6, 34].

While the research on biomolecules and on living organisms was taking place, the methods of making UV filters, transmitting UVA light (and possibly some UVB), and absorbing visual light were developed [58]. These developments allowed studies on the properties of chemicals such as monoalkyl substituted benzene, absorbing UVA light, and the development of hyperconjugation and Baker-Nathan effects [41]. Analysis was also carried out on other photosensitive biochemicals and proteins such as glyceraldehayde-3-phosphate dehydrogenase [55] and in 1989 Hoeter and Eisenstark showed that *E .coli* mutant lacking manganese or iron superoxide dismutase (Mn-SOD or Fe-SOD respectively) was causing 9-fold increase in mutation frequency employing the role of superoxide ($O_2^{\cdot-}$) in causing the damage to DNA.

1.3 The Types of UV Light

There are two major sources of Ultraviolet (UV) light – our sun and the man-made UV lamps. The UV light which comes from the sun ranges from 180–400 nm. In the earlier days, this range of UV light was classified in three types: UVA (also called near UV or black-light) ranging 320–400 nm, UVB ranging 290–320 nm and UVC ranging 180–290 nm. At a later stage, this classification was further sub divided into narrowband UVB of 311–313 nm, man-made Excimer laser of 380 nm, UVA2 of 320–340 nm and UVA 340–400 nm. Beyond 400 nm is blue light [56]. It is interesting to note that these synonymous words have been used: UV light, UV radiation and UV rays, which mean the same, but for consistency in this book, we wish to keep the word "UV radiation or sometimes UV light as needed."

1.4 Man-Made UV Generating Lamps

1.4.1 Arc Lamp

Arc lamps, also known as gas discharge lamps, are composed of two electrodes with plasma (the arc) between them; these are sealed within a transparent envelope with gas (e.g. mercury or xenon). When high voltage is applied, the electrons become excited and the light is emitted when they return to the ground state [39]. Different gases and pressures lead to different spectral output. Xenon lamps are used as solar simulators. The arc lamp was the first effective artificial light source and is currently used in photodynamic therapy for treatment of conditions such as acne and actinic keratosis [17].

1.4.2 Cold-Light UV-A1

A filter is installed in this lamp to eliminate the light of the 530 nm wavelength and it dissipates the excessive heat load generated by the UV-A1 generator. This type of UV light has been found to be more effective than UV-A alone or conventional UV-A1 at clearing lesions and reducing duration of atopic dermatitis (AD) flares [63].

1.4.3 Excimer

This lamp consists of a coherent single-wavelength light source of UV-B light at 308 nm. The excimer is a complex of excited gases which upon decomposition gives off excess energy in the form of ultraviolet radiation (UVR). The excimer exists both as a lamp and a laser. The lamp is a polychromatic (wavelengths 306–310 nm), non-targeted (incoherent) light used to treat a range of body surface areas. The laser, on the other hand, is a monochromatic (308 nm), intermittent (pulsing) light, which can deliver targeted therapy (coherent). It is commonly used for the treatment of psoriasis, atopic dermatitis, and vitiligo [49, 50], and has also been used in the treatment of early stage mycosis fungoides [45]. Excimer Laser exposure for 10 weeks has been shown to yielded good results in the prurigo form of atopic dermatitis (AD) versus clobetasol propionate [8].

Also a far UVC light with single wavelength of 207 nm, generated by excimer lamp, has been produced which can kill microbes of all kinds. An advantage of this germicidal lamp over conventional UVC lamp is that the former lamp is a human health hazard, is catastrophic, and carcinogenic; in contrast, this single wavelength generating excimer lamp is safer and shown in laboratory conditions to be totally harmless to human skin tissues. A biological explanation is that, due to its extremely short range, this light cannot penetrate the human stratum corneum or even the cytoplasm of human cells [9].

1.4.4 Fluorescent Lamp

Fluorescent lamps are cylindrical glass tubes coated with phosphors and containing mercury. When current is applied to the ends of the tubes, the mercury is vaporized to a higher state and radiation is produced when the mercury falls to its ground state. The phosphors coating the tube act as a fluorophores (i.e. chromophores for fluorescence), absorbing the light and then re-emitting it at longer wavelengths [39]. Different phosphors lead to UVA, UVB, or visible light. Fluorescent lamps can be used for both treatment and diagnosis of dermatologic conditions. In fact, they are the most frequently used sources of therapeutic UVR in dermatology. They can be used for full body treatment or targeted treatment such as for hands and feet. The Wood's lamp, commonly used for diagnosis of vitiligo, fungal infections, and erythema, emits UVA, which is absorbed by skin fluorophores (collagen, elastin, and porphyrins) and then re-emitted at a longer wavelength as visible light [5].

1.4.5 Full Spectrum Light

The Full Spectrum light (FSL) is a newly developed phototherapy device. It generates full spectral light with a continuous wavelength ranging from 320–5000 nm and has been used in conjunction with an emollient, demonstrating greater improvement in AD severity scores at 4 weeks as compared to the emollient alone [10].

1.4.6 UV Gas Discharge Lamps

These are specialized arc lamps which are loaded with gases such as argon or deuterium and may be used without windows or with magnesium fluoride window; they produce UV light at particular spectral lines. These kinds of lamps are often used in UV spectroscopy equipment for chemical analysis. They are also suitable for generating several lined spectra of hydrogen (85–160 nm) and the Hopefield continuum helium (60–100 nm) [47]. Also available are gas-discharged bulbs to be used in headlight and better visibility of colored road signs [62].

1.4.7 Narrowband UVB

Narrowband ultraviolet B (NBUVB) (311–312 nm) was developed for the treatment of psoriasis. It has been shown to be more effective than broadband UVB [48, 64] and allows for a lower dose of UV to be used. NBUVB is also effective for the treatment of AD, early stage mycosis fungoides, vitiligo, and pityriasis rosea [15]. NBUVB can also be used in combination with topical agents such as calcipotriol, or oral agents such as retinoids, to augment efficacy and promote faster resolution of skin disease [11, 15].

NBUVB is safe enough to give to children and women during pregnancy [15]. The most common side effect from UVB exposure is an acute phototoxic reaction manifesting as erythema [5]. Other side effects may include conjunctivitis and keratitis (if adequate eye protection is not used during treatment) [25] and an increased long term risk of skin cancer [15].

1.4.8 Narrow Band Ultraviolet B and C Lamp

This lamp is a combination of UVB/UVC with a ratio of 8.7 or Wolff Helarium lamps with UVB/UVC greater than 1300. These lamps have been found safe to treat AD although leaving a risk of melanoma [27].

1.4.9 Near UV or UVA Light

The near UV light, also known as black light and UVA light, emits UV light of 315–400 nm. This lamp was invented in 1935 and this fluorescent lamp uses a phosphor on the inner glass surface of the tube which absorbs the visible light and allows the UVA light to go through. Certain other lamps make use of the deep bluish purple Wood's glass optical filter that blocks the visible light of the wavelength longer than 400 nm. Other cheaper versions of UVA lamps are available in the market.

This light on its own has weak damaging effects on the biological system. This is evidenced by studies on *E. coli Sod* A or *sod* B mutant which could not be mutated when carrying an individual mutation but when the double mutant *Sod* A/*Sod* B was exposed to NUV a 9-fold increased mutation was seen, suggesting the weakness of the UVA light [23]. Also, results from a number of studies on other systems, *Salmonella typhimurium, Streptomyces gricius,* embryonic fibroblasts, fertilized sea urchin eggs, and mouse ocular tissues supported that UVA alone can affect these organisms more weakly than UVC [13, 29, 65, 67]. ROS have been implicated in this effect [37, 52].

Although UVA alone affects living organisms more weakly than UVC, when this energy is combined with a photosensitive agent, its potency is significantly enhanced known as synergistic action [21]. An earlier study on the reactivity of NUV with hydrogen peroxide showed that a combination of these two can produce the ROS, superoxide anion killing phage T7 synergistically [1]. Several subsequent studies revealed that a large number of other agents can be reactivated by UVA generating ROS [2].

1.4.10 Plasma and Synchrotron Sources of Extreme UV

Non-coherent extreme UV (EUV) light at 13.5 nm has been produced and used for extreme ultraviolet lithography (From Wikipedia, the free encyclopedia: 19th October, 2015).

1.4.11 Microwave Discharge Electrodeless Lamp

A novel Microwave discharge electrodeless lamp (MDEL) has been produced consisting of a three layered cylindrical structure that is effective for the removal of 2,4-D herbicide and near total sterilization of bacteria (*E. coli* and other microbes inactivated up to 99%) in waste water through photolysis with the emitted vacuum-UV (185 nm) and UVC (254 nm) light. In this process, the chemical oxidation with ROS is produced by the photolysis of dioxygen and air oxygen through one of the photoreactors. The integrated UV/ROSO2 and UV/ROSair methods have been found to be more effective than either the UV alone or ROSO2 and ROSair [22].

1.4.12 TL-01 Ultraviolet Lamp

This lamp emits a narrowband UV spectrum of 311 nm and has shown to be most effective for the treatment of psoriasis [6]. A modified version of this lamp (Narrow-band, TL-01 UVB air conditioned) has been developed and tried to treat a number of patients suffering from AD. This narrow band UVB phototherapy was found to be an effective and sparing steroid treatment for chronic severe atopic dermatitis, offering long term benefits in the majority of the patients treated [16].

1.4.13 TL-12 Philips Lamp

This near UV broadband lamp emits light 270–350 nm covering small amount of UVC, all UVB, and certain wavelength of UVA. It use is limited [18].

1.4.14 Black-Light ILTs

This UVA lamp which was originally developed in 1935 and used in agriculture since then, is also used as the insect light traps (ILTs). Now this lamp is universally used all over, in industry, in food production, in hospitals, and as indoor settings [57].

1.4.15 UV Light-Emitting Diodes

Light emitting diodes (LEDs) are semiconductors that convert electrical current into narrow band light in wavelengths ranging from 274 to 1300 nm. LEDs can deliver the same wavelengths as lasers but at lower energy output. Therefore, LEDs provide a more gentle delivery of light and do not carry the same risk of tissue damage as lasers do [4]. Because LEDs can be made into panels, they can cover greater body surface areas compared to lasers [12] resulting in faster treatment times.

LEDs have been used safely in the accelerated healing of wounds, both traumatic and iatrogenic, inflammatory acne and the patient-driven application of skin rejuvenation [30]. Other wide range of applications of this lamp include: treating skin wound [36], colonoscopy [54] and reducing stethoscope contamination [42]. Also in a recent study a pulsed 405 nm light emitting diode (LED) light has been tested for its efficacy as antimicrobial photodynamic inactivator [19].

1.4.16 Ultraviolet Lasers

Laser is an acronym for light amplification by stimulated emission of radiation. In lasers, the incident and emitted photons are of the same wavelength, phase, and direction; this characteristic gives lasers their monochromatic (i.e. single wavelength) and coherent spectral output [39]. The wavelength is determined by a lasing medium (e.g. solid, liquid, gas) in the optical cavity of the laser through which the light passes. It is selected based on the depth and absorption characteristics of the target chromophore. Hemoglobin, melanin, artificial pigment (tattoos), and collagen are some of the chromophores targeted by lasers.

Once laser energy is absorbed in the skin, photothermal, photochemical, or photomechanical effects are possible. Photothermal and photomechanical reactions are the most commonly observed effects in current laser practice. Photothermal effects occur when a chromophore absorbs a specific wavelength and the conversion of absorbed energy into heat leads to destruction of the target (chromophore). Rapid thermal expansion can lead to acoustic waves and subsequent photomechanical destruction of the absorbing tissue [60].

Currently, lasers work on the principle of selective photodermolysis (i.e. selective thermal damage), in which a wavelength is chosen that will be preferentially absorbed by the target tissue (chromophore) and cause its destruction with minimal thermal damage to the surrounding tissue. To limit the amount of thermal energy deposited within the skin, the exposure duration of tissue to light (pulse duration) must be adequate and is chosen based on the size of the target; smaller targets require shorter pulse [39]. Lastly, the energy density (fluence, measured in joules per square centimeter) must be sufficient to achieve destruction within the allotted time [3]. Lasers have been used in the treatment of benign vascular and pigmented birthmarks, hypertrophic scars and keloids, removal of facial or body hair, tattoos, and rhytides.

1.4.17 UVA-1, UVA-2 and a Combination of Both

Although UVA lamp (see above) is a broad spectrum light emitter covering light of 320,400 nm, now UVA-l lamp has been produced which makes use of the lower frequencies of UV-A light spectrum (340–400 nm) and filters out UV-A2 radiation (320–340 nm) and hence its adverse effects when used [53].

UVA and UVB lamps exist on their own, for the treatment of certain skin diseases, such as atopic eczema, and although UV-A had been found to be quite effective for the treatment of AD, its long exposure time remains unacceptable. This problem was overcome with the development of UV-A1 lamps. UV-A1 can be administered either employing a high dose (80–130 J/cm^2), medium dose (40–80 J/cm^2), or low dose (<40 J/cm^2) [27]. An issue with UV-A1 is the high dose producing excessive heating of the equipment making its use intolerable in many situations [51]. Although these lamps have been used independly, a combination of these two lights has been shown to result in the most effective treatment when compared to either lamp used alone [43]. Additional importance of combination therapy is the reduction in exposure time required to treat diseases by UVA which decreases potential side effects.

1.4.18 Short Wave Ultraviolet Lamp

This is also known as UVC or germicidal lamp emitting radiation of 180–290 nm. This is one of the most widely used lamps in biological research. It has played the most important roles in determining the DNA damage and its repair, as UV of 254 nm is maximally absorbed by DNA causing a number of different types of damage. As mentioned above in Introduction section that according to Pub Med 82,617 research papers have been published in which UV light has been used. Another PubMed search (November, 2016) shows that 11,774 papers have been published when papers on UVC light searched. This shows the importance of UVC light on human life and importance of UVC light in research. To produce more accurate results from short wavelength lamps, cut off-filters were developed for the UV radiation [7].

1.5 Phototherapies

1.5.1 UVA1 Phototherapy

By penetrating deep into the dermis, UVA1 phototherapy has led to therapeutic responses without the usual side effects caused by less penetrating UVB wavelengths and UVB-like wavelengths in the UVA-2 region [26]. UVA1 has been shown to induce apoptosis of skin-infiltrating T-helper cells [44] and their depletion from affected skin, eventually leading to clearing of atopic dermatitis [20, 32]. It has also been shown to be an effective treatment for sclerosis by inducing collagenase I expression in a dose dependent manner [33, 35].

1.5.2 PUVA Therapy

PUVA, also known as photochemotherapy, is the combination of 8-methoxypsoralens (MOP) plus UVA light. MOP are compounds found in plants that when taken orally or applied topically, absorb light and produce photochemical reactions in skin cells resulting in a therapeutic effect [38]. Some of the skin disorders treated with PUVA include psoriasis, dermatitis, vitiligo, polymorphic light eruption, and early stage cutaneous T-cell lymphoma. Immediate side effects include burning, itching, and nausea. Cumulative high-dose exposure to PUVA causes photoaging and increases the risk for skin aging, and skin cancer, in particular squamous cell carcinoma and possibly, melanoma [25, 66]).

1.5.3 Photodynamic Therapy

Photodynamic therapy (PDT) uses photosensitizing agents to amplify the effects of visible light or lasers. A photosensitizer agent such as 5-aminolevulinic acid is applied to the skin, where it accumulates in the target cells. These cells absorb light and along with oxygen, lead to formation of reactive oxygen species and selective cell apoptosis [59]. PDT is effective in treating neoplasms such as actinic keratosis and superficial nonmelanoma skin cancers (i.e, superficial basal carcinoma and Bowen's disease). Topical PDT can also be used for other non-neoplastic indications such as psoriasis, localized scleroderma, acne, and skin rejuvenation [24, 28, 61]. Most common side effects include phototoxic reactions and pain. The pain is often referred to as smarting reaction and it may require analgesia for control [66].

1.5.4 Extracorporeal Photopheresis

Extracorporeal photopheresis (ECP), also known as extracorporeal photochemotherapy, refers to a type of systemic light treatment, in which leukocytes are separated from the patient's blood, combined with MOP and irradiated with UVA (PUVA). The treated white cells are then re-infused into the patient. ECP is a first-line of treatment for Sézary syndrome (leukemic CTCL). Other indications include graft-versus host disease and systemic scleroderma, among others [31].

1.6 Conclusion

Multiple UV light sources in the form of lamps and lasers are currently being used in the treatment of dermatosis such as psoriasis, atopic dermatitis, cutaneous T-cell lymphomas, morphea, and vitiligo, and despite the development of newer therapies, phototherapy remains, in many cases, the first line treatment for these dermatologic conditions.

References

1. Ahmad SI (1981) Synergistic action of near ultraviolet radiation and hydrogen peroxide on the killing of coliphage T7: possible role of superoxide radical. Photobiochem Photobiophys 2:173–180
2. Ahmad SI (2016) Ultraviolet light, chromophores, reactive oxygen species and human health. In: Ahmad SI (ed) Taylor and Francis Publication, chapter 3, 25–40
3. Anderson RR, Parrish JA (1983) Selective photothermolysis: precise microsurgery by selective absorption of pulsed radiation. Science 220:524–527
4. Barolet D (2008) Light-emitting diodes (LEDs) in dermatology. Semin Cutan Med Surg 27(4):227–238
5. Baron ED, Suggs AK (2014) Introduction to photobiology. Dermatol Clin 32(3):255–266
6. Barth J, Pinzer B (1990) Therapy of psoriasis with the Philips TL01 ultraviolet lamp. Dermatol Monatsschr 176(11):707–710
7. Bass AM (1948) Short wave-length cut-off filters for the ultraviolet. J Opt Soc Am 38(11):997–999
8. Brenninkmeijer EE, Spuls PI, Lindeboom R et al (2010) Excimer laser vs. clobetasol propionate 0.05% ointment in prurigo form of atopic dermatitis: a randomized controlled trial, a pilot. Br J Dermatol 163(4):823–831
9. Buonanno M, Stanislauskas M, Ponnaiya B et al (2016) 207 nm light-A promising tool for safe low-cost reduction of surgical site infections: in veivo safety study. PLoS One 11(6.) more to be added later
10. Byun HJ, Lee HI, Kim B et al (2011) Full-spectrum light phototherapy for atopic dermatitis. Int J Dermatol 50(1):94–101

11. Coven TR, Burack LH, Gilleaudeau R et al (1997) Narrowband UV-B produces superior clinical and histopathological resolution of moderate-to-severe psoriasis in patients compared with broadband UV-B. Arch Dermatol 133(12):1514–1522
12. Dierickx C, Anderson RR (2002) Visible light treatment of photoaging. Dermatol Ther 18(3):191–208
13. Ferron WL, Eisenstark A, Mackay D (1972) Distinction between far- and near-ultraviolet light killing of recombinationless (recA) Salmonella typhimurium. Biochim Biophys Acta 277(3):651–658
14. Gamlen HE (1903) Treatment of lupus by X-rays and ultraviolet rays. Br Med J 1(2214):1310–1313
15. Gawkrodger DJ, Ardern-Jones MR (2012) Dermatology: an illustrated colour text. Churchill Livingstone/Elsevier, Edinburgh
16. George SA, Bilsland DJ, Johnson BE et al (1993) Narrow-band (TL-01) UVB air-conditioned phototherapy for chronic severe adult atopic dermatitis. Br J Dermatol 128(1):49–56
17. Gilbert DJ (2011) How I perform ALA-photodynamic therapy in my practice. In: Gold MH (ed) Photodynamic therapy in dermatology. Springer, New York, pp 161–172
18. Gibbs NK, Norval M, Traynor NJ et al (1993) Comperative potency of broad band and narrow band phototherapy sources to induce edema, sunburn cells and urocanic acid photoisomerization in hairless mouse skin. Photochem Photobiol 58(5):643–647
19. Gillespie JB, Maclean M, Given MJ et al (2016) Efficacy of pulsed 405-nm light emitting diodes for antimicrobial photodynamic inactivation: effects of intensity, frequency and duty cycle. Photomed Laser Surg. EPUB ahead of print
20. Grewe M, Bruijnzeel-Koomen CA, Schöpf E et al (1998) A role for Th1 and Th2 cells in the immunopathogenesis of atopic dermatitis. Immunol Today. 19(8):359–361
21. Hartman P, Eisenstark A (1978) Synergistic killing of Escherichia coli killing by near-UV radiation and hydrogen peroxide: distinction between rec A-repairable and rec A-nonrepairable damage. J Bacteriol 133(2):769–774
22. Horikoshi S, Tsuchida A, Shinomiya T et al (2015) Microwave discharge electrodeless lamps(MDELs). A novel MDEL photoreactor for the photolytic and chemical oxidation treatment of contaminated wastewater. Photochem Photobiol Sci 14(12):2187–2194
23. Hoerter J, Eisenstark A, Touati D (1989) Mutation by near-ultraviolet radiation in *Escherichia coli* strains lacking superoxide dismutase. Mutat Res 215(2):161–165
24. Ibbotson SH (2002) Topical 5-aminolaevulinic acid photodynamic therapy for the treatment of skin conditions other than non-melanoma skin cancer. Br J Dermatol 146(2):178–188
25. Iordanou E, Berneburg M (2010) Phototherapy and photochemotherapy. J Dtsch Dermatol Ges 8(7):533–541
26. Jekler J, Larkö O (1990) Combined UVA-UVB versus UVB phototherapy for atopic dermatitis: a paired-comparison study. J Am Acad Dermatol 22(1):49–53
27. Jekler J, Diffey B, Larko O (1990) Ultraviolet radiation dosimetry in phototherapy for atopic dermatitis. J Am Acad Dermatol 23(1):49–51
28. Karrer S, Abels C, Landthaler M et al (2000) Topical photodynamic therapy for localized scleroderma. Acta Derm Venereol 80(1):26–27
29. Kelner A (1951) Action spectra for photoreactivation of ultraviolet-irradiated *Escherichia coli* and Streptomyces griseus. J Gen Physiol 34(6):835–852
30. Kim WS, Calderhead RG (2011) Is light-emitting diode phototherapy (LED-LLLT) really effective. Laser Ther 20(3):205–2015
31. Knobler R, Berlin G, Calzavara-Pinton P et al (2014) Guidelines on the use of extracorporeal photopheresis. JEADV 28(Suppl 1):1–37
32. Krutann J, Czech W, Diepgen T et al (1992) High-dose UVA1 therapy in the treatment of patients with atopic dermatitis. J Am Acad Dermatol 26:225–230
33. Krutman J (1999) Therapeutic photomedicine: phototherapy. In: Freedberg IM, Az E, Wolff H et al (eds) Fitzpatrick's Dermatology in General Medicine. McGraw-Hill, New York
34. Larko O (1989) Treatment of psoriasis with a new UVB lamp. Acta Derm Venereol 69(4):357–359
35. LeRoy EC (1974) Increased collagen synthesis by scleroderma skin fibroblasts in vitro: a possible defect in the regulation or activation of the scleroderma fibroblast. J Clin Invest 54(4):880–889
36. Li Y, Zhang J, Xu Y et al (2016) The histopathological investigation of red and blue light emitting diod on traeting skin wound in Japanese big-ear white rabbit. PLoS One 11(6)………….more to come
37. Linetsky M, Chemoganskiy VG, Hu F et al (2003) Effect of UVA light on the activity of several aged human lense enzymes. Invest Opthalmol Vis Sci 44(1):264–274
38. Ling TC, Clayton TH, Crawley J et al (2016) British Association of Dermatologists and British Photodermatology Group guidelines for the safe and effective use of psoralen-ultraviolet A therapy 2015. Br J Dermatol 174(1):24–55. https://doi.org/10.1111/bjd.14317
39. Lui H, Anderson RR (2007) Radiation sources and interaction with the skin. In: Lim H, Honigsmann H, Hawk JL (eds) Photodermatology. Informa Healthcare, New York, pp 29–40
40. Lutz FE, Richtmyer FK (1922) The reaction of Drosophila to ultraviolet. Science 55(1428):519
41. Martsen FA, Robertson WW, Chuoke RL (1947) The near-ultraviolet absorption spectra of monoalkyl-substituted benzene: hyperconjugation and Baker-Nathan effect. Chem Rev 41(2):273–279
42. Messina G, Fattorini M, Nante N et al (2016) Time effectiveness of ultraviolet C light (UVC) emitted by light emitting diodes (LEDs) in reducing stetho-

scope contamination. Int J Environ Res Public Health 13(10):940
43. Midelfart K, Stenvold S-E, Volden G (1985) Combined UVB and UVA phototherapy of atopic eczema. Dermatology 171(2):95–98
44. Morita A, Werfel T, Stege H et al (1997) Evidence that singlet oxygen-induced human T helper cell apoptosis is the basic mechanism of ultraviolet-A radiation phototherapy. J Exp Med 17 186(10):1763–1768
45. Nistico A, Costanzo A, Saraceno R et al (2004) Efficacy of monochromatic excimer laser radiation (308 nm) in the treatment of early stage mycosis fungoides. Br J Dermatol 151:877–879
46. Nicholls AG (1928) The ultraviolet rays in therapeutics. Can Med Assoc J 18(3):321–322
47. Nicholson AJ (1970) A gas discharge lamp for the extreme ultraviolet. Appl Opt 9(5):1155–1158
48. Parrish JA, Jaenicke KF (1981) Action spectrum for phototherapy of psoriasis. J Invest Dermatol 76:359–362
49. Park KK, Swan J, Koo J (2012a) Effective treatment of etanercept and phototherapy-resistant psoriasis using excimer laser. Dermatol Online J 18(3):2
50. Park KK, Liao W, Maurase JE (2012b) A review of monochromatic excimer light in vitiligo. Br J Dermatol 167(3):468–478
51. Patrizi A, Raone B, Ravaioli GM (2015) Management of atopic dermatitis: safety and efficacy of phototherapy. Clin Cosmet Investig Dermatol 8:511–520
52. Petersen AB, Gniadecki R, Vicanova J et al (2000) Hydrogen peroxide is responsible for UVA-induced DNA damage measured by alkaline comet assay in HaCaT keratinocytes. J Photochem Photobiol B 59(1–3):123–131
53. Polderman MCA, Wintzen M, Cessie S et al (2005) UVA-1 cold light therapy in the treatment of atopic dermatitis: 61 patients treated in the Leiden University Medical Center. Photodermatol Photoimmunol Photomed 21(2):93–96
54. Sasaki S, Nishikawa J, Yanai H (2016) Image quality of a novel light-emitting diod (LED)- illuminated colonoscope. Endoscopy 48(10):934–938
55. Shugard D (1951) Photoreactivation in the near ultraviolet of d-glyceraldehyde-3-phosphate dehydrogenase. Experientia 7(1):26–28
56. Situm M, Bulat V, Majcen K et al (2014) Benefit of controlled ultraviolet radiation in the treatment of dermatological diseases. Coll Antropol 38(4):1249–1253
57. Sliney DH, Gilbert DW, Lyon T (2016) Ultraviolet safety assessments of insect light traps. J Occup Environ Hyg 13(6):413–424
58. Smith B (1946) Method of making Filters Transmitting the Near Ultraviolet and absorbing visual light. Science 104(2708):490–491
59. Szeimies RM, Karrer S, Abels C et al (2001) Photodynamic therapy in dermatology. In: Krutmann J, Honigsmann H, Elmets CA et al (eds) Dermatological Phototherapy and Photodiagnostic Methods. Springer, Berlin
60. Tanzi EL, Lupton JR, Alster TS (2003) Lasers in dermatology: four decades of progress. J Am Acad Dermatol 49(1):1–31
61. Touma D, Yaar M, Whitehead S et al (2004) A trial of short incubation, broad-area photodynamic therapy for facial actinic keratoses and diffuse photodamage. Arch Dermatol 140(1):33–40
62. Venkatachalam K, Smith G (2000) Gas discharge headlights and visibility of coloured road sign. Clin Exp Optom 83(5):246–256
63. Von Kobyletzki G, Pieck C, HOxtermann S et al (1999) Circulating activation markers of severe atopic dermatitis following ultraviolet A1 cold light phototherapy: eosinophil cationic protein, soluble interleukin 2 receptor and soluble interleukin-4 receptor. Br J Dermatol 140(5):966–968
64. Walters IB, Burack LH, Coven TR et al (1999) Suberythemogenic narrow-band UVB is markedly more effective than conventional UVB in treatment of psoriasis vulgaris. J Am Acad Dermatol 40:893–900
65. Yoakum G, Ferron W, Eisenstark A et al (1974) Inhibition of replication gap closure in Escherichia coli by near ultraviolet light photoproducts of L-tryptophan. J Bacteriol 119(1):62–69
66. Zanolli M (2003) The modern paradigm of phototherapy. Clin Dermatol 21(5):398–406
67. Zigman S, Hare JD (1976) Inhibition of cell growth by near ultraviolet light photoproducts of tryptophan. Mol Cell Biochem 10(3):131–135

Part II

UV Photoproducts, Damage to DNA and Mutagenesis

Ultraviolet Light Induced Generation of Reactive Oxygen Species

2

T.L. de Jager, A.E. Cockrell, and S.S. Du Plessis

Abstract

As ultraviolet (UV) radiation is naturally and ubiquitously emitted by the sun, almost everyone is exposed to it on a daily basis, and it is necessary for normal physiological function. Human exposure to solar UV radiation thus has important health implications. The generation of reactive oxygen species (ROS) by UV radiation is one of the mechanisms through which UV light can manifest its possible detrimental effects on health. When an imbalance develops due to ROS generation exceeding the body's antioxidant defence mechanisms, oxidative stress can develop. Oxidative stress can lead to cellular damage (e.g. lipid peroxidation and DNA fragmentation), apoptosis and cell death. Broadly UV can induce ROS by affecting the cellular components directly or by means of photosensitization mechanisms. More specifically UV light can induce ROS by affecting the enzyme catalase and up-regulating nitric oxide synthase (NOS) synthesis. It may also cause a decrease in protein kinase C (PKC) expression leading to increased ROS production. UVR is capable of modifying DNA and other chromophores resulting in elevated ROS levels. The effects of raised ROS levels can vary based on the intracellular oxidant status of the cell. It is therefore important to protect yourself against the potentially harmful effects of UV light as it can lead to pathological UV-induced ROS production.

Keywords

Reactive oxygen species • Oxidant • Antioxidants • Catalase • Nitric oxide • Protein kinase C

T.L. de Jager • A.E. Cockrell • S.S. Du Plessis (✉)
Division of Medical Physiology, Faculty of Medicine and Health Sciences, Stellenbosch University, Tygerberg, South Africa
e-mail: ssdp@sun.ac.za

S. Ahmad (ed.), *Ultraviolet Light in Human Health, Diseases and Environment*, Advances in Experimental Medicine and Biology 996, https://doi.org/10.1007/978-3-319-56017-5_2

2.1 Introduction

As ultraviolet (UV) radiation is naturally emitted by the sun and thus regarded as ubiquitous, almost everyone is exposed to it on a daily basis. Human exposure to solar ultraviolet radiation has important health implications. Suitable amounts of UV radiation exposure has beneficial effects, however many studies have demonstrated evidence in support of harm associated with overexposure to UV. Adequate exposure is vital for UV-induced vitamin D synthesis, while a number of health effects (e.g. skin cancer, malignant melanoma) have been identified as a result of excess exposure.

The generation of reactive oxygen species (ROS) by UV radiation is one of the mechanisms through which UV light can manifest its detrimental effects on health. ROS are free radicals and can be defined as an unstable chemical species possessing an unpaired electron. When an imbalance develops due to ROS generation exceeding the body's antioxidant defence mechanisms, oxidative stress can develop. Oxidative stress can lead to cellular damage (e.g. lipid peroxidation and DNA fragmentation), apoptosis and cell death [22].

Due to industrialization there is a dramatic increase in chlorofluorocarbons (CFCs), with resulting loss of stratospheric ozone. This situation subsequently can lead to increased levels of specially UVC radiation reaching the earth's surface (which otherwise cannot) and therefore growing human exposure to UV radiation. The burden on human health is more noticeable with increased UV radiation induced pathologies and thus the need to explore this phenomenon in more detail. The chapter aims to provide a detailed relationship between UV and ROS as well as the associated burden.

2.2 Ultraviolet Light

Ultraviolet (UV) light is part of the solar emissions spectrum which falls between the electromagnetic radiation spectrum of X-rays and visible light with wavelengths ranging from 100 nm to 400 nm. Based on their wavelengths, UV light can be subdivided into several categories with three bands, namely UVA, UVB and UVC [3, 15]. Environmental and dermatological photobiologists commonly use slightly different divisions, which are more closely associated with the biological effect of the different wavelengths [19].

UVA 320–400 nm: The most commonly encountered UV light as it passes through the atmospheric ozone with little change. Initially UVA cause pigment darkening (tanning) followed by sunburn when over exposed. UVA is necessary for Vitamin D production in humans but over exposure could result in epidermal hardening, immune system suppression and formation of cataracts. UVA is vastly used in the cosmetic industry (sunbeds or tanning booths).

UVB 290–320 nm: UVB is the key factor in photochemical damage to cellular DNA. UVB is also essential for production of Vitamin D in humans; however, over exposure may hold harmful effects to the human body. These harmful effects include sunburn, cataracts as well as the initiation of the carcinogenic process in the skin.

UVC 220–290 nm: UVC is almost completely absorbed by the atmospheric ozone and has little effect on human health. Germicidal lamps emit UVC in order to kill microbes. Humans get exposed to UVC accidentally may cause corneal burns and snow blindness. UVC is absorbed by the dead outer layer of the dermis thus exposure may cause severe pain but clears up within a few days [3, 15].

2.2.1 The Global Solar UV Index (UVI)

The global solar UV index [28] provides a description of the level of solar UV radiation at the Earth's surface [19]. The values are reported as a number from zero upwards; the higher the number the greater the risk for UV induced damage. The UV index (UVI) can be used as a guideline and educational tool to determine the risk for potential UV damage to the skin and eye when exposed to solar UV radiation. This could be a valuable tool to inform individuals of the increased risk of skin cancers associated with excessive UV radiation exposure, and to encourage individuals to adopt protective measures [19].

UVI is presented usually as the maximum UV radiation levels on a given day, which occurs around solar noon. Geographical location plays a determining factor but none the less, solar noon takes place between local noon and 2 pm [19].

2.3 Sources of UV Light

Humans are exposed to UV radiation through outdoor sun exposure or due to artificial sources.

2.3.1 Outdoor

Outdoor exposure by the sun is either through deliberate activities (sun tanning) or as a result of recreational or occupational activities.

2.3.2 Artificial Sources

Exposure includes sources from medical and cosmetic treatments. The following sources are major contributors to artificial UV exposure:

- *Phototherapy/ sunbeds*: They expose human skin to UVA and some UVB radiation. Black light, which is also referred to as a UV-A light, is predominately used by tanning booths and phototherapies
- *Medical*: Exposure will depend upon treatment type but, include a majority of diagnostic and treatment apparatus.
- *Industrial/ commercial*: Arc welding is a potential source of UV exposure which can cause severe damage to the eyes and skin.
- *Lighting:* Fluorescent lamps emit minor amount of UV light which only contributes a small percentage to the total yearly exposure [3].

2.4 Tissue Exposure to UV

Vitamin D binds to steroid vitamin D receptors in the body. These receptors are directly involved in proliferation and differentiation of cells. The vitamin D dependent renewal process takes place in the keratinocytes. Keratinocytes accounts for a large portion of the cells in the epidermis.

Vitamin D is able to regulate the low resting levels of cell signalling components such as ROS by means of controlling the expression of these components. Therefore, vitamin D is vital in the replenishment of new cells for the skin surface. Vitamin D production is dependent on the amount of melanin in the skin [4, 5]. The amount of melanin the skin produces depends how fair or dark the skin colour is and the Vitamin D production is dependent on the amount of melanin in the skin. Darker skins have more melanin which allows less UVB to enter the skin. Due to less UVB entering the skin, less vitamin D is produced; therefore dark skinned individuals require more sun exposure.

These factors make it difficult to determine the amount of sun exposure time one requires in order to get adequate vitamin D and avoid damage. A good rule of thumb for sun exposure is to expose your skin for half the time before it turns pink. For fair skinned individuals it could be only a few minutes, but darker skinned person requires a longer exposure time [9].

A small amount of UV radiation is required for the production of vitamin D in humans. The amount of vitamin D one gets from exposing the bare skin to the sun depends on the intensity of the sun's UV rays reaching the earth's surface which is influenced by a number of different factors [19]. Vitamin D's role in cell proliferation and differentiation could possibly have a protective role against UV-induced ROS damage.

Prolonged human exposure to solar UV radiation may result in acute and chronic health effects on the skin, eye and immune system. Sunburn and tanning are the best known acute effects of excessive UV radiation exposure; in the long term, UV radiation-induced degenerative changes in cells, fibrous tissue and blood vessels lead to premature skin ageing. UV radiation can also cause inflammatory reactions of the eye, such as photokeratitis [19].

2.5 UV Light Induced Reactive Oxygen Species

Human skin provides continuous protection against chemicals, radiation and infection. The skin is composed of different layers namely the epidermis (composed of mainly dead cells), the dermis and subcutaneous tissue. The epidermis is constantly renewed and is separated from the dermis by a layer of continuously dividing cells namely keratinocytes and melanocytes. Melanin pigment is produced by melanocytes, which is the precursor of melanoma [17], and is transferred to the neighbouring keratinocytes. Keratinocytes create a highly effective physical barrier; they accumulate melanin pigments as they mature, and epidermal melanin functions to potently block UV penetration into the skin. A third cell type is the Langerhans cells which are present under the stratum corneum. They play a role in immunological reactions of the skin and their actions are highly sensitive to UV. The dermis contains collagen fibres which assists in the skins' elasticity and provides supportive strength. Collagen fibres are broken down by high levels of UV which leads to premature aging as skin loses its elasticity [3, 12].

As the human skin is the largest organ and covers the body's whole surface area it is prone to continuous UV light exposure [11]. Melanin synthesis is stimulated by sun exposure [11], artificial sources [3] and inflammation. This results in post inflammatory hyperpigmentation. Epidermal melanocytes are thus susceptible to ROS which is induced through excessive sun exposure [11]. If homeostasis is disrupted by the increased production of ROS this could possibly drive the process of malignant transformation of cells [11, 21].

2.5.1 What Are Reactive Oxygen Species?

Free radicals are defined as molecules containing one or more unpaired electrons in their electron orbitals. Electrons are considered to be more stable when paired free radicals are generally more reactive than non-radical species [13]. Radicals can combine their unpaired electrons by forming a covalent bond. Radical interactions with non-radicals often involve the radical donating its electron (a reducing action) or it could accept an electron from the non-radical molecule (an oxidising action) or it could simply join onto a non-radical [13]. This results in the non-radical becoming a radical and can have various physiological repercussions.

ROS is the umbrella term used to describe oxygen derived free radicals as well as non-radicals such as hydrogen peroxide. Organisms that are dependent on the reduction of oxygen for energy, aerobic organisms, are the most susceptible to the potentially damaging actions of ROS that are released during this process [6] (Table 2.1).

ROS is constitutively produced in the body and sources of ROS can either be exogenous or endogenous.

Endogenous ROS are predominantly produced by the mitochondria in the cell, as this is where oxygen is reduced for ATP production by the addition of four electrons to produce water as the product. The largest portion of oxygen is reduced in the electron transport chain (ETC). No remaining intermediates are produced. However approximately 5% of oxygen is reduced by the

Table 2.1 Examples of ROS

Superoxide	$^{\bullet}O_2^-$
Hydroxyl anion	$^{\bullet}OH^-$
Hydrogen peroxide	H_2O_2
Peroxyl	$ROO^{\bullet-}$
Alkoxyl radicals	$RO^{\bullet}$
Radicals of nitric oxide	$^{\bullet}NO$
Peroxynitrite	$^{\bullet}ONOO^-$
Ozone	O_3
Oxygen singlet	O^1O_2

Adapted from Ref. [6]

univalent pathway thereby leading to free radical production. ROS producing capabilities are often dependent on the composition of the mitochondrial membrane, the species of animal and the age of the animal [6].

Most reactions that take place in the mitochondrial ETC consist of redox reactions. These reactions involve the exchange and movement of electrons, with the enzyme cytochrome oxidase being the only enzyme involved in a reaction that uses oxygen. It has been found that the redox reactions in the ETC 'leak' electrons and subsequently generate superoxide ($O_2^{\cdot-}$) with the majority of the ROS formation taking place at complex III and to a lesser extent at complex I. ROS is also formed in the endoplasmic reticulum (ER) and peroxisomes of the cell. Extracellular or exogenous ROS sources are vast and are responsible for a large percentage of ROS present in the body. Common air pollutants and industrial contaminants, exhaust fumes and the smoke generated by cigarettes have been implicated in ROS generated in the body. ROS generated in this manner has been found to both directly and indirectly cause $O_2^{\cdot-}$ formation and various types of nitric oxide derivatives, either by direct contact with the skin or by inhalation. Some drugs, narcotic substances and anesthetizing gases are also thought to contribute to ROS production. Certain food and alcohol consumption have also been implicated in ROS formation. Other environmental agents and non-genotoxic carcinogens such as gamma irradiation can also induce ROS formation [6].

2.5.2 UV – Light Induced ROS

Even exposure to the ultraviolet light spectrum can indirectly lead to the production of a variety of ROS including $O_2^{\cdot-}$, singlet oxygen, hydroxyl radicals and hydrogen peroxide through various mechanisms [17] (Fig. 2.1).

2.5.2.1 Catalase

The first mechanism that will be discussed is the UV induced ROS by the enzyme catalase. Catalase is known to be able to degrade hydrogen peroxide via the reaction $2H_2O_2 \rightarrow 2H_2O + O_2$ through a process known as catalytic activity. The enzyme catalase is also capable of exhibiting peroxidatic activity when low concentrations of hydrogen peroxide (H_2O_2) are present. However, it was observed that UVB light caused a marked increase in ROS in keratinocytes, more so in glutathione depleted cells. Upon further experimentation it was found that the oxidant generating protein was catalase [14].

This is thought to be a result of the ability of the short wave UV (Likely UVC) light to alter the H_2O_2 binding site on the catalase enzyme, thus allowing for water molecules to access the heme iron. This essentially allows for the water molecules to act as a source for the generation of protons. These protons are then able to interact with diatomic oxygen to generate ROS such as reactive peroxides. The activity of the charge relay network of the enzyme was also found to be of importance on the effect that UV light mediates on catalase. The effects of UVB light on catalase were found to be pH-sensitive and oxygen dependent, this and the intracellular oxidant status influences whether the catalase activity in response to UV light is cytotoxic or protective [14].

2.5.2.2 Nitric Oxide Synthase

UVB has also been implicated in the up-regulation of membrane-bound nitric oxide synthase (NOS) in human keratinocytes, thereby resulting in increased production of NO [10]. While UVA has been shown by [2] to generate increased superoxide levels at moderate UVR levels in real time, $O_2^{\cdot-}$ is able to inactivate iron-sulphur proteins leading to release of the reducing ferrous iron, thereby resulting in further production of ROS [23]. The exact mechanisms as to how UV causes these effects are not fully elucidated, however the simultaneous production of $O_2^{\cdot-}$ and NO within proximity of each other can lead to the spontaneous generation of peroxynitrite, a highly reactive free radical [10].

2.5.2.3 Signalling Pathways

Various signaling pathways are activated by UV-mediated ROS generation, especially in the pathophysiology of skin diseases. MAPKs have

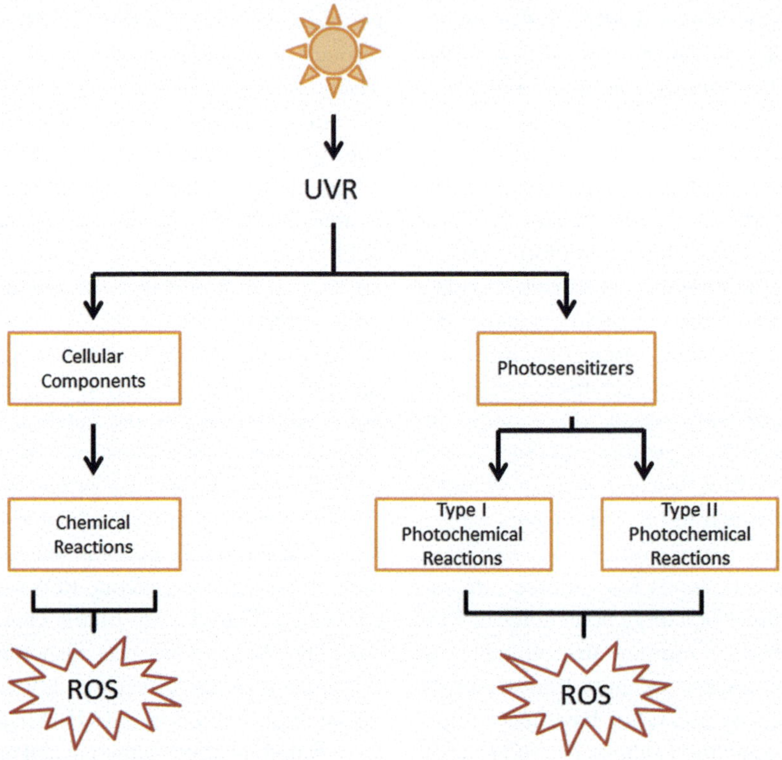

Fig. 2.1 UV radiation is able to mediate damage to cellular components in two ways: The first mechanism is by the direct absorption of incident rays by the cell and its components. This results in the formation of an excited state of the components and subsequent chemical reactions. The second mechanisms are by means of photosensitization. Incident rays are absorbed by endogenous or exogenous photosensitizers such as bilirubin. This results in an excitation of the sensitizers to their triplicate states. The excited photosensitizers exert their effects by two mechanisms. Type I photochemical reactions involve electron transfer and the process of hydrogen abstraction to form free radicals. Type II photochemical reactions involve the transfer of energy with O_2 to yield reactive state singlet oxygen ($^1O_2$) [20, 22]

been shown to be a target of oxidative stress such the oxidative stress induced by solar UV radiation, the UV radiation influences the pathways in a manner that closely resembles ROS [7, 16]. The ROS produced lead to the activation of the MAPKs such as ERK and JNK. These MAPKs play a pivotal role in the recruitment of factors that lead to the downstream activation of transcription factor AP-1. The factor p38 and the inhibitory kappa kinase activation are vital in the process of transcriptional activation of NF-kB. UVA irradiation of the fibroblast cells in the skin result in the release of labile iron, which is implicated in the activation of NF-kB [7, 24]. AP-1 and NF-kB play essential roles in the regulation of a diverse array of genes involved in processes such as the cell cycle, proliferation and apoptosis [7].

2.5.2.4 Protein Kinase C

UV irradiation was found to have an effect on protein kinase C (PKC) in murine fibroblasts. Several different isoforms of this serine/threonine kinase exist and have been implicated in UV-induced signal transduction pathways [8]. Low doses of UV exposure result in the adhesion of cultured fibroblasts to the collagen matrices by means of PKC isoform activation and integrins. PKC α has also been found to be irreversibly inhibited in UVA irradiated cells while PKC ε is known to act as an endogenous photosensitizer that plays a role in inducing UV-induced cutaneous damage. The accumulation of ROS associated with skin aging is thought to be linked to UV radiations effect on PKC. In a study done by Bossi et al. [8] it was observed that UVA irradiation causes an increase in ROS levels and lipid peroxidation in both young and aging fibroblasts; however basal ROS levels were much higher in the aged cell population and they also exhibited a much slower response to UV irradiation. A decline in PKC δ expression in these fibroblasts was also observed to be closely linked to the translocation of PKCδ to the nucleus due to the exposure to UV light. At the same time it also resulted in an increase in ROS production. This observation is suggestive of the role that PKC δ may have in regulating ROS production in ROS-induced states and lean towards the idea that loss of PKC δ expression can cause the activation and elevation of ROS levels [8]. It was also found that PKC α expression post UV irradiation is not linked to the elevation in ROS levels, regardless of increased PKC α expression following UV irradiation.

2.5.2.5 DNA Damage

Exposure of cells to UVA, UVB or UVC is capable of causing 8-oxo-7-8-dihydro-2′-deoxyguanosine (8-oxodGuo). This is an oxidatively modified DNA base that is often used as a marker of possible oxidative DNA damage, both in vivo and in vitro in calf thymus DNA as well as cultured cells [28]. Various ROS have been implicated in the UV-induced oxidation of 8-oxodGuo. Singlet oxygen ($^1O_2$) is proposed to be the only source of ROS involved in UVC-induced production of 8-oxodGuo. While for UVA and UVB $^1O_2$ is a large source of the ROS involved. Both H_2O_2 and hydroxyl radicals (·OH) were also found to play a significant role as well. Surprisingly it was observed that the $O_2^{\cdot-}$, usually a highly reactive free radical, was not involved in UV-induced oxidative damage in these cells [28]. These observations were also noted in a study conducted by Wei et al. [27] where they confirmed the role of $^1O_2$ in oxidative DNA damage induced by UV exposure.

2.5.3 Biochemical Actions of UV in the Skin

Chromophores absorb UV energy which allows for biochemical reactions in human skin [26]. They are molecules that absorb certain wavelengths of visible light and transmit to others. Chromophores could either be an exogenous agent or endogenous compound [1]. Cutaneous chromophores include DNA, urocanic acid, aromatic amino acids, retinoids, carotenoids, bilirubin, flavins, haemoglobin, melanin and NAD(P)H [26]. Chromophores may be damaged directly or they can act as photosensitizers. This results in the generation of ROS in the presence of molecular oxygen [1].

If the epidermis is exposed to direct UV radiation from the sun it can lead to oxidative stress through activating the enzyme NADPH oxidase or by promoting lipid peroxidation. When molecular oxygen is reduced to $O_2^{\cdot-}$ the production of ROS is initiated. This process can be enzymatic through NADPH oxidase or xanthine oxidase catalysed reactions or non-enzymatic. Glutathione is recycled by glutathione reductase to the detriment of NADPH, which is recycled by glucose 6- phosphate dehydrogenase after glucose has been phosphorylated by hexokinase [26].

UV radiation also causes oxidative stress in the skin by inducing the release of mediators of inflammation. Leukocytes produce the radical anion $O_2^{\cdot-}$, catalysed by the enzyme NADPH oxidase. $O_2^{\cdot-}$ then undergoes dismutation to oxygen and H_2O_2, catalysed by superoxide dismutase. Myeloperoxidase then converts H_2O_2 to hypochlorite. Another pathway for oxidative stress caused by UV is the removal of a proton and an electron from lipid molecules which produces lipid radicals. Lipid radicals can interact with molecular oxygen giving rise to lipid peroxidase and new lipid molecules [26].

2.6 UV, ROS and Pathophysiological Skin Effects

The best known effect of excessive UV radiation exposure is erythema, which is reddening of the skin due to sunburn. Chronic exposure to UV radiation can cause a series of degenerative skin conditions.

Photoaging is a consequence of exposure to UVA and UVB. Abnormal elastic fibres in the dermis and a decrease in collagen fibres are distinct characteristics of photo aged skin. Photoaged skin displays an increased degradation of collagen and elastic fibres in the dermis which is caused by an increase in proteolytic activation and abnormal Extra Cellular Matrix (ECM) turnover. UV rays can also contribute to the generation of ROS that stimulates the inflammatory process in the skin. This causes reduced levels of natural enzymatic and non-enzymatic antioxidant defence mechanisms of the photoaged skin as well as an increased neutrophil infiltration into skin and increased inflammation [3].

UVA mainly drives the production of ROS. UVA has the ability to penetrate the deep dermal layer, inducing changes and driving the process and progression of photoaging. Once UVA penetrates the skin it is absorbed by cellular chromophores which comprise of molecules like riboflavins, melanin, bilirubin, but DNA is not included. The absorption of photons/energy results in the excitation of chromophores, referred to as the singlet exited state. The photon/energized molecule then falls back to the ground state and emits either heat or fluorescence or second an intersystem crossing leading to a triplet excited state. The triplet state could react with both DNA and molecular oxygen which can induce changes in DNA or lead to ROS production. Photoaging could also drive DNA damage, particularly mtDNA damage. ROS production in mitochondrial DNA increases as UV radiation causes a well-recognized 4977 base pair long deletion of the mtDNA [25].

UVB has the ability to penetrate the epidermis but not the deeper dermal layer. Damage is thus limited to keratinocytes and melanocytes in the epidermal layer. However, UVA passes the through the epidermal layer inducing damage which leads to damage in the deeper dermal tissue [25].

2.7 Conclusion

Humans are exposed to UV light on a daily basis, with the skin bearing the brunt of this exposure due to its large surface area. The constant UV radiation is imperative for normal physiological function. However over-exposure to UV light is known to have detrimental physiological effects. It is believed that these effects are a result of an increase in ROS in the cells. UV light can induce ROS in various ways, such as affecting the enzyme catalase and up-regulating NOS synthesis. UV radiation may also cause a decrease in PKC expression leading to increased ROS production. It can also modify DNA and other chromophores resulting in elevated ROS levels. The effects of raised ROS levels can vary based on the intracellular oxidant status of the cell. However, it is still important to protect yourself against the potentially harmful effects of UV light as it can lead to pathological UV-induced ROS production.

References

1. Ainbinder D, Touitou E (2010) Skin photodamage prevention: state of the art and new prospects. In: Textbook of aging skin. Springer, Berlin
2. Aitken GR, Henderson JR, Chang SC, Mcneil CJ, Birch-Machin MA (2007) Direct monitoring of UV-induced free radical generation in HaCaT keratinocytes. Clin Exp Dermatol 32:722–727
3. Belkin M, Césarini J, Diffey B, Hietanen M, Kojima M, Mariutti G, Mckinlay A, Repacholi M, Roy C, Rubenstein R (1994) Protection against exposure to ultraviolet radiation. World Health Organization/United Nations Environment Programme, Geneva
4. Berger, U., Wilson P, Mcclelland RA, Colston K, Haussler MR, Pike JW, Coombes RC (1988) Immunocytochemical detection of 1,25-dihydroxyvitamin D receptors in normal human tissues. J Clin Endocrinol Metab 67(3):607–613
5. Berridge MJ (1988) Vitamin D, reactive oxygen species and calcium signalling in ageing and disease. LID - 10.1098/rstb.2015.0434 [doi] LID - 20150434 [pii]

6. Bhattacharyya S, Saha J (2015) Tumour, oxidative stress and host T cell response: cementing the dominance. Scand J Immunol 82(6):477–488
7. Bickers DR, Athar M (2006) Oxidative stress in the pathogenesis of skin disease. J Invest Dermatol 126:2565–2575
8. Bossi O, Gartsbein M, Leitges M, Kuroki T, Grossman S, Tennenbaum T (2008) UV irradiation increases ROS production via PKCδ signaling in primary murine fibroblasts. J Cell Biochem 105:194–207
9. Chen TC, Lu Z, Holick MF (2010) Photobiology of vitamin D. Vitamin D. Springer, Berlin
10. Deliconstantinos G, Villiotou V, Stavrides JC (1996) Alterations of nitric oxide synthase and xanthine oxidase activities of human keratinocytes by ultraviolet B radiation: potential role for peroxynitrite in skin inflammation. Biochem Pharmacol 51:1727–1738
11. Denat L, Kadekaro AL, Marrot L, Leachman SA, Abdel-Malek ZA (2014) Melanocytes as instigators and victims of oxidative stress. J Investig Dermatol 134:1512–1518
12. D'orazio J, Jarrett S, Amaro-Ortiz A, Scott T (2013) UV radiation and the skin. Int J Mol Sci 14:12222–12248
13. Halliwell B (1991) Reactive oxygen species in living systems: source, biochemistry, and role in human disease. Am J Med 91:S14–S22
14. Heck DE, Vetrano AM, Mariano TM, Laskin JD (2003) UVB light stimulates production of reactive oxygen species unexpected role for catalase. J Biol Chem 278:22432–22436
15. Hussein A, Elhassaneen Y (2014) Natural dye from red onion skins and applied in dyeing cotton fabrics for the production of women's headwear resistance to ultraviolet radiation (UVR). J Am Sci 10(3):129–139
16. Kim AL, Labasi JM, Zhu Y, Tang X, Mcclure K, Gabel CA (2005) Role of p38 MAPK in UVB-induced inflammatory responses in the skin of SKH-1 hairless mice. J Invest Dermatol 124:1318–1325
17. Kimeswenger S, Schwarz A, Födinger D, Müller S, Pehamberger H, Schwarz T, Jantschitsch C (2016) Infrared A radiation promotes survival of human melanocytes carrying ultraviolet radiation-induced DNA damage. Exp Dermatol 25:447–452
18. Liu S, Mizu H, Yamauchi H (2010) Photoinflammatory responses to UV-irradiated ketoprofen mediated by the induction of ROS generation, enhancement of cyclooxygenase-2 expression, and regulation of multiple signaling pathways. Free Radic Biol Med 48:772–780
19. Lucas R, Mcmichael T, Smith W, Armstrong B (2006) Solar ultraviolet radiation. Assessing the environmental burden of disease at national and local levels, Environmental burden of disease series, vol 13. World Health Organization, Geneva
20. Pattison DI, Davies MJ (2006) Actions of ultraviolet light on cellular structures. EXS 96:131–157
21. Picardo M, Grammatico P, Roccella F, Roccella M, Grandinetti M, Del Porto G, Passi S (1996) Imbalance in the antioxidant pool in melanoma cells and normal melanocytes from patients with melanoma. J Investig Dermatol 107:322–326
22. Poljšak B, Dahmane R (2012) Free radicals and extrinsic skin aging. Dermatol Res Prac, pp. 1–4
23. Raha S, Robinson BH (2000) Mitochondria, oxygen free radicals, disease and ageing. Trends Biochem Sci 25:502–508
24. Reelfs O, Tyrrell RM, Pourzand C (2004) Ultravioleta radiation-induced immediate iron release is a key modulator of the activation of NF-kappaB in human skin fibroblasts. J Invest Dermatol 122:1440–1447
25. Rinnerthaler M, Bischof J, Streubel MK, Trost A, Richter K (2015). Oxidative stress in aging human skin. Biomol Ther 5:545–589
26. Sorg O, Antille C, Saurat J-H (2004) Retinoids, other topical vitamins, and antioxidants. Basic Clin Dermatol 28:89–116
27. Wei H, Cai Q, Rahn R, Zhang X (1997) Singlet oxygen involvement in ultraviolet (254 nm) radiation-induced formation of 8-hydroxy-deoxyguanosine in DNA. Free Radic Biol Med 23:148–154
28. World Health Organization (2002) Global Solar UV Index: A Practical Guide. WHO, Geneva
29. Zhang X, Rosenstein BS, Wang Y, Lebwohl M, Wei H (1997) Identification of possible reactive oxygen species involved in ultraviolet radiation-induced oxidative DNA damage. Free Radic Biol Med 23:980–985

Part III

UV Light and Human Diseases

UV-Induced Molecular Signaling Differences in Melanoma and Non-melanoma Skin Cancer

3

Feng Liu-Smith, Jinjing Jia, and Yan Zheng

Abstract

There are three major types of skin cancer: melanoma, basal cell carcinoma (BCC) and squamous cell carcinoma (SCC). BCC and SCC are often referred to as non-melanoma skin cancer (NMSC). NMSCs are relatively non-lethal and curable by surgery, hence are not reportable in most cancer registries all over the world. Melanoma is the deadliest skin cancer. Its incidence rate (case number) is about 1/10th of that for NMSC, yet its death toll is ~8 fold higher than NMSC.

Melanomas arise from melanocytes which are normally located on the basement membrane with dendrites extending into the epidermal keratinocytes. A major known function of melanocytes is to produce pigments which are enclosed by lipid membrane (termed melanosomes) and distribute them into keratinocytes, thus give different shade of skin colors. BCCs arise from basal cells, which are a layer of cells located at the deepest part of epidermis. Basal cells are recently considered to be skin stem cells as they are constantly proliferating and generating keratinocytes which are continuously pushed to the surface and eventually become a dead layer of stratum corneum. Squamous cells are the keratinocytes which resembles fish scale shape, ie, those initiated from basal cells and differentiated into squamous cells. Both basal cells and squamous cells belong to keratinocytes, therefore sometimes BCC and SCC are termed keratinocyte cancer.

These three types of cancer share many characteristics, yet they are very different from etiology to progression. One shared characteristic of skin cancer is that, according to the current views, they all are caused by

F. Liu-Smith (✉)
Department of Epidemiology, University of California, Irvine, CA, USA

University of California, Irvine, CA 92697, USA
e-mail: liufe@uci.edu

J. Jia • Y. Zheng
Department of Dermatology, The Second Affiliated Hospital, School of Medicine, Xi'an Jiaotong University, Xi'an, China

S. Ahmad (ed.), *Ultraviolet Light in Human Health, Diseases and Environment*, Advances in Experimental Medicine and Biology 996, https://doi.org/10.1007/978-3-319-56017-5_3

solar or artificial ultraviolet radiation (UVR). UVA and UVB from solar UVR are the major UV bands reaching the earth surface. Both UV types cause DNA damage and immune suppression which play crucial roles in skin carcinogenesis. UVB can be directly absorbed by DNA molecules and thus causes UV-signature DNA damages; UVA, on the other hand, may function through inducing cellular ROS which then causes oxidative DNA damages [1–4]. This chapter will discuss the molecular signaling differences of UVR in melanoma and NMSC.

Keywords
Ultraviolet • Melanoma • Basal cell carcinoma • Squamous cell carcinoma • Non-melanoma skin cancer • DNA damage • Oncogenes • Tumor suppressors

3.1 UVR Impact on NMSC

3.1.1 Introduction

NMSC refers to non-melanoma malignant growths that involve the skin and its appendages, which mainly include basal cell carcinoma (BCC) and squamous cell carcinoma (SCC). Their incidence varies in race and region, and mainly occurs in head, neck and other exposure sites of the elderly [5]. NMSC is the most frequently diagnosed cancer in the United States.

We receives UVR every day, and long-term UV exposure will lead to photo-aging, such as rough, loose and sagging skin, wrinkles and freckles, and even the occurrence of benign or malignant tumors. It has been well recognized that UVR is the main predisposition of NMSC. However, the molecular events leading to transformation is not fully understood. Now we attempt to summarize the current understanding of UVR in the occurrence and development of NMSC.

UV-induced skin carcinogenesis is a complex and continuous biological procedure, caused by the different wavelengths of UV. As stated above, UVB can be directly absorbed by DNA and thus causes DNA damage and changes in gene expression by intracellular signaling transduction, which induces skin cancer. UVA irradiation can produce reactive oxygen species (ROS), which cause secondary damage to DNA, and thus induce skin cancer, and often require chronic and cumulative exposure. Cells are able to repair the damaged DNA, however, DNA repair can go wrong due to genetic or environmental factors, therefore mutations in proto-oncogenes and tumor suppressor genes may occur, which lead to the formation of tumor. The various immunological reaction (often decreased after UV exposure, termed immune suppression), individual basic level antioxidant protection, virus or genetic predisposition may all participate in the regulation of DNA repair efficiency, thus affecting tumorigenesis in NMSC [1–4].

3.1.2 UVB-Caused DNA Damage in Keratinocytes and Basal Cells

UVB can penetrate several cell layers into the dermal layer of skin, and perhaps also the basal cell layer. The forms of DNA photodamage induced by UVB include pyrimidine dimer, purine and pyrimidine dimer, purine photoproduct and protein-DNA cross link and single strand break, among which the most important form of DNA photodamage is pyrimidine dimer.

The mechanisms of DNA photo-damage include: (1) after the DNA bases directly absorb UVB photons, the photoproducts, namely cyclobutane pyrimidine dimers (CPD, the majority) and 6-pyrimidine-4-pyrimidone photoproducts (6–4(PP), the minority), are formed between two adjacent pyrimidine sites (TT, CT, TC, CC) in the same DNA strand. Both of them are important premises for increased mutation frequency, and the basis of the UVR-induced skin cancer. Cells can repair (6–4) PP more effectively than CPD dimers. The thymine-cytosine (TC) and cytosine-cytosine (CC) dimers of CPD are the most mutagenic, because in UV-induced skin cancers, the C → T and CC → TT mutations can be often seen in p53 gene, thus termed UV signature mutation [1]. The main photoproduct thymine-thymine (TT) dimer rarely mutate due to the repair efficacy of DNA polymerase. (2) UVB can induce formation of dimers by adenine residues and thymine residues in DNA strands. Although the amount is limited, it has been confirmed that such dimers can cause mutations [1]. (3) UVB irradiation also induced oxidation of guanine, generating purine photoproduct, 8-hydroxy-2-deoxyguanosines (8-OHdG). 8-OHdG is the general marker of oxidative stress. Although it only accounts for a small part of DNA damages caused by UVB, it induces G → T translocation thus leading to DNA mutation. In addition, UVB radiation can also cause other types of DNA damage, such as protein-DNA cross-linking and single strand DNA breaks [6].

3.1.3 The DNA Repair Mechanisms

Normal human keratinocytes contain an effective DNA damage repair system, which can prevent a variety of gene mutations caused by UVR damage. DNA can be repaired through a number of signal pathways including DNA double strand break repair (DSB), nucleotide excision repair (NER), base excision repair (BER), mismatch repair (MMR), and post replication repair (PRR) [7]. For UV-induced lesions two major types of DNA repair systems play crucial roles: The NER and BER systems.

3.1.3.1 The NER System

It is the main repair pathway for DNA photoproducts. A large variety of DNA damages, such as CPD and (6–4) PP, can be repaired by this system. NER pathway is generally divided into the following steps: (1) recognition damage site; (2) cut the injury sites on both sides of the nucleotide chains; (3) remove short nucleotide fragments (24–32 bp) at damaged sites; (4) synthesize the gap in DNA chains; (5) join the synthesized nucleotide chain and the parent chain. The NER system is well demonstrated by a group of genetically compensated mutations for Xeroderma pigmentosum (XP), which is a rare autosomal recessive disease. Its characteristics are the high-risk occurrence of skin cancers such as SCC, BCC or epidermal benign tumors seborrheic keratosis at exposure sites. The molecular basis is the congenital defects of NER, i.e., mutations in a series of XP genes including XPA, XPB, XPC, XPD, XPE, XPF, and XPG, all of which play roles in the NER pathway. Mutations in these genes led to unrepaired damaged DNA and results in mutations in other genes [7–9].

In addition, the endonuclease V of bacteriophage T4 (T4 N5) is a special kind of cyclobutane pyrimidine dimer NER enzyme. Nishigori et al. [10] used the T4 N5 liposome to apply to UV-irradiated mice, and found that the delayed hypersensitivity (DTH) and the contact hypersensitivity (CHS) were inhibited, and the inhibitory T lymphocytes was induced after UVR. Using minimal erythema dose of UV to irradiate buttock skins of 15 patients with skin cancers, and topically applicate T4 N5 liposome after 2, 4 and 5 hours, the endonuclease could be found by biopsy after additional 6 hours. Immunohistochemical study on the damaged DNA showed that T4 N5 could significantly enhance the repair after DNA damage. Although the active T4 N5 could not significantly inhibit the formation of erythema and sunburn cells after UVR, it could almost completely inhibit UV mediated upregulation of IL-10, TNF-α and IL-10 protein levels.

3.1.3.2 The BER System

Single base lesions such as 8-OHdG rely on this mechanism to be repaired. BER system includes enzymatic glycosylation, replication protein A (RPA), proliferating cell nuclear antigen (PCNA), apurinic/apyrimidinic site endonuclease (APEX1/REF-1) [11]. Usually the glycosylation enzymes recognize the lesion and cut off the glycosylic bond between the base and the main DNA strand, thus the lesion presents as an abasic site (AP site), which is then recognized by the APEX1/REF-1 protein. The APEX1/REF-1 protein generates a single-strand break (a nick site) which is then repaired by DNA polymerase I and DNA ligase III [11]. NER pathway repairs endogenous DNA damages caused by ROS, active hydrolytic substance and alkylating agents. The activity of BER system is regulated by HOGG1, and inhibited by nitric oxide (NO) [12]. Therefore, the NO induced by UVR not only can cause DNA damage but also inhibit the activity of BER system, resulting in the increase of gene mutation and the risk of skin cancer [11].

3.1.4 UVR and Mutations

Usually, cells can accurately and effectively repair the CPD and (6–4) PP lesions after UVR. Defects in repair system or occasional error-prone repairs can lead to gene mutations. Mutations alter cell cycle regulation, which further lead to the cloning hyperplasia and immortalized growth, and eventually the occurrence of skin cancers. According to their different functions, these genes can be divided into: (1) caretaker genes: with the execution of DNA damage repair and maintenance of genome integrity, such as the repair gene XPA → XPG of XP. (2) Gatekeeper genes: regulate signaling transduction pathways, cell proliferation, differentiation and apoptosis, such as p53, patched (PTCH) and RAS gene. In NMSC gatekeepers are closely related to the occurrence of some skin cancers. Their main mutations are CC → TT at the site of pyrimidine-pyrimidine sequences, which are the same with the main mutations of CPD and (6–4) PP induced by UVB [13–16], i.e. the UV-signature mutations. The major tumor suppressors and oncogenes are discussed in details below.

3.1.4.1 The p53 Gene

The tumor suppressor gene p53 can regulate cell cycle, and its products play essential roles for the induction of apoptosis and the maintenance of normal cells. p53 monitors the integrity of the genome. If cellular DNA is damaged, p53 activates cell cycle check points and allow sufficient time to repair the damaged DNA. If repair fails, p53 can cause programmed cell death to prevent the generation of mutant cells. When a cell is affected by the external environment carcinogens, various reasons will lead to inactivation of p53 gene, so that DNA damage could not be repaired, resulting in cellular transformation. UVR played an essential role in causing mutations in p53 gene. Mutations in p53 gene can be detected in more than 90% of SCC and nearly 50% of BCC. The biopsy of sun exposed skin showed C → T conversion in DNA sequences [5]. The results clearly indicate that after daily exposure to sunlight, DNA mutations may exist in the skin without any pathological abnormalities. Brash et al. [17] suggested that the mutant p53 led to inability of apoptosis, resulting in clonal amplification of the p53-mutant cells in sun-exposed skin, resulting in accumulation of gene mutations. Mutations in p53 can be found in skin tumors such as BCC, SCC, and precancerous lesions such as Xeroderma pigmentosum [18] (Fig. 3.1).

UVR can also cause excessive expression of heat shock proteins (HSP) in skin. Experiments by Zhou et al. [19] cannot be replaced showed that when human keratinocytes exposed to 300 J/m^2 UVB in the culture medium, the heat shock protein expression was gradually increased and reached a peak at 6 h after irradiation. This kind of stress reaction was thought to be a double-edged sword, because it is extremely important to protect the skin from UVB, but also had certain potential carcinogenic potential. The immunofluorescence and immunoprecipitation analysis showed that HSP 27, HSP 70, HSP 90 and mutant p53 were

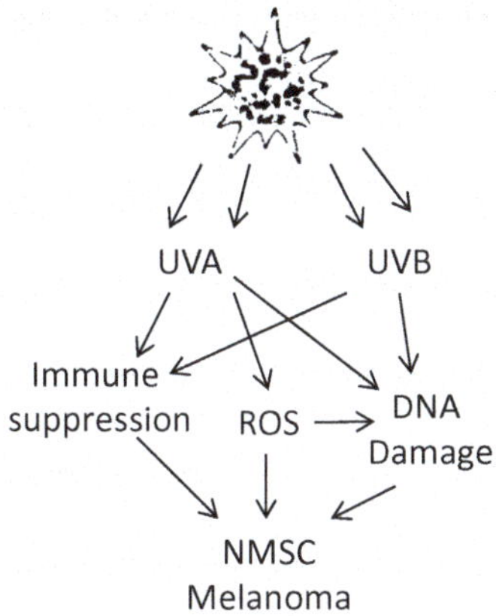

Fig. 3.1 Overview of UVR and skin cancer: The earth surface solar UVR mainly are UVA and UVB types, both of which cause immune suppression and DNA damage in human skin. UVA causes DNA damage via ROS while UVB directly cause DNA crosslink. UVR is the single most important known etiological factor for NMSC (both BCC and SCC) and melanoma

co-localized in the squamous cell carcinoma cell line A431, indicating that the occurrence of molecular chaperone complexes HSP and p53 had combined effect of tumorigenesis [20].

3.1.4.2 PTCH (PATCHED1)

PTCH is a tumor suppressor gene initially isolated from *Drosophila*. The gene product PATCHED1 is a membrane receptor for sonic hedgehog signaling pathway. The hedgehog-PTCH pathway is defective in over 70% of BCC according to one early report [21], and over 85% BCC in a recent study [22, 23]. In the absence of ligands, PTCH inhibits expression of smoothened (SMO) gene, which activates a transcriptional factor Gli1 and lead to tumor formation [21, 24] (Fig. 3.2). PTCH point mutations were found in 50% ~ 60% BCC and XP patients with DNA repair defects. Most of the PTCH mutations bore the UV fingerprint (i.e., C → T and CC → TT transitions at dipyrimidine sites); and their mutations were associated with irradiation dose. These mutations inactivate PTCH function and allow constitutive activation of the SMO-Gli pathway, which seems to be sufficient for BCC development [25]. But mutations could not be found in SCC, indicating that the PTCH gene mutation is a crucial step for BCC but not for SCC tumorigenesis.

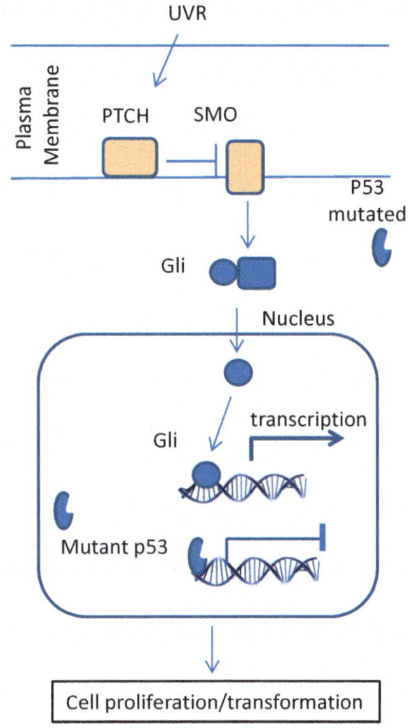

Fig. 3.2 The major oncogenic pathway for BCC: Mutations in tumor suppressor p53 and PTCH are all UV-caused mutation types. PTCH functions through SMO and Gli pathway for tumorigenesis while p53 mutations function through loss of genome integrity

3.1.4.3 PTEN

PTEN is a classic tumor suppressor inhibiting the pro-proliferating PI3K/AKT pathway. PTEN has recently evolved to show activities in DNA repair including DSB and NER pathways [7]. Targeted down-regulation of PTEN in mice epidermis predisposed these mice to skin lesions, and a possible mechanism is that PTEN loss caused XPC down-regulation, a crucial component for NER pathway [26, 27]. In addition, a recent genomic profiling of DNA mutations in BCC showed that a small percentage (2%) of BCC tumors harbor PIK3CA mutations which is the direct PTEN target [22].

3.1.4.4 RAS Oncogenes

There are three major RAS oncogenes: HRAS (Harvey-RAS), KRAS (Kristen-RAS) and NRAS (Neuronal RAS), all up-stream of MAPK (mitogen activated protein kinase) pathway which is a crucial cell proliferation pathway. The three RAS genes share sequence homology and mutation spectrums; most oncogenic mutations are found in codon 12, 13 and 61. It seems that RAS gene mutations are not common in BCCs and SCCs as published data is very limited. However, ectopic expression of HRAS in immortalized keratinocytes introduced an invasive phenotype [28].

3.1.4.5 Cell Cycle Genes and HPV Virus Infection

The cell cycle regulation genes include a number of cell division cycle (CDC) genes, kinases that regulate the CDC genes (CDKs) and inhibitors that inhibit the CDK genes (cyclin-dependent kinase inhibitors, CKIs) such as $p16^{INK4A}$, $p14^{ARF}$, $p19^{ARF}$ (all three proteins are encoded by the CDKN2A locus), $p21^{CIP1/WAF1}$ and $p27^{Kip1}$ (CDKN2B), and regulatory genes including retinoblastoma (RB) family genes, checkpoint kinase 2 (CHK2), ataxia telangiectasia mutated (ATM) protein, and p53. Viral oncogenes such as E6, E7 genes carried by HPV virus are able to interact with cell cycle genes and therefore interfere their function, leading to aberrant cell cycling. A recent study revealed that $p16^{INK4A}$ staining can serve as a biomarker for HPV infection [29]. Mutations in the above mentioned genes are often found in BCCs and SCCs and are correlated with UVR [30]. On the other hand, UVR also induces cell cycle arrest through these genes, for example, several studies showed that UVB induced $p16^{INK4A}$ upregulation; and CDKN2A mutations were found in 24% of SCC and 3.5% of BCC samples [31].

3.1.5 UVR-Induced Immunosuppression

Skin tumors are the result of both gene mutations and the loss of immune surveillance. Numerous studies revealed that UVR triggered immune suppression response in skin through different signaling molecules, including the Fas/FasL system, TNF, interleukins, or even simple induction of apoptosis of T cell.

3.1.5.1 The Fas/FasL System

Fas, also known as Apo-1, now named CD95, belongs to the tumor necrosis factor receptor and nerve growth factor receptor family and is a type I cell membrane protein. Fas gene is expressed in various types of tissues, and skin tissues express relatively higher levels of Fas. Fas and its ligand (FasL) together can induce apoptosis and play important roles in maintaining the stability and homeostasis of skin micro- environment, but its excessive expression may contribute to immune escape of some tumor cells. During tumor development, tumor cells can actively express FasL to kill the infiltrating T cells, thus weakening the immune surveillance system. Fas was normally found only in the cell membranes and intercellular bridges of keratinocytes in the basal layers of the epidermis [32]. After long-term exposure to sunlight, Fas was up-regulated in whole layers of the epidermis [32], thus induced necessary apoptosis if cells were beyond repair. This UV-induced up-regulation of Fas can be inhibited by protein kinase C epsilon (PKC) [33], resulting in proliferation of cells with mutations, thus enhancing UV-induced carcinogenesis.

3.1.5.2 Langerhans Cells

UVR also elevates the TNF-α, IL-6 and IL-10 levels in epidermal cells, which are involved in down-regulating the activities of Langerhans cells (LC), thereby inhibiting the immune reaction of the local skin [34]. Clinical and experimental results show that the LC of the skin is the main target of UVR. LC is the most important antigen presenting cell, and is the key of UVB induced skin immune suppression [35]. The main manifestation of UVB induced skin immune suppression is the defects of antigen presentation ability. The mechanism may include the following aspects: (1) UVR directly damages the epidermal LC, reduces its number and makes alters their morphology and function; (2) UVR interferes the expression of membrane co-stimulating molecules of LC and influences its function of antigen presenting to T lymphocytes; (3) UVR interferes the migration, differentiation and maturation of LC from epidermis to draining lymph nodes; (4) light-induced DNA damage may be another cause of UVR induced immunosuppression; (5) UVR induces secretion of IL-10 and TNF-α by keratinocytes, which may be related to the inhibition of delayed hypersensitivity and contact allergy [36].

Although LC is the core cell of skin immune suppression after UVR, the morphological and functional changes of LC are not the only mechanisms of UV impact on immune function. The study on T lymphocytes, macrophages, mast cells and cytokines such as IL-1, IL-10 and IL-12 has become a hot topic in recent years.

T cells play an important role in anti-tumor immunity. T cells can be divided into two subgroups: CD4+ and CD8+. CD4+ is the T helper cell and CD8+ is the suppressor T cell. Many studies on infiltrating T cell subtypes of tumor tissues suggest that the ratio of CD4/CD8 is often negatively correlated with the degree of malignancy, but positively correlated with the prognosis. The experimental results showed that the CD3+ T cells in human skins only expressed CD4 but not CD8 1 week after UVR, and the ratio of CD4/CD8 was up-regulated [37, 38], indicating that UVR could activate inhibitory CD4+ T cell and then mediate the proliferation of antigen specific CD8+ T cells, thus executing the inhibitory effect. If the changes of T cell subtypes cannot recover to a normal state and thus persist, it will result in a state of low immunity, which promotes tumor development. In addition, UVR may also inhibit Th1 type of cellular immune response, with no significant effect on Th2 type immune response. But these results need further confirmation as there was a contradictory report [39]. The exact effect of UVB irradiation on local changes of T cells has not yet been fully explained.

3.1.5.3 Natural Killer Cells and Master Cells

Natural killer cells (NK) are important immune cells of the innate immune system. NKs not only are related to anti-tumor, anti-viral infection and immune regulation, but also in some cases participate in the occurrence of hypersensitivity and autoimmune diseases. UVR can produce dose-dependent inhibition on NK cells. After UVB irradiation, the number of NK was decreased, and this inhibitory effect was positively correlated with the UVB dose. Long-term exposure to UVR can decrease the number of NK and inhibit its activity [40].

Animal experiments found that mast cells played key roles in the systemic immune modulation induced by UVB in mice. UVB dose required to complete the 50% immunosuppression was linearly related to the number of mast cells, significant increase of mast cells was also found in the unexposed buttocks of BCC patients [41]. Studies of histamine receptor antagonists demonstrated that histamine was the main product of mast cells, and played a role in the prostaglandin dependent pathways. Almost 50% of normal people and more than 90% of BCC or SCC patients are prone to produce inhibitory contact hypersensitivity reaction. This susceptibility was not related to the polymorphism of cytokines, nor the trans-urocanic acid levels induced by UVB in the exposure sites. It is speculated that the mast cells in the human body can induce immune suppression and create an environment for the development of tumors [41].

3.1.5.4 Cyclooxygenase (COX-2)

Cyclooxygenase (COX-2) is the key enzyme to catalyze the initial step of arachidonic acid into active prostaglandin. A large number of studies have indicated that COX-2 is involved in the occurrence and development of many kinds of tumors in addition to the inflammatory response. UVB exposure caused excessive expression of COX-2 in mouse skin, which was confined to the superficial layer of epidermis, dominantly on the stratum granulosum and stratum spinosum of SCC [42]. Upon UVR, mice deficient in the COX-2 enzyme or those treated with pharmacological inhibitors of COX-2 developed significantly fewer tumors than control mice [43]. Furthermore, COX-2 inhibitor indomethacin (which also inhibits COX-1) diminished the UV-induced immune suppression response [44].

3.2 The UV Impact on Melanoma

Melanomas arise from the pigment-producing cells called melanocytes, which normally reside in the basal cell layer and extend their dendrites into the keratinocytes for melanin transportation and other cell-cell contacts. Usually one melanocyte is corresponding to about 40 keratinocytes. Melanocytes also interact with skin fibroblasts and basal cells, which make this type of cell an excellent model for studying cell-cell interaction. Many researchers believe that the microenvironment of melanocytes plays an important role in melanoma development. In this chapter we will only focus on the UV impact on melanomagenesis.

3.2.1 The UVR-Induced DNA Damage in Melanoma

Because melanocytes are located relatively in the deeper layer of the skin, it has been under debate for the past decades as whether UVA or UVB caused melanoma. UVA is able to penetrate to, but UVB can hardly reach, the melanocytes layer. Although UVA accounts for most of the solar UV radiation (90–95%), the relative smaller portion of UVB exhibits higher energy which is more potent in inducing DNA damage. As discussed above for NMSC, UVR causes similar mutations at the DNA nucleotide levels in melanoma, namely the UV-signature mutations. A distinct difference is that the target genes are very different. For example, mutations in p53 are common in NMSC but very rare in melanoma.

3.2.1.1 The MAP Kinase Pathway: NRAS and BRAF Oncogenes

Mutations in NRAS were found in 15–30% of melanomas and mutations in BRAF were found in ~60% of melanomas [45]. Mutations in these two oncogenes are generally mutually exclusive, i.e., if a tumor harbors an NRAS mutation, it would not contain a BRAF mutation [46]. This is mainly because both of these genes are upstream of the crucial cell proliferation pathway: the mitogen activated protein kinase (MAPK) pathway (Fig. 3.3). Quite interestingly, BRAF mutations were found mostly on intermittently sun-exposed body sites [47], while NRAS mutations are more frequently found on chronic sun damaged body sites. All mutations are activating mutations which enhance the MAPK pathway signaling. Most frequently mutated codon (>90%) in BRAF is codon 600 (nucleotide 1799 T > A; codon GTG > GAG) from a valine to a glutamic acid. This mutation, however, is not a typical UV-signature mutation, hence the cause of the most common mutation in melanoma is actually unknown. $BRAF^{V600E}$ mutation is also frequently found in many other cancer types such as colon cancer and thyroid cancer, which are not related to UV exposure.

In contrast to BRAF mutation, NRAS mutations are more frequently found in sun exposed body sites [48, 49]. The most frequent mutation of NRAS found in melanoma is codon 61 mutation from CAA (glutamine) to AAA (Lysine) or CGA (Argenine) (Q61K or Q61R) [48]. Opposite strain of this codon is TTG to TTT or TCG. Again this is a C > A or A > G point mutation on the plus strand, and G > T or T > C on the minus strand, and are not the typical C > T or CC > TT type of UV signature mutation. These mutations are also found in other cancer types, not specific for melanoma. Therefore the mutations on NRAS, like those found in BRAF codon 600, are not typical UV signature mutations. However, when the entire genome mutation profile was analyzed, the UV signature mutations still ranked number one in melanoma among all different signatures mutations [50]. Nevertheless, comparing these mutations to mutation types in bacteria in a previous study [51], we believe that NRASQ61K or Q61L mutations may be derived from oxidative DNA damage, which could be a result from UVA radiation.

3.2.1.2 The Newly Discovered Oncogenes: RAC1and PREX2

The current genomics technology allowed for large scale genomic profiling of all types of cancer. Such studies revealed many previously unknown oncogenes in melanoma, including RAC1, PREX2

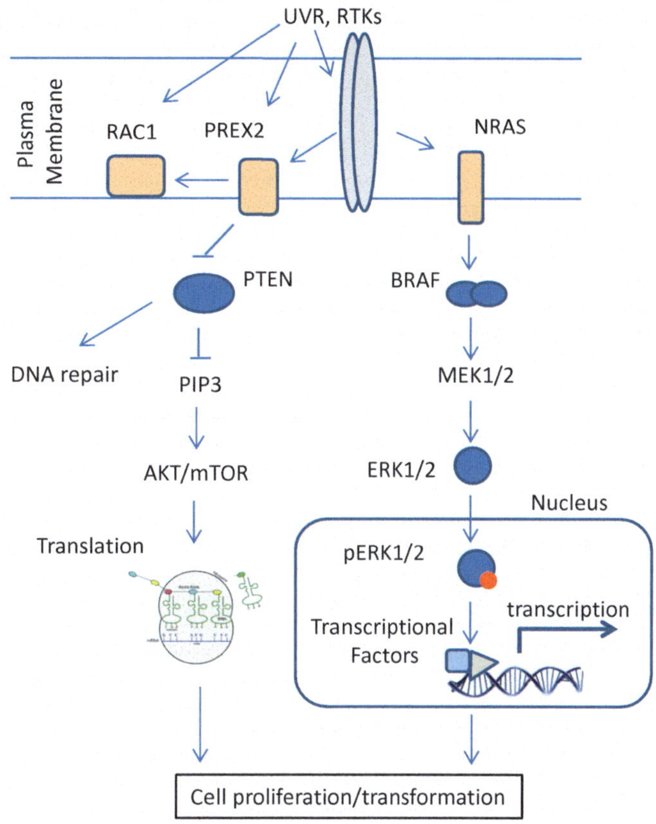

Fig. 3.3 The major oncogenic pathway for melanoma: For MAPK pathway UV promotes cellular transformation for cells carrying BRAF or NRAS oncogenic mutations already. For the PI3K pathway, UV may function through activating RAC1 and PREX2 oncogenes. PIP3: Phosphatidylinositol (3,4,5)-trisphosphate

and ERBB4 [52–54]. Evidence for RAC1 and PREX2 is solid and confirmed by multiple studies, but evidence for ERBB4 is contradictory [55]. In fact PREX2 and RAC1 are functionally linked because they interact with each other to promote PI3K/AKT pathway [56, 57], which ultimately cross talks with the MAPK pathway for promoting melanoma cell proliferation [58, 59] (Figure 3.3). PREX2 interacts with PTEN and inhibits PTEN function which is an inhibitor for PI3K pathway [60]. PREX2 full name is "phosphatidylinositol-3,4,5-trisphosphate-dependent RAC exchange factor 2". The PREX2 mutations in melanoma are also more frequently found in sun-exposed body sites [53]. 11 of 25 melanomas harbored at least one non-synonymous mutations, the mutation spectrum included three truncating mutations (K278*, E824* and Q1430*) and non-synonymous point mutations. Patients with these mutations showed worse survival as compared to patients with wild-type PREX2 [53].

The most studied mutation in RAC1 is an activating mutation (P29S), which ranks as the third most common mutations in sun-exposed melanomas (the other two were $NRAS^{Q61}$ and $BRAF^{V600}$ mutations) [52, 61–64]. More importantly, $RAC1^{P29S}$ is a typical UV-signature mutation (C- > T transition, unlike those in the $NRAS^{Q61}$ and $BRAF^{V600}$ mutations), suggesting a UV radiation origin of this mutation [61]; and it is a driving mutation for melanoma [52].

3.2.1.3 MiTF

MiTF-m (microphthalmia transcriptional factor, melanocytic specific) is a melanocytes lineage transcriptional factor, which is one of the several transcripts derived from the MiTF locus driven by a melanocytic-specific promoter [65]. MiTF is phosphorylated by ERK1/2, resulting in proteasome degradation of this protein [66, 67]. Our results showed that UVB did not affect MiTF-m expression but UVA induced a transient MiTF degradation [68]. This observation seemingly contradicts the role of MiTF as a master regulator of melanin synthesis, because UVR induces melanin synthesis. The exact role of MiTF in UVR induced signal cascade is still to be defined. MiTF mutations including gene amplification were found in a small set of melanoma samples [69]. Although MiTF played a pro-survival role and serves as a specific and sensitive melanoma diagnostic biomarker [70], evidence from many research laboratories suggested that the role of MiTF was much more complex than an oncogene because over-expression of MiTF could directly leading to cell cycle arrest and differentiation [69, 71]. Therefore the cellular dose of MiTF seems to be crucial for its function.

3.2.1.4 Cell Cycle Genes: CDK4 and CDKN2A

Of all the cell cycle genes and regulators, p16^{INK4A}, CCND1 (cyclin D1) and CDK4 are the three major genes which were found in familial melanoma cohorts and sporadic melanomas [72]. CDK4 and CCND1 amplification was found in many melanoma tumors without mutations in BRAF or NRAS [73], and loss of p16^{INK4A}, which enhances the CDK4 activity and promotes cell proliferation, was identified in most familial melanoma cases [74]. As described above in this chapter, CDKN2A locus encodes both p16^{INK4A} and p19^{ARF} proteins. Loss of either gene product sensitized melanocytes to UVR induced transformation [75, 76]. On the other hand, UVR also induced decreased expression of p16^{INK4A} in skin cancer patients but not in the normal healthy control individuals [77]. However, it seemed loss of p16^{INK4A} or amplification of CDK4 did not impact the cellular DNA repair capacity [78]; therefore the melanogenesis through these cell cycle genes may be more through promoting cell proliferation.

CDKN2A mutations are mostly gene deletion, i.e., 9p21 deletions [79]. Point mutations and short sequence duplications within the coding region are also reported [80]. As these are all familial cases, therefore the mutations are germline and not related to solar UVR. Quite interestingly, ambient UVR seems not to promote melanoma development in individuals carrying these mutations [81]. CDK4 mutation spectrum includes gene amplification, R24C and R24H point mutations. Both of the point mutations impair the CDK4 binding with p16^{INK4A}, thus evading the p16^{Ink4A} inhibitory effect [82].

3.2.2 UV Radiation and Melanoma Oncogenes

Although mutations in NRAS and BRAF were found in most of human melanoma samples, these same mutations were also found in benign nevi, and the percentage of these mutations were even higher than that in melanoma [83]. In general, it is believed that activating mutations of these oncogenes in fact led to senescence, which is a typical feature for benign nevi. New evidence from mouse models showed that UV played a crucial role in stimulating these cells into cycling and hence promoting oncogenesis. In a BRAF^{V600E} mouse model, under normal condition without other mutation background, only skin hyperplasia was induced, without melanoma formation [84]. But when PTEN was also deleted, this model would produce melanoma [84]. Furthermore, one time UVR greatly enhanced the melanoma formation in these mice, and the mechanism was through p53 pathway [74]. Therefore in the mouse model, the BRAF mutations rely on loss of tumor suppressor p53 and/or PTEN to induce melanoma. The p53 gene was in general considered not to be crucial for melanoma development as mutations in this gene was rare; however updated data suggest that other mutations in the p53 pathway still played a prominent role.

3.2.3 UV-Induced Immune Suppression in Melanoma

The pioneer work by Kripke in 1974 demonstrated that UV carcinogen induced much higher degree of antigenicity compared with skin tumors induced by chemical carcinogens [85]. Up to date, the triggers for UV-induced immune suppression include three major classes: DNA damage, urocanic acid (UCA) and membrane lipids

[86, 87]. UCA is a skin-rich histamine deamination product which is a well-known activator for immune suppression [88]. UCA antibody was shown to delay skin tumor formation upon UVR [89]. The local immune suppression, as described in NMSC section, was through local immune cells, mainly Langerhans cells and suppressor T cells, while the systemic immune suppression was through cytokines [86]. A more recent study indicated that solar simulated UVR did not induce systemic immune changes in both men and women, while the local response was prominent [90]. Quite interestingly, the solar simulated UVR caused more severe immune suppression in men than in women [90], which may provide a partial explanation why melanoma incidence in men is higher than that in women.

Further evidence that UVR-induced immune suppression plays a role in melanomagenesis stems from epidemiological studies using immune-suppressed populations. A comprehensive study summarized melanoma and NMSC incidence rates in organ transplant patients (ORPs) in the literature [91]. ORPs with a pre-transplant melanoma history showed a 19% rate of recurrence, which was much greater than the recurrence rate for general population [92]. ORPs also showed higher de novo melanoma incidence rates than the general population [91].

3.3 Conclusions

There are shared and distinct components in the mutation spectrum and signal transduction in UVR-induced NMSC and melanoma. The difference and share characteristics of NMSC and melanoma are summarized in Figs. 3.2 and 3.3. Specifically, the oncogenic mutation spectrum and signal transduction pathways are quite different in NMSC and melanoma, although all are related or induced by UVR. The well- studied tumor suppressors in NMSC include PTCH, p53 and PTEN. Oncogenes for BCC and SCC are much less defined than those in melanoma [93]. In melanoma a major oncogenic pathway is the MAPK pathway including NRAS, BRAF oncogenes and PTEN tumor suppressor. New evidence suggested that p53 and RB pathway was also important, despite the classical point mutations in p53 or RB were very rare in melanoma. PTCH mutation was also rare in melanoma.

The UVR-induced immune suppression pathway is very similar in both NMSC and melanoma. Both local and systemic immune responses are involved, with similar cell types (Langerhans cells, T cells and other immune cells) and cytokines involved. The antigen-presenting function of each cell type (basal cells, keratinocytes and melanocytes) may be slightly different due to different cellular components.

Many other shared components in oncogenesis in NMSC and melanoma such as redox regulating genes including APEX1/REF-1, the AP-1 gene family, the NFE2L1/NFE2L2 (nuclear factor erythroid 2 [NF-E2]-related factor 1 or 2, NRF1/NRF2) gene family, NADPH oxidase gene family, nitric oxide synthase gene family are not included in this chapter because the mutations in these genes seem not to be very common in both cancer types. However they all showed certain activities in UV-induced oncogenesis. It is important to understand whether some sequence variations in the population affect the skin cancer susceptibility.

Overall, although UVR is the single most important etiological environmental factor for both NMSC and melanoma, the oncogenic pathway, mutation spectra and tumor suppression genes are quite cell-specific. For this reason it was proposed that cancers may exhibit lineage specific oncogene addition which is crucial for tumor development and growth [94]. In melanoma the MAPK pathway plus the MiTF-dependent lineage specific oncogenesis play the pro-survival role for cancer cells [69], while in BCC the hedgehog-PTCH-SMO-Gli pathway is crucial. Thus the targeted therapy and prevention methods may require different approaches for these different cancer types.

References

1. Seebode C, Lehmann J, Emmert S (2016) Photocarcinogenesis and skin cancer prevention strategies. Anticancer Res 36(3):1371–1378
2. Battie C, Verschoore M (2012) Cutaneous solar ultraviolet exposure and clinical aspects of photodamage. Indian J Dermatol Venereol Leprol 78(Suppl 1):S9–S14
3. Huang XX, Bernerd F, Halliday GM (2009) Ultraviolet A within sunlight induces mutations in the epidermal basal layer of engineered human skin. Am J Pathol 174(4):1534–1543
4. Karran P, Brem R (2016) Protein oxidation, UVA and human DNA repair. DNA Repair (Amst) 44:178–185
5. Kumar R, Deep G, Agarwal R (2015) An Overview of Ultraviolet B Radiation-Induced Skin Cancer Chemoprevention by Silibinin. Curr Pharmacol Rep 1(3):206–215
6. Afaq F et al (2007) Delphinidin, an anthocyanidin in pigmented fruits and vegetables, protects human HaCaT keratinocytes and mouse skin against UVB-mediated oxidative stress and apoptosis. J Invest Dermatol 127(1):222–232
7. Kim Y, He YY (2014) Ultraviolet radiation-induced non-melanoma skin cancer: regulation of DNA damage repair and inflammation. Genes Dis 1(2):188–198
8. Zhang L, Gong F (2016) The emerging role of deubiquitination in nucleotide excision repair. DNA Repair (Amst) 44:118–122
9. Sugasawa K (2016) Molecular mechanisms of DNA damage recognition for mammalian nucleotide excision repair. DNA Repair (Amst) 44:110–117
10. Nishigori C et al (1996) The immune system in ultraviolet carcinogenesis. J Investig Dermatol Symp Proc 1(2):143–146
11. Quinones JL, Demple B (2016) When DNA repair goes wrong: BER-generated DNA-protein crosslinks to oxidative lesions. DNA Repair (Amst) 44:103–109
12. Suzuki T et al (1998) Misincorporation of 2′-deoxyoxanosine 5′-triphosphate by DNA polymerases and its implication for mutagenesis. Biochemistry 37(33):11592–11598
13. de Gruijl FR, van Kranen HJ, Mullenders LH (2001) UV-induced DNA damage, repair, mutations and oncogenic pathways in skin cancer. J Photochem Photobiol B 63(1–3):19–27
14. Norgauer J et al (2003) Xeroderma pigmentosum. Eur J Dermatol 13(1):4–9
15. Ehrhart JC et al (2003) UVB-induced mutations in human key gatekeeper genes governing signalling pathways and consequences for skin tumourigenesis. Photochem Photobiol Sci 2(8):825–834
16. Besaratinia A et al (2011) Wavelength dependence of ultraviolet radiation-induced DNA damage as determined by laser irradiation suggests that cyclobutane pyrimidine dimers are the principal DNA lesions produced by terrestrial sunlight. FASEB J 25(9):3079–3091
17. Brash DE (2006) Roles of the transcription factor p53 in keratinocyte carcinomas. Br J Dermatol 154(Suppl 1):8–10
18. Giglia-Mari G, Sarasin A (2003) TP53 mutations in human skin cancers. Hum Mutat 21(3):217–228
19. Zhou X et al (1998) Heat shock transcription factor-1 regulates heat shock protein-72 expression in human keratinocytes exposed to ultraviolet B light. J Invest Dermatol 111(2):194–198
20. Kindas-Mugge I et al (2002) Characterization of proteins associated with heat shock protein hsp27 in the squamous cell carcinoma cell line A431. Cell Biol Int 26(1):109–116
21. Boukamp P (2005) Non-melanoma skin cancer: what drives tumor development and progression? Carcinogenesis 26(10):1657–1667
22. Bonilla X et al (2016) Genomic analysis identifies new drivers and progression pathways in skin basal cell carcinoma. Nat Genet 48(4):398–406
23. Mizuno T et al (2006) Molecular basis of basal cell carcinogenesis in the atomic-bomb survivor population: p53 and PTCH gene alterations. Carcinogenesis 27(11):2286–2294
24. Denef N et al (2000) Hedgehog induces opposite changes in turnover and subcellular localization of patched and smoothened. Cell 102(4):521–531
25. Rahnama F et al (2006) Inhibition of GLI1 gene activation by Patched1. Biochem J 394(Pt 1):19–26
26. Ming M et al (2011) PTEN positively regulates UVB-induced DNA damage repair. Cancer Res 71(15):5287–5295
27. Ming M et al (2011) UVA induces lesions resembling seborrheic keratoses in mice with keratinocyte-specific PTEN downregulation. J Invest Dermatol 131(7):1583–1586
28. Vaughan MB et al (2009) H-ras expression in immortalized keratinocytes produces an invasive epithelium in cultured skin equivalents. PLoS One 4(11):e7908
29. Michaelsen SH, Larsen CG, von Buchwald C (2014) Human papillomavirus shows highly variable prevalence in esophageal squamous cell carcinoma and no significant correlation to p16INK4a overexpression: a systematic review. J Thorac Oncol 9(6):865–871
30. Conscience I et al (2006) P16 is overexpressed in cutaneous carcinomas located on sun-exposed areas. Eur J Dermatol 16(5):518–522
31. Soufir N et al (1999) P16 UV mutations in human skin epithelial tumors. Oncogene 18(39):5477–5481
32. Filipowicz E et al (2002) Expression of CD95 (Fas) in sun-exposed human skin and cutaneous carcinomas. Cancer 94(3):814–819
33. de Gruijl F (2009) Protein kinase Cepsilon reveals importance of extrinsic apoptosis in preventing UV carcinogenesis. J Invest Dermatol 129(8):1853–1856
34. Petit-Frere C et al (1998) Induction of interleukin-6 production by ultraviolet radiation in normal human epidermal keratinocytes and in a human keratinocyte cell line is mediated by DNA damage. J Invest Dermatol 111(3):354–359

35. Bacci S, Alard P, Streilein JW (2001) Evidence that ultraviolet B radiation transiently inhibits emigration of Langerhans cells from exposed epidermis, thwarting contact hypersensitivity induction. Eur J Immunol 31(12):3588–3594
36. Nishisgori C (2015) Current concept of photocarcinogenesis. Photochem Photobiol Sci 14(9):1713–1721
37. Kasahara S, Wago H, Cooper EL (2002) Dissociation of innate and adaptive immunity by UVB irradiation. Int J Immunopathol Pharmacol 15(1):1–11
38. Boehm T et al (2012) VLR-based adaptive immunity. Annu Rev Immunol 30:203–220
39. Schmitt DA, Ullrich SE (2000) Exposure to ultraviolet radiation causes dendritic cells/macrophages to secrete immune-suppressive IL-12p40 homodimers. J Immunol 165(6):3162–3167
40. Miyauchi-Hashimoto H et al (2005) Ultraviolet radiation-induced impairment of tumor rejection is enhanced in xeroderma pigmentosum a gene-deficient mice. J Invest Dermatol 124(6):1313–1317
41. Hart PH, Grimbaldeston MA, Finlay-Jones JJ (2001) Sunlight, immunosuppression and skin cancer: role of histamine and mast cells. Clin Exp Pharmacol Physiol 28(1–2):1–8
42. Athar M et al (2001) Ultraviolet B(UVB)-induced cox-2 expression in murine skin: an immunohistochemical study. Biochem Biophys Res Commun 280(4):1042–1047
43. Elmets CA, Ledet JJ, Athar M (2014) Cyclooxygenases: mediators of UV-induced skin cancer and potential targets for prevention. J Invest Dermatol 134(10):2497–2502
44. Soontrapa K et al (2011) Prostaglandin E2-prostaglandin E receptor subtype 4 (EP4) signaling mediates UV irradiation-induced systemic immunosuppression. Proc Natl Acad Sci U S A 108(16):6668–6673
45. Davies H et al (2002) Mutations of the BRAF gene in human cancer. Nature 417(6892):949–954
46. Brose MS et al (2002) BRAF and RAS mutations in human lung cancer and melanoma. Cancer Res 62(23):6997–7000
47. Curtin JA et al (2005) Distinct sets of genetic alterations in melanoma. N Engl J Med 353(20):2135–2147
48. Omholt K et al (2003) NRAS and BRAF mutations arise early during melanoma pathogenesis and are preserved throughout tumor progression. Clin Cancer Res 9(17):6483–6488
49. Pavey S et al (2004) Microarray expression profiling in melanoma reveals a BRAF mutation signature. Oncogene 23(23):4060–4067
50. Alexandrov LB et al (2013) Signatures of mutational processes in human cancer. Nature 500(7463):415–421
51. Cheng KC et al (1992) 8-Hydroxyguanine, an abundant form of oxidative DNA damage, causes G----T and A----C substitutions. J Biol Chem 267(1):166–172
52. Hodis E et al (2012) A landscape of driver mutations in melanoma. Cell 150(2):251–263
53. Berger MF et al (2012) Melanoma genome sequencing reveals frequent PREX2 mutations. Nature 485(7399):502–506
54. Lee MKT, Sharma A, Czerniecki BJ (2010) It's all in for the HER family in tumorigenesis. Expert Rev Vaccin 9(1):29–34
55. Brockhoff G et al (2011) No evidence for ErbB4 gene amplification in malignant melanoma. Acta Derm Venereol 91(4):488–490
56. Mense SM et al (2015) PTEN inhibits PREX2-catalyzed activation of RAC1 to restrain tumor cell invasion. Sci Signal 8(370):ra32
57. Barrows D et al (2015) p21-activated Kinases (PAKs) Mediate the Phosphorylation of PREX2 Protein to Initiate Feedback Inhibition of Rac1 GTPase. J Biol Chem 290(48):28915–28931
58. Graells J et al (2004) Overproduction of VEGF concomitantly expressed with its receptors promotes growth and survival of melanoma cells through MAPK and PI3K signaling. J Invest Dermatol 123(6):1151–1161
59. Paluncic J et al (2016) Roads to melanoma: key pathways and emerging players in melanoma progression and oncogenic signaling. Biochim Biophys Acta 1863(4):770–784
60. Fine B et al (2009) Activation of the PI3K pathway in cancer through inhibition of PTEN by exchange factor P-REX2a. Science 325(5945):1261–1265
61. Halaban R (2015) RAC1 and melanoma. Clin Ther 37(3):682–685
62. Krauthammer M et al (2012) Exome sequencing identifies recurrent somatic RAC1 mutations in melanoma. Nat Genet 44(9):1006–1014
63. Mar VJ et al (2014) Clinical and pathological associations of the activating RAC1 P29S mutation in primary cutaneous melanoma. Pigment Cell Melanoma Res 27(6):1117–1125
64. Watson IR et al (2014) The RAC1 P29S hotspot mutation in melanoma confers resistance to pharmacological inhibition of RAF. Cancer Res 74(17):4845–4852
65. Shibahara S et al (2001) Microphthalmia-associated transcription factor (MITF): multiplicity in structure, function, and regulation. J Investig Dermatol Symp Proc 6(1):99–104
66. Weilbaecher KN et al (2001) Linkage of M-CSF signaling to Mitf, TFE3, and the osteoclast defect in Mitf(mi/mi) mice. Mol Cell 8(4):749–758
67. Hemesath TJ et al (1998) MAP kinase links the transcription factor Microphthalmia to c-Kit signalling in melanocytes. Nature 391(6664):298–301
68. Liu F et al (2010) MiTF links Erk1/2 kinase and p21 CIP1/WAF1 activation after UVC radiation in normal human melanocytes and melanoma cells. Mol Cancer 9:214
69. Garraway LA et al (2005) Integrative genomic analyses identify MITF as a lineage survival oncogene amplified in malignant melanoma. Nature 436(7047):117–122

70. King R et al (1999) Microphthalmia transcription factor. A sensitive and specific melanocyte marker for MelanomaDiagnosis. Am J Pathol 155(3):731–738
71. Wellbrock C, Arozarena I (2015) Microphthalmia-associated transcription factor in melanoma development and MAP-kinase pathway targeted therapy. Pigment Cell Melanoma Res 28(4):390–406
72. Lee B, Sandhu S, McArthur G (2015) Cell cycle control as a promising target in melanoma. Curr Opin Oncol 27(2):141–150
73. Kwong LN et al (2012) Oncogenic NRAS signaling differentially regulates survival and proliferation in melanoma. Nat Med 18(10):1503–1510
74. Viros A et al (2014) Ultraviolet radiation accelerates BRAF-driven melanomagenesis by targeting TP53. Nature 511(7510):478–482
75. Luo C et al (2013) Loss of ARF sensitizes transgenic BRAFV600E mice to UV-induced melanoma via suppression of XPC. Cancer Res 73(14):4337–4348
76. Tsao H et al (2012) Melanoma: from mutations to medicine. Genes Dev 26(11):1131–1155
77. Krahn G et al (2001) UVB-induced decrease of p16/CDKN2A expression in skin cancer patients. Pigment Cell Res 14(3):201–205
78. Shannon JA et al (1999) Normal repair of ultraviolet radiation-induced DNA damage in familial melanoma without CDKN2A or CDK4 gene mutation. Melanoma Res 9(2):133–137
79. Fountain JW et al (1992) Homozygous deletions within human chromosome band 9p21 in melanoma. Proc Natl Acad Sci U S A 89(21):10557–10561
80. Wadt KA et al (2015) Molecular characterization of melanoma cases in Denmark suspected of genetic predisposition. PLoS One 10(3):e0122662
81. Cust AE et al (2011) Melanoma risk for CDKN2A mutation carriers who are relatives of population-based case carriers in Australia and the UK. J Med Genet 48(4):266–272
82. Zuo L et al (1996) Germline mutations in the p16INK4a binding domain of CDK4 in familial melanoma. Nat Genet 12(1):97–99
83. Bastian BC (2014) The molecular pathology of melanoma: an integrated taxonomy of melanocytic neoplasia. Annu Rev Pathol 9:239–271
84. Dankort D et al (2009) Braf(V600E) cooperates with Pten loss to induce metastatic melanoma. Nat Genet 41(5):544–552
85. Kripke ML (1974) Antigenicity of murine skin tumors induced by ultraviolet light. J Natl Cancer Inst 53(5):1333–1336
86. Ullrich SE (2005) Mechanisms underlying UV-induced immune suppression. Mutat Res 571(1–2):185–205
87. Muller HK et al (2008) Effect of UV radiation on the neonatal skin immune system - implications for melanoma. Photochem Photobiol 84(1):47–54
88. De Fabo EC, Noonan FP (1983) Mechanism of immune suppression by ultraviolet irradiation in vivo. I. Evidence for the existence of a unique photoreceptor in skin and its role in photoimmunology. J Exp Med 158(1):84–98
89. Beissert S et al (2001) IL-12 prevents the inhibitory effects of cis-urocanic acid on tumor antigen presentation by Langerhans cells: implications for photocarcinogenesis. J Immunol 167(11):6232–6238
90. Damian DL et al (2008) UV radiation-induced immunosuppression is greater in men and prevented by topical nicotinamide. J Invest Dermatol 128(2):447–454
91. Kubica AW, Brewer JD (2012) Melanoma in immunosuppressed patients. Mayo Clin Proc 87(10):991–1003
92. Penn I (1996) Malignant melanoma in organ allograft recipients. Transplantation 61(2):274–278
93. Marionnet C et al (2003) Differential molecular profiling between skin carcinomas reveals four newly reported genes potentially implicated in squamous cell carcinoma development. Oncogene 22(22):3500–3505
94. Weinstein IB, Joe AK (2006) Mechanisms of disease: oncogene addiction--a rationale for molecular targeting in cancer therapy. Nat Clin Pract Oncol 3(8):448–457

Xeroderma Pigmentosa Group A (XPA), Nucleotide Excision Repair and Regulation by ATR in Response to Ultraviolet Irradiation

4

Phillip R. Musich, Zhengke Li, and Yue Zou

Abstract

The sensitivity of *Xeroderma pigmentosa* (XP) patients to sunlight has spurred the discovery and genetic and biochemical analysis of the eight XP gene products (XPA-XPG plus XPV) responsible for this disorder. These studies also have served to elucidate the nucleotide excision repair (NER) process, especially the critical role played by the XPA protein. More recent studies have shown that NER also involves numerous other proteins normally employed in DNA metabolism and cell cycle regulation. Central among these is ataxia telangiectasia and Rad3-related (ATR), a protein kinase involved in intracellular signaling in response to DNA damage, especially DNA damage-induced replicative stresses. This review summarizes recent findings on the interplay between ATR as a DNA damage signaling kinase and as a novel ligand for intrinsic cell death proteins to delay damage-induced apoptosis, and on ATR's regulation of XPA and the NER process for repair of UV-induced DNA adducts. ATR's regulatory role in the cytosolic-to-nuclear translocation of XPA will be discussed. In addition, recent findings elucidating a non-NER role for XPA in DNA metabolism and genome stabilization at ds-ssDNA junctions, as exemplified in prematurely aging progeroid cells, also will be reviewed.

Keywords

UV irradiation • Nucleotide excision repair (NER) • *Xeroderma pigmentosum* Group A (XPA) • Ataxia telangiectasia and Rad3-related (ATR) • ATR-XPA interaction • Regulation of XPA by ATR • XPA nuclear import • XPA phosphorylation and acetylation • XPA cytoplasmic interactions and functions • Non-NER functions of XPA

P.R. Musich • Y. Zou (✉)
Department of Biomedical Sciences, Quillen College of Medicine, East Tennessee State University, Johnson City, TN 37614, USA
e-mail: zouy@etsu.edu

Z. Li
Department of Cancer Genetics and Epigenetics, City of Hope Comprehensive Cancer Center, 1500 E Duarte Rd, Duarte, CA 91007, USA

S. Ahmad (ed.), *Ultraviolet Light in Human Health, Diseases and Environment*, Advances in Experimental Medicine and Biology 996, https://doi.org/10.1007/978-3-319-56017-5_4

4.1 Introduction

Individuals with mutations in *Xeroderma pigmentosa* (XP) genes are especially sensitive to the ultraviolet (UV) rays (180–315 nm) in sunlight [1, 2]. These individuals accumulate DNA damage in their skin cells after solar irradiation, primarily as a cyclobutane pyrimidine dimer (CPD) and, to a lesser extent, as a (6–4) photoproduct ((6–4) PP). Normally, these intrastrand cross-links of adjacent pyrimidine bases are removed from the DNA by nucleotide excision repair (NER) [1, 3–5]. In addition to other repair factors, seven XP gene products are involved in the NER process: *Xeroderma pigmentosa* complementation groups A through G (XPA–XPG). Mutations in any of these XP gene products reduces the efficiency of this repair process with XPA and XPC mutations being the most frequent [6] and XPA deficiency showing the highest sensitivity to UV [7]. If adducts persist they may be bypassed by error-prone translesion synthesis using DNA polymerase eta (Pol η), a product of the XPV (pol H) gene [8]. The structure and mutational features, plus post-translational modifications of these XP proteins have been reviewed recently by Feltes and Bonatto [9]. XPA mutation is the most severe XP deficiency since this protein is required in both the global genomic NER (GG-NER) and the transcription-coupled NER (TC-NER) sub-pathways of nucleotide excision repair [10–14]. XPC mutations, though relatively frequent, are less severe since this protein is primarily involved in GG-NER [10, 15]. Though not an XP protein, the DNA damage checkpoint protein ataxia telangiectasia and Rad3-related (ATR) also is essential for initiation and regulation of the NER process [16, 17]. Thus, this review will focus on new information from the last decade on the biochemical roles and cellular mechanisms of XPA and ATR in the nucleotide excision repair process and cell death, and discuss recent findings on possible non-NER functions of XPA in both the nucleus and in the cytoplasm.

4.2 ATR Signaling Mediates the Cellular Response to DNA Damaged Induced by Ultraviolet Radiation

The presence of UV-induced CPD and (6-4) PP adducts in mammalian nuclear DNA generates a cascade of events as part of the DNA damage response (DDR). Generally, these helix-distorting, replication- and transcription-blocking DNA adducts induce activation of the DNA repair process and arrest the cell cycle to allow for repair of the damaged DNA. ATR, a key regulator of these processes, is a member of the phosphatidylinositol 3-kinase (PI3K) family. The PI3K family of protein kinases also includes the other stress-responsive protein kinases *ataxia telangiectasia* mutated (ATM), DNA-dependent protein kinase (DNA-PK) and mammalian target of rapamycin (mTOR) [18, 19]. Although it functions in multiple DDR processes [20] ATR is the primary regulator of the nucleotide excision repair pathway due to its ability to detect the replicative and transcriptional stresses caused by UV-induced damage and other bulky DNA adducts resulting from chemical toxins and some chemotherapeutic agents [21–24].

Induction of CPDs and (6-4) PPs in DNA generates obstacles to DNA replication and transcription. The resulting replicative and transcriptional stresses stall DNA polymerization during replication and pol II progression in RNA synthesis [11, 12], respectively, leading to an accumulation of stretches of single-strand DNA (ssDNA), which become coated with the ssDNA-binding replication protein A (RPA) [25]. ATR in complex with its nuclear binding partner ATR-interacting protein (ATRIP) binds to this RPA-coated ssDNA via an ATRIP-RPA interaction. ATRIP also serves to activate the checkpoint kinase activity of ATR [4, 26–29]. Activated ATR kinase phosphorylates many downstream mediators/effectors which include checkpoint kinase 1 (Chk1), A-kinase-anchoring protein 12 (AKAP12), p38/mitogen-activated protein kinase-activated protein (MAPKAP) kinase 2 (MK2), the tumor suppressor protein p53, ATRIP and XPA [26, 30–32]. Phosphorylation activates these downstream proteins resulting in arrest of cell cycle progression, activation of DNA repair and, in cases of severe damage, apoptotic cell death [21, 33, 34]. ATR is an essential gene for the initiation and regulation of NER and for genome maintenance [16, 35, 36].

Historically, ATR has been described as a necessary protein kinase which functions in the cell nucleus to regulate DNA replication and various responses to DNA damage and cellular stress [37, 38]. Possible non-nuclear roles for ATR have received little attention. However, a recent study described an anti-apoptotic, cytoplasmic role for ATR [39, 40]. It was demonstrated that in mammals a small fraction of cellular ATR normally exists in the cytoplasm (cytoATR) and that, in response to DNA damaging agents, the amount of this cytoATR increases and changes conformation, resulting in a slower-migrating, higher electrophoretic band (ATR-H) as compared with the faster-migrating, lower electrophoretic band (ATR-L). The most efficient induction of ATR-H formation was by UV irradiation, though it also was induced by camptothecin and hydroxyurea, agents which cause DNA double-strand breaks (DSBs). Interestingly, the increase in cytoATR appears to result from nuclear export and not from new protein biosynthesis [40]. This nuclear export of ATR-L and its conversion to cytoplasmic ATR-H by UV irradiation was observed in normal human fibroblasts, transformed skin keratinocytes, multiple human cancer cell lines, and in transformed mouse embryonic fibroblasts [39].

It was found that the ATR-L is a prolyl *trans*-isomer of cytoplasmic ATR while ATR-H is the *cis*-isomer [39]. The formation of cytoplasmic ATR-L (*trans*-ATR) from ATR-H (*cis*-ATR) is mediated by peptidylprolyl cis/trans isomerase NIMA-interacting 1 (Pin1) [39]; this enzyme is a critical regulator of many biological processes in both normal and diseased cells [41–47]. Since ATR is naturally more stable in its *cis*-isomeric form, newly-synthesized ATR is in the ATR-H isoform but is quickly converted to the ATR-L isoform by Pin1 isomerization of the phospho-$Ser^{428}Pro^{429}$ site of the ATR protein [39]. This isomerization converts Pro^{429} from the *cis*- (ATR-H) to the *trans*-isoform (ATR-L). Surprisingly, this conformational change of only one out of 2644 amino acids is sufficient to reduce the electrophoretic mobility of the ATR protein in 3–8% gradient SDS-polyacrylamide gels, similar to adding ~10 kilodaltons, to generate a clearly distinguishable higher band (ATR-H). The mechanism of this protective response stems from UV-induced changes in the phosphorylation status of the $Ser^{428}Pro^{429}$ site in ATR and the Ser^{71} residue in Pin1. UV irradiation induces DAPK1 to phosphorylate Pin1 at Ser^{71}, thus inactivating the isomerase activity [48, 49]. The UV irradiation also induces a dephosphorylation of the phospho-$Ser^{428}Pro^{429}$ site in ATR, rendering it a non-recognizable Pin1 site [39]. Together, these changes in phosphorylation status allow cytoATR to assume the *cis* isoform, ATR-H. Although the details of the UV-induced changes in DAPK1 kinase and the unknown phosphatase activities remain to be elucidated these observations reveal a very sensitive cellular sensor for ultraviolet damage and ATR isomeric conversion.

Upon UV irradiation-induced DNA damage ATR initiates the nuclear NER process to repair the genome. To allow time for completion of this repair the cell needs to stall two processes: cell cycle progression, especially through S phase, and the initiation of damage-induced cell death. Cell cycle arrest is needed to allow time for DNA repair and, thus, prevent the introduction of mutations by replication through unrepaired CPD and (6–4) PP damage sites. A classic feature of ATR in response to UV damage is its phosphorylation of Chk1 kinase, which then phosphorylates other proteins to arrest cell cycle progression [50]. UV-induced damage also can activate the intrinsic cell death pathway through the release of mitochondrial cytochrome C into the cytosol which activates caspase cleavage and eventual apoptosis [51]. But how does ATR stall the onset of apoptotic cell death to allow sufficient time for cell recovery by repair of the CPD and (6–4) PP damage? The answer lies in the interaction of cytosolic ATR-H with the proapoptotic protein tBid (truncated BH3 interacting-domain death agonist) as described by Hilton et al. [39]. In response to damage tBid promotes polymerization of proapoptotic proteins Bax (bcl-2-

associated X) and Bak (bcl-2 homologous antagonist-killer) at the mitochondrial surface, which induces cytochrome C release leading to apoptotic cell death [51]. Hilton et al. surprisingly found that ATR contains a BH3-like domain which allows it to function like a prosurvival Bcl-2 family protein. In the nucleus, ATR remains in the form of ATR-L, regardless of UV, whose BH3 domain appears to be masked in a folded N-terminal region of the *trans*-isoform protein; however, the N-terminus is unfolded in the cytosolic *cis*-isoform which exposes this BH3 domain, allowing ATR-H to bind to and sequester tBid protein, thus delaying initiation of the intrinsic cell death pathway [39]. Figure 4.1 illustrates how the *cis*- and *trans*-isoforms may affect these changes in the accessibility of the BH3 domain in ATR-L vs. ATR-H isoforms, and how the ATR-L form is necessary for the regulation of XPA nuclear import and NER efficiency.

Nuclear ATR is well known for its association with ATRIP, a necessary interaction which activates the kinase activity of ATR in addition to localizing it to the RPA-coated ssDNA at damage sites [4, 26–29]. This kinase activity is essential for ATR's activation of downstream proteins during the DDR. In contrast, Hilton et al. found that cytoATR is free of ATRIP, which remains sequestered in the nucleus after UV irradiation. Also, the anti-apoptotic function of mitochondrial ATR-H is independent of its checkpoint kinase activity [39, 40]. Thus, the regulated *cis*- *vs.* *trans*-isoform switching between ATR-H and ATR-L allows distinct prosurvival functions of ATR in the cytoplasm versus those in the nucleus in response to UV irradiation. Particularly, the cytoplasmic ATR-H prevents premature cell death at mitochondria. This coordination of the cytoplasmic anti-apoptotic and the nuclear cell cycle arrest/DNA repair roles provides time for damage repair before any decision on programmed cell death needs to be made. Note that, once formed, ATR-H reaches a maximum within 2 h but persists in the cytoplasm for over 8 h, sufficient time for most NER-competent cells to repair all the (6-4) PP adducts and most, if not all, of the CPD adducts [39, 40, 52]. Thus, this slow re-isomerization of ATR-H to ATR-L may serve as an internal timer of repair efficiency and death.

The novel finding of the cytoplasmic role of ATR as an anti-apoptotic protein at mitochondria highlights that much remains to be discovered about the signaling molecules involved in the DNA damage responses. These observations support previous findings that prolyl isomerization of a single residue in a large protein may have

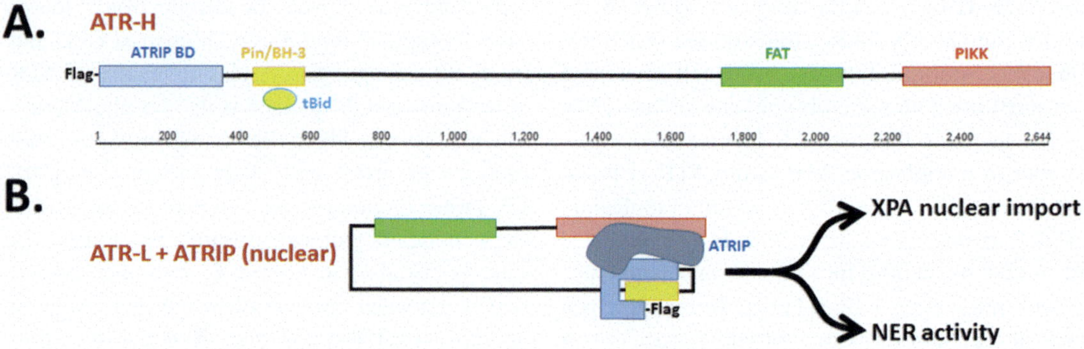

Fig. 4.1 Possible alternative folding conformations of ATR-H vs. ATR-L. There currently are no 3-demensional structures described for ATR. The diagrammatic representations presented here are based on the predictions of Hilton et al. for the N-terminal regions of ATR-H vs. ATR-L [39]. A. The N-terminal region of ATR-H, which has the *cis*-Pro^{429} isomer and an unphosphorylated Ser^{428}, is accessible to both tBid binding and to Flag antibody binding. Thus, ATR-H is presented in an open conformation. B. In ATR-L, which contains a phosphorylated Ser^{428} and a *trans*-Pro^{429}, the BH3 domain is inaccessible to tBid binding as is the Flag tag [39]. Thus, ATR-L is drawn with a folded N-terminal region. The N-terminus of ATR contains the ATRIP binding site; binding of ATRIP leads to activation of the ATR kinase *via* interaction with the C-terminal PIKK region [26–29]. Although speculative, the lower diagram of ATR-L illustrates this folding of the N-terminal region onto the C-terminal region, perhaps mediated by ATRIP binding

pleotropic effects on a protein's structure and function [47, 53]. Also, these cytoplasmic pro-survival functions are not only novel for ATR since ATM also displays similar stress functions at peroxisomes in response to increased levels of reactive oxygen species [54–58] and at mitochondria in response to DNA damage [59, 60].

4.3 ATR-XPA Interactions Are Necessary for the Nuclear Import of XPA and for Efficient Nucleotide Excision Repair

The data sheets accompanying nearly all commercial anti-XPA antibodies recommended for immunofluorescence studies by the suppliers indicate that XPA is a protein located in the nucleus only. This discrepancy stems from the early studies in which formalin (2–4% *para*-formaldehyde) was used for cell fixation [61, 62]. More recent immunofluorescence studies of XPA's subcellular distribution confirmed that in *para*-formaldehyde-fixed cells the endogenous protein was observed to be nuclear [63–65]. However, biochemical fractionation of millions of cells into nuclear vs. cytoplasmic fractions revealed that XPA occurs predominantly in the cytoplasm of normal mammalian cells and that it is translocated to the nucleus in response to DNA damage, especially from UV irradiation [66–69]. These biochemical findings were confirmed by immunofluorescence observations of methanol-fixed cells [66–69]. We have observed that with either fixative the anti-XPA antibodies revealed XPA in the nucleus, but antibody detection of the cytosolic XPA occurred only in cells fixed with cold methanol. Methanol fixation extracts lipids, dehydrates and permeablizes cells causing proteins to denature and precipitate onto the cellular architecture. In contrast, *para*-formaldehyde fixation cross-links proteins and other macromolecules in place [70]. A possible explanation, then, for the reported differences in the subcellular localization of XPA with these two methods is that methanol fixation disrupts the cloaking interaction between XPA and an as yet undescribed cytosolic XPA sequestration protein (cXSP) which sequesters XPA in the cytoplasm; the methanol fixation with denaturation then exposes XPA's antigenic site; in contrast, *para*-formaldehyde fixation locks this XPA-cXSP complex in place, thus masking the XPA epitopes in the cytosol. UV irradiation induces a disruption of this cytosolic XPA-cXSP complex, releasing XPA for nuclear import and detection in nuclei of cells fixed with methanol or *para*-formaldehyde. This also could be true for other so-called nuclear proteins.

Wu et al. reported that ATR regulated XPA nuclear import in response to UV radiation [66, 71]. More recent studies by Li et al. have revealed further important details of the cytosol-nuclear translocation of XPA. The tumor suppressor protein p53 is a major downstream effector molecule and phosphorylation substrate in the ATR-mediated DDR. In support of earlier observations [52, 66], Li et al. demonstrated that the nuclear import of XPA in response to UV irradiation or cisplatin treatment is ATR-dependent in normal fibroblasts and in cancer cells that are p53 proficient; XPA import also is dependent on the transcriptional activity of p53 in these cells [68, 69]. In addition, this dependence on ATR checkpoint activity is cell-cycle phase dependent, occurring only during the S phase [68]. Most XPA remained sequestered in the cytosol in the G_1 phase even after UV treatment; in contrast, in G2-phase cells the nucleus contained the majority of the XPA molecules irrespective of UV irradiation. Consistently, NER recently was found to recruit ATR to the UV-damage sites and to activate ATR in G1-phase but not in S-phase [72–75]. Regulation of S-phase cytosolic XPA translocation into the nucleus by ATR is consistent with previous findings that the peak activity of this checkpoint kinase occurs in S phase as part of normal DNA replication and also in response to DNA damage [33, 76]. Li et al. observed that the maximum UV-induced phosphorylation of Ser^{15} of p53 occurred in S phase and that the NER removal of CPD adducts also was most efficient in S phase [68]. Recall that ATR binds to XPA *via* the Lys^{188} and Ser^{196} residues in its HTH motif

[52] and that these residues are important for the efficient repair of CPD adducts.

Interestingly, the p53 status of cells significantly influences the role of ATR in regulating DNA repair after UV or cisplatin damage. Although efficient NER removal of the damage was dependent on ATR kinase activity in p53-proficient ($p53^{+/+}$) cells the repair process seemed to be ATR-independent in p53-deficient ($p53^{-/-}$) cells [68, 69, 76]. Consistently, nuclear import of cytosolic XPA is dependent on p53 transcriptional activity in $p53^{+/+}$ cells and occurs much slower in $p53^{-/-}$ cells, but the import still occurs [69]. Thus, damage-induced ATR activation of the p53 tumor suppressor protein appears to be a primary but not the sole mediator of XPA nuclear import in $p53^{+/+}$ *vs.* $p53^{-/-}$ cells in S phase. The cell cycle checkpoint kinases ATM, Chk1 and MK2 appear not to have a role in XPA nuclear import in $p53^{+/+}$ nor $p53^{-/-}$ cells [68, 69].

The phosphorylation of XPA by ATR is essential for the NER function of XPA [71]. Shell et al. found that ATR binds XPA *via* a specific helix-turn-helix motif in the minimal DNA-binding domain (DBD) and that this XPA motif contains an ATR phosphorylation site (Ser^{196}) [52]. In addition, disruption of this phosphorylation site in XPA with a Ser^{196}Ala mutation significantly reduced the repair efficiency of CPDs but not the repair of (6-4) PPs. The nucleotide excision repair of (6-4) PPs is generally much more efficient than the repair of CPDs [77, 78] and the above finding indicates that ATR's phosphorylation of Ser^{196} in XPA is mechanistically important in the repair of the more prevalent CPDs which represent persistent UV damage. The phosphorylation of Ser^{196} in XPA by ATR appears to stabilize XPA against HERC2-mediated ubiquitinylation and degradation [79].

Shell's structure-function studies also found that the Lys^{188} residue, which is nearby in the same helical DBD of XPA, was critical since a Lys^{188}Ala mutation disrupted the ATR-XPA interaction, thus significantly reducing DNA repair efficiency [80]. Moreover, the normal UV-induced nuclear translocation of cytosolic XPA was lost with the Lys^{188}Ala mutation. However, the Ser^{196}Ala mutation had no effect on XPA's nuclear translocation. The targeting of XPA to the nucleus occurs *via* its nuclear localization sequence (NLS) which contains basic residues located at positons 30-34 of the 273 amino acid protein [61, 62, 67, 81]. This raises the interesting and important question of how XPA is normally held in the cytoplasm if it contains a NLS sequence and its normal NER function is in the nucleus. One possibility is that XPA is sequestered in the cytosol in normal cells *via* association with cXSP, from which it is released for nuclear import after a DNA damaging event such as UV irradiation. Perhaps the stability of the XPA-cXSP complex is disrupted by the phosphorylation of XPA at Ser^{196} and/or by a post-translational modification of the necessary Lys^{188} (i.e., acetylation). Note that highly over-expressed XPA mutants lacking the NLS site can be detected in the cytoplasm by immunofluorescence microscopy in *para*-formaldehyde fixed cells [62], indicating that cXSP may occur is physiologically limiting amounts. This as yet uninvestigated cytosolic XPA sequestration and release could be one of the dynamic components of the UV-induced damage response. Also, note that AKAP12 is normally a cytosolic protein associated with protein kinase A (PKA) but becomes phosphorylated by ATR after UV irradiation and then is transported into the nucleus in association with ATR [30].

It is obvious that the DNA damage-induced import of cytosolic XPA into the nucleus is a highly regulated process. Mechanistic features of this import process have been resolved in additional studies by Li et al. [67]. It was shown that the NLS in the N-terminal region of XPA was required for nuclear localization. In addition, siRNA knockdown revealed that nucleo-cytoplasmic transport proteins importin-α4 and -α7 were required for XPA nuclear import, but not the other importin-α proteins. Co-immunoprecipitation studies demonstrated that importin-α4 and importin-α7 mediate this nuclear import by direct physical interactions with XPA. However, these two carrier proteins appear to serve different functions during the cell cycle. Importin-α4 transport of XPA was activated by UV radiation and required functional

ATR kinase activity, consistent with importin-α4 being responsible for the nuclear import of XPA during the S-phase DNA damage response. In contrast, importin-α7 functioned independent of DNA damage and ATR kinase activation, perhaps reflecting the observed nuclear import of XPA in the G_2 phase irrespective of UV exposure [68]. These features of XPA cytosolic localization and cell cycle-dependent nuclear import in response to UV irradiation are diagrammatically summarized in Fig. 4.2.

Nuclear import of proteins requires a GTPase to coordinate protein-protein interactions [82–84]. XAB1 was observed in a yeast two-hybrid system to be an XPA-binding protein with GTPase activity [85]. However, Li et al. demonstrated that XAB1 is not the GTPase involved in XPA nuclear import [67]. Also, questions remain on how XPA is released from cXSP in the cytosolic sequestration complex to bind to the importin-α4 in S phase cells exposed to UV. These authors demonstrated that there was an increase in the XPA available for importin-α4 binding within 30 min after UV exposure; however, the mechanistic details of the cytosolic DDR remain to be resolved. In addition, how the cytosolic XPA sequestered by cXSP during G_1 and S phases is released in non-irradiated cells for importin-α7-mediated nuclear import in the G_2 phase also remains to be elucidated.

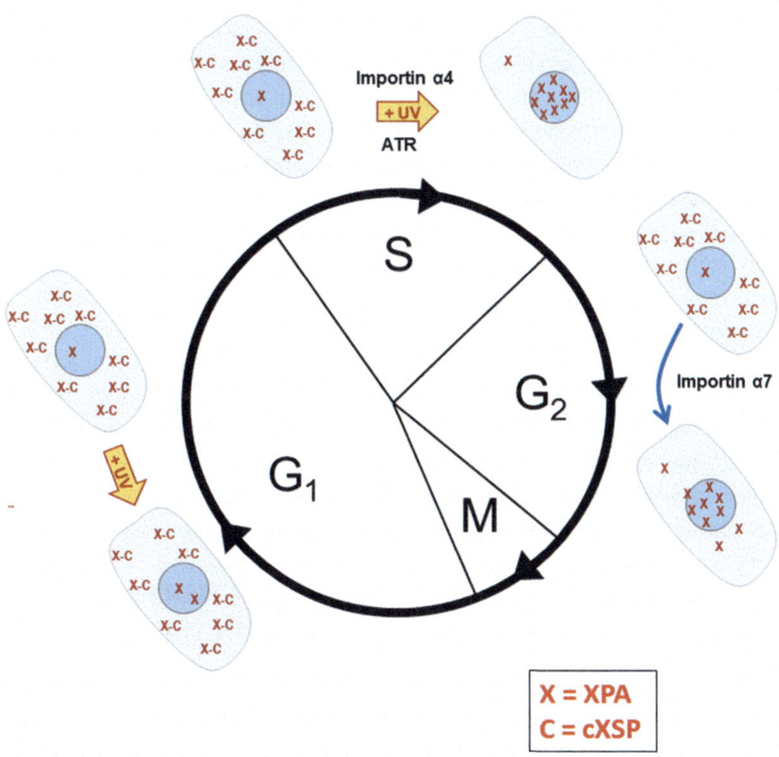

Fig. 4.2 Normal and UV-induced redistribution during progression through the cell cycle in p53-competent human cells. This model is based on the studies of Li et al. [67–69] In non-damaged cells in the G_1 phase XPA (X) is mostly located in the cytosol, likely bound to cXSP (C), a hypothetical cytosolic XPA sequestration protein. Exposure of G_1 cells to UV does not change this distribution. Likewise, in S phase cells XPA is mostly cytosolic; however, UV exposure induces a release of XPA from cXSP and a translocation of XPA into the nucleus. This XPA nuclear translocation in S phase requires the importin α4 transport protein and is ATR kinase- and p53-dependent in p53-competent cells. XPA is primarily located in the nucleus in G_2 phase cells, transported there via importin α7 in a process independent of UV exposure. The XPA redistributes to the cytosol during the M-G_1 phase transition and reassociates with cXSP

4.4 Does XPA Have a Cytosolic Function Outside of Nucleotide Excision Repair?

Why is the XPA protein localized in the cytosol of normal (non-DNA damaged) cells during G_1 and S phases of the cell cycle, but not in the G_2 phase? Does its complex with cXSP provide a cytosolic function in G_1 and S phases, and/or is it sequestered there to prevent interference with ongoing nuclear processes?

In addition to high dermatological sensitivity to sunlight XP patients, especially those with an XPA deficiency, often suffer from neurological deficiencies and an early-aging phenotype [2], likely due to non-NER mechanisms as exogenous, genotoxin-induced bulky adducts would not be a concern. XPA interacts with a variety of XP and other proteins during the DNA repair process in the nucleus [67–69, 81, 86–88], but interactions with cytosolic proteins have not been described. Are these non-NER features of XPA deficiency related to XPA's cytosolic location, especially in the G_0/G_1 phase status typical of neurons, cardiomyocytes or other differentiated cell types? Other than the descriptions of its UV-induced cytoplasmic-to-nuclear translocation [52, 66–69], possible XPA binding partners and/or functional roles in the cytosol have received little to no attention. One possibility might be that cXSP, the proposed cytoplasmic sequestration factor to which XPA is bound in normal G_1 and S phase cells, influences abnormal, dis-regulatory activity in $XPA^{-/-}$ cells leading to deleterious metabolic events. Using a bioinformatics analysis Fang et al. observed that the $XPA^{-/-}$ phenotype includes neurological features similar to mitochondrial diseases, and results in abnormal mitochondrial energy metabolism, even though cytoplasmic XPA in XPA-proficient cells was absent from the mitochondrial matrix [89]. They also reported increased poly(ADP-ribose) polymerase 1 (PARP1) activity, resulting in higher parsylation of cellular proteins resulting in NAD^+ depletion, thus reducing mitochondrial energy generation. They observed that the reduced level of NAD^+ downregulated SIRT1, a NAD^+-dependent deacetylase involved in regulating mitochondrial homeostasis and XPA repair activity [90–92]. Fang et al. assumed that the PARP1 was activated in $XPA^{-/-}$ cancer cells and neurons by an increased level of basal nuclear DNA damage [89]. However, the presence of a basal level and the type of DNA damage occurring in the $XPA^{-/-}$ cells was not demonstrated. In addition, as reviewed by Weaver and Yang [93], PARP1 activation can be induced by stress responses other than DNA damage, including the ERK-1 [94, 95] and Notch/HES-1 [96] signaling pathways and intracellular calcium overload [97]. In addition, XPA and PARP1 appear to have regulatory interactions which would be upset in the $XPA^{-/-}$ cells [98]. These studies and their interpretation are complicated further by the observed cell-type specificity of PARP1 activation [95–97]. Resolution of these ambiguities rest, in part, on an elucidation of the cXSP cytosolic binding partner of XPA which sequesters this NER protein in the cytosol in normal G_1 and S phases of the cycling cell and in the G_1/G_0 states of the non-cycling, highly differentiated cells. There are multiple possibilities since XPA has been described as a highly flexible scaffold protein capable of interacting with numerous proteins simultaneously [81, 88]. Future studies also are needed to elucidate XPA's possible cytosolic binding partner(s) in the G_1 and S phase cells, their biochemical properties, and possible normal function after XPA dissociation in G_2 and M phases.

4.5 Non-NER Functions of XPA in the Nucleus

XPA functions as an essential component of the DNA damage repair complexes for both GG-NER and TC-NER. In addition, XPA binds to ds-ssDNA junctions with a significantly higher affinity (Kd = 49.1 ± 5.1 nM) [99] than it's specific binding to bulky DNA lesions (Kd = 200 nM) [100]. This suggests that, in addition to DNA damage recognition/verification, XPA may bind independently to and stabilize such ds-ssDNA junctions during the NER process and/or during other types of DNA metabolism. Hilton et al.

recently demonstrated that in binding to ds-ssDNA junctions XPA employs a larger DNA-binding domain [101] than was previously described for repair substrates [102, 103].

How might this essential biochemical affinity for ds-ssDNA junctions relate to XPA's cytoplasmic restriction during S phase and XPA's performance of non-NER functions in cells? Hutchison-Gilford progeria syndrome (HGPS) patients suffer from a variety of laminopathy ailments due to a sporadic deficiency in the proteolytic processing of the precursor form of lamin A into the mature protein. The aberrantly processed protein produced is called progerin, a truncated form of lamin A with a hydrophobic farnesylated C-terminal [104–111]. HGPS cells with progerin accumulation exhibit a reduced replicative lifespan plus a deficiency in the repair of endogenous, laminopathy-induced DNA DSBs, which increase with age [112, 113]. These DNA metabolism deficiencies also correlate with a proteolytic truncation of replication factor C1 (RFC1) [114] and a sequestration of proliferating cell nuclear antigen (PCNA) in a complex with progerin [115]. Both the intact RFC complex and PCNA are essential replication factors and are needed for loading the replicative polymerase onto DNA [116, 117], thus accounting for the reduced replicative lifespan of HGPS cells [112, 118]. Interestingly, cellular nucleotide excision repair protein XPA misaccumulates at the DSB sites consisting of ds-ssDNA junctions even though XPA never has had a documented role in DSB repair, causing these breaks to become progressively devoid of DSB repair proteins [113]. Those DSBs appear to be generated from stalled and collapsed replication forks in HGPS. Depletion of XPA in these aging HGPS cells significantly relieves the deficiency in DSB repair, possibly by shifting the binding of available free PCNA to these XPA-free junctions [115]. These observations suggest that as HGPS cells age progerin accumulates and sequesters PCNA, resulting in collapsed replication forks with DSBs and ds-ssDNA junctions to which XPA binds. Although this XPA binding may limit access to DNA DSBs repair proteins, it appears that the binding could stabilize the forks and prevent the HGPS cells from progerin-induced apoptosis [115].

These potential non-NER roles allow for interesting speculation concerning XPA's pleiotropic functions and those of it's as-yet undescribed binding partners (i.e., cXSP) and will lead to many interesting experimental studies.

4.6 Conclusions

XPA is indispensable for both transcription-coupled repair and global genomic repair, and, thus, has a central and critical role in the NER process. Recent studies have revealed that XPA is kept in the cytosol in non-UV irradiated cells where it may be sequestered by a cytosolic XPA-binding protein, here termed cXSP. This subcellular distribution can be easily detected by immunofluorescence microscopy if the cells are fixed in cold methanol but not in cells fixed with *p*-formaldehyde. In the S phase UV irradiation induces a translocation of XPA into the nucleus for NER of UV-induced adducts. This S phase nuclear import is facilitated by XPA binding to by the transport protein importin-α4 (Fig. 4.2). In contrast, cells in G_1 phase retain XPA in the cytosol while XPA is mostly located in the nucleus in the G_2 phase; both the G_1 and G_2 phase distributions are largely independent of UV irradiation. Importin-α7 facilitates the G_2 phase nuclear import of XPA. The S phase nuclear import of XPA is dependent on the kinase activity of ATR and on the tumor suppressor protein p53, which also is activated by the ATR kinase.

The ATR protein has multiple roles in regulating the NER process. In response to UV damage ATR regulates the NER process *via* its phosphorylation of numerous cell cycle control and DNA repair proteins. One of these is XPA; its phosphorylation by ATR is required for its essential role in NER of persistent CPD adducts. In addition, ATR kinase activity is required for the cytosolic-to-nuclear translocation of XPA by importin-α4 during S phase, the period when ATR kinase activity is at its highest. In addition to these kinase-dependent DDR nuclear functions a recent study reports an important cytosolic, kinase-independent role for ATR in moderating the intrinsic cell death response induced by UV irradiation. Surprisingly, newly formed ATR is a *cis*-conformer (ATR-H) at the Pro^{429} residue but the nuclear ATR is isomerized into the *trans*-isomer (ATR-L) by the proline isomerase Pin1. It is likely that the prolyl isomerization of ATR may change the conformation of ATR between an unfolded structure to expose its BH3 domain and a folded structure making BH3 inaccessible; the former is able to bind to and sequester the pro-apoptotic factor tBid at the mitochondrial surface to prevent initiation of the intrinsic apoptosis, thus allowing time for DNA repair.

XPA binds to ds-ssDNA junctions, such as those found at exposed replication forks and DNA regions undergoing repair. This binding, which is not necessarily unrelated to XPA's NER activity, is stronger than its binding to bulky DNA adducts. Prematurely-aging progeroid cells accumulate progerin, an abnormal form of lamin A and suffer from an accumulation of DNA DSBs and stalled replication forks. Interestingly, these sites are exposed due to sequestration of PCNA by progerin, allowing XPA to bind to these DSB sites and stalled forks.

These studies have revealed several potential sites for therapeutic intervention to enhance the chemotherapy of cancer cells and/or the survival of progeroid cells.

References

1. Cleaver JE, Lam ET, Revet I (2009) Disorders of nucleotide excision repair: the genetic and molecular basis of heterogeneity. Nat Rev Genet 10:756–768
2. DiGiovanna JJ, Kraemer KH (2012) Shining a light on xeroderma pigmentosum. J Invest Dermatol 132:785–796
3. Sancar A (2016) Mechanisms of DNA repair by photolyase and excision nuclease (Nobel lecture). Angew Chem 55:8502–8527
4. Sancar A, Lindsey-Boltz LA, Unsal-Kacmaz K, Linn S (2004) Molecular mechanisms of mammalian DNA repair and the DNA damage checkpoints. Annu Rev Biochem 73:39–85
5. Sugasawa K (2016) Molecular mechanisms of DNA damage recognition for mammalian nucleotide excision repair. DNA Repair 44:110–117
6. Hengge UR, Emmert S (2008) Clinical features of xeroderma pigmentosum. Adv Exp Med Biol 637:10–18
7. Niedernhofer LJ, Garinis GA, Raams A, Lalai AS, Robinson AR, Appeldoorn E, Odijk H, Oostendorp R, Ahmad A, van Leeuwen W, Theil AF, Vermeulen W, van der Horst GT, Meinecke P, Kleijer WJ, Vijg J, Jaspers NG, Hoeijmakers JH (2006) A new progeroid syndrome reveals that genotoxic stress suppresses the somatotroph axis. Nature 444:1038–1043
8. Auclair Y, Rouget R, Belisle J, Costantino S, Drobetsky EA (2010) Requirement for functional DNA polymerase eta in genome-wide repair of UV-induced DNA damage during S phase. DNA Repair (Amst) 9:754–764
9. Feltes BC, Bonatto D (2015) Overview of xeroderma pigmentosum proteins architecture, mutations and post-translational modifications. Mutat Res Rev Mutat Res 763:306–320
10. Gillet LC, Scharer OD (2006) Molecular mechanisms of mammalian global genome nucleotide excision repair. Chem Rev 106:253–276
11. Hanawalt PC, Spivak G (2008) Transcription-coupled DNA repair: two decades of progress and surprises. Nat Rev Mol Cell Biol 9:958–970
12. Lagerwerf S, Vrouwe MG, Overmeer RM, Fousteri MI, Mullenders LH (2011) DNA damage response and transcription. DNA Repair (Amst) 10:743–750
13. Camenisch U, Nageli H (2008) XPA gene, its product and biological roles. Adv Exp Med Biol 637:28–38
14. Ding D, Zhang Y, Yu H, Guo Y, Jiang L, He X, Ma W, Zheng W (2012) Genetic variation of XPA gene and risk of cancer: a systematic review and pooled analysis. Int J Cancer 131:488–496
15. Sugasawa K (2008) XPC: its product and biological roles. Adv Exp Med Biol 637:47–56
16. Cimprich KA, Cortez D (2008) ATR: an essential regulator of genome integrity. Nat Rev Mol Cell Biol 9:616–627
17. Cortez D (2015) Preventing replication fork collapse to maintain genome integrity. DNA Repair (Amst) 32:149–157
18. Weber AM, Ryan AJ (2015) ATM and ATR as therapeutic targets in cancer. Pharmacol Ther 149:124–138
19. Perry J, Kleckner N (2003) The ATRs, ATMs, and TORs are giant HEAT repeat proteins. Cell 112:151–155
20. Nam EA, Cortez D (2011) ATR signalling: more than meeting at the fork. Biochem J 436:527–536
21. Abraham RT (2001) Cell cycle checkpoint signaling through the ATM and ATR kinases. Genes Dev 15:2177–2196
22. Flynn RL, Zou L (2011) ATR: a master conductor of cellular responses to DNA replication stress. Trends Biochem Sci 36:133–140

23. Fong YW, Cattoglio C, Tjian R (2013) The Intertwined roles of transcription and repair proteins. Mol Cell 52:291–302
24. Maréchal A, Zou L (2013) DNA damage sensing by the ATM and ATR kinases. Cold Spring Harb Perspect Biol 5:a012716
25. Zeman MK, Cimprich KA (2014) Causes and consequences of replication stress. Nat Cell Biol 16:2–9
26. Cortez D, Guntuku S, Qin J, Elledge SJ (2001) ATR and ATRIP: partners in checkpoint signaling. Science 294:1713–1716
27. Choi JH, Sancar A, Lindsey-Boltz LA (2009) The human ATR-mediated DNA damage checkpoint in a reconstituted system. Methods 48:3–7
28. Zou L, Elledge SJ (2003) Sensing DNA damage through ATRIP recognition of RPA-ssDNA complexes. Science 300:1542–1548
29. Zou L, Liu D, Elledge SJ (2003) Replication protein A-mediated recruitment and activation of Rad17 complexes. Proc Natl Acad Sci U S A 100:13827–13832
30. Jarrett SG, Wolf Horrell EM, D'Orazio JA (2016) AKAP12 mediates PKA-induced phosphorylation of ATR to enhance nucleotide excision repair. Nucleic Acids Res. 44:10711–10726
31. Andrés-León E, Cases I, Arcas A, Rojas AM (2016) DDRprot: a database of DNA damage response-related proteins. Database. doi:10.1093/database/baw1123
32. Matsuoka S, Ballif BA, Smogorzewska A, McDonald ER, Hurov KE, Luo J, Bakalarski CE, Zhao Z, Solimini N, Lerenthal Y, Shiloh Y, Gygi SP, Elledge SJ (2007) ATM and ATR substrate analysis reveals extensive protein networks responsive to DNA damage. Science 316:1160–1166
33. Dart DA, Adams KE, Akerman I, Lakin ND (2004) Recruitment of the cell cycle checkpoint kinase ATR to chromatin during S-phase. J Biol Chem 279:16433–16440
34. Roos WP, Kaina B (2013) DNA damage-induced apoptosis: from specific DNA lesions to the DNA damage response and apoptosis. Cancer Lett 332:237–248
35. Brown EJ, Baltimore D (2000) ATR disruption leads to chromosomal fragmentation and early embryonic lethality. Genes Dev 14:397–402
36. Brown EJ, Baltimore D (2003) Essential and dispensable roles of ATR in cell cycle arrest and genome maintenance. Genes Dev 17:615–628
37. Buisson R, Boisvert JL, Benes CH, Zou L (2015) Distinct but concerted roles of ATR, DNA-PK, and Chk1 in countering replication stress during S phase. Mol Cell 59:1011–1024
38. Cho YJ, Liang P (2011) S-phase-coupled apoptosis in tumor suppression. Cell Mol Life Sci 68:1883–1896
39. Hilton BA, Li Z, Musich PR, Wang H, Cartwright B, Serrano MA, Zhou XZ, Lu KP, Zou Y (2015) ATR plays a direct antiapoptotic role at mitochondria which is regulated by prolyl isomerase Pin1. Mol Cell 60:35–46
40. Li Z (2013) New insights into the roles of human DNA damage checkpoint protein ATR in the regulation of nucleotide excision repair and DNA damage-induced cell death. In: Biomedical sciences. East Tennessee State University, Johnson City
41. Hunter T (1998) Prolyl isomerases and nuclear function. Cell 92:141–143
42. Lu KP, Liou YC, Zhou XZ (2002) Pinning down proline-directed phosphorylation signaling. Trends Cell Biol 12:164–172
43. Lu Z, Hunter T (2014) Prolyl isomerase Pin1 in cancer. Cell Res 24:1033–1049
44. Yaffe MB, Schutkowski M, Shen M, Zhou XZ, Stukenberg PT, Rahfeld J-U, Xu J, Kuang J, Kirschner MW, Fischer G, Cantley LC, Lu KP (1997) Sequence-specific and phosphorylation-dependent proline isomerization: a potential mitotic regulatory mechanism. Science 278:1957–1960
45. Zheng H, You H, Zhou XZ, Murray SA, Uchida T, Wulf G, Gu L, Tang X, Lu KP, Xiao ZX (2002) The prolyl isomerase Pin1 is a regulator of p53 in genotoxic response. Nature 419:849–853
46. Zhou XZ, Lu KP (2016) The isomerase PIN1 controls numerous cancer-driving pathways and is a unique drug target. Nat Rev Cancer 16:463–478
47. Brichkina A, Nguyen NT, Baskar R, Wee S, Gunaratne J, Robinson RC, Bulavin DV (2016) Proline isomerisation as a novel regulatory mechanism for p38MAPK activation and functions. Cell Death Differ 23:1592–1601
48. Kim BM, You MH, Chen CH, Lee S, Hong Y, Kimchi A, Zhou XZ, Lee TH (2014) Death-associated protein kinase 1 has a critical role in aberrant tau protein regulation and function. Cell Death Dis 5:e1237
49. Lee TH, Chen CH, Suizu F, Huang P, Schiene-Fischer C, Daum S, Zhang YJ, Goate A, Chen RH, Zhou XZ, Lu KP (2011) Death-associated protein kinase 1 phosphorylates Pin1 and inhibits its prolyl isomerase activity and cellular function. Mol Cell 42:147–159
50. Enders GH (2008) Expanded roles for Chk1 in genome maintenance. J Biol Chem 283:17749–17752
51. Ichim G, Tait SWG (2016) A fate worse than death: apoptosis as an oncogenic process. Nat Rev Cancer 16:539–548
52. Shell SM, Zou Y (2009) Protein-protein interactions in Ataxia telangiectesia. In: Ahmad S, Hanaoka F (eds) Molecular mechanisms of Ataxia Telangiectasia. Landes Bioscience, Austin, pp 42–51
53. Lu KP, Zhou XZ (2007) The prolyl isomerase PIN1: a pivotal new twist in phosphorylation signalling and disease. Nat Rev Mol Cell Biol 8:904–916
54. Alexander A, Cai S-L, Kim J, Nanez A, Sahin M, MacLean KH, Inoki K, Guan K-L, Shen J, Person MD, Kusewitt D, Mills GB, Kastan MB, Walker CL (2010) ATM signals to TSC2 in the cytoplasm

to regulate mTORC1 in response to ROS. Proc Natl Acad Sci U S A 107:4153–4158
55. Alexander A, Kim J, Walker CL (2010) ATM engages the TSC2/mTORC1 signaling node to regulate autophagy. Autophagy 6:672–673
56. Alexander A, Walker CL (2010) Differential localization of ATM is correlated with activation of distinct downstream signaling pathways. Cell Cycle 9:3685–3686
57. Tripathi DN, Zhang J, Jing J, Dere R, Walker CL (2016) A new role for ATM in selective autophagy of peroxisomes (pexophagy). Autophagy 12:711–712
58. Zhang J, Tripathi DN, Jing J, Alexander A, Kim J, Powell RT, Dere R, Tait-Mulder J, Lee JH, Paull TT, Pandita RK, Charaka K, Pandita TK, Kastan MB, Walker CL (2015) ATM functions at the peroxisome to induce pexophagy in response to ROS. Nat Cell Biol 17:1259–1269
59. Barroso-Gonzalez J, Auclair S, Luan S, Thomas L, Atkins KM, Aslan JE, Thomas LL, Zhao J, Zhao Y, Thomas G (2016) PACS-2 mediates the ATM and NF-[kappa]B-dependent induction of anti-apoptotic Bcl-xL in response to DNA damage. Cell Death Differ 23:1448–1457
60. Valentin-Vega YA, MacLean KH, Tait-Mulder J, Milasta S, Steeves M, Dorsey FC, Cleveland JL, Green DR, Kastan MB (2012) Mitochondrial dysfunction in ataxia-telangiectasia. Blood 119:1490–1500
61. Miura N, Miyamoto I, Asahina H, Satokata I, Tanaka K, Okada Y (1991) Identification and characterization of xpac protein, the gene product of the human XPAC (xeroderma pigmentosum group A complementing) gene. J Biol Chem 266:19786–19789
62. Miyamoto I, Miura N, Niwa H, Miyazaki J, Tanaka K (1992) Mutational analysis of the structure and function of the xeroderma pigmentosum group A complementing protein. Identification of essential domains for nuclear localization and DNA excision repair. J Biol Chem 267:12182–12187
63. Bomgarden RD, Lupardus PJ, Soni DV, Yee MC, Ford JM, Cimprich KA (2006) Opposing effects of the UV lesion repair protein XPA and UV bypass polymerase eta on ATR checkpoint signaling. EMBO J 25:2605–2614
64. Rademakers S, Volker M, Hoogstraten D, Nigg AL, Mone MJ, Van Zeeland AA, Hoeijmakers JH, Houtsmuller AB, Vermeulen W (2003) Xeroderma pigmentosum group A protein loads as a separate factor onto DNA lesions. Mol Cell Biol 23:5755–5767
65. Solimando L, Luijsterburg MS, Vecchio L, Vermeulen W, van Driel R, Fakan S (2009) Spatial organization of nucleotide excision repair proteins after UV-induced DNA damage in the human cell nucleus. J Cell Sci 122:83–91
66. Wu X, Shell SM, Liu Y, Zou Y (2007) ATR-dependent checkpoint modulates XPA nuclear import in response to UV irradiation. Oncogene 26:757–764
67. Li Z, Musich PR, Cartwright BM, Wang H, Zou Y (2013) UV-induced nuclear import of XPA is mediated by importin-alpha4 in an ATR-dependent manner. PLoS One 8:e68297
68. Li Z, Musich PR, Serrano MA, Dong Z, Zou Y (2011) XPA-mediated regulation of global nucleotide excision repair by ATR Is p53-dependent and occurs primarily in S-phase. PLoS One 6:e28326
69. Li Z, Musich PR, Zou Y (2011) Differential DNA damage responses in p53 proficient and deficient cells: cisplatin-induced nuclear import of XPA is independent on ATR checkpoint in p53-deficient lung cancer cells. Int J Biochem Mol Biol 2:138–145
70. Schnell U, Dijk F, Sjollema KA, Giepmans BNG (2012) Immunolabeling artifacts and the need for live-cell imaging. Nat Meth 9:152–158
71. Wu X, Shell SM, Yang Z, Zou Y (2006) Phosphorylation of nucleotide excision repair factor Xeroderma pigmentosum group A by ataxia telangiectasia mutated and Rad3-related-dependent checkpoint pathway promotes cell survival in response to UV irradiation. Cancer Res 66:2997–3005
72. Ray A, Blevins C, Wani G, Wani AA (2016) ATR- and ATM-Mediated DNA damage response is dependent on excision repair assembly during G1 but bot in S phase of cell cycle. PLoS One 11:e0159344
73. Ray A, Milum K, Battu A, Wani G, Wani AA (2013) NER initiation factors, DDB2 and XPC, regulate UV radiation response by recruiting ATR and ATM kinases to DNA damage sites. DNA Repair 12:273–283
74. Lindsey-Boltz LA, Kemp MG, Reardon JT, DeRocco V, Iyer RR, Modrich P, Sancar A (2014) Coupling of human DNA excision repair and the DNA damage checkpoint in a defined in vitro system. J Biol Chem 289:5074–5082
75. Kemp MG, Gaddameedhi S, Choi JH, Hu J, Sancar A (2014) DNA repair synthesis and ligation affect the processing of excised oligonucleotides generated by human nucleotide excision repair. J Biol Chem 289:26574–26583
76. Auclair Y, Rouget R, Affar el B, Drobetsky EA (2008) ATR kinase is required for global genomic nucleotide excision repair exclusively during S phase in human cells. Proc Natl Acad Sci U S A 105:17896–17901
77. Rouget R, Auclair Y, Loignon M, Affar EB, Drobetsky EA (2008) A sensitive flow cytometry-based nucleotide excision repair assay unexpectedly reveals that mitogen-activated protein kinase signaling does not regulate the removal of UV-induced DNA damage in human cells. J Biol Chem 283:5533–5541
78. Lima-Bessa KMD, Armelini MG, Chiganças V, Jacysyn JF, Amarante-Mendes GP, Sarasin A, Menck CFM (2008) CPDs and 6-4PPs play different roles in UV-induced cell death in normal and NER-deficient human cells. DNA Repair (Amst) 7:303–312

79. Lee TH, Park JM, Leem SH, Kang TH (2012) Coordinated regulation of XPA stability by ATR and HERC2 during nucleotide excision repair. Oncogene 33:19–25
80. Shell SM, Li Z, Shkriabai N, Kvaratskhelia M, Brosey C, Serrano MA, Chazin WJ, Musich PR, Zou Y (2009) Checkpoint kinase ATR promotes nucleotide excision repair of UV-induced DNA damage via physical interaction with Xeroderma Pigmentosum Group A. J Biol Chem 284:24213–24222
81. Sugitani N, Sivley RM, Perry KE, Capra JA, Chazin WJ (2016) XPA: a key scaffold for human nucleotide excision repair. DNA Repair (Amst) 44:123–135
82. Stewart M (2007) Molecular mechanism of the nuclear protein import cycle. Nat Rev Mol Cell Biol 8:195–208
83. Wickner W, Schekman R (2005) Protein translocation across biological membranes. Science 310:1452–1456
84. Knudsen NO, Andersen SD, Lutzen A, Nielsen FC, Rasmussen LJ (2009) Nuclear translocation contributes to regulation of DNA excision repair activities. DNA Repair (Amst) 8:682–689
85. Nitta M, Saijo M, Kodo N, Matsuda T, Nakatsu Y, Tamai H, Tanaka K (2000) A novel cytoplasmic GTPase XAB1 interacts with DNA repair protein XPA. Nucl Acids Res 28:4212–4218
86. Shell SM, Zou Y (2008) Other proteins interacting with XP proteins. Adv Exp Med Biol 637:103–112
87. Jiang G, Zou Y, Wu X (2012) Replication mediated disassociation of replication protein A-XPA complex upon DNA damage: Implications for RPA handing off. Cell Biol Int 36:713–720
88. Fadda E (2016) Role of the XPA protein in the NER pathway: a perspective on the function of structural disorder in macromolecular assembly. Comput Struct Biotech J 14:78–85
89. Fang EF, Scheibye-Knudsen M, Brace LE, Kassahun H, SenGupta T, Nilsen H, Mitchell JR, Croteau DL, Bohr VA (2014) Defective mitophagy in XPA via PARP-1 hyperactivation and NAD(+)/SIRT1 reduction. Cell 157:882–896
90. Fan W, Luo J (2010) SIRT1 regulates UV-induced DNA repair through deacetylating XPA. Mol Cell 39:247–258
91. Mouchiroud L, Houtkooper RH, Moullan N, Katsyuba E, Ryu D, Canto C, Mottis A, Jo YS, Viswanathan M, Schoonjans K, Guarente L, Auwerx J (2013) The NAD(+)/sirtuin pathway modulates longevity through activation of mitochondrial UPR and FOXO signaling. Cell 154:430–441
92. Mouchiroud L, Houtkooper RH, Auwerx J (2013) NAD(+) metabolism: a therapeutic target for age-related metabolic disease. Crit Rev Biochem Mol Biol 48:397–408
93. Weaver AN, Yang ES (2013) Beyond DNA repair: additional functions of PARP-1 in cancer. Front Oncol 3:290
94. Cohen-Armon M (2007) PARP-1 activation in the ERK signaling pathway. Trends Pharmacol Sci 28:556–560
95. Cohen-Armon M, Visochek L, Rozensal D, Kalal A, Geistrikh I, Klein R, Bendetz-Nezer S, Yao Z, Seger R (2007) DNA-independent PARP-1 activation by phosphorylated ERK2 increases Elk1 activity: a link to histone acetylation. Mol Cell 25:297–308
96. Kannan S, Fang W, Song G, Mullighan CG, Hammitt R, McMurray J, Zweidler-McKay PA (2016) Notch/HES1-mediated PARP1 activation: a cell type – specific mechanism for tumor suppression. Blood 117:2891–2900
97. Geistrikh I, Visochek L, Klein R, Miller L, Mittelman L, Shainberg A, Cohen-Armon M (2011) Ca^{+2}-induced PARP-1 activation and *ANF* expression are coupled events in cardiomyocytes. Biochem J 438:337–347
98. Fischer JM, Popp O, Gebhard D, Veith S, Fischbach A, Beneke S, Leitenstorfer A, Bergemann J, Scheffner M, Ferrando-May E, Mangerich A, Burkle A (2014) Poly(ADP-ribose)-mediated interplay of XPA and PARP1 leads to reciprocal regulation of protein function. FEBS J 281:3625–3641
99. Yang Z, Roginskaya M, Colis LC, Basu AK, Shell SM, Liu Y, Musich PR, Harris CM, Harris TM, Zou Y (2006) Specific and efficient binding of Xeroderma pigmentosum complementation group A to double-strand/single-strand DNA junctions with 3′- and/or 5′-ssDNA branches. Biochemistry 45:15921–15930
100. Liu Y, Liu Y, Yang Z, Utzat C, Wang G, Basu AK, Zou Y (2005) Cooperative interaction of human XPA stabilizes and enhances specific binding of XPA to DNA damage. Biochemistry 44:7361–7368
101. Hilton B, Shkriabai N, Musich PR, Kvaratskhelia M, Shell S, Zou Y (2014) A new structural insight into XPA-DNA interaction. Biosci Rep 34:831–840
102. Buchko GW, Daughdrill GW, de Lorimier R, Rao BK, Isern NG, Lingbeck JM, Taylor JS, Wold MS, Gochin M, Spicer LD, Lowry DF, Kennedy MA (1999) Interactions of human nucleotide excision repair protein XPA with DNA and RPA70 Delta C327: chemical shift mapping and 15N NMR relaxation studies. Biochemistry 38:15116–15128
103. Ikegami T, Kuraoka I, Saijo M, Kodo N, Kyogoku Y, Morikawa K, Tanaka K, Shirakawa M (1998) Solution structure of the DNA- and RPA-binding domain of the human repair factor XPA. Nat Struct Mol Biol 5:701–706
104. Dechat T, Adam SA, Taimen P, Shimi T, Goldman RD (2010) Nuclear Lamins. Cold Spring Harb Perspect Biol 2:1–22
105. Goldman RD, Shumaker DK, Erdos MR, Eriksson M, Goldman AE, Gordon LB, Gruenbaum Y, Khuon S, Mendez M, Varga R, Collins FS (2004) Accumulation of mutant lamin A causes progressive changes in nuclear architecture in Hutchinson-Gilford progeria syndrome. Proc Natl Acad Sci U S A 101:8963–8968

106. Kudlow BA, Kennedy BK, Monnat RJ Jr (2007) Werner and Hutchinson-Gilford progeria syndromes: mechanistic basis of human progeroid diseases. Nat Rev Mol Cell Biol 8:394–404
107. Misteli T, Scaffidi P (2005) Genome instability in progeria: when repair gets old. Nat Med 11:718–719
108. Musich PR, Zou Y (2009) Genomic instability and DNA damage responses in progeria arising from defective maturation of prelamin A. Impact Aging 1:28–37
109. Pereira S, Bourgeois P, Navarro C, Esteves-Vieira V, Cau P, De Sandre-Giovannoli A, Lévy N (2008) HGPS and related premature aging disorders: from genomic identification to the first therapeutic approaches. Mech Ageing Dev 129:449–459
110. Smith ED, Kudlow BA, Frock RL, Kennedy BK (2005) A-type nuclear lamins, progerias and other degenerative disorders. Mech Ageing Dev 126:447–460
111. Wiesel N, Mattout A, Melcer S, Melamed-Book N, Herrmann H, Medalia O, Aebi U, Gruenbaum Y (2008) Laminopathic mutations interfere with the assembly, localization, and dynamics of nuclear lamins. Proc Natl Acad Sci U S A 105:180–185
112. Liu Y, Rusinol A, Sinensky M, Wang Y, Zou Y (2006) DNA damage responses in progeroid syndromes arise from defective maturation of prelamin A. J Cell Sci 119:4644–4649
113. Liu Y, Wang Y, Rusinol AE, Sinensky MS, Liu J, Shell SM, Zou Y (2008) Involvement of Xeroderma pigmentosum group A (XPA) in progeria arising from defective maturation of prelamin A. FASEB J 22:603–611
114. Tang H, Hilton B, Musich PR, Fang DZ, Zou Y (2011) Replication factor C1, the large subunit of replication factor C, is proteolytically truncated in Hutchinson-Gilford Progeria Syndrome. Aging Cell 11:363–365
115. Hilton BA, Liu J, Cartwright BM, Liu Y, Breitman M, Wang Y, Jones R, Tang H, Rusinol A, Musich PR, Zou Y (2017) Progerin sequestration of PCNA promotes replication fork collapse and mislocalization of XPA in laminopathy-related progeroid syndromes. FASEB J:fj.201700014R
116. Thompson JA, Marzahn MR, O'Donnell M, Bloom LB (2012) Replication factor C is a more effective PCNA opener than the checkpoint clamp loader, RAD24-RFC. J Biol Chem 287:2203–2209
117. Wang S-C (2014) PCNA: a silent housekeeper or a potential therapeutic target? Trends Pharmacol Sci 35:178–186
118. Musich PR, Zou Y (2011) DNA-damage accumulation and replicative arrest in Hutchinson-Gilford progeria syndrome. Biochem Soc Trans 39:1764–1769

Impact of Ultraviolet Light on Vitiligo

5

Rasnik K. Singh

Abstract

Vitiligo is a disorder of the melanocytes that results in a dynamic spectrum of skin depigmentation. Its etiology is complex and multifactorial, with data supporting several different hypotheses. Given its prominent phenotype, vitiligo has a significant negative impact on quality of life. Coupled with the chronic and incurable nature of the disease, this presents a formidable treatment challenge. Several treatment modalities have been instituted over the years, with varying efficacy. This chapter focuses on the use of ultraviolet light in vitiligo as an established therapeutic option.

Keywords

Vitiligo • Inflammatory • Ultraviolet • Light • Phototherapy • UVB • PUVA • Narrowband • Repigmentation • VASI

5.1 Introduction

Vitiligo is a chronic persistent disease of the skin's pigment-producing cells with an estimated worldwide prevalence of up to 2%. Half of all cases have their onset before age 20 [1]. Disease onset before 3 years of age, in particular, has been associated with a more extensive and progressive course [2]. Epidemiologic studies suggest that all skin types and races, as well as both sexes, are equally affected [3, 4]. Although the disease phenotype of patchy depigmentation is more noticeable in darker skin types, vitiligo has a significant impact on quality of life for all those affected [5, 6]. Multiple treatment options currently exist, including lasers, topical corticosteroids, calcineurin inhibitors, and vitamin D analogues, as well as phototherapy. Camouflaging with dermatological-grade makeup is another option for more immediate results.

R.K. Singh (✉)
David Geffen School of Medicine at UCLA,
University of California,
Los Angeles, CA 90095, USA
e-mail: Rasnik.s@gmail.com

S. Ahmad (ed.), *Ultraviolet Light in Human Health, Diseases and Environment*, Advances in Experimental Medicine and Biology 996, https://doi.org/10.1007/978-3-319-56017-5_5

5.2 Vitiligo Types

Vitiligo is classified broadly into non-segmental (NSV) and segmental (SV) types. NSV, which accounts for 85–90% of all cases, manifests with multifocal symmetrically distributed lesions. Several subphenotypes of NSV have been described, including mucosal, acrofacial, generalized, and universal [7]. SV accounts for 10–15% of cases and is characterized by a unifocal lesion in a band-shaped distribution [8]. Studies suggest that in general, SV has earlier onset and tends to stabilize over the course of a few months. On the other hand, NSV more often exhibits long-term disease progression [9].

5.3 Vitiligo Pathophysiology

Several hypotheses have been generated to define the pathophysiology of vitiligo; however, the disease remains poorly understood. The genetic component is thought to be complex and multifactorial, with many susceptibility loci identified through genome-wide association studies [10–13]. Monozygotic twin studies reveal a relatively low concordance rate of 23%, indicating an additional strong environmental component [14]. Various groups have investigated a neural hypothesis, in which neurogenic inflammatory mediators are released that are directly toxic to melanocytes [15, 16]. A redox hypothesis suggests that overproduction of reactive oxygen species (ROS) in the skin may be toxic to critical cell components, leading to melanocyte destruction [17]. Finally, an autoimmune hypothesis with contributions from cell-, antibody-, and cytokine-mediated mechanisms is also supported. Specifically, evidence suggests a role for cytotoxic T cells, autoantibodies against tyrosine hydroxylase and pigment cell-surface antigens, and the Th1, Th2, and Th17 programs [18–20]. Interestingly, vitiligo has also been significantly associated with several autoimmune conditions, including thyroid disease, alopecia areata, inflammatory bowel disease, pernicious anemia, systemic lupus erythematosus, myasthenia gravis, and others [21].

5.4 Ultraviolet Light Therapy (Phototherapy)

The ultraviolet (UV) light spectrum is divided into ultraviolet A (UVA) (400-320-nm), ultraviolet B (UVB) (320-280 nm), and ultraviolet C (UVC) (280-200 nm) portions. UVB is subcategorized as either narrowband (313-311 nm) or broadband (320-280 nm). Phototherapy involves consistent cutaneous exposure to specific wavelengths of ultraviolet light under medical supervision for treatment of an underlying condition.

5.4.1 Cellular and Molecular Changes Associated with Phototherapy

Phototherapy is currently a well-accepted therapeutic modality for several skin diseases, including psoriasis, atopic dermatitis, and vitiligo [22]. Ultraviolet light is thought to work by multiple mechanisms to modulate disease phenotype. For instance, it has been shown to regulate antigen presentation, with marked reduction in numbers of Langerhans cells [23]. It has also been shown to induce apoptosis and regulate cytokine secretion of macrophages [24]. The downstream effects include dampening of cytotoxic and helper T-cell activity with concomitant induction of regulatory T cells [25]. In particular, narrowband UVB has been shown to decrease inflammatory Th17 cell abundance as well as serum and tissue IL-17 levels [20].

Finally, phototherapy is also thought to act on the innate arm of the immune system by antagonizing both neutrophil and NK cell activity [26].

Phototherapy is thought to affect repigmentation in vitiligo by stimulating follicular melanocytes to migrate upwards to the epidermis [27]. This results in the perifollicular repigmentation pattern classically seen after UV light treatment. Narrowband UVB has also been shown to promote release of keratinocyte-derived fibroblast growth factor and endothelin-1, which in turn leads to local proliferation of melanocytes [28].

5.5 Phototherapy in Vitiligo

Vitiligo is evaluated quantitatively by first dividing the body into 6 regions: face/neck, trunk, upper extremities, lower extremities, hands, and feet. For each region, the affected body surface area measured in 'hand units' is multiplied by the extent of depigmentation (e.g. 0.25, 0.5, 0.75, 1.0). The summation of these products across all body sites is known as the Vitiligo Area Scoring Index (VASI) [29]. Treatment response is often reported as a change in VASI score or more generally, as an assessed percent value of repigmentation. More than 75% repigmentation is thought to be cosmetically acceptable. Changes in VASI score are characterized on a wide scale from "very much worse" to "very much improved", with several intermediate milestones. At the extremes are changes of –50 and +50, respectively.

Narrowband UVB or UVA in combination with an oral or topical photosensitizing drug such as 8-methoxypsoralen (PUVA) have been used for decades in patients with vitiligo. Phototherapy is typically administered two to three times per week at a fixed starting dose ranging from 100 to 280 mJoules/cm^2. Dosage can be increased by 10-20% at each subsequent visit, as tolerated, until development of a mild erythema. For PUVA therapy, the photosensitizing agent is given prior to light exposure. Eye and genital photoprotection with goggles, shields, and/or towels is always recommended to prevent unnecessary exposure to UV radiation [22]. 100-200 treatment sessions are typically required to maximize and optimize pigment induction.

5.5.1 Efficacy

Prior to development of the most recent guidelines for the treatment of vitiligo in 1999, PUVA was a mainstay therapy for the generalized type. PUVA has been shown to induce repigmentation [30]. With the development of narrowband UVB and subsequent data supporting its application in psoriasis, several studies were conducted to additionally determine its efficacy in vitiligo. Rates of cosmetically acceptable repigmentation have been reported to range from 12.5% [31] to 75% of patients receiving narrowband UVB. The higher response rates were observed in Indian cohorts [32, 33]. Along those lines, evidence suggests that narrowband UVB is more likely to lead to satisfactory repigmentation in patients with darker skin types, facial lesions, and whom demonstrate treatment response within the first month [34]. Although unclear, facial lesions may repigment better than acral sites due to a higher native density of hair follicles, which in turn are melanocyte reserves [35]. Relapse after discontinuation of narrowband UVB therapy is reported in up to 44% of patients, however, the disease usually responds to an additional course of treatment [29].

Many studies have directly compared the efficacy profiles of narrowband UVB and PUVA. These found narrowband UVB to be as or more efficacious with more stable repigmentation and fewer side effects than PUVA [36–39].

Narrowband UVB has also been investigated in combination with several other agents, including antioxidants, folic acid, vitamin B12, topical vitamin D analogues (calcipotriene, calcipotriol, and tacalcitol), and tacrolimus. Addition of folic acid, vitamin B12, or calcipotriene was not shown to increase efficacy over UVB monotherapy in any studies; however, addition of tacalcitol, calcipotriol, or antioxidants had variable results. Combination therapy with PUVA and calcipotriol has also been studied, with variable results [40].

5.5.2 Safety

Adverse effects of phototherapy can be characterized as acute or chronic. Acute adverse effects include erythema, pruritus, tanning, xerosis, and rarely burning or blistering [22]. Chronic adverse effects include photoaging [41]. Skin cancer, as a chronic adverse effect of phototherapy in vitiligo remains, controversial. Interestingly, on the whole patients with vitiligo have been shown to have a decreased risk of melanoma and non-melanoma skin cancer (NMSC) than the general population [42, 43]. This is thought to be due to the anti-melanocyte immune response and increased wild type p53 tumor suppressor gene levels that occur in the disease [44]. In one study, subgroup analyses of patients treated with narrowband UVB or topical or oral PUVA did not demonstrate increased prevalence than untreated patients [43]. A separate study, however, demonstrated that vitiligo patients receiving phototherapy did have a higher risk of both melanoma and NMSC than their untreated counterparts [42]. More data need to be collected to clarify the risk for carcinogenesis associated with phototherapy in vitiligo. A number of reports are available, however, suggesting that PUVA therapy for other skin disorders including psoriasis and mycosis fungoides leads to increased incidence of nonmelanoma skin cancer [45, 46]. Therefore, vitiligo patients receiving long-term treatment with ultraviolet light should be followed closely.

5.6 Conclusion

Vitiligo is a chronic persistent disease of melanocytes with a complex etiology thought to derive from genetic, autoimmune, neural, and redox mechanisms. There are multiple treatment options with varying efficacy, including targeted lasers, topical corticosteroids, vitamin D analogues, and calcineurin inhibitors, as well as phototherapy. Phototherapy is particularly useful for the generalized subtype of vitiligo given its ease of widespread administration compared to topical therapy. The use of phototherapy in vitiligo has evolved from PUVA to largely narrowband UVB over the past couple of decades and is a mainstay of treatment given its favorable efficacy and safety profiles, as well as reasonable cost. Long-term risks of phototherapy in vitiligo—in particular for melanoma and non-melanoma skin cancer—are still unclear, necessitating further studies.

References

1. Ezzedine K, Silverberg N (2016) A practical approach to the diagnosis and treatment of vitiligo in children. Pediatrics. doi:10.1542/peds.2015-4126
2. Mu EW, Cohen BE, Orlow SJ (2015) Early-onset childhood vitiligo is associated with a more extensive and progressive course. J Am Acad Dermatol 73:467–470
3. Boisseau-Garsaud AM, Garsaud P, Calès-Quist D, Hélénon R, Quénéhervé C, Claire RC (2000) Epidemiology of vitiligo in the French West Indies (Isle of Martinique). Int J Dermatol 39:18–20
4. Howitz J, Brodthagen H, Schwartz M, Thomsen K (1977) Prevalence of vitiligo. Epidemiological survey on the Isle of Bornholm, Denmark. Arch Dermatol 113:47–52
5. Linthorst Homan MW, Spuls PI, de Korte J, Bos JD, Sprangers MA, van der Veen JPW (2009) The burden of vitiligo: patient characteristics associated with quality of life. J Am Acad Dermatol 61:411–420
6. Porter JR, Beuf AH (1991) Racial variation in reaction to physical stigma: a study of degree of disturbance by vitiligo among black and white patients. J Health Soc Behav 32:192–204
7. Ezzedine K, Le Thuaut A, Jouary T, Ballanger F, Taieb A, Bastuji-Garin S (2014) Latent class analysis of a series of 717 patients with vitiligo allows the identification of two clinical subtypes. Pigment Cell Melanoma Res 27:134–139
8. Ezzedine K, Lim HW, Suzuki T et al (2012) Revised classification/nomenclature of vitiligo and related issues: the Vitiligo Global Issues Consensus Conference. Pigment Cell Melanoma Res 25:E1–13
9. Mazereeuw-Hautier J, Bezio S, Mahe E et al (2010) Segmental and nonsegmental childhood vitiligo has distinct clinical characteristics: a prospective observational study. J Am Acad Dermatol 62:945–949
10. Jin Y, Birlea SA, Fain PR et al (2010) Variant of TYR and autoimmunity susceptibility loci in generalized vitiligo. N Engl J Med 362:1686–1697
11. Jin Y, Birlea SA, Fain PR et al (2011) Genome-wide analysis identifies a quantitative trait locus in the mhc class ii region associated with generalized vitiligo age of onset. J Invest Dermatol 131:1308–1312
12. Quan C, Ren Y-Q, Xiang L-H et al (2010) Genome-wide association study for vitiligo identifies susceptibility loci at 6q27 and the MHC. Nat Genet 42:614–618
13. Cheong KA, Kim N-H, Noh M, Lee A-Y (2013) Three new single nucleotide polymorphisms identified by a genome-wide association study in korean patients with vitiligo. J Korean Med Sci 28:775
14. Alkhateeb A, Fain PR, Thody A, Bennett DC, Spritz RA (2003) Epidemiology of vitiligo and associated autoimmune diseases in Caucasian probands and their families. Pigment Cell Res 16:208–214
15. Peters EMJ, Handjiski B, Kuhlmei A, Hagen E, Bielas H, Braun A, Klapp BF, Paus R, Arck PC (2004)

Neurogenic inflammation in stress-induced termination of murine hair growth is promoted by nerve growth factor. Am J Pathol 165:259–271
16. Norris DA (2001) In this issue. J Invest Dermatol 117:1025–1026
17. Khan R, Satyam A, Gupta S, Sharma VK, Sharma A (2009) Circulatory levels of antioxidants and lipid peroxidation in Indian patients with generalized and localized vitiligo. Arch Dermatol Res 301:731–737
18. You S, Cho Y-H, Byun J-S, Shin E-C (2013) Melanocyte-specific CD8 + T cells are associated with epidermal depigmentation in a novel mouse model of vitiligo. Clin Exp Immunol 174:38–44
19. Taher ZA, Lauzon G, Maguiness S, Dytoc MT (2009) Analysis of interleukin-10 levels in lesions of vitiligo following treatment with topical tacrolimus. Br J Dermatol 161:654–659
20. Singh RK, Lee KM, Vujkovic-Cvijin I et al (2016) The role of IL-17 in vitiligo: a review. Autoimmun Rev 15:397–404
21. Gill L, Zarbo A, Isedeh P, Jacobsen G, Lim HW, Hamzavi I (2016) Comorbid autoimmune diseases in patients with vitiligo: a cross-sectional study. J Am Acad Dermatol 74:295–302
22. Singh RK, Lee KM, Jose MV, Nakamura M, Ucmak D, Farahnik B, Abrouk M, Zhu TH, Bhutani T, Liao W (2016) The patient's guide to psoriasis treatment. Part 1: UVB phototherapy. Dermatol Ther (Heidelb) 6:307–313
23. Duthie MS, Kimber I, Norval M (1999) The effects of ultraviolet radiation on the human immune system. Br J Dermatol 140:995–1009
24. Sethi G, Sodhi A (2004) Role of p38 mitogen-activated protein kinase and caspases in UV-B-induced apoptosis of murine peritoneal macrophages. Photochem Photobiol 79:48–54
25. Okamoto H, Horio T, Maeda M (1987) Alteration of lymphocyte functions by 8-methoxypsoralen and long-wave ultraviolet radiation. II. The effect of in vivo PUVA on IL-2 production. J Invest Dermatol 89:24–26
26. Weitzen ML, Bonavida B (1984) Mechanism of inhibition of human natural killer activity by ultraviolet radiation. J Immunol 133:3128–3132
27. Cui J, Shen LY, Wang GC (1991) Role of hair follicles in the repigmentation of vitiligo. J Invest Dermatol 97:410–416
28. Wu C-S, Yu C-L, Wu C-S, Lan C-CE YH-S (2004) Narrow-band ultraviolet-B stimulates proliferation and migration of cultured melanocytes. Exp Dermatol 13:755–763
29. Nicolaidou E, Antoniou C, Stratigos A, Katsambas AD (2009) Narrowband ultraviolet B phototherapy and 308-nm excimer laser in the treatment of vitiligo: a review. J Am Acad Dermatol 60:470–477
30. Whitton ME, Ashcroft DM, González U (2008) Therapeutic interventions for vitiligo. J Am Acad Dermatol 59:713–717
31. Chen G-Y, Hsu MM-L, Tai H-K, Chou T-C, Tseng C-L, Chang H-Y, Lan C-CE, Sheu H-M (2005) Narrow-band UVB treatment of vitiligo in Chinese. J Dermatol 32:793–800
32. Kanwar AJ, Dogra S (2005) Narrow-band UVB for the treatment of generalized vitiligo in children. Clin Exp Dermatol 30:332–336
33. Kanwar AJ, Dogra S, Parsad D, Kumar B (2005) Narrow-band UVB for the treatment of vitiligo: an emerging effective and well-tolerated therapy. Int J Dermatol 44:57–60
34. Nicolaidou E, Antoniou C, Stratigos AJ, Stefanaki C, Katsambas AD (2007) Efficacy, predictors of response, and long-term follow-up in patients with vitiligo treated with narrowband UVB phototherapy. J Am Acad Dermatol 56:274–278
35. Falabella R (2009) Vitiligo and the melanocyte reservoir. Indian J Dermatol 54:313
36. Parsad D, Kanwar AJ, Kumar B (2006) Psoralen-ultraviolet A vs. narrow-band ultraviolet B phototherapy for the treatment of vitiligo. J Eur Acad Dermatol Venereol 20:175–177
37. Bhatnagar A, Kanwar AJ, Parsad D, De D (2007) Comparison of systemic PUVA and NB-UVB in the treatment of vitiligo: an open prospective study. J Eur Acad Dermatol Venereol 21:638–642
38. El Mofty M, Mostafa W, Esmat S, Youssef R, Azzam O, Hunter N, El Hanafi G, Fawzi M (2006) Narrow band Ultraviolet B 311 nm in the treatment of vitiligo: two right-left comparison studies. Photodermatol Photoimmunol Photomed 22:6–11
39. Yones SS, Palmer RA, Garibaldinos TM, Hawk JLM (2007) Randomized double-blind trial of treatment of vitiligo: efficacy of psoralen-UV-A therapy vs Narrowband-UV-B therapy. Arch Dermatol 143:578–584
40. Prabhu S, Shenoi S (2014) Photochemotherapy (PUVA) in psoriasis and vitiligo. Indian J Dermatol Venereol Leprol 80:497
41. Choi CP, Kim YI, Lee JW, Lee MH (2007) The effect of narrowband ultraviolet B on the expression of matrix metalloproteinase-1, transforming growth factor-?1 and type I collagen in human skin fibroblasts. Clin Exp Dermatol 32:180–185
42. Paradisi A, Tabolli S, Didona B, Sobrino L, Russo N, Abeni D (2014) Markedly reduced incidence of melanoma and nonmelanoma skin cancer in a nonconcurrent cohort of 10,040 patients with vitiligo. J Am Acad Dermatol 71:1110–1116
43. Teulings HE, Overkamp M, Ceylan E, Nieuweboer-Krobotova L, Bos JD, Nijsten T, Wolkerstorfer AW, Luiten RM, van der Veen JPW (2013) Decreased risk of melanoma and nonmelanoma skin cancer in patients with vitiligo: a survey among 1307 patients and their partners. Br J Dermatol 168:162–171
44. Schallreuter KU, Behrens-Williams S, Khaliq TP, Picksley SM, Peters EMJ, Marles LK, Westerhof W, Miehe B, Fanghänel J (2003) Increased epidermal

functioning wild-type p53 expression in vitiligo. Exp Dermatol 12:268–277
45. Querfeld C (2005) Long-term follow-up of patients with early-stage cutaneous T-Cell Lymphoma who achieved complete remission with psoralen plus UV-a monotherapy. Arch Dermatol 141:305–311
46. Maiorino A, De Simone C, Perino F, Caldarola G, Peris K (2016) Melanoma and non-melanoma skin cancer in psoriatic patients treated with high-dose phototherapy. J Dermatolog Treat 27:443–447

Polymorphous Light Eruption

6

Marcella Guarrera

Abstract

Polymorphous light eruption (PLE) is the commonest immuno-mediated photodermatosis. It occurs after solar or artificial UV-light exposure and affects only the sun-exposed areas with preference of the V-area of the chest, of arms and forearms, legs, upper part of the back, and rarely the face. The lesions are itching or burning, and vary morphologically from erythema to papules, vesico-papules and occasionally blisters, plaques, sometimes erythema multiforme-like, insect bite-like wheals and purpura. The clinical manifestations befall within a few hours to days from light exposure, last a few days, and subside in about a week without sequelae. Its diagnosis is based on history, morphology and phototests. PLE is considered as a delayed hypersensitivity response to newly UV induced, but still unidentified, antigen(s). Usually, MED is normal, but the provocative phototests with UVA or UVB reproduce the spontaneous lesions in about 50% of the patients. Broad spectrum sunscreens and antioxidants, photo-hardening with PUVA or narrow band UVB may be beneficial to prevent the disease. Therapy is based mainly on topical or systemic corticosteroids.

Keywords

Polymorphous light eruption • Idiopathic photodermatosis • Immunomediated photodermatosis • UV light • Phototests • Minimal erythema dose • Photoprovocation tests

6.1 Introduction

Polymorphous light eruption (PLE), once called idiopathic, is the commonest immuno-mediated photodermatosis. The first description dates back to the nineteenth century, when Bateman [1]

M. Guarrera (✉)
University of Genoa, Genoa, Italy
e-mail: guarrera@unige.it

S. Ahmad (ed.), *Ultraviolet Light in Human Health, Diseases and Environment*, Advances in Experimental Medicine and Biology 996, https://doi.org/10.1007/978-3-319-56017-5_6

defined as *eczema solare* some recurrent non scarring eczematous lesions provoked by sun-exposure. PLE has been labeled also solar dermatitis, summer prurigo and sun allergy. The present title is due to Rasch in 1900 [2].

6.2 Epidemiology

PLE affects both genders, but women are most affected (f/m ratio 3:1–7:1) [3], and all ages, mostly adulthood [4]. Apparently, the prevalence depends on the latitude: about 21% in Scandinavia [5], 10–15% in northern United States [6] and United Kingdom [7], but only 5% in Australia [8], 1% in Singapore [5], and 0.6% in India [9]. In Italy, it occurs in 6% of the population [10], less than in some other European Countries (18%) [11]. It affects all skin types, preferring the fair ones, and all races with an apparently paradoxical prevalence (86%) in African-Americans [12, 13]. A positive family history can be found in one-sixth of patients [6] or even more [14].

6.3 Clinical Manifestations

PLE lesions occur always after solar or artificial UV-light exposure and affect only the sun-exposed areas with preference of the V-area of the chest, arms and forearms, legs, upper part of the back, and in the severest forms also the face. They are always itching or burning, but vary morphologically (explaining the adjective "polymorphous" or "polymorphic") from erythema to papules, vesico-papules and occasionally blisters, plaques, sometimes erythema multiforme-like, insect bite-like wheals and purpura [5, 15] (Figs. 6.1, 6.2, 6.3, 6.4 and 6.5). In the same patient, however, the lesions are monomorphic. Often, itching or burning shortly herald the appearance of the lesions.

The clinical manifestations befall within a few hours to days from light exposure, last a few days, and subside in about a week without sequelae but rare small hyper- or hypopigmentations. PLE may last for many years in several patients, often recurring annually in the same season, improves over the years in others, and sometimes remits spontaneously [16]. Usually, there are no systemic symptoms. Chills, headache, fever and nausea

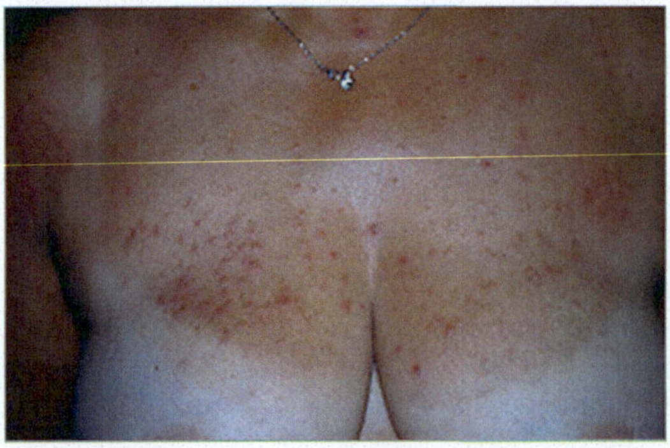

Fig. 6.1 Papular PLE of the chest

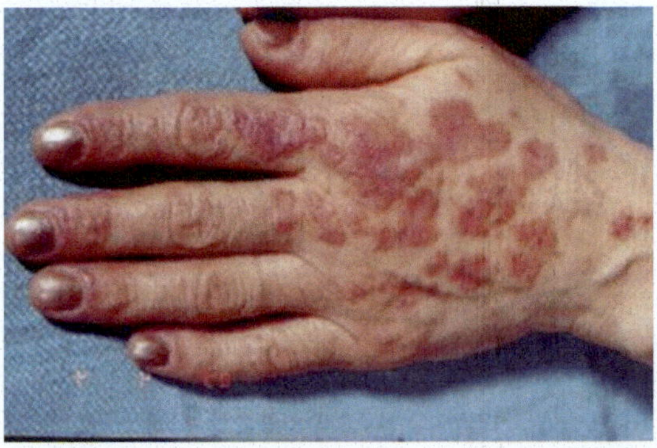

Fig. 6.2 PLE plaques on the hand

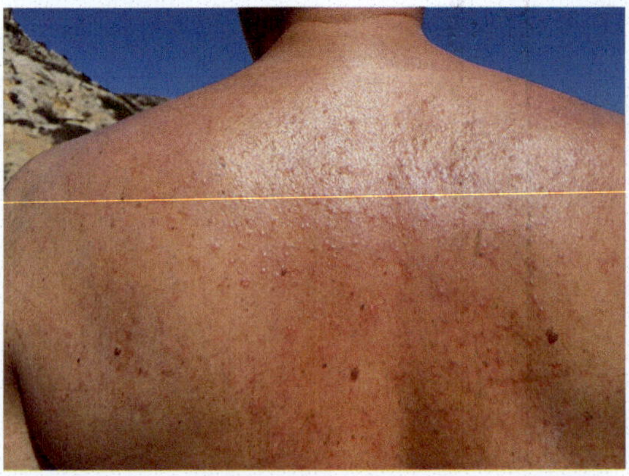

Fig. 6.3 Vesico-papular PLE of the dorsum

have been described, but they probably result from heatstroke or sunburn [5].

Particular forms have been reported, such as PLE sine eruption [17], pinpoint papular eruption [18, 19] especially in individuals with skin type IV-VI, characterized by 1–2 mm pinpoint papules, similar to the pinhead papular eruption form [20], persistent PLE after UVA1 therapy [21], PLE of the elbows [22, 23], solar brachioradial pruritus [24], that needs to be distinguished by the cervical spine disease form [25].

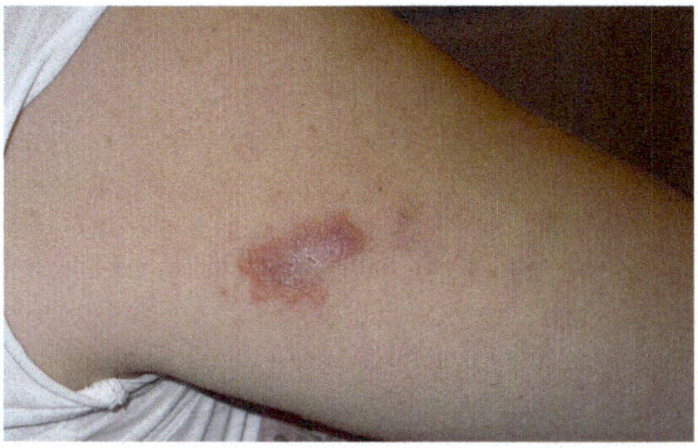

Fig. 6.4 Erythema multiforme-like PLE of the arm

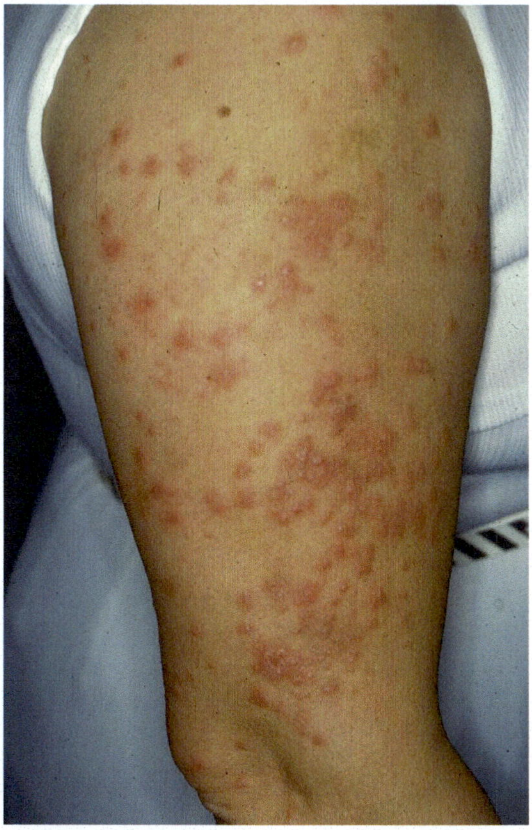

Fig. 6.5 Vesico-papular PLE of the arm

There are variants that should be distinguished, such as Juvenile spring eruption [26, 27] characterized by itching papules and vesicles of the ears in young boys occurring in spring, and Actinic prurigo, which is characterized by persistent, pruritic, excoriated, papular or nodular eruption of sun-exposed and unexposed areas in childhood often present also in winter. Actinic prurigo is a typical manifestation of native American people, in whom it is often hereditary with the presence of HLA DR4 in about 90% of cases and, in particular, the subtype DRB1*0407 present in 60% of cases [28–30]. Hydroa aestivale-vacciniforme is characterized by groups of vesicles with crusts that leave vacciniform scars on the face, chest and dorsum of the hands, more often in women in their first decade of life [31]. Another variant is the Benign summer light eruption (BSLE) [32], which affects mostly young women on the upper chest with small pruritic papules, shortly after an intense UV-light exposure and without annual relapse. In 2011, an Italian multicenter study [33] tried to distinguish BSLE and classical PLE enrolling 346 patients with typical clinical history and/or presentation of PLE, on the basis of some clinical and laboratory criteria. The studied criteria distinguished only a minority of BSLE patients. BSLE may be considered as a mild form of PLE [33], it is always positive to UVA-induced phototest, and is probably more frequent, but often goes unobserved by dermatologists because its mildness simply prevents their visit. This conclusion has been shared by others [34, 35].

6.4 Histology and Immunohistochemistry

The histology of PLE is not specific in accordance with the polymorphic clinical patterns, and depending also on the timing of the biopsy. Characteristic feature is a moderate to intense

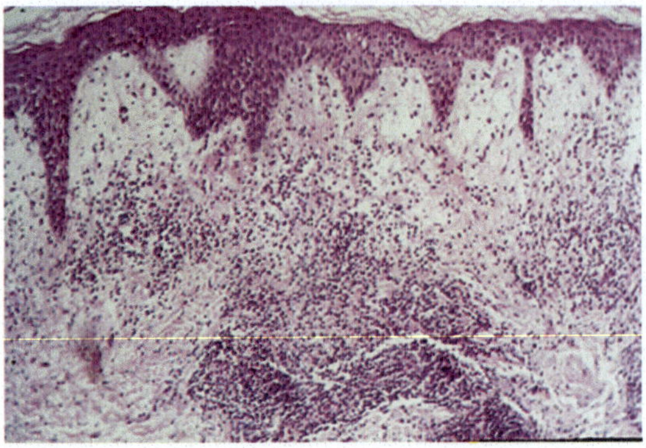

Fig. 6.6 Histopathology of PLE showing papillary edema and middermis lymphocytic infiltrate

perivascular T cell infiltrate [36] and the edema in the upper part of the dermis (Fig. 6.6), even though the latter may be observed in LE and dermatomyositis [37]. In the papular form, the edema of the papillary dermis is common, focal dyskeratotic cells and slight vacuolar alteration of the basal layer can be observed. The plaque-type PLE exhibits also a band like mononuclear cell infiltrate. The papulovesicular form shows spongiotic microvesicles, marked subepidermal edema, extravasation of erythrocytes and a mixed, mainly lymphocytic, dermal infiltrate. Lastly, the eczematous form shows parakeratosis, acanthosis, spongiosis and sporadic dyskeratosis. Immunohistochemistry shows an increase of Langerhans cells (OKT6) in the epidermis. The direct immunofluorescence is not contributory.

6.5 Diagnosis

Diagnosis is not difficult. Taking the history of sun or artificial light exposure (either professional or not), excluding the possible responsibility of photosensitizing cosmetics or drugs, and the clinical examination (morphology and the sun-exposed site of the pruritic lesion are highly suggestive). The age of the first manifestation, the interval from the light exposure (latency time), the duration and the seasonality are also helpful data, and a series of phototests may be confirmatory. MED (minimal erythema dose), the provocative UV phototests, patch and photopatch tests, the porphyrins blood levels and the antinuclear antibodies (ANA) assessment are mandatory. Biopsy may be helpful.

6.6 Differential Diagnosis

The commonest differential diagnoses are solar urticaria, which develops just a few minutes after sun-exposure, the rare photosensitive erythema multiforme, which occurs after intake of drugs like carbanilides, phenylbutazone and aflaqualone and affects the oral mucosa as well [38], and the photocontact allergic dermatitis, in which the photopatch tests are diagnostic. It should not be forgotten that PLE patients may also be suffering from photocontact allergic dermatitis.

The most important disease to be considered in the differential diagnosis, however, is systemic lupus erythematosus (SLE). Classically, SLE lesions last more than 2 weeks, and are accompanied by positive serology and direct immunofluorescence. Although ANA may be present in PLE as well, they do not exceed the 1:80 dilution. Ro (SSa) and LA (SSb) antibodies, which characterize the photosensitive LE subset (SCLE), are absent in PLE. Nonetheless, a relationship between the two diseases probably exists, though denied by some Authors [16, 39–42]. In fact, about 10% of PLE patients with positive ANA develop SLE over time [43], PLE symptoms have been reported in 50% of LE patients and LE diagnosis has been done in PLE patients up to 7 years after the PLE onset [44]. ANA may already be present many years before in 78% of PLE patients who are destined to develop LE, though there is no way to predict such an outcome [45]. Lastly, PLE symptoms have been described in 60% of DLE or SCLE patients and are more frequent in LE patients' relatives [46].

6.7 Pathogenesis

After the great intuition of Epstein [47], many details help consider PLE as a delayed hypersensitivity response to UV induced, but still unidentified, antigen(s). The delayed occurrence of the lesions, the HLA-DR expression at least in 50% of patients, the pro-inflammatory cytokines and adhesion molecules expression are indicative findings. In addition, the presence in the dermis of T CD4+ cells within 72 h and, later, T CD8+, the presence of macrophages 1–5 h after irradiation, the increasing numbers of Langerhans cells in 5 h after UV exposure, the improvement after immunosuppression therapy, all justify such a conclusion.

The abnormal immune response has been attributed to the resistance of the PLE patients towards the immune suppressive effects of sunlight [48]. The exact UV-induced immunosuppression mechanism and the relative contribution of UVB and UVA in healthy subjects are as yet unclear, but the expression of TNF-alpha, IL-4 and IL-10 and the Langerhans cell depletion seem to be crucial phenomena [49]. In PLE, the resistance to immunosuppression is documented by a reduced expression of TNF-alpha, IL-4 and IL-10 and by an impaired Langerhans cell and neutrophil migration into the epidermis [50]. Incidentally, the UV-induced immunosuppression is lower in healthy women [51], possibly via 17β-estradiol [52] or estrogen receptors [53], explaining the disproportionate prevalence of PLE in women. Moreover, the disease can be favored by oral contraceptives and pregnancy [54], and, personally, I observed that, usually, PLE may occur during the first pregnancy (unpublished data).

Genetic factors also play a role, though with poor penetrance [55, 56]. There is no difference in the prevalence of the disease between monozygotic and dizygotic twins [7, 56], and a reverse link to glutathione-S-transferase1 allele, which is protective against the pathogenetic role of ROS, was advocated [57], but not confirmed [58]. However no gene has been identified till now. Heat-shock protein immunoreactivity has been suggested [59]. In fact, the heat-shock protein expression increases in keratinocytes and endothelial cells of dermal blood vessels in experimental PLE, 1 h to 6 days after UVR exposure [7].

Moreover, abnormalities in arachidonic acid metabolism, especially in the severest forms, and in prostaglandins have been reported [7]. The mentioned role of ROS would be confirmed by the decreased levels of epidermal (by 30%) [60] and blood catalase, superoxide dismutase and vitamin E levels [61] and of the global serum antioxidant capacity [62, 63].

Lastly, the 25-OH-vitamin-D3 serum level is lower than in controls, but may be increased after a prophylactic treatment with narrow band UVB [64]. PLE, like other severe photosensitive diseases, would be at high risk of low vitamin D status [65], which would contribute to the autoimmune process [66].

6.8 Photobiological Investigations

Most patients (almost 50% according to some Authors) [67] have a normal MED both to UVB or UVA, and do not react to Visible and Infrared lights [5, 33, 67]. Researchers do not agree as for the prevalence of the response to the provocative phototest and, furthermore, the action spectrum is still unclear. The positive reactions range from 47% up to 90% [15, 33, 42, 67–72]. The discrepancy may depend on numbers of variables, such as the different light sources, the number of the UV exposures or different UV light doses, the size of the irradiated skin area (exposed or not exposed, previously affected by the lesions or not) and the season in which the phototest is done. In brief, it depends on the lack of standardized phototest protocol. By irradiating an area divided into three parts, one receiving only UVA, one receiving only UVB and the middle one receiving both UVA and UVB (which is more similar to the natural sunlight irradiation) (Fig. 6.7), a positive reaction in the middle has been obtained in 10% of patients, a reaction that otherwise would be missed [33]. Generally, the provocative phototest, preferably on a previously involved skin area, yields positive response to UVA light in about 50% of patients. The best total doses may be 0.75–1.5 UVB MED, and 30–50 J/cm^2 for UVA for 3–5 consecutive days. The reading should be done the

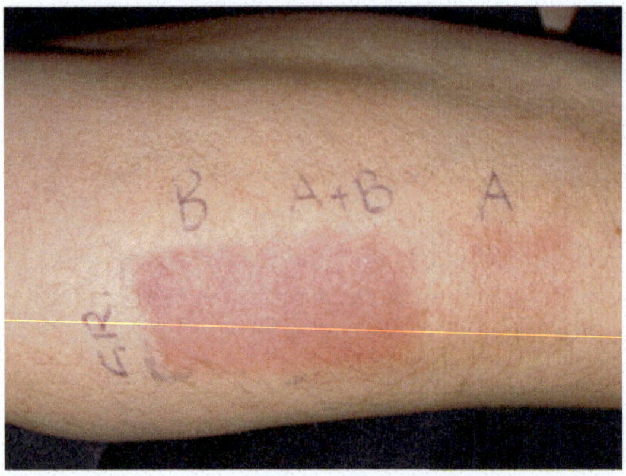

Fig. 6.7 Positive phototest to UVB and with less intensity to UVA. The figure shows the three areas irradiated with UVA, with UVB and in the middle one with both UVA and UVB

same day of irradiatons and repeated daily for up to 1 week. UVC as well may provoke PLE [73] as it has been described in welders [74]. Patch and photopatch tests are positive in about 7–10% of PLE patients [33, 75], although higher rates of positive results have also been reported [67]. Sunscreens are mostly responsible because of their large use.

6.9 Prognosis

PLE lasts for many years, often improving over time. In a study conducted for 7 years, 11% of the patients completely cleared [41], 24% in a 32-years follow up study [16] and 9% in a study in Mediterranean area [76]. The improvement is often obtained by educating the patients to avoid sunlight or to use topical and systemic photoprotection.

About 22% of PLE patients, mostly women, develop an autoimmune disease including thyroid dysfunction [16] especially autoimmune thyroiditis (8.7%) [76], and, as mentioned above, SLE in 2–10% [16, 33, 43]. On the contrary, PLE bears less risk of skin cancer [77].

Co-morbidities of PLE are respiratory allergy, such as asthma and allergic rhino-conjunctivitis [76], atopic eczema (19.8%) [33] or other photosensitive diseases like solar urticaria [78].

6.10 Quality of Life

In 40% of PLE patients, the psychosocial impact (greater in women) [79] leads to discomfort and loss of quality of life in spring and summer [80] and to high levels of anxiety and depression [81].

6.11 Prevention

The best prevention of PLE is avoiding UV light, but practically, especially in the southern Countries, such prescription is unrealistic. However, wearing clothing and hats and using broad spectrum sunscreens is useful. Broad spectrum sunscreens with a high UVA and UVB protection factor, may be beneficial in mild forms, even with only 1 mg/cm^2 (a minor thickness than guidelines suggest)[82]. Topical vitamin D3 analogs such as calcipotriol [4] may be useful. Sunscreens containing liposomal DNA repair enzymes, such as photolyase from *Anacystis nidulans* and T4 endonucleases from *Micrococcus luteus* lysate [4], proved to be effective. In addition, ectoin, a natural substance from halophilic bacteria, which protects Langerhans cells from UV-impairment, proved to reduce the *sunburn cells* and to counteract UVA-induced cell damage [83]. The low level of antioxidants [60–63] may suggest the use of topical and oral antioxidants such as beta-carotene (75–100 mg/day) [84] or oral nicotinamide (3 g/d for 2 weeks) to correct a possible error in the tryptophan pathway [85]. Results are however, controversial [86, 87]. The extract of *Polypodium leucotomas*, a fern from Central and South America, containing polyphenolic compounds, would be helpful both topically and orally [88–90]. More helpful is the desensitization treatment (photohardening). This procedure should be done in early spring or at least one month before the intense sun-exposure. Photohardening includes PUVA, the carcinogenic risk [91] notwithstanding, broad and narrow band UVB [92], the latter being more effective with less adverse effects [42, 93–97]. The starting dose should be 50% of the minimal phototoxic dose for PUVA or 75% of MED for UVB, followed by 20% increments three times a week for 4–5 weeks [78]. Photohardening [4] increases the thickening of stratum corneum and the melanin production, depletes neoantigen(s) and the Langerhans cells, whose UV-induced less migration from and to epidermis is displayed in PLE [5, 98]. In any case the natural photohardening is preferable.

6.12 Therapy

Topical corticosteroids can be used in the milder forms, systemic corticosteroids in the severe one (prednisone 40–60 mg/d, tapered within 10–14 days) and even so in short-course therapy [99]. Antimalarial drugs (chloroquine or OH-chloroquine 125–500 mg/d) [100] as immunosuppressive agents are of benefit only in selected forms, always considering their adverse effects especially the ocular ones. Azathioprine (50–100 mg/d) has been used in severe forms [101]. Cyclosporin (3.3 mg/Kg/d) was reported to be effective in a single case of PLE associated to psoriasis [102] and (3–4 mg/Kg/d) in three cases of PLE without psoriasis who profited of it also as a preventing measure [103]. Thalidomide had good to excellent results in 88% of 25 patients. There are doubts however about the correct diagnosis of the treated patients [104]. Omega-3 polyunsaturated fatty acids may act modulating inflammatory and immune response [104], while antihistamine should be used only to reduce itching [4, 7]. In conclusion, PLE is the most common photodermatosis, affecting mostly young women. Although the relationship of PLE and SLE is unclear, the assessment of ANA is highly recommendable and positive patients should be monitored over time. Prevention with topical and oral photoprotection (sunscreens and antioxidants) associated with photohardening is advisable.

References

1. Bateman D (1817) Delineations of cutaneous disease. Longman, London
2. Rasch C (1900) Om et polymorft [erythematost, vesikulost og ekzematoidt] lysudslet. Hospitalstid 43:478–480
3. Tutrone WD, Spann CT, Scheinfeld N et al (2003) Polymorphic light eruption. Dermatol Ther 16:28–39
4. Gruber-Wackernagel A, Byrne SN, Wolf P (2014) Polymorphous light eruption: clinic aspects and pathogenesis. Dermatol Clin 32:315–334
5. Honigsmann H (2008) Polymorphous light eruption. Photodermatol Photoimmunol Photomed 24:155–161
6. Morison WL, Stern RS (1982) Polymorphous light eruption: a common reaction uncommonly recognized. Acta Derm Venereol 62:237–240
7. Stratigos AJ, Antonoiu C, Katsambas AD (2002) Polymorphous light eruption. J Eur Acad Dermatol Venereol 16:193–206
8. Pao C, Norris PG, Corbett M et al (1994) Polymorphic light eruption: prevalence in Australia and England. Br J Dermatol 130:62–64
9. Sharma L, Basnet A (2008) A clinicoepidemiological study of polymorphic light eruption. Indian J Dermatol Venereol Leprol 74:15–17
10. Procaccini EM, Fabbrocini G, Affaticati V et al (2006) Epidemiologic data about polymorphous light eruption in Italy. G Ital Dermatol Venereol 141:215–219
11. Rhodes LE, Bock M, Janssens AS et al (2010) Polymorphic light eruption occurs in 18% of Europeans and does not show higher prevalence with increasing latitude: multicenter survey of 6,895 individuals residing from the Mediterranean to Scandinavia. J Invest Dermatol 130:626–628
12. Nakamura M, Henderson M, Jacobsen G et al (2014) Comparison of photodermatoses in African-Americans and Caucasians: a follow-up study. Photodermatol Photoimmunol Photomed 30:231–236
13. Kerr HA, Lim HW (2007) Photodermatoses in African Americans: a retrospective analysis of 135 patients over a 7-year period. J Am Acad Dermatol 57:638–643
14. Ros AM (1988) Solar purpura-an unusual manifestation of polymorphous light eruption. Photo-Dermatology 5:47–48
15. Hölzle E, Plewig G, von Kries R et al (1987) Polymorphous light eruption. J Invest Dermatol 88(3 Suppl):32s–38s
16. Hasan T, Ranki A, Jansen CT et al (1998) Disease associations in polymorphous light eruption. A long-term follow-up study of 94 patients. Arch Dermatol 134:1081–1085
17. Dover JS, Hawk JL (1988) Polymorphic light eruption sine eruption. Br J Dermatol 118:73–76
18. Kontos AP, Cusack CA, Chaffins M et al (2002) Polymorphous light eruption in African Americans: pinpoint papular variant. Photodermatol Photoimmunol Photomed 18:303–306
19. Bansal I, Kerr H, Janiga JJ et al (2006) Pinpoint papular variant of polymorphous light eruption: clinical and pathological correlation. J Eur Acad Dermatol Venereol 20:406–410

20. Isedeh P, Lim HW (2013) Polymorphous light eruption presenting as pinhead papular eruption on the face. J Drugs Dermatol 12:1285–1286
21. AlJasser MI, Lui H, Ball NJ et al (2013) Persistent polymorphous light eruption after ultraviolet A1 phototherapy. Photodermatol Photoimmunol Photomed 29:52–54
22. Goitre M, Roncarolo G (1987) Unusual photodermatitis. Presentation of 2 cases. G Ital Dermatol Venereol 122:261
23. Molina-Ruiz AM, Sanmartin O, Santonja C et al (2013) Spring and summer eruption of the elbows: a peculiar localized variant of polymorphous light eruption. J Am Acad Dermatol 68:306–312
24. Knight TE (1994) Solar [brachioradial] pruritus. Int J Dermatol 33:206–209
25. Shumway NK, Cole E, Fernandez KH (2016) Neurocutaneous disease: Neurocutaneous dysesthesias. J Am Acad Dermatol 74:215–228
26. Hawk J (1996) Juvenile spring eruption is a variant of polymorphic light eruption. N Z Med J 109:389
27. Lava SA, Simonetti GD, Ragazzi M et al (2013) Juvenile spring eruption: an outbreak report and systematic review of the literature. Br J Dermatol 168:1066–1072
28. Norris PG, Hawk JLM (1999) The idiopathic photodermatoses: polymorphic light eruption, actinic prurigo and hydroa vacciniforme. In: Hawk JLM (ed) Photodermatology. Arnold, London, pp 106–108
29. Ross G, Foley P, Baker C (2008) Actinic prurigo. Photodermatol Photoimmunol Photomed 24:272–275
30. Grabczynska SA, McGregor JM, Kondeatis et al (1999) E Actinic prurigo and polymorphic light eruption: common pathogenesis and the importance of HLA-DR4/DRB1*0407. Br J Dermatol 140:232–236
31. Norris PG, Hawk JLM (1999) The idiopathic photodermatoses: polymorphic light eruption, actinic prurigo and hydroa vacciniforme. In: Hawk JLM (ed) Photodermatology. Arnold, London, pp 108–109
32. Thomas P, Amblard P (1988) Lucite idiopathiques in Photodermatologie et Phototherapie, Paris, Masson, 49–53
33. Guarrera M, Cardo P, Rebora AE et al (2011) Polymorphous light eruption and benign summer light eruption in Italy. Photodermatol Photoimmunol Photomed 27:35–39
34. Leroy D, Dompmartin A, Verneuil L et al (2002) La lucite estivale bénigne existe-t-elle? Ann Dermatol Venereol 129:855–858
35. Hawk J (2004) Benign summer light eruption and polymorphic light eruption: genetic and functional studies suggest that a revised nomenclature is required. J Cosmet Dermatol 3:173–175
36. Farmer ER, Hood AF (2000) Pathology of the skin, 2nd edn. McGraw Hill, New York, p 331
37. Pincus LB, LeBoit PE, Goddard DS et al (2010) Marked papillary dermal edema–an unreliable discriminator between polymorphous light eruption and lupus erythematosus or dermatomyositis. J Cutan Pathol 37:416–425
38. Calzavara Pinton PG, Venturini M, Capezzera R, Zane C, Facchetti F (2003) Photosensitive erythema multiforme and erythema multiforme-like polymorphous light eruption. Photodermatol Photoimmunol Photomed 19:157–159
39. Tzaneva S, Voltc-Platzer B, Kittler H et al (2008) Antinuclear antibodies in patients with polymorphic light eruption: a long-term follow-up study. Br J Dermatol 158:1050–1054
40. Cahn M, Levy EJ, Shaffer B (1963) Polymorphous light eruption. A ten-year follow-up and evaluation. Arch Dermatol 88:756–758
41. Jansén CT, Karvonen J (1984) Polymorphous light eruption. A seven-year follow-up evaluation of 114 patients. Arch Dermatol 120:862–865
42. Mastalier U, Kerl H, Wolf P et al (1998) Clinical, laboratory, phototest and phototherapy findings in polymorphic light eruptions: a retrospective study of 133 patients. Eur J Dermatol 8:554–559
43. Murphy GM, Hauk JL (1991) The prevalence of antinuclear antibodies in patients with apparent polymorphic light eruption. Br J Dermatol 125:448–451
44. Nyberg F, Hasan T, Puska P et al (1997) Occurrence of polymorphous light eruption in lupus erythematosus. Br J Dermatol 136:217–221
45. Arbuckle MR, McClain MT, Rubertone MV et al (2003) Development of autoantibodies before the clinical onset of systemic lupus erythematosus. N Engl J Med 349:1526–1533
46. Millard TP, Lewis CM, Khamashta MA et al (2001) Familial clustering of polymorphic light eruption in relatives of patients with lupus erythematosus: evidence of a shared pathogenesis. Br J Dermatol 144:334–338
47. Epstein S (1942) studies in abnormal human sensitivity to light. IV. Photoallergic concept of prurigo aestivalis. J Invest Dermatol 5:289–298
48. Wolf P, Byrne SN, Gruber-Wackernagel A (2009) New insights into the mechanisms of polymorphic light eruption: resistance to ultraviolet radiation-induced immune suppression as an aetiological factor. Exp Dermatol 18:350–356
49. van de Pas CB, Kelly DA et al (2004) Ultraviolet-radiation-induced erythema and suppression of contact hypersensitivity responses in patients with polymorphic light eruption. J Invest Dermatol 122:295–299
50. Kölgen W, van Meurs M, Jongsma M et al (2004) Differential expression of cytokines in UV-B-exposed skin of patients with polymorphous light eruption: correlation with Langerhans cell migration and immunosuppression. Arch Dermatol 140:295–302
51. Damian DL, Patterson CR et al (2008) UV radiation-induced immunosuppression is greater in men and prevented by topical nicotinamide. J Invest Dermatol 128:447–454

52. Aubin F (2004) Why is polymorphous light eruption so common in young women? Arch Dermatol Res 296:240–241
53. Widyarini S, Domanski D et al (2006) Estrogen receptor signaling protects against immune suppression by UV radiation exposure. Proc Natl Acad Sci U S A 103:12837–12842
54. Boonstra H, Boer J (1989) Polymorphic light eruption induced by oral contraceptives and pregnancy? Photo-Dermatol 6:56–57
55. Millard TP, Bataille V, Snieder H et al (2000) The heritability of polymorphic light eruption. J Invest Dermatol 115:467–470
56. McGregor JM, Grabczynska S, Vaughan R et al (2000) Genetic modeling of abnormal photosensitivity in families with polymorphic light eruption and actinic prurigo. J Invest Dermatol 115:471–476
57. Millard TP, Fryer AA, McGregor JM (2008) A protective effect of glutathione-AS-transferas GSTP1*Val [105] against polymorphic light eruption. J Invest Dermatol 128:1901–1905
58. Zirbs M, Pümer C, Buters JTM et al (2013) GSTM1,GSTT1 and GSTP1 gene polymorphism in polymorphous light eruption. J Eur Acad Dermatol Venereol 27:157–162
59. McFadden JP, Norris PG, Cerio R et al (1994) Heat shock protein 65 immunoreactivity in experimentally induced polymorphic light eruption. Acta Derm Venereol 74:283–285
60. Guarrera M, Ferrari P, Rebora A (1998) Catalase in the stratum corneum of patients with polymorphic light eruption. Acta Derm Venereol 78:335–336
61. Briganti S, Cristaudo A (1998) Guarrera M et al Alteration of systemic antioxidant levels is correlated with the manifestation of polymorphohic light eruption? Exp Dermatol 7:423
62. Guarrera M, Rebora A (2007) Serum antioxidant capacity in polymorphic light eruption. Acta Derm Venereol 87:228–230
63. Giardini R, Cardo PP (2008) Decreased hydrosoluble antioxidant capacity in women: comment on the paper by Guarrera & Rebora on polymorphic light eruption. Acta Derm Venereol 88:204
64. Gruber-Wackernagel A, Obermayer-Pietsch B, Byrne SN et al (2012) Patients with polymorphic light eruption have decreased serum levels of 25-hydroxyvitamin-D3 that increase upon 311 nm UVB photohardening. Photochem Photobiol Sci 11:1831–1836
65. Reid SM, Robinson M, Kerr AC et al (2012) Prevalence and predictors of low vitamin D status in patients referred to a tertiary photodiagnostic service: a retrospective study. Photodermatol Photoimmunol Photomed 28:91–96
66. Cantorna MT (2000) Vitamin D and autoimmunity: is vitamin D status an environmental factor affecting autoimmune disease prevalence? Proc Soc Exp Biol Med 223:230–233
67. Boonstra HE, van Weelden H, Toonstra J et al (2000) Polymorphous light eruption: A clinical, photobiologic, and follow-up study of 110 patients. J Am Acad Dermatol 42:199–207
68. Guarrera M, Micalizzi C, Rebora A (1993) Heterogeneity of polymorphous light eruption: a study of 105 patients. Arch Dermatol 129:1060–1061
69. Hölzle E, Plewig G, Hofmann C et al (1982) Polymorphous light eruption. Experimental reproduction of skin lesions. J Am Acad Dermatol 7:111–125
70. Leroy D, Dompmartin A, Verneuil L et al (2002) La sensibilité du phototest polychromatique supérieure à celle du phototest UVA dans les lucites. Ann Dermatol Venereol 129:860–864
71. Salomon N, Messer G, Dick D et al (1997) Phototesting for polymorphic light eruption [PLE] with consecutive UVA1/UVB-irradiation. Photodermatol Photoimmunol Photomed 13:72–74
72. Das S, Lloyd JJ, Walshaw D et al (2004) Provocation testing in polymorphic light eruption using fluorescent ultraviolet [UV] A and UVB lamps. Br J Dermatol 151:1066–1070
73. Schmutz JL, Trechot P (2011) Polymorphous light eruption caused by ultraviolet C light. Ann Dermatol Venereol 138:639
74. Majoie IM, van Weelden H, Sybesma IM et al (2010) Polymorphous light eruption-like skin lesions in welders caused by ultraviolet C light. J Am Acad Dermatol 62:150–151
75. Gudmundsen KJ, Murphy GM, O'Sullivan D et al (1991) Polymorphic light eruption with contact and photocontact allergy. Br J Dermatol 124:79–82
76. Matekovits A, Dalamaga M, Stratigos A et al (2016) Polymorphous light eruption under the Mediterranean sun: a clinico-epidemiological and photobiological study. Eur J Dermatol 26:304–306
77. Lembo S, Fallon J, O'Kelly P et al (2008) Polymorphic light eruption and skin cancer prevalence: is one protective against the other? Br J Dermatol 159:1342–1347
78. Ferguson J (2003) Diagnosis and treatment of the common idiopathic photodermatoses. Aust J Dermatol 44:90–96
79. Richards HL, Ling TC, Evangelou G et al (2007) Psychologic distress in polymorphous light eruption and its relationship to patients' beliefs about their condition. J Am Acad Dermatol 56:426–431
80. Ling TC, Richards HL, Janssens AS et al (2006) Seasonal and latitudinal impact of polymorphic light eruption on quality of life. J Invest Dermatol 126:1648–1651
81. Richards HL, Ling TC, Evangelou G et al (2008) Evidence of high levels of anxiety and depression in polymorphic light eruption and their association with clinical and demographic variables. Br J Dermatol 159:439–444
82. Bissonnette R, Nigen S, Bolduc C (2012) Influence of the quantity of sunscreen applied on the ability to protect against ultraviolet-induced polymorphous light eruption. Photodermatol Photoimmunol Photomed 28:240–243

83. Buenger J, Driller H (2004) Ectoin: an effective natural substance to prevent UVA-induced premature photoaging. Skin Pharmacol Physiol 17:232–237
84. Swanbeck G, Wennersten G (1972) Treatment of polymorphous light eruptions with beta-carotene. Acta Derm Venereol 52:462–466
85. Neumann R, Rappold E, Pohl-Markl H (1986) Treatment of polymorphous light eruption with nicotinamide: a pilot study. Br J Dermatol 115:77–80
86. Corbett MF, Hawk JL, Herxheimer A, Magnus I (1982) A Controlled therapeutic trials in polymorphic light eruption. Br J Dermatol 107:571–581
87. Ortel B, Wechdorn D, Tanew A, Hönigsmann H (1988) Effect of nicotinamide on the phototest reaction in polymorphous light eruption. Br J Dermatol 118:669–673
88. Del Rosso JQ (2016) Use of Polypodium leucotomas Extract in Clinical Practice: A Primer for the Clinician. J Clin Aesthet Dermatol 9:37–42
89. Winkelmann RR, Del Rosso J, Rigel DS (2015) Polypodium leucotomos extract: a status report on clinical efficacy and safety. J Drugs Dermatol 14:254–261
90. Tanew A, Radakovic S, Gonzalez S et al (2012) Oral administration of a hydrophilic extract of Polypodium leucotomos for the prevention of polymorphic light eruption. J Am Acad Dermatol 66:58–62
91. Stern RS, Nichols KT, Väkevä LH (1997) Malignant melanoma in patients treated for psoriasis with methoxsalen [psoralen] and ultraviolet A radiation [PUVA]. The PUVA Follow-Up Study. N Engl J Med 336:1041–1045
92. Horkay I, Bodolay E, Kósa A (1986) Immunological aspects of prophylactic UVB and PUVA therapy in polymorphic light eruption. Photo-Dermatology 3:47–49
93. Murphy GM, Logan RA, Lovell CR et al (1987) Prophylactic PUVA and UVB therapy in polymorphic light eruption–a controlled trial. Br J Dermatol 116:531–538
94. Rücker BU, Häberle M, Koch HU et al (1991) Ultraviolet light hardening in polymorphous light eruption – a controlled study comparing different emission spectra. Photodermatol Photoimmunol Photomed 8:73–78
95. Naleway AL (2002) Polymorphous light eruption. Int J Derm 41:377–383
96. Santoro FA, Lim HV (2011) Update on photodermatoses. Semin Cutan Med Surg 30:229–238
97. Bilsland D, George SA, Gibbs NK, Aitchison T, Johnson BE, Ferguson JA (1993) Comparison of narrow band phototherapy [TL-01] and photochemotherapy [PUVA] in the management of polymorphic light eruption. Br J Dermatol 129:708–712
98. Janssens AS, Pavel S, Out-Luiting JJ et al (2005) Normalized ultraviolet [UV] induction of Langerhans cell depletion and neutrophil infiltrates after artificial UVB hardening of patients with polymorphic light eruption. Br J Dermatol 152:1268–1273
99. Patel DC, Bellaney GJ, Seed PT, McGregor JM, Hawk JL (2000) Efficacy of short-course oral prednisolone in polymorphic light eruption: a randomized controlled trial. Br J Dermatol 143:828–831
100. Murphy GM, Hawk JL, Magnus IA (1987) Hydroxychloroquine in polymorphic light eruption: a controlled trial with drug and visual sensitivity monitoring. Br J Dermatol 116:379–386
101. Norris PG, Hawk JL (1989) Successful treatment of severe polymorphous light eruption with azathioprine. Arch Dermatol 125:1377–1379
102. Shipley DR, Hewitt JB (2001) Polymorphic light eruption treated with cyclosporin. Br J Dermatol 144:446–447
103. Lasa O, Trebol I, Gardeazabal J, Diaz-Perez JL (2004 Nov) Prophylactic short-term use of cyclosporin in refractory polymorphic light eruption. J Eur Acad Dermatol Venereol 18(6):747–748
104. Ling TC, Gibbs NK, Rhodes LE (2003) Treatment of polymorphic light eruption. Photodermatol Photoimmunol Photomed 19:217–227

Ultraviolet Radiations: Skin Defense-Damage Mechanism

7

Dheeraj Mohania, Shikha Chandel, Parveen Kumar, Vivek Verma, Kumar Digvijay, Deepika Tripathi, Khushboo Choudhury, Sandeep Kumar Mitten, and Dilip Shah

Abstract

UV-radiations are the invisible part of light spectra having a wavelength between visible rays and X-rays. Based on wavelength, UV rays are subdivided into UV-A (320–400 nm), UV-B (280–320 nm) and UV-C (200–280 nm). Ultraviolet rays can have both harmful and beneficial effects. UV-C has the property of ionization thus acting as a strong mutagen, which can cause immune-mediated disease and cancer in adverse cases. Numbers of genetic factors have been identified in human involved in inducing skin cancer from UV-radiations. Certain heredity diseases have been found susceptible to UV-induced skin cancer. UV radiations activate the cutaneous immune system, which led to an inflammatory response by different mechanisms. The first line of defense mechanism against UV radiation is melanin (an epidermal pigment), and UV absorbing pigment of skin, which dissipate UV radiation as heat. Cell surface death receptor (e.g. Fas) of keratinocytes responds to UV-induced injury and elicits apoptosis to avoid malignant transformation. In addition to the formation of photo-dimers in the genome, UV also can induce mutation by generating ROS and nucleotides are highly susceptible to these free radical injuries. Melanocortin 1 receptor (MC1R) has been known to be implicated in different UV-induced damages such as

D. Mohania (✉)
Dr. Rajendra Prasad Centre for Ophthalmic Sciences, All India Institute of Medial Sciences (AIIMS), Ansari Nagar, New Delhi, India
e-mail: dmohania@gmail.com

S. Chandel • K. Digvijay • D. Tripathi
K. Choudhury
Department of Research, Sir Ganga Ram Hospital (SGRH), New Delhi, India

P. Kumar • V. Verma
Dr. B.R. Ambedkar Centre for Biomedical Research, University of Delhi, (North Campus), New Delhi, India

S.K. Mitten
Department of Neurology, PGIMER, Ram Manohar Lohia Hospital, New Delhi, India

D. Shah
Drexel University College of Medicine, Philadelphia, PA, USA

S. Ahmad (ed.), *Ultraviolet Light in Human Health, Diseases and Environment*, Advances in Experimental Medicine and Biology 996, https://doi.org/10.1007/978-3-319-56017-5_7

pigmentation, adaptive tanning, and skin cancer. UV-B induces the formation of pre-vitamin D3 in the epidermal layer of skin. UV-induced tans act as a photoprotection by providing a sun protection factor (SPF) of 3–4 and epidermal hyperplasia. There is a need to prevent the harmful effects and harness the useful effects of UV radiations.

Keywords
UV radiations • Skin cancer • Melanin • Melanoma • Vitamin-D

7.1 Introduction

Light is referred to all electromagnetic radiations with a particular wavelength, energy, and frequency. The wavelength of electromagnetic radiations is inversely proportional to energy and frequency. These electromagnetic waves are organized in light spectra of decreasing order of wavelengths as radio waves, microwaves, terahertz radiations, infra-red radiations (IR), visible light, UV rays, X-rays, and γ-rays. Some among these radiations are emitted by the sun including visible light, IR and UV Rays. UV radiations have both adverse as well as the beneficial effect on all terrestrial living organisms.

7.2 What Are Ultra-Violet Radiations?

The term ultraviolet is described as "beyond violet" which according to light spectra is referred to wavelengths longer than X-rays and shorter than visible violet light. Furthermore, these radiations have broadly been sub-divided into different types based on wavelength differences as: UV-A (320–400 nm), UV-B (280–320 nm) and UV-C (200–280 nm). Ultraviolet radiations with an important property of ionizing molecules and inducing chemical reactions renders it separable from visible rays and hence, is acting as a strong environmental mutagen through damaging cellular components, which, lead to immune-mediated diseases and adversely it can cause fatal disease like Cancer. Although some UV-induced reactions are not harmful, for instance, in the epidermal layer of skin, UV-B induces the formation of pre-vitamin D3 from 7-dehydrocholestrol during an electrocyclic reaction.

In the atmosphere, compounds like oxygen (O_2), ozone (O_3), and water (H_2O) vapors are acting as selective filters for both UV-C and UV-B. This makes 95% of UV-A to reach the earth and almost no UV-C can penetrate earth's atmosphere. Although, UV-A and some UV-B can reach, but its UV-B which burns the skin, causes damage to cellular components and cancer. UV-A, on another hand, has a higher penetration (penetration is directly proportional to the wavelength), thus penetrates deeper into the dermis, while UV-B is absorbed superficially [69].

7.3 Skin – Vulnerability to UV Radiations

The skin plays a significant interface between humans and their biological, chemical, and physical atmosphere which provides an important access for entry of potentially harmful agents. Further, it is the most susceptible target tissue for damage by environmental hazardous agents. Skin with its large surface and in direct contact with the environment (i.e. natural as well as man-made), makes it among the most vulnerable of organ systems. Thus, it is not surprising to find that skin diseases of environmental origin, including chemical stimuli, physical stresses, and infectious agents collectively comprise the majority of skin illness, which in numeric, a most important segment of disabling acute and chronic skin disease. Even, as a clue to the relative importance of environmental factors, if, one just considers that segment of skin diseases provoked by occupational environment alone, the statistics are quite convincing [1]. It becomes essential to understand the physiological and biochemical aspects of percutaneous absorption and the factors which enhance cutaneous penetration, the pathologic patterns in response to environmental injury and the agents, as well as the preventive approaches to reduce cutaneous hazard. A specific pattern of adverse response can be characterized by morphologic, physiologic and biochemical features and these reactions range in intensity and quality from simple itching to metastatic neoplasia [1].

The epidemiological demonstrations have shown that sunlight plays a role in malignant melanoma. A significant number of genetic factors have been recognized in human skin involved in inducing cancer from UV rays, and mutation in any one of these is characterized by hypersusceptibility [2, 5]. The phenotypic characters associated with susceptibility to skin cancer are: fair complexion, poor ability to tan, frequent sunburn or erythema, light eye and hair color. Certain hereditary diseases are characterized by high susceptibility to developing UV radiation-induced skin tumors. They include albinism, Xeroderma pigmentosum, and erythropoetic protoporphyria. One of the genetic defects in Xeroderma pigmentosum is the incapability to restore UV-induced DNA injury in cutaneous cells and the non-lethal damage caused to DNA allows for persistence of transformed cells. In addition to ultraviolet radiation, ionizing radiation, burns, skin trauma, and chronic infections have a connection with the development of squamous cell carcinoma.

7.4 Cutaneous Immune Response

The ionization energy of UV radiation can be absorbed by an array of molecules in the skin when it hits the skin. DNA is one of these molecules. The absorption of ionizing rays by DNA leads mutation and which ultimately results in malignant transformation. The immune system destroys UV-induced skin malignancies [69].

In addition, skin is an important component of the innate immune system and also compromised with several adaptive immune cells such as Langerhans cell, antigen presenting cell (APC), and unique T cells having skin-homing receptors. The specific receptors like cutaneous lymphocyte antigen (CLA) and chemokine receptors (i.e. CCR6 and CCR4) target T cells for migration to the skin. Keratinocytes also get activated on interaction with immune cells and act as a fundamental part of the cutaneous immune system. Keratinocytes possess major histocompatibility complex class II molecules (MHC-II) which help them to present antigen to T cells. Further, they express ICAM-1 as costimulatory molecules to secrete cytokines and chemokines [70]. UV radiations activates components of immune system and induces an inflammatory response by different mechanisms: (i) activating keratinocytes and other cells to release inflammatory chemokines and cytokines, (ii) release of sequestered auto antigens from UV-damaged cells and its reorganization, (iii) altering self-proteins to become more immunogenic, (iv) increasing the immunogenicity of externally applied molecules, and (v) chemical alterations of systemically introduced medicines whose allocation involves skin [69] (Fig. 7.1).

7.5 Melanin

Melanin is a skin pigment which provides a first line of defense against UV radiation injury by blocking UV radiation and dissipates it as harmless heat. This pigment is synthesized in the melanin producing cells called the melanocytes, which transports them in granular form to adjacent cells of the basal layer named the prickle cells, and is finally desquamated with the horny shingles at the surface.

Normal skin pigment serves as a protecting agent against damage by actinic radiation and under the suitable stimulus of UV radiation, the rate of pigment formation increases and already existing pigment appears to darken. This pigmentary response to UV light, as well as corneal thickening, provides additional protection against further actinic radiation injury. In recent years, new and revealing information has been pre-

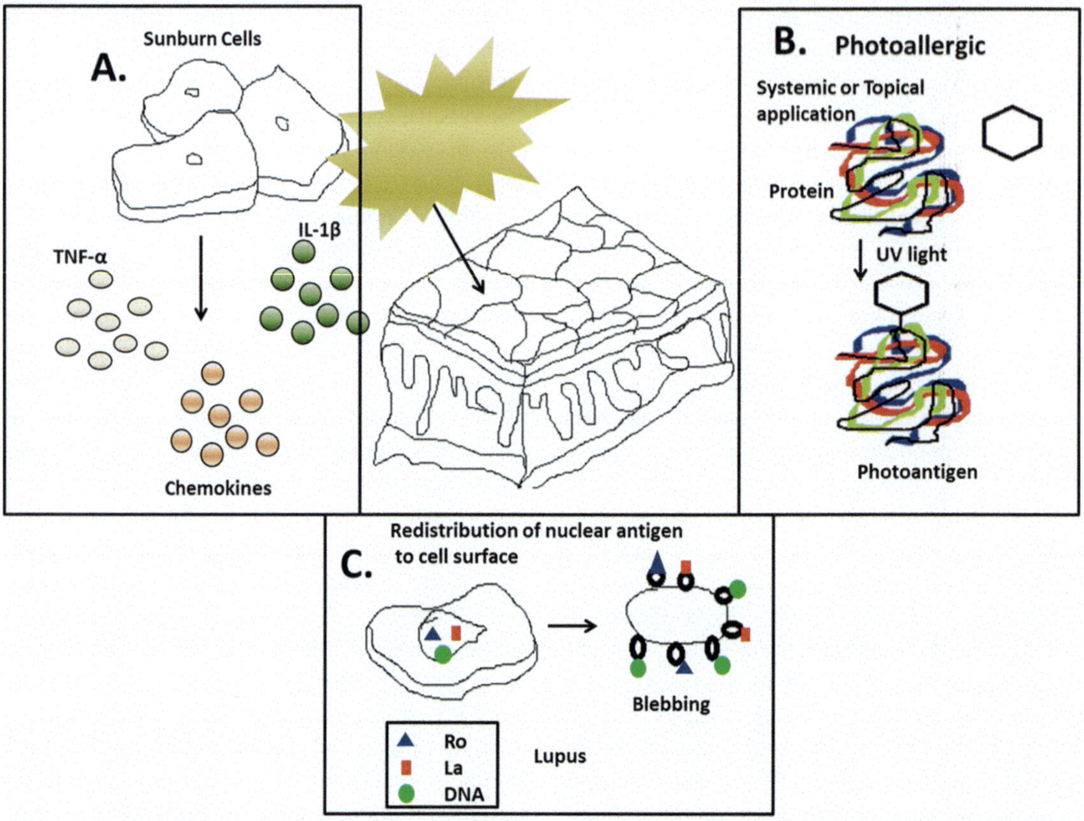

Fig. 7.1 Mechanisms of activation of the immune system by UV radiation. (**a**) IL-1β, TNF-α, and chemokines are released by keratinocytes in response to UV irradiation. (**b**) UV induced chemical alterations in the skin. Described here is a drug (i.e. Hapten) which forms a covalent bond with self-protein and represents a novel antigen. (**c**) Keratinocytes undergo apoptosis to avert malignant alteration in UV-induced DNA damage. Following UV irradiation, nuclear antigens undergoes reorganization to the surface of the keratinocytes

sented about the complex biochemical synthesis of melanin. It has proposed a much better understanding of pigment disturbances in the skin, especially the ones provoked by occupational exposures [69].

The melanin synthesis is started with tyrosine in the presence of the copper-protein complex tyrosinase and oxygen which is then transformed into dihydroxyphenylalanine (DOPA). In turn, this initial product acts as an additional catalyst to increase the rate of transformation of the precursor tyrosine, into DOPA and then to melanin (the final product) there is a series of intermediate steps in which substances such as dopaquinone, 5, 6-dihydroxyindole, dopachrome, and indole-5,6-quinone are sequentially produced before the melanin polymerization occurs. There are variable physiologic factors regulating the formation of melanin, and at particular environmental stimuli there is an altered pigmentation in the skin. Some of the most dramatic pigment changes occur as a consequence of occupational and other environmental exposures [1].

The UV radiations which escape melanin absorption leads DNA damage by generating either reactive oxygen species (ROS) or by directly causing DNA breaks by chemical reaction e.g. inducing ionization or by creating cyclobutane pyrimidine dimers (CPD) within DNA. Further, these led to the development of 6-4 photoproducts (6-4PP).

The UV-B is a common cause of 6-4PPs. It is reported that patients with XP (a recessive genetic skin disorder) can develop multiple skin cancers because of defective nucleotide excision repair system which is normally used for recognition and repairs of UV induced damage [71]. Malignant transformations can eventually occur with an accumulation of more DNA damage. The direct link between skin cancer and CPD formation has been established. For example, in the squamous cell carcinomas and their precancerous precursors, there are several mutational hotspots presents within a p53 gene that corresponds to UV-induced CPD dimer formation sites [72, 73]. UV-induced CPDs and 6-4PPs have various harmful impacts on cells. These molecules increased melanin production in melanocytes and invoke tanning response [74]. 6-4PPs can induce apoptosis and CPDs can cause cell cycle arrest [75]. Therefore, UV-induced DNA injury can guide the cells to end proliferation and undergo apoptosis if required (Fig. 7.2).

7.6 Keratinocytes Protective Mechanism – "Apoptosis"

In UV induced damage, cell surface death receptors trigger apoptosis in keratinocytes which initiate apoptosis even without the presence of its ligand [76, 77]. During initiation of apoptosis in keratinocyte in DNA damage, cytochrome C and other pro-apoptotic factors are released from the mitochondria which lead to the formation of apoptosome by capsase-9 activation and sequentially activates caspases-3/7. Apoptotic keratinocytes are referred to as the "sunburn cells" [78, 79]. Thus, it may be interpreted apoptosis is the escape pathway of keratinocytes to avoid cancer (Fig. 7.3).

In sunburn, the majority of keratinocytes undergoes apoptosis to prevent malignant transformation. Dying keratinocytes release lot of auto-antigens which are picked by Langerhans cells. Therefore, every time the disposal of UV-damaged cells ignites an autoimmune response. As keratinocytes undergo apoptosis, in

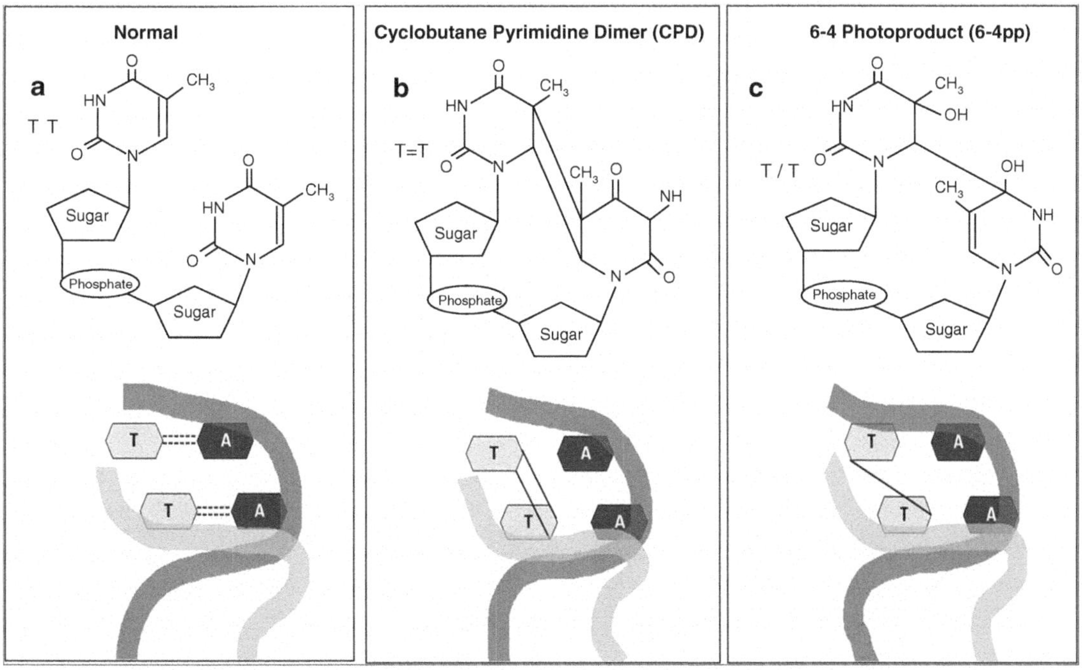

Fig. 7.2 Formation of 6–4 photoproducts and cyclobutane pyrimidine dimmers. (**a**) Two normal thymidine residues. (**b, c**) Absorption of ionization radiation of UV rays by DNA undergoes chemical modifications including the formation of 6-4 photoproducts (6-4PPs) or cyclobutane pyrimidine dimers (CPD)

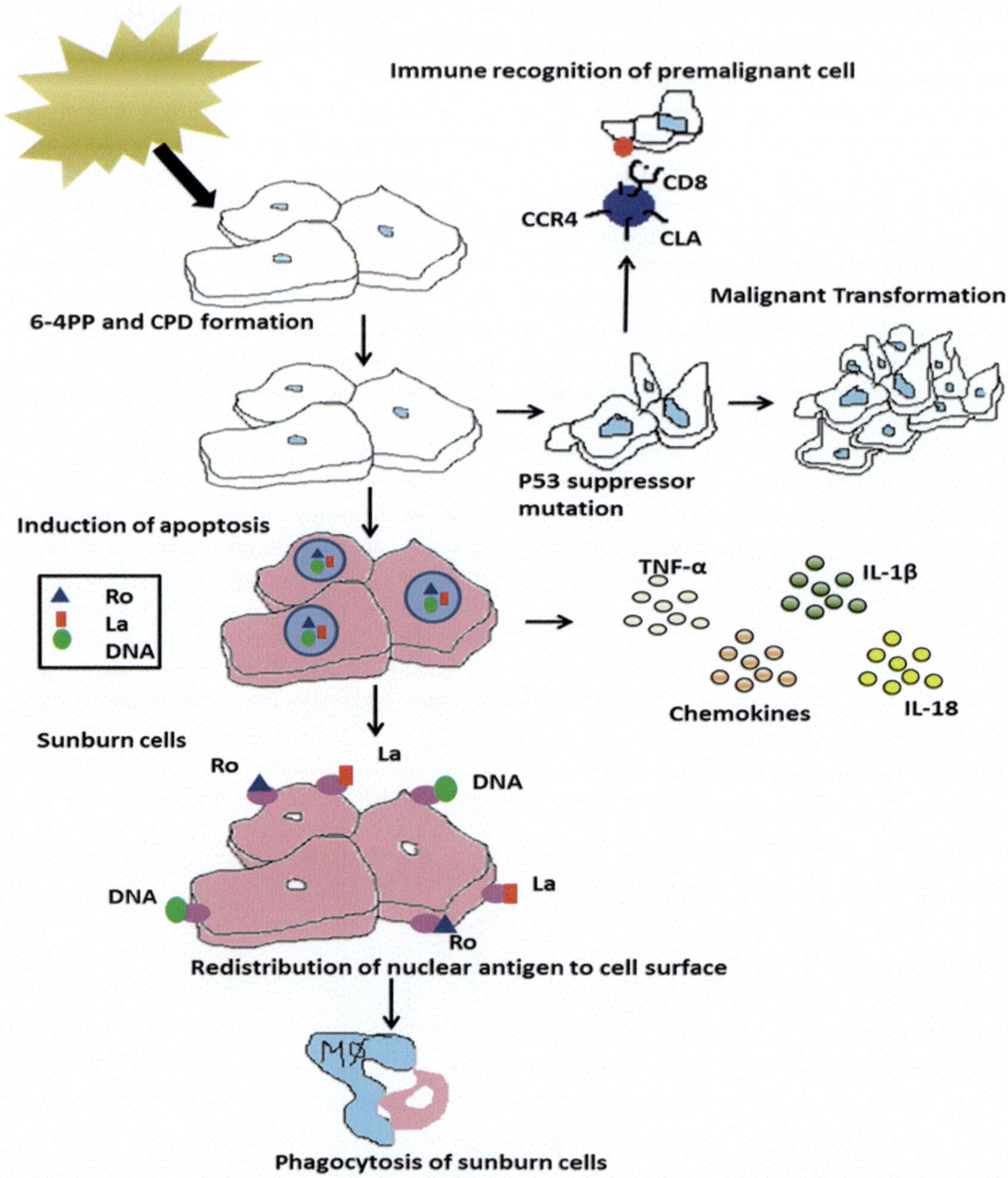

Fig. 7.3 Apoptosis of UV induced DNA-injured keratinocytes. UV injured cells undergo apoptosis followed by engulfment by antigen presenting cells (i.e. macrophages). This process is facilitated by secretion of pro-inflammatory cytokines. These cytokines help in the recruitment of the leukocytes to the skin, which facilitates the clearance of the dying keratinocytes. T cells are targeted towards skin by the receptors such as the chemokine receptors (i.e. CCR6 and CCR4) and cutaneous lymphocyte antigen (CLA). If a damaged keratinocytes avoid apoptosis, they may develop into malignancy which is mostly characterized by mutations in the p53 gene

first 24 h, these cells discharge minute concentration of nuclear materials [80], which gives time for self-processing of their cellular contents and clearance by phagocytosis. Autoimmunity occurs in the case of defective cellular processing of the auto-antigens [81], increased nucleosomal antigen release [82] or decreased clearance of the apoptotic cells [83]. UV-induced apoptosis involves exposure to potential auto-antigens; however, their antigenicity is reduced after processing and disposal of auto-antigens. UV rays also induce the reorganization of nuclear auto-antigens to the surface of keratinocytes [84].

7.7 Effect of UV Radiations on Melanomas

Ultraviolet radiations have been known to cause many adverse effects on the skin and other organs leading to various diseases. Among these, cancers, especially of the skin, are the most common form of cancer mostly seen with Caucasian populations [2]. The most common inciting factor of skin cancer in humans is UV radiation mostly due to excessive sunlight exposure. There are different types of skin cancer due to solar radiation [3–9] i.e. melanomas, squamous cell carcinomas, basal cell carcinomas and kerato-acanthomas also known as Muir-Torre syndrome.

An association has been found between cancers and skin pigmentation in different races. The skin's response to sunlight has been quoted as the major factor affecting the rate and incidence of melanoma. Due to variation in the melanin content of skin content, incidences of melanoma are more prevalent in lightly pigmented persons than deeply pigmented persons [11, 12]. As discussed earlier, melanin has a photo-protective effect on skin in the form of a neutral density filter, which corresponds to an attenuation of radiations by scattering and absorption of the UV radiation. Additionally, it can undergo rapid oxidation as stable free radical which also acts as a biological exchange polymer [13]. These impacts of UV radiations, without influencing UV penetration, may block chemical reactions leading to malignant transformation [10].

Incidences of melanoma are influenced by individual pigment characteristics predominantly in lightly pigmented populations. Different studies have shown that incidences of myeloma are least in those with black hair and highest in red hair, less in olive or dark skin, while more in the fair or light skin; decreased in brown eyes and increased in blue eyes [14]. Also burned or freckled persons are at increased risk of melanoma than those who tend to tan in response to sunlight [15]. Further, the correlation between skin pigmentation and melanoma is supported by reports showing direct measurement of the skin sensitivity to sunlight [16].

Among environmental factors, short-wave UV-B (280-320 nm) in sunlight is mainly accountable for melanoma and other skin tumors [14]. The major effects of this band of UV radiation are the production of pyrimidine dimers (i.e. T = T), cross-linking of nucleoproteins and DNA bases, and DNA breaks [17]. This band of UV can also induce basal cell carcinoma, sarcomas, and squamous cell carcinomas of the skin [18] but use a different mechanism to induce melanoma.

7.8 Molecular Mechanism of UVR to Induce Skin Cancer

The UV exposure is a well established risk factor for skin cancer which has been increasing at an alarming rate and accentuating the importance to understand the molecular mechanism of UV-induced skin cancer [19]. The specific pattern of mutation in p53 gene has been recognized as a powerful biomarker in UV induced skin cancer. UV exposure has been related to increased risk for squamous cell carcinomas (SCC), basal cell carcinoma (BCC) and melanoma development [23, 24]. Dipyramidine sites in DNA are prone to UV exposure where it primes to CC > TT double base substitutions and C > T substitutions. These sites are the signatures of UV exposure, as these sorts of alterations in DNA are rarely resulting from other mutagens [25, 26]. Mutations in p53 have been found mainly at dipyrimidine sites (frequently C > T substitutions) in different skin cancers, particularly SCC [20] and BCC [21, 22], suggesting a close association between UV exposure, p53 mutation, and cancer. This tight interrelationship is further evident from the finding that non-sun exposed sites of the body do not possesses p53 mutations [21, 27].

7.9 Inflammatory Response

Skin modulates the immune response to UV-B irradiation is the main cause of sunburn and leads neutrophils accumulation in the skin. Mast cells are present abundantly in the skin and play a major role in the UV-B-induced skin inflammation by inducing a variety of pro-inflammatory mediators [28]. UV-B exposure of skin considerably enhanced the release of IL-8 while there was a modest increase in IL-1 production, but other cytokines such as IL-2, IL-4, IL-6, IL-10, IL-12, IL-13, TNF-α and IFN-γ remained unchanged. Additionally, a reduction in the UV-B induced increase of IL-8 by up to 40%, with cycloheximide suggesting that *de-novo* protein production is contributing to IL-8 synthesis. Furthermore, UV-B exposure of mast cells significantly induced IL-8 expression, with its contrasting effect on IL-8 secretion.

UV-B-induced sunburn is characterized by an acute inflammatory response is a hallmark features for vascular permeability change, erythema, dermal blood vessels dilation and epithelial hyperplasia [29]. It is well established that accumulation of mononuclear cells and neutrophils within the dermis are acute inflammatory changes in the skin after the excessive short-term UV-B irradiation [30]. The exposure of UV-B irradiation (6–26 h) on skin leads to increase the expression of ICAM-1 and E-selection as a sign of vascular endothelial activation and promote binding and transmigration of leukocytes to the injured site in skin [31–33, 34, 35]. The neutrophil accumulation in the skin peak at 12 h of UV-B irradiation [30], however, the specific mechanism for the neutrophil recruitment to skin after sunburn are not clearly defined. Existing evidence showed that UV-B causes the functional defects in cell mediated immunity through the alteration in epidermal inflammatory cytokines profiles, phagocytosis suppression and enhances ROS production by keratinocytes, reduction in antigen presentation by Langerhans cells, induction of early lymphocyte depletion and late T cell proliferation [33–36, 37]. Further, UV-B exposure to the skin causes an increase in generation of ROS which may damge DNA and intiate the apoptosis and cell injury [37–40, 41]. Although there is a clear evidence of the role of various cytokines such as IL-1, IL-8, IL-10, IL-15 and TNF-α in UV-B induced dermal inflammation, yet cellular sources and production kinetics of these cytokines have not been fully understood [41–44, 45–49].

7.10 Oxidative Injury

In addition to the formation of photo-dimers in the genome, UV also can induce mutation by generating ROS which includes hydroxyl radical (OH), hydrogen peroxide (H_2O_2) and superoxide anion (O_2^-) and nucleotides are prone to these ROS. Oxidation of nucleotide bases promotes base pair mismatch leading to mutagenesis [84, 85]. For example, the transversion of G → T (guanine to thymine) is a well known base mispairing caused by ROS due to guanine oxidation at the 8th position to form 8-hydroxy-2′-deoxyguanine (8-OHdG) [85, 86, 87] where 8-OHdG tends to pair with an adenine (A) rather than cytosine (C) and leads to G/C to A/T in the second round of replication. This transversion in base pairing can lead to mutation and such mutations are evident in skin cancers indicating carcinogenic properties of oxidative injuries [87, 88].

To repair the DNA damage occurred due to oxidative injuries genome maintenance pathways have been present from a long time. DNA damage caused by ROS can be repaired by Base excision repair (BER) mechanism to avoid mutagenesis induced by reactive oxygen species. In this mechanism, DNA is scanned for specific changes like oxidized, deaminated and alkylated bases, by glycosylases enzymes. After recognition of chemically altered base, glycosylases cleaves of the nitrogenous base by breaking the N-glycosidic bond of the base from the phosphodiester backbone of a deoxyribose sugar, which creates an abasic or apurinic/apyrimidinic (AP) site. Then the gap created (AP site) is repaired by using the complementary strand to maintain the fidelity.

Cellular machinery possesses a lot of antioxidants to neutralize the excessive ROS and prevents DNA and other biomolecules from ROS induced damage. For example, Glutathione is a natural antioxidant present abundantly inside the cells to scavenge ROS by oxidizing itself through electron transfer to reactive species. During this process, oxidized glutathione is recycled through gaining electron from NADPH reverting back to its original state. Thus, glutathione is found in both oxidized and reduced state in a cell their ratio could possibly predict the oxidative load on the cellular machinery. Some other major antioxidants are catalase which eradicates hydrogen peroxide species and superoxide dismutase (SOD) is involved in inactivating superoxide anion (O_2^-), [91, 92] and regulation of this antioxidant machinery is an important area of study due to its crucial role in regulating UV induced skin cancer [88–90, 91, 92, 93, 94, 127, 128].

7.11 Nucleotide Excision Repair and Skin Cancer

Other than free radical production, UV exposure influences the base pairing of DNA [94, 95]. Pyrimidines are particularly more prone to UV induced chemical alterations. Particularly UV-B and UV-C which have the shorter wavelength photons are able to cleave the double bonds between fifth and sixth position. This leads to the formation of abnormal covalent bonds between adjacent pyrimidines leading to an altered 3D structure of DNA. The most anticipated photolesions- CPD and 6-4PP formed during UV exposure are highly mutagenic [96]. It has been observed that one-day exposure to the sun around 105 photolesions in every skin cell could arise as a result of UV-damage [96]. Abnormal base pairing resulted from UV induced photolesions block DNA replication, and it also influences transcription machinery in a negative manner. Additionally, it causes peculiar transition mutations which are signature mutations of UV exposure like thymidine to cytosine transition. These types of transitions are abundant in genes associated with cancer-regulation reported in various primary skin cancer patients strongly supporting the role of UV as a mutagen leading to malignancies [97–101].

UV-induced bulky DNA lesions and photoproducts are repaired by an evolutionarily conserved mechanism known as Nucleotide excision repair (NER) [102]. The significance of NER in protection from cancer is best represented by Xeroderma pigmentosum (XP), a rare condition in which subject is hypersensitive towards UV exposure resulting from the homozygous genetic defect in one of the eight proteins involved in NER: *ERCC1*, *XPA*, *XPC*, *ERCC2 (XP-D)*, *ERCC3 (XP-B)*, *ERCC4 (XP-F)*, *ERCC5 (XP-G)*, *DDB2 (XP-E)*, and *POLH*.

XP patients are highly sensitive to UV exposure which leads to the changes in skin characteristics at very early ages like capillary telangiectasias, pigment abnormalities, and atrophied skin which has been exposed to UV. Patients of XP develop premalignant lesions and skin cancer at higher frequencies in early stages of life compared to unaffected individuals. In the second decade of life BCC, SCC and melanomas are evident at uneven rates [103]. Also, XP-related skin cancers have shown UV signature transition mutations, apparently demonstrating the significance of NER in the disease resistance [104]. NER mechanism involves well-organized enzymatic reactions that work in cooperation for repairing bulky lesions altering the 3D structure of DNA.

In this pathway, damaged DNA is recognized and an oligomer of proteins is recruited, which nicks the damaged strand far from a lesion in both directions. After nick formation damaged strand is excised by exonuclease and polymerase starts filling the gap by using non damaged strand as a template [105–107]. In addition to basic components of DNA repair pathway, there are a lot of other accessory factors, which have an important role in maintaining the genome integrity. Common characteristics of XP patients have clearly demonstrated the impact of NER mechanism in UV and cancer resistance, NER polymorphisms associated with UV hypersensitivity and skin cancer incidence have been studied a lot in random populations [127, 128].

7.12 Melanocortin 1 Receptor (MC1R) and Skin Cancer Susceptibility

Melanocortin 1 receptor (MC1R) is a basic hereditary locus required for pigmentation, versatile tanning reaction and skin cancer vulnerability [108–115]. MC1R is present on melanocytes where α-melanocyte stimulating hormone (MSH) binds to it and shows the downstream signal for differentiation through activation of adenylyl cyclase and cAMP generation [116–118]. This signaling activates the protein kinase A (PKA) cascade leading to elevated levels and increased activity of melanin synthesizing enzymes resulting in increased amount of melanin and its localization [119, 120]. In melanocytes, UV induced mutations could be decreased by boosting DNA repair machinery [121]. Defective signaling due to MC1R polymorphisms is common in fair-skinned, sun-sensitive and skin cancer-prone people (e.g. Northern Europeans). The most common MC1R mutations: R151C, R160W, D84E and D294H, known as RHC (red hair color) because they are associated with red color of hair, spotted skin and increased skin burns on UV exposure [122, 123], along with an increased risk of melanoma and other types of skin cancers [124–126]. On UV exposure, MC1R signaling protects our skin via two mechanisms; one is through enhanced pigment (melanin) and MC1R production which in turn increases the rate of eumelanin deposition underneath the epidermis. This epidermal localization of melanin reduces the dosage of UV by absorbing it which in turn lowers down the risk of cancer through decreased mutagenesis. Additionally, MC1R signaling enhances nucleotide excision repair and oxidative resistance, so potent therapeutics could be used to influence cAMP levels for decreasing the UV vulnerability in decrease the risk of skin cancer [127].

7.13 Health Benefits

Along with harmful effect of UV exposure to skin, the controlled exposure of UV to the skin has shown numberous health benefits such as improved mood, enhanced appearance and elevated levels of vitamin D.

Indoor Tanning Association (ITA) has claimed that "catching some rays may lengthen your life" [49]. Sunlight exposure is connected with elevated levels of energy and mood upliftment. According to a report on the tanning attitudes of young adults, it was reported that 81% people in 2007, felt that a tan enhances appearance, while just 58% of them in 1968 held a similar conviction [50]. On exposure to sunlight, individuals with seasonal affective disorder were reported to have improved mood status [51]. According to ITA, a base tan can act as "the body's natural protection against sunburn" [49]. Although the specific mechanism for UV induced photoprotection remains unclear, it has been proposed that employing a sun protection factor (SPF of 3–4) induces the hyperpigmentation, such as epidermal hyperplasia, is likely to play a role in UV induced photoprotection [52, 53].

7.13.1 Vitamin D Production and UV Radiation

Skin synthesizes previtamin D3 and its production is induced by UV-B radiations. Vitamin D levels could vary in healthy individuals according to the season; therefore vitamin D deficiency could be spotted in the winter season [54, 55]. Lower levels of vitamin D are linked with increased risk of various types of cancer, coronary artery diseases (CAD) and bone ailments [56–61, 62, 64]. There is a significant role of vitamin D in autoimmune disease [62]. According to a new study carried by ITA on how the sunshine deceases infection in Western Africa where more tuberculosis (TB) patients are prevalant. Their finding showed that TB patients have low levels of Vitamin D, defined as hypo-vitaminosis of Vitamin D, 25(OH) D3 $\leq$ 75 nmol/L as compared to controls [84]. But, the causal relation of these associations is still not known.

Recommended Dietary Allowance (RDA) of Vitamin D is 400–600 IU, to retain blood levels of 25-hydroxyvitamin D greater than 75 nmol/L daily intake is to be increased upto 800–2000 IU

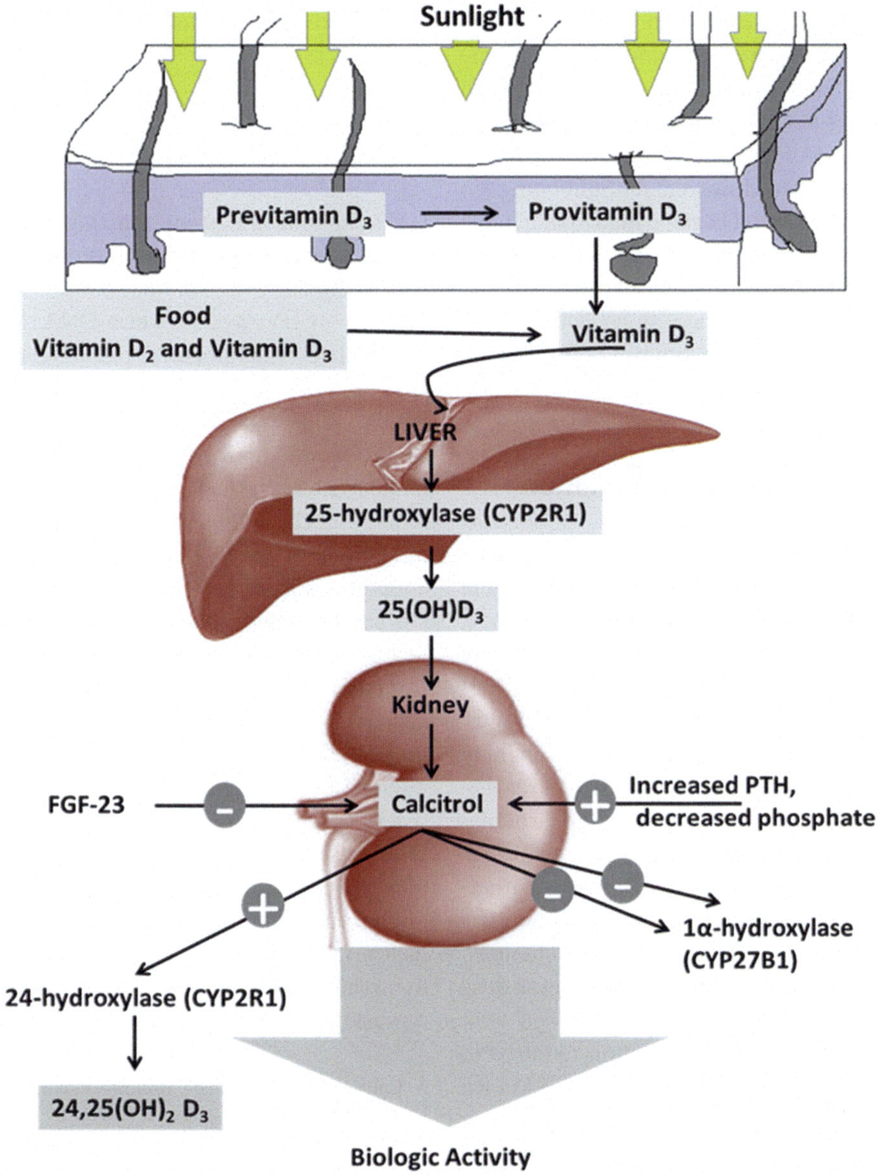

Fig. 7.4 7-dehydrocholestrol or Pro-vitamin D3 is mainly found in the spinous and basal cell layers of the epidermis, undergoes a series of photochemical changes to form pre-vitamin D3. In dark skinned population, UV light blocking the function of melanin leads to a greater requirement for UV exposure to produce equivalent amounts of Vitamin D3. In addition to its action on the kidneys, vitamin D binding protein bound to calcitriol acts by both genetic and non-genetic mechanisms on other target tissues like intestine, bone, and parathyroid gland which express the vitamin D receptors

[64, 65]. Endogenous synthesis of previtamin D3 is due to UV tanning, but according to the previous studies amount of previtamin D3 plateaus with exposure time and further exposure does not increase the total amount of previtamin D3 [66]. In light skinned persons daily moderate quantity of sun exposure to the hands, face, and arms generates adequate cutaneous previtamin D3 to meet day-to-day needs, even with the increased daily requirement to 1000 IU [67, 68]. For a fair skinned person (Type I-III), 5–20 min of sun light exposure is enough to meet a daily requirement

of previtamin D3 depending upon the season. At higher latitudes where sun induced vitamin D synthesis is the less efficient same amount of time is needed [68]. A moderate sun exposure is as efficient as prolonged sun exposure for previtamin D production. However, sunlight being the only source of vitamin D can be inadequate in the winter time and for people with darker skin types [68]. Thus, a moderate sunlight exposure should be made under consideration along with combinatorial diet, fortified with vitamin D for optimal vitamin D requirement (Fig. 7.4).

7.13.2 Phototherapy (UV Radiation Therapy)

UV radiations could also be used in the treatment of various skin diseases known as phototherapy. Based on the wavelength (λ) used it could be divide into following types: broadband UV-B (280–320 nm), narrowband UV-B (311–313 nm), UV-A (320–400 nm), and combinatorial therapy of psoralen plus UV-A (PUVA). To decide the module used for the treatment of a particular patient, following issues needs to be considered: (a) type of disease, (b) depth of disease pathology, and (c) risks of skin diseases. After choosing the required modality, treatment is individualized according to the patient's minimal erythema dose (MED).

The minimal amount of radiation required to produce erythema after 24 h of exposure is known as minimal erythema dose (MED). Phototherapy (especially PUVA therapy) has revolutionized the severe psoriasis patient treatment. In PUVA photochemotherapy psoralens used are of very high quality and their mode of action is similar to that of environmental psoralens, which are responsible for phytophotodermatitis. Physicians can maximize the immunosuppressive effects and minimize the risk of photo-burns by deciding the precise dose of psoralen through titration and the extent and type of UV radiation used by MED. Despite being the most powerful phototherapy against many immune mediated skin diseases like cutaneous T cell lymphoma, PUVA could also give rise to skin cancers like melanoma [41].

7.14 Conclusion

UV radiations are strong ionizing rays responsible for a huge number of skin problems and cancer, particularly melanoma. This is due to the property of UV rays to induce DNA damage and accumulate inflammatory pathologies in skin. Our skin has an excellent pigment (melanin) which could dampen the effect of UV rays by absorbing them. In addition, DNA damage is looked after by nucleotide and base excision repair mechanisms at molecular level. The inflammatory response is taken care either by apoptotic machinery or cell cycle checkpoints. On another hand, pigmentation is an important factor which could reduce the risk factor for the formation of melanoma and other forms of skin cancer. UV radiations are not always harmful, they are beneficial also. Vitamin D production is induced by UV rays, which in turn reduces the risk of coronary artery diseases, cancer, and bone problems. Even short exposure to UV rays can play a significant role in providing photoprotection through tanning, employing a sun protection factor. Even controlled UV exposure is used in the treatment of disease by knowing the disease type, depth, and risk for a particular patient.

Acknowledgements The authors thank Dr. Parveen Kumar for excellent assistance in editing and reviewing this manuscript.

References

1. Raymond R, Suskind RR (1977) Environment and the skin. Environ Health Perspect 20:27–37
2. Emmett EA (1975) Occupational skin cancer: a review. J Occup Med 17:44
3. Fitzpatrick TB et al (1974) An introduction to the problem of normal and abnormal responses of an's skin to solar radiation. In: Pathak MA et al (eds) Sunlight and man: normal and abnormal photobiologic responses. University of Tokyo Press, Tokyo, pp 3–14

4. Suskind RR (1974) Ultraviolet radiation carcinogenesis: UVR and atmospheric contaminants. In: Pathak MA et al (eds) Sunlight and man: normal and abnormal photobiologic responses. University of Tokyo Press, Tokyo, pp 285–298
5. Emmett EA (1973) Ultraviolet radiation as a cause of skin tumors. Crit Rev Toxicol 2:211
6. Suskind RR, Horton AW (1959) Etiologic factors and the pathogenesis of premalignant and malignant lesions. In: Rothman S (ed) The human integument, normal and abnormal. American Association for the Advancement of Science, Washington, DC, pp 171–192
7. Emmett EA (1974) Occupational skin cancer: a review. J Occup Med 17:44. 1975
8. Epstein JH (1974) Xeroderma pigmentosum and UVL carcinogenesis. In: Pathak MA et al (eds) Sunlight and Man: Normal and Abnormal Photobiologic Responses. University of Tokyo Press, Tokyo, pp 299–315
9. Epstein JH (1970) Ultraviolet carcinogenesis. In: Giese AC (ed) Photophysiology, vol 5. Academic Press, New York, pp 235–273
10. Amstrong BK, Holman CDJ (1987) Malignant melanoma of the skin. Bull World Health Organ 65:245–252
11. Crombie IK (1979) Racial differences in melanoma incidence. Br J Cancer 40:185–193
12. Sober AJ et al (1979) The melanin pigmentary system in man. In: Clark W et al (eds) Human malignant melanoma. Grune & Stratton, New York, pp 3–13
13. Pathak MA (1982) The role of natural photoprotective agents in human skin. In: Pathak M et al (eds) Sunlight and man. University of Tokyo Press, Tokyo, pp 725–750
14. Lee JAH (1982) Melanoma. In: Schottenfeld D et al (eds) Cancer epidemiology and prevention. W. B. Saunders, Philadelphia, pp 984–995
15. Holman, CDJ et al (1984) Pigmentary traits, ethnic origin, benign nevi, and family history as risk factors for cutaneous malignant melanoma. J Natl Cancer Inst 72: 257–266
16. Beitner H et al (1981) Further evidence for increased light sensitivity in patients with malignant melanoma. Br J Dermatol 104:289–294
17. World Health Organization (1979) Ultraviolet radiation Geneva. Environ Health Criter 14
18. Freeman RG (1978) Action spectrum for ultraviolet carcinogenesis. Natl Cancer Inst Monogr 50:27–29
19. Semenza JC, Weasel LH (1997) Molecular epidemiology in environmental health: the potential of tumor suppressor gene p53 as a biomarker. Environ Health Perspect 105 (1):155–163
20. Brash DE, Rudolph JA, Simon JA, Lin A, GJ MK, Baden HP, Halperin AJ, Ponten J (1991) A role for sunlight in skin cancer: UV-induced p53 mutations in squamous cell carcinoma. Proc Natl Acad Sci U S A 88:10124–10128
21. Moles JP, Moyret C, Guillot B, Jeanteur P, Guilhou JJ, Theillet C, Basset-Seguin N (1993) p53 gene mutations in human epithelial skin cancers. Oncogene 8:583–588
22. Ziegler A, Leffell DJ, Kunala S, Sharma HW, Gailani M, Simon JA, Halperin AJ, Baden H, Shapiro PE, Bale AE, Brash DE (1993) Mutation hot spots due to sunlight in the p53 geneof nonmelanoma skin cancers. Proc Natl Acad Sci U S A 90:4216–4220
23. Council on Scientific Affairs (1989) Harmful effects of ultraviolet radiation. JAMA 262:380–384
24. Stretch JR, Gatter KC, Ralfkiaer E, Lane DP, Harris AL (1991) Expression of mutant p53 in melanoma. Cancer Res 51:5976–5979
25. Reid TM, Loeb LA (1992) Mutagenic specificity of oxygen radicals produced by human leukemia cells. Cancer Res 52:1082–1086
26. Reid TM, Loeb LA (1993) Tandem double CC−+TT mutations are produced by reactive oxygen species. Proc Natl Acad Sci U S A 90:3904–3907
27. Nakazawa H, English D, Randell P, Nakazawa K, Martel N, Armstrong BK, Yamasaki H (1994) UV and skin cancer: specific p53 gene mutation in normal skin as a biologically relevant exposure measurement. Proc Natl Acad Sci U S A 91:360–364
28. Endoh I, Di Girolamo N, Hampartzoumian T, Cameron B, Geczy CL, Tedla N (2007) Ultraviolet B irradiation selectively increases the production of interleukin-8 in human cord blood-derived mast cells. British society for immunology. Clin Exp Immunol 148:161–167
29. Walsh LJ (1995) Ultraviolet B. Irradiation of skin induces mast cell degranulation and release of tumour necrosis factor-alpha. Immunol Cell Biol 73:226–233
30. Hawk JL, Murphy GM, Holden CA (1988) The presence of neutrophils in human cutaneous ultraviolet-B inflammation. Br J Dermatol 118:27–30
31. Rhodes LE, Joyce M, West DC, Strickland I, Friedmann PS (1996) Comparison of changes in endothelial adhesion molecule expression following UVB irradiation of skin and a human dermal microvascular cell line (HMEC-1). Photodermatol Photoimmunol Photomed 12:114–121
32. Dosquet C, Weill D, Wautier JL (1992) Molecular mechanism of blood monocyte adhesion to vascular endothelial cells. Nouv Rev Fr Hematol 34:55–59
33. Heck DE, Vetrano AM, Mariano TM, Laskin JD (2003) UVB light stimulates production of reactive oxygen species: unexpected role for catalase. J Biol Chem 278:22432–22436
34. Schwarz T (2002) Photoimmunosuppression. Photodermatol Photoimmunol Photomed 18:141–145
35. Di Nuzzo S, de Rie MA, van der Loos CM, Bos JD, Teunissen MB (1966) Solar-simulated ultraviolet irradiation induces selective influx of CD4+ T lymphocytes in normal human skin. Photochem Photobiol 64:988–993

36. Kulms D, Schwarz T (2002) Independent contribution of three different pathways to ultraviolet-B-induced apoptosis. Biochem Pharmacol 64:837–841
37. Duthie MS, Kimber I, Norval M (1999) The effects of ultraviolet radiation on the human immune system. Br J Dermatol 140:995–1009
38. Nicolo C, Tomassini B, Rippo MR, Testi R (2001). UVB-induced apoptosis of human dendritic cells: contribution by caspase-dependent and caspase-independent pathways. Blood 97:1803–1808
39. Horio T, Miyauchi-Hashimoto H, Okamoto H (2005) DNA damage initiates photobiologic reactions in the skin. Photochem Photobiol Sci 4:709–714
40. Strickland I, Rhodes LE, Flanagan BF, Friedmann PS (1997) TNF-alpha and IL-8 are upregulated in the epidermis of normal human skin after UVB exposure: correlation with neutrophil accumulation and E-selectin expression. J Invest Dermatol 108:763–768
41. Saade NE, Nasr IW, Massaad CA, Safieh-Garabedian B, Jabbur SJ, Kanaan SA (2000) Modulation of ultraviolet-induced hyperalgesia and cytokine upregulation by interleukins 10 and 13. Br J Pharmacol 131:1317–1324
42. Ding W, Beissert S, Deng L et al (2003) Altered cutaneous immune parameters in transgenic mice overexpressing viral IL-10 in the epidermis. J Clin Invest 111:1923–1931
43. Mohamadzadeh M, Takashima A, Dougherty I, Knop J, Bergstresser PR, Cruz PD Jr (1995) Ultraviolet B radiation up-regulates the expression of IL-15 in human skin. J Immunol 155:4492–4496
44. Gordon JR, Burd PR, Galli SJ (1990) Mast cells as a source of multifunctional cytokines. Immunol Today 11:458–464
45. Parikh SA, Cho SH, Oh CK (2003) Preformed enzymes in mast cell granules and their potential role in allergic rhinitis. Curr Allergy Asthma Rep 3:266–272
46. Malaviya R, Ikeda T, Ross E, Abraham SN (1996) Mast cell modulation of neutrophil influx and bacterial clearance at sites of infection through TNF-alpha. Nature 381:77–80
47. Clydesdale GJ, Dandie GW, Muller HK (2001) Ultraviolet light induced injury: immunological and inflammatory effects. Immunol Cell Biol 79:547–568
48. Raja KS, Lori AC, Robert PD (2009) The benefits and risks of ultraviolet (UV) tanning and its alternatives: the role of prudent sun exposure. Dermatol Clin 27:149–156
49. Robinson JK, Kim J, Rosenbaum S et al (2008) Indoor tanning knowledge, attitudes, and behavior among young adults from 1988–2007. Arch Dermatol 144:484–488
50. Wirz-Justice A, Graw P, Krauchi K et al (1996) 'Natural' light treatment of seasonal affective disorder. J Affect Disord 37:109–120
51. Hillhouse J, Stapleton J, Turrisi R (2005) Association of frequent indoor UV tanning with seasonal affective disorder. Arch Dermatol 141:1465
52. Cripps DJ (1981) Natural and artificial photoprotection. J Invest Dermatol 77:154–157
53. Nash JF, Tanner PR, Matts PJ (2006) Ultraviolet a radiation: testing and labeling for sunscreen products. Dermatol Clin 24:63–74
54. Tangpricha V, Pearce EN, Chen TC et al (2002) Vitamin D insufficiency among free-living healthy young adults. Am J Med 112:659–662
55. Giovannucci E (2007) Epidemiological evidence for vitamin D and colorectal cancer. J Bone Miner Res 22(2):V81–V85
56. Mitka M (2008) Vitamin D deficits may affect heart health. JAMA 299:753–754
57. Wang TJ, Pencina MJ, Booth SL et al (2008) Vitamin D deficiency and risk of cardiovascular disease. Circulation 117:503–511
58. Cranney A, Horsley TO'D et al (2007) Effectiveness and safety of vitamin D in relation to bone health. Evid Rep Technol Assess 158:1–235
59. Li H, Stampfer MJ, Hollis JB et al (2007) A prospective study of plasma vitamin D metabolites, vitamin D receptor polymorphisms, and prostate cancer. PLoS Med 4(3), e103
60. Dobnig H, Pilz S, Scharnagl H et al (2008) Independent association of low serum 25-hydroxyvitamin D and 1,25-dihydroxyvitamin D levels with all-cause and cardiovascular mortality. Arch Intern Med 168:1340–1349
61. Lips P (2006) Vitamin D physiology. Prog Biophys Mol Biol 92:4–8
62. Wejse C, Olesen R, Rabna P et al (2007) Serum 25-hydroxyvitamin D in a west African population of tuberculosis patients and unmatched healthy controls. Am J Clin Nutr 86:1376–1383
63. Heaney RP (2005) The vitamin D requirement in health and disease. J Steroid Biochem Mol Biol 97:13–19
64. Holick MF, Chen TC (2008) Vitamin D deficiency: a worldwide problem with health consequences. Am J Clin Nutr 87:1080S–1086S
65. Holick MF, MacLaughlin JA, Doppelt SH (1981) Regulation of cutaneous previtamin D3 photosynthesis in man: skin pigment is not an essential regulator. Science 211:590–593
66. Webb AR, Engelsen O (2008) Ultraviolet exposure scenarios: risks of erythema from recommendations on cutaneous vitamin D synthesis. Adv Exp Med Biol 624:72–85
67. Webb AR, Engelsen O (2006) Calculated ultraviolet exposure levels for a healthy vitamin D status. Photochem Photobiol 82:1697–1703
68. Tangpricha V, Turner A, Spina C et al (2004) Tanning is associated with optimal vitamin D status (serum 25-hydroxyvitamin D concentration) and higher bone mineral density. Am J Clin Nutr 80:1645–1649

69. Maverakis E, Miyamura Y, Bowen MP, Correa G, Ono Y, Goodarzi H (2010) Light, including ultraviolet. J Autoimmun 34:J247–J257
70. Cleaver JE (1968) Defective repair replication of DNA in xeroderma pigmentosum. Nature 218:652–656
71. Tommasi S, Denissenko M, Pfeifer G (1997) Sunlight induces pyrimidine dimers preferentially at 5- methylcytosine bases. Cancer Res 57:4727–4730
72. Ziegler A, Jonason AS, Leffell DJ, Simon JA, Sharma HW, Kimmelman J, Remington L, Jacks T, Brash DE (1994) Sunburn and p53 in the onset of skin cancer. Nature 372:773–777
73. Eller MS, Yaar M, Gilchrest BA (1994) DNA damage and melanogenesis. Nature 372:413–414
74. Lo HL, Nakajima S, Ma L, Walter B, Yasui A, Ethell DW, Owen LB (2005) Differential biologic effects of CPD and 6-4PP UV-induced DNA damage on the induction of apoptosis and cell-cycle arrest. BMC Cancer 5:135
75. Leverkus M, Yaar M, Gilchrest BA (1997) Fas/Fas ligand interaction contributes to UV-induced apoptosis in human keratinocytes. Exp Cell Res 232:255–262
76. Aragane Y, Kulms D, Metze D, Wilkes G, Poppelmann B, Luger TA, Schwarz T (1998) Ultraviolet light induces apoptosis via direct activation of CD95 (Fas/APO-1) independently of its ligand CD95L. J Cell Biol 140:171–182
77. Olson RL, Everett MA (1975) Epidermal apoptosis: cell deletion by phagocytosis. J Cutan Pathol 2:53–57
78. Kerr JF, Wyllie AH, Currie AR (1972) Apoptosis: a basic biological phenomenon with wide-ranging implications in tissue kinetics. Br J Cancer 26:239–257
79. van Nieuwenhuijze AE, van Lopik T, Smeenk RJ, Aarden LA (2003) Time between onset of apoptosis and release of nucleosomes from apoptotic cells: putative implications for systemic lupus erythematosus. Ann Rheum Dis 62:10–14
80. Napirei M, Karsunky H, Zevnik B, Stephan H, Mannherz HG, Moroy T (2000) Features of systemic lupus erythematosus in Dnase1-deficient mice. Nat Genet 25:177–181
81. Botto M (1998) C1q knock-out mice for the study of complement deficiency in autoimmune disease. Exp Clin Immunogenet 15:231–234
82. Furukawa F, Kashihara-Sawami M, Lyons MB, Norris DA (1990) Binding of antibodies to the extractable nuclear antigens SS-A/Ro and SS-B/la is induced on the surface of human keratinocytes by ultraviolet light (UVL): implications for the pathogenesis of photosensitive cutaneous lupus. J Invest Dermatol 94:77–85
83. Meyskens FL Jr, Farmer P, Fruehauf JP (2001) Redox regulation in human melanocytes and melanoma. Pigment Cell Res 14:148–154
84. Schulz I, Mahler HC, Boiteux S, Epe B (2000) Oxidative DNA base damage induced by singlet oxygen and photosensitization: recognition by repair endonucleases and mutagenicity. Mutat Res 461:145–156
85. Nishimura S (2002) Involvement of mammalian OGG1 (MMH) in excision of the 8-hydroxyguanine residue in DNA. Free Radic Biol Med 32:813–821
86. Kunisada M, Sakumi K, Tominaga Y, Budiyanto A, Ueda M, Ichihashi M, Nakabeppu Y, Nishigori C (2005) 8-Oxoguanine formation induced by chronic UVB exposure makes *Ogg1* knockout mice susceptible to skin carcinogenesis. Cancer Res 65:6006–6010
87. Agar NS, Halliday GM, Barnetson RS, Ananthaswamy HN, Wheeler M, Jones AM (2004) The basal layer in human squamous tumors harbors more UVA than UVB fingerprlnt. mutations: a role for UVA in human skin carcinogenesis. Proc Natl Acad Sci U S A 101:4954–4959
88. Schallreuter KU, Moore J, Wood JM, Beazley WD, Gaze DC, Tobin DJ, Marshall HS, Panske A, Panzig E, Hibberts NA (1999) *In vivo* and *in vitro* evidence for hydrogen peroxide (H_2O_2) accumulation in the epidermis of patients with vitiligo and its successful removal by a UVB-activated pseudocatalase. J Investig Dermatol Symp Proc 4:91–96
89. Song X, Mosby N, Yang J, Xu A, Abdel-Malek Z, Kadekaro AL (2009) Alpha-MSH activates immediate defense responses to UV-induced oxidative stress in human melanocytes. Pigment Cell Melanoma Res 22:809–818
90. Kadekaro AL, Chen J, Yang J, Chen S, Jameson J, Swope VB, Cheng T, Kadakia M, Abdel-Malek Z (2012) Alpha-melanocyte-stimulating hormone suppresses oxidative stress through a p53-mediated signaling pathway in human melanocytes. Mol Cancer Res 10:778–786
91. Krol ES, Kramer-Stickland KA, Liebler DC (2000) Photoprotective actions of topically applied vitamin E. Drug Metab Rev 32:413–420
92. Bickers DR, Athar M (2006) Oxidative stress in the pathogenesis of skin disease. Invest Dermatol 126:2565–2575
93. Kokot A, Metze D, Mouchet N, Galibert MD, Schiller M, Luger TA, Bohm M (2009) Alpha-melanocyte-stimulating hormone counteracts the suppressive effect of UVB on Nrf2 and Nrf-dependent gene expression in human skin. Endocrinology 150:3197–3206
94. Cleaver JE, Crowley E (2002) UV damage, DNA repair and skin carcinogenesis. Front Biosci 7:d1024–d1043
95. Wei Q, Lee JE, Gershenwald JE, Ross MI, Mansfield PF, Strom SS, Wang LE, Guo Z, Qiao Y, Amos CI et al (2003) Repair of UV light-induced DNA damage and risk of cutaneous malignant melanoma. J Natl Cancer Inst 95:308–315

96. Sarasin A (1999) The molecular pathways of ultraviolet-induced carcinogenesis. Mutat Res 428:5–10
97. Hoeijmakers JH (2009) DNA damage, aging, and cancer. N Engl J Med 361:1475–1485
98. Kanjilal S, Pierceall WE, Cummings KK, Kripke ML, Ananthaswamy HN (1993) High frequency of p53 mutations in ultraviolet radiation-induced murine skin tumors: evidence for strand bias and tumor heterogeneity. Cancer Res 53:2961–2964
99. Sato M, Nishigori C, Zghal M, Yagi T, Takebe H (1993) Ultraviolet-specific mutations in p53 gene in skin tumors in xeroderma pigmentosum patients. Cancer Res 53:2944–2946
100. Daya-Grosjean L, Dumaz N, Sarasin A (1995) The specificity of p53 mutation spectra in sunlight induced human cancers. J Photochem Photobiol B 28:115–124
101. Hodis E, Watson IR, Kryukov GV, Arold ST, Imielinski M, Theurillat JP, Nickerson E, Auclair D, Li L, Place C et al (2012) A landscape of driver mutations in melanoma. Cell 150:251–263
102. Nouspikel T (2009) DNA repair in mammalian cells: nucleotide excision repair: variations on versatility. Cell Mol Life Sci 66:994–1009
103. DiGiovanna JJ, Kraemer KH (2012) Shining a light on xeroderma pigmentosum. J Invest Dermatol 132:785–796
104. Daya-Grosjean L (2008) Xeroderma pigmentosum and skin cancer. Adv Exp Med Biol 637:19–27
105. Sancar A, Lindsey-Boltz LA, Unsal-Kacmaz K, Linn S (2004) Molecular mechanisms of mammalian DNA repair and the DNA damage checkpoints. Annu Rev Biochem 73:39–85
106. Reed SH (2005) Nucleotide excision repair in chromatin: the shape of things to come. DNA Repair 4:909–918
107. Leibeling D, Laspe P, Emmert S (2006) Nucleotide excision repair and cancer. J Mol Histol 37:225–238
108. Valverde P, Healy E, Jackson I, Rees JL, Thody AJ (1995) Variants of the melanocyte-stimulating hormone receptor gene are associated with red hair and fair skin in humans. Nat Genet 11:328–330
109. Valverde P, Healy E, Sikkink S, Haldane F, Thody AJ, Carothers A, Jackson IJ, Rees JL (1996) The Asp84Glu variant of the melanocortin 1 receptor (MC1R) is associated with melanoma. Hum Mol Genet 5:1663–1666
110. Koppula SV, Robbins LS, Lu D, Baack E, White CR Jr, Swanson NA, Cone RD (1997) Identification of common polymorphisms in the coding sequence of the human MSH receptor (MCIR) with possible biological effects. Hum Mutat 9:30–36
111. Rees JL, Healy E (1997) Melanocortin receptors, red hair, and skin cancer. J Investig Dermatol Symp Proc 2:94–98
112. Abdel-Malek ZA, Kadekaro AL, Kavanagh RJ, Todorovic A, Koikov LN, JC MN, Jackson PJ, Millhauser GL, Schwemberger S, Babcock G, Haskell-Luevano C, Knittel JJ (2006) Melanoma prevention strategy based on using tetrapeptide alpha-MSH analogs that protect human melanocytes from UV-induced DNA damage and cytotoxicity. FASEB J 20:1561–1563
113. Landi MT, Bauer J, Pfeiffer RM, Elder DE, Hulley B, Minghetti P, Calista D, Kanetsky PA, Pinkel D, Bastian BC (2006) MC1R germline variants confer risk for BRAF-mutant melanoma. Sci Signal 313:521
114. Roberts DW, Newton RA, Leonard JH, Sturm RA (2008) Melanocytes expressing MC1R polymorphisms associated with red hair color have altered MSH-ligand activated pigmentary responses in coculture with keratinocytes. J Cell Physiol 215:344–355
115. Abdel-Malek ZA, Ruwe A, Kavanagh-Starner R, Kadekaro AL, Swope V, Haskell-Luevano C, Koikov L, Knittel JJ (2009) Alpha-MSH tripeptide analogs activate the melanocortin 1 receptor and reduce UV-induced DNA damage in human melanocytes. Pigment Cell Melanoma Res 22:635–644
116. Suzuki I, Cone RD, Im S, Nordlund J, Abdel-Malek ZA (1996) Binding of melanotropic hormones to the melanocortin receptor MC1R on human melanocytes stimulates proliferation and melanogenesis. Endocrinology 137:1627–1633
117. Sturm RA, Duffy DL, Box NF, Newton RA, Shepherd AG, Chen W, Marks LH, Leonard JH, Martin NG (2003) Genetic association and cellular function of MC1R variant alleles in human pigmentation. Ann N Y Acad Sci 994:348–358
118. Kadekaro AL, Kanto H, Kavanagh R, Abdel-Malek ZA (2003) Significance of the melanocortin 1 receptor in regulating human melanocyte pigmentation, proliferation, and survival. Ann N Y Acad Sci 994:359–365
119. Bertolotto C, Abbe P, Hemesath TJ, Bille K, Fisher DE, Ortonne JP, Ballotti R (1998) Microphthalmia gene product as a signal transducer in cAMP-induced differentiation of melanocytes. J Cell Biol 142:827–835
120. Newton RA, Roberts DW, Leonard JH, Sturm RA (2007) Human melanocytes expressing MC1R variant alleles show impaired activation of multiple signaling pathways. Peptides 28:2387–2396
121. Hauser JE, Kadekaro AL, Kavanagh RJ, Wakamatsu K, Terzieva S, Schwemberger S, Babcock G, Rao MB, Ito S, Abdel-Malek ZA (2006) Melanin content and MC1R function independently affect UVR-induced DNA damage in cultured human melanocytes. Pigment Cell Res 19:303–314
122. Sturm RA (2002) Skin colour and skin cancer—MC1R, the genetic link. Melanoma Res 12:405–416
123. Beaumont KA, Shekar SN, Cook AL, Duffy DL, Sturm RA (2008) Red hair is the null phenotype of MC1R. Hum Mutat 29:E88–E94
124. Landi MT, Kanetsky PA, Tsang S, Gold B, Munroe D, Rebbeck T, Swoyer J, Ter-Minassian M, Hedayati

M, Grossman L et al (2005) MC1R, ASIP, and DNA repair in sporadic and familial melanoma in a Mediterranean population. J Natl Cancer Inst 97:998–1007
125. Fargnoli MC, Spica T, Sera F, Pellacani G, Chiarugi A, Seidenari S, Carli P, Chimenti S, Peris K (2006) Re: MC1R, ASIP, and DNA repair in sporadic and familial melanoma in a Mediterranean population. J Natl Cancer Inst 98:144–145
126. Wong SS, Ainger SA, Leonard JH, Sturm RA (2012) MC1R variant allele effects on UVR-induced phosphorylation of p38, p53, and DDB2 repair protein responses in melanocytic cells in culture. J Invest Dermatol 132:1452–1461
127. D'Orazio J, Jarrett S, Amaro-Ortiz A, Scott T (2013) UV radiation and the skin. Int J Mol Sci 14:12222–12248
128. Ahmad SI, Hanaoka F (eds) (2008) Molecular mechanisms of Xeroderma pigmentosum. In: Advances in experimental medicine and biology. Springer Science + Business Media, LLC, Landes Bioscience. ISBN: 978–0–387-09598-1

Ultraviolet Photobiology in Dermatology

8

Luisa Christensen, Amanda Suggs, and Elma Baron

Abstract

The effects of ultraviolet radiation on human skin have been studied for years, and both its harmful and therapeutic effects are well known. Exposure to UV light can lead to sunburn, immunosuppression, skin aging, and carcinogenesis, and photoprotection is strongly advocated. However, when used under controlled conditions, UV radiation can also be helpful in the diagnosis and treatment of many skin conditions.

Keywords

UV radiation • Photocarcinogenesis • Photoaging • Photoprotection • Skin disease

8.1 Introduction

In Dermatology, photobiology refers to the interactions of light, mainly in the ultraviolet radiation (UVR) spectrum, and skin. For years, studies have demonstrated both the harmful and therapeutic effects of UVR on human skin. Photoprotection is widely and strongly advocated in order to prevent the deleterious effects of UVR such as sunburn, premature skin aging, and UVR-induced skin cancers. Nonetheless, phototherapy continues to play a significant role in the treatment of cutaneous diseases such as atopic dermatitis, psoriasis, and cutaneous T-cell lymphoma. More recently, the science of photobiology has enabled the development of diagnostic tools to help with the approach to various diseases. This chapter provides a basic overview of the effects of UVR on human skin and its use in therapy and diagnosis of cutaneous disease.

8.2 Light Properties

Sunlight at the earth's surface consists mostly of short wavelength ionizing radiation (cosmic, gamma, and X-rays) and long wavelength non-ionizing radiation (UV, visible, and infrared) [32, 61]. UVR is the area of the electromagnetic

L. Christensen (✉) • A. Suggs • E. Baron
Department of Dermatology, UH Hospitals Cleveland Medical Center/Case Western Reserve University, Cleveland, OH 44106, USA
e-mail: luisa.christensen@uhhospital.org

S. Ahmad (ed.), *Ultraviolet Light in Human Health, Diseases and Environment*, Advances in Experimental Medicine and Biology 996, https://doi.org/10.1007/978-3-319-56017-5_8

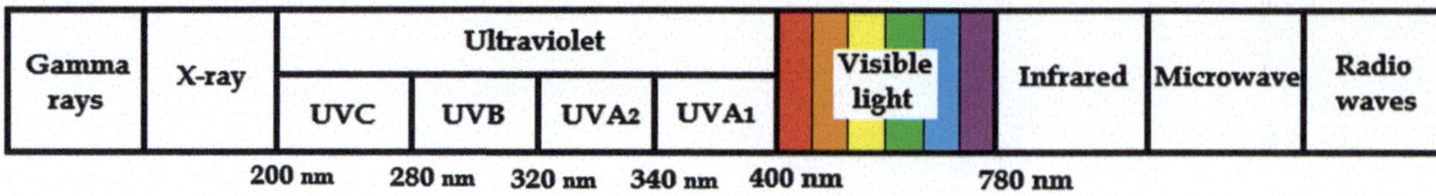

Fig. 8.1 Electromagnetic spectrum

spectrum that is considered most biologically active and therefore of greatest impact on health and disease [7]. UVR spans the wavelengths 100–400 nm and is subdivided into UVA (400–315 nm), UVB (315–280 nm), and UVC (280–100 nm) [7, 32]. Of note, some authors use 320 nm as the demarcation between UVA and UVB [73]. UVA is further subdivided into UVA2 (315–340) and UVA1 (340–400 nm). UVB has greater capacity to induce erythema than UVA, and UVA 2 is more erythemagenic than UVA1. Beyond ultraviolet radiation is visible light (400–760 nm), infrared radiation, micro waves, and radio waves (Fig. 8.1).

Most UV radiation that reaches the earth's surface is UVA (95%), only a small percentage is UVB (approximately 5%). UVC, which is extremely hazardous to skin, is absorbed by the ozone layer [25, 32, 101]. UV radiation peaks around noon and is increased by reflection from snow, water, and sand [32, 77]. UVA, but not UBV, can penetrate glass [12].

8.3 UV Radiation and Skin Interactions

Light has the properties of both waves and particles (photons). When a photon of light reaches the skin surface it can be reflected, scattered, or absorbed [37]. It is only absorbed light that can elicit cellular changes and lead to clinical response [73, 101]. Reflection of light provides the means by which we can diagnose skin disease but it does not itself lead to a biologic effect [37]. The depth of light propagation in the skin is influenced by the degree to which its direction is scattered by structures in the skin. Most scattering occurs in the dermis due to the properties of collagen [37, 58]. Scattering of UV is wavelength dependent; longer wavelengths scatter less and can therefore penetrate deeper into the skin [2, 14, 37]. Absorption is an important event as it leads to energy production. Without energy, no biologic or therapeutic effect occurs [37]. When a light-absorbing molecule (chromophore) absorbs a photon, it changes to a transient, excited state. When the chromophore returns to its ground state, energy is released. Such reaction can change the chromophore or indirectly change other molecules via energy transfer, leading to cellular changes and a biologic response [6, 73]. Light absorption depends on both the wavelength and the absorption profile of the chromophore. Chromophores can be endogenous cellular and molecular components such as DNA, hemoglobin, melanin, 7-dehydrocholesterol, or exogenous, such as tattoo pigments or photosensitizing drugs like psoralen [37].

Scattering and absorption determine the depth to which light penetrates the skin [37]. UVA penetrates deeply into the dermis, while UVB is absorbed in the epidermis [32] and only a small amount enters the superficial dermis [32, 95] (Fig. 8.2). Depth of light penetration and the absorption profile of a given chromophore are important for therapeutic intervention and form the basis of phototherapy. In general, shorter wavelength light can be used to treat epidermal disease while longer wavelengths can be used to target deeper structures such as sebaceous glands and thicker lesions [6].

8.4 Acute and Chronic Effects of UV Radiation

8.4.1 Acute Changes

UV radiation induces a multitude of acute responses in the skin including sunburn, tanning,

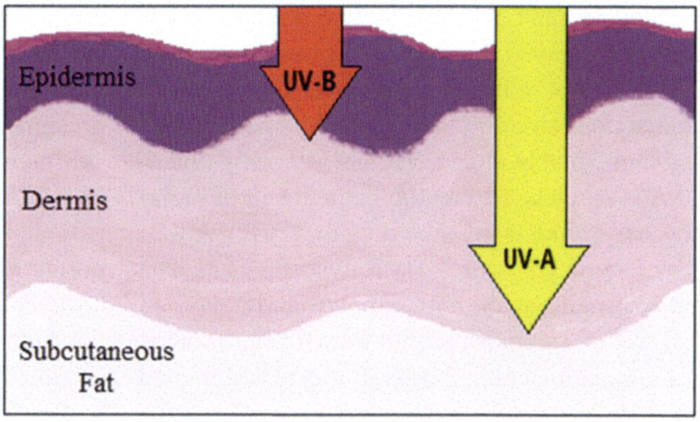

Fig. 8.2 UV wavelength and depth of skin penetration

Table 8.1 Fitzpatrick skin phototypes

Type	Unexposed skin color	Reaction to sun exposure
I	White	Always burns, never tans
II	White	Always burns, sometimes tans
III	White to olive	Sometimes burns, gradually tans
IV	Light brown	Sometimes burns, tans well
V	Brown	Rarely burns, always tans
IV	Dark brown to black skin	Never burns, always tans

epidermal thickening, vitamin D production, and immunologic effects [53].

A sunburn is an acute inflammatory response to UV exposure caused by penetration of primarily UVB into the epidermis and superficial dermis. It stimulates the production and release of prostaglandins, leukotrienes, histamine, interleukin 1 (IL-1), and tumor necrosis factor α (TNF-α), which cause pain and lead to the production of nitric oxide (NO), which in turn causes dermal dilation and redness [95]. The redness is maximal at 24 h and fades over the next 2–3 days to leave desquamation and pigmentation [32]. Sunburn cells (apoptotic keratinocytes) can be observed histologically as early as 1 h after UVR exposure [72]. Cells are more vulnerable to apoptosis when they are undergoing DNA synthesis. Keratinocytes are more vulnerable than melanocytes due to their frequent turnover from stem cells to desquamation. A severely damaged keratinocyte is likely to undergo apoptosis while a melanocyte is more likely to survive [33].

An individual's inherent tendency to burn or tan after UV radiation exposure has been used to categorize skin phototypes [73], also known as Fitzpatrick skin types, of which there are six categories (Table 8.1). Light-skinned individuals, or lower Fitzpatrick skin types, suffer the most damage and burn more than individuals of darker skin color, or higher Fitzpatrick skin types, who typically burn less and tan more [57, 66].

Tanning occurs when UVR stimulates melanin production via oxidation of melanin precursors, redistribution of melanocytes, and melanogenesis (ie, synthesis of new melanin) [12, 57, 73]. Melanocytes are dendritic cells found in the stratum basale of the epidermis. Melanin (pigment), a complex polymer of tyrosine derivatives stored in melanosomes, is transferred to keratinocytes along their dendritic arms. The melanin in the keratinocytes give the skin its natural pigment. Melanosomes have eumelanin (brown/black pigment) and pheomelanin (yellow/red pigment). Diversity of skin color is independent of melanocyte number. Instead, it is determined by variation in melanogenesis, amount of pigment within the melanosome, and the size and density of melanosomes [13, 31, 57]. When keratinocytes and melanocytes are exposed to UVR, α-melanocyte stimulating hormone and adrenocorticotropic hormone are secreted. These stimulate the melanocortin 1 receptor (MC1R), located in the melanocyte surface and melanogenesis occurs [57].

Tanning can be divided in two phases: immediate and delayed. Immediate pigment darkening is the result of redistribution of melanosomes and photooxidation of melanin [57]. It occurs during and immediately after UVR and is mostly due to UVA [12]. Delayed tanning is the result of melanogenesis and is mostly due to UVB [73]. It peaks about 3 days after UVR exposure [12]. The tan is usually at its maximum from 10 days to 3–4 weeks and fades within the ensuing months as the melanin within the keratinocyte is sloughed off [13].

New melanin forms a physical barrier over the keratinocytes, which protects DNA from further damage. It scatters and absorbs 50–75% of UVR [31, 53] and is also an antioxidant and scavenger of free radicals [13]. It is important to note that a tan from only UVA light, such as that from a tanning bed, provides 5–10 times less protection from subsequent UVR exposure, when compared to a tan from UVB, most likely because UVB also induces epidermal hyperplasia [12]. Whereas melanogenesis is a protective response against further UV damage [32], intentional tanning is not recommended as a means of sun protection, as it entails its own risks.

After UVR exposure, skin cells undergo cell cycle arrest and apoptosis as a protective mechanism to prevent propagaton of damaged DNA [53]. UVB can lead to G1 and G2 phase cycle arrest [31] and keratinocyte hyperproliferation, which can be observed as epidermal hyperplasia histologically [95]. The epidermis may double in thickness following UVR exposure [53].

Synthesis of vitamin D from the skin is the major source of the vitamin in humans. Limited vitamin D is obtained from the diet as only a few varieties of food contain it. Vitamin D is responsible for calcium absorption and bone maintenance and is therefore helpful in preventing rickets and osteomalacia [53]. For Vitamin D to exert a physiological effect, it must be converted into its active form 1α25-dehydroxyvitamine D3 (1,25(OH)2D3), also known as calcitriol. Synthesis of vitamin D starts in keratinocytes with photolysis of 7-dehydrocholesterol by UVB in sunlight, leading to formation of Vitamin D3 (cholecarciferol). Although keratinocytes are capable of metabolizing D3 to the metabolically active 1,25(OH)2D3, very little appears to enter the circulation, in fact, most of the circulating 1,25(OH)2D3 is produced by the kidney. It is presumed that 1,25(OH)2D3 produced by the skin is used for autocrine or paracrine purposes [9, 10, 30, 54, 62]. In the circulation, Vitamin D3 is bound to vitamin D binding protein and transported to the liver, where it is hydroxylated into 25-hydroxyvitamin D3. This form is then sent to the kidney, where its conversion will result in production of 1,25(OH)2D3 [13, 31, 46, 57].

Exposure to UV radiation has a profound effect on the skin's immune system. It has both, pro-inflammatory as well as immunosuppressive effects and it involves both innate and adaptive immunity. UV radiation releases several pro-inflammatory mediators such as serotonin, prostaglandins, IL-1, IL-6, IL-8, and TNF-α. Examples of pro-inflammatory responses observed clinically include sunburn, photodermatosis, and exacerbation of systemic disease in patients with systemic lupus erythematosus, among others [73].

Immunosuppression by UV radiation is a complex mechanism that involves multiple pathways. One of the most studied paradigms starts with UV-induced DNA damage to Langerhans cells and keratinocytes in the epidermis, trans to cis isomerization of urocanic acid in the stratum corneum, and peroxidation of membrane lipids, resulting in production of anti-inflammatory mediators and suppression of skin immunity [73, 80]. The immunosuppressive effects of UV light may not only involve areas directly exposed to radiation, but also non-irradiated sites, depending on the UV dose and other factors [26, 51, 65]. This immunosuppressive effect is mainly mediated by soluble immunosuppressive cytokines such as IL-10. Furthermore, interactions between damaged Langerhans cells, lymph node antigen presenting cells, and suppressor B lymphocytes lead to the eventual activation of UV-induced regulatory T cells [73, 80]. Urocanic acid (UCA), a metabolic product of histidine which accumulates in keratinocytes due to their inability to metabolize it [22, 67], is an important epidermis chromophore. UV radiation causes trans to cis isomerization of UCA and lipid peroxidation which leads to production of platelet activating factor (PAF) with eventual production of immunoregulatory factors including prostaglandin E2, IL-4, IL-5, and IL-10 and systemic immunosuppression [90]. Both UVA and UVB contribute to immunosuppression, and it is believed that the combined spectrum makes sunlight more suppressive than either UVA or UVB alone [71].

8.4.2 Chronic Changes

DNA is regarded as the major chromophore for the most critical biologic effects of UVB, including immunosuppression and carcinogenesis [99]. UVB and UVC, and to a lesser extent, UVA, are absorbed directly by DNA, leading to formation of DNA photoproducts. DNA photoproducts are thymne dimers formed by covalently binding two adjacent pyrimidines in the same polynucleotide chain [73]. Cyclobutane-pyrimidine dimers (CPDs) and pyrimidine-pyrimidone (6–4) photoproducts are considered the signature lesions of UVR-induced DNA damage [19, 33]. CPD is the most common photoproduct formed when damage from UVR occurs [57, 73]. A CPD is formed when linkage at the 5 and 6 positions between adjacent pyrimidine bases (thymine or cytosine) occurs and a cyclobutyl ring is formed [73]. The 5,6 double bond of pyrimidine bases is the most effective area of DNA that absorbs UVR [7]. If a covalent bond is instead formed between the 6 position of one pyrimidine and the 4 position of the adjacent pyrimidine, it is referred to as a pyrimidine-pyramidone (6–4) photoproduct [57] (Fig. 8.3).

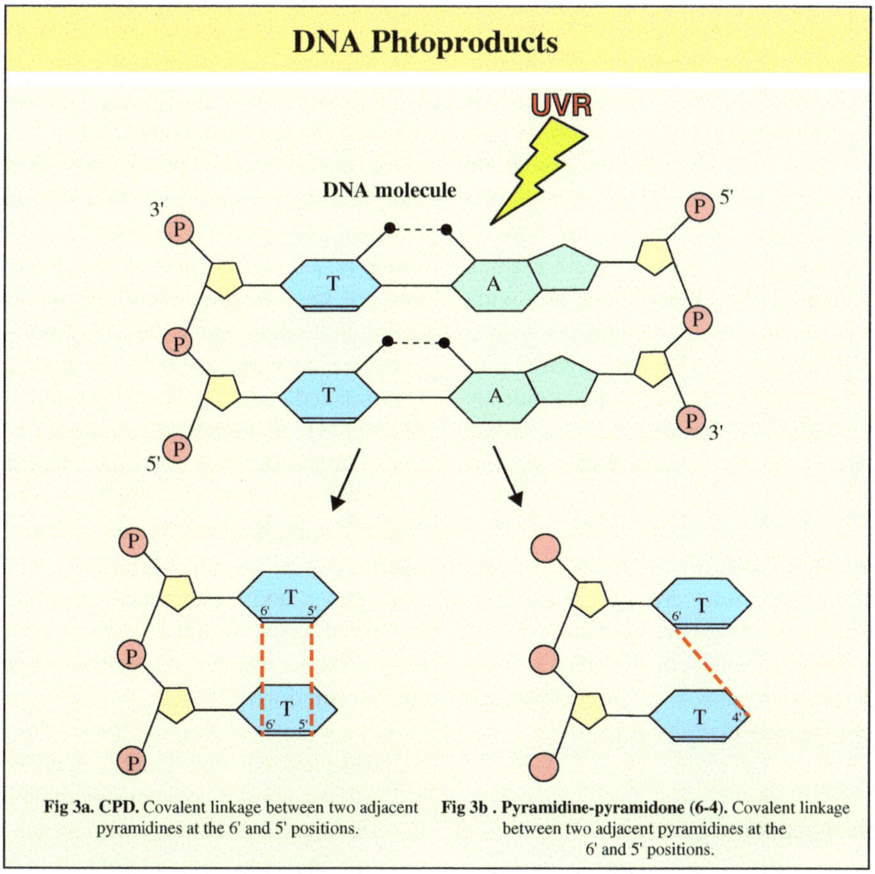

Fig. 8.3 UVR induced DNA photoproducts. (**a**) **CPD.** Covalent linkage between two adjacent pyramidines at the 6' and 5' positions. (**b**) **Pyramidine-pyramidone (6-4).** Covalent linkage between two adjacent pyramidines at the 6' and 4' position

UVA can also induce pyrimidine dimers, although at a much lower frequency than UVB [73]. The effects of UVA on DNA are believed to be indirect and involve induction of oxidative stress [19]. UV-induced reactive oxygen species (free radicals) include singlet oxygen and other oxygen species; such as hydrogen peroxide and the superoxide radical [21]. These oxygen species react predominantly with guanine and generate several DNA changes, including the formation of 7,8-dihydro-8-oxoguanosine (8-oxoG). Oxidative stress not only affects DNA, but also membranes and proteins [73]. Although UVA has been shown to be responsible for almost all oxidative DNA damage, UVB also plays a role in their production [52]. Despite the effects of oxidative stress on cell damage, pyrimidine dimers are still the most common product of DNA damage from UVR with oxidative DNA damage playing a minor role in mutagenesis [74, 75].

Photoproducts are structurally damaging. They distort the DNA helix, which results in halting of RNA polymerase and inhibition of gene expression [7]. To counteract the potentially mutagenic and cytotoxic effects of UV radiation, several defenses and DNA repair mechanisms exist, such as inducible melanin, epidermal thickening, enzymatic and non-enzymatic antioxidative defenses, repair processes, and removal of damaged cells [73]. DNA damage requires excision and replacement of damaged nucleotides by DNA repair pathways. For example, nucleotide excision repair is used to repair bulky products such as CPD [25] and base excision repair is used to repair modified bases such as 8oG [73]. If these methods fail, DNA mutations are the result. Most of these mutations are not catastrophic because the genetic code is redundant and large portions of DNA are not used. However, if mutations occur at oncogenes or tumor suppressor genes, carcinogenesis may occur [33].

8.4.3 Photocarcinogenesis

It is well established that chronic exposure of the skin to UV light is a major risk factor for the development of skin cancers [26]. These cancers are of epidermal origin and can be divided into three main types: malignant melanoma (MM), basal cell carcinoma (BCC), and squamous cell carcinoma (SCC). BCC and SCC are commonly known as nonmelanoma skin cancers (NMSC). Actinic keratosis is a precancerous lesion also caused by UV radiation, which if left untreated, may progress into SCC [94]. Malignant melanoma is one of the most aggressive cancers with early metastatic capacity. The vast majority of deaths from skin cancer result from MM. SCCs can also metastasize but at much lower rates. BCC very rarely metastasizes [100]. The risk of developing skin cancer is closely related to the number of sunburn episodes in a person's life and is influenced by skin color. Very light skin individuals have an increased incidence of skin cancer when compared to dark skin individuals [88].

Basal and squamous cell carcinomas develop from epidermal keratinocytes and hair follicles. [7]. They are linked to cumulative exposure to UVR and occur most often on chronically sun exposed areas of the body (ie, head, neck, forearms, dorsal hands), particularly in those with daily UVR exposure over their lifetime such as farmers, sailors, and fishermen [33]. Chronically sun-exposed skin harbors many clonal proliferations of keratinocytes with *p53* mutations, indicating that mutations in the *p53* tumor suppressor gene are an early event in the pathogenesis of UV-induced actinic keratosis and cutaneous SCCs [73]. Although SCC appears to be more associated with UVB exposure, the role of UVA in formation of SCC can't be excluded [23]. Natural sunlight contains much more UVA than UVB. Also, UVA is less filtered by window glass or clothing, and it penetrates more effectively to the basal layer of the epidermis, these factors may serve to increase its relative contribution to SCC formation [73].

Cutaneous melanomas arise from epidermal melanocytes. All individuals, regardless of skin color, have similar numbers of melanocytes; as described earlier, the major difference lies in the size, number, and pigment content of their melanosomes [72, 100]. In addition to genetic factors, melanoma formation is associated with intense, intermittent UVR exposure. It has been suggested that melanocytes might be more prone to UV-induced mutagenesis following a single high dose of UV light, because of their relative inability to undergo apoptosis [33], possibly due to high levels of anti-apoptotic proteins such as Bcl-2. Unlike keratinocytes, which readily undergo apoptosis, damaged melanocytes are more likely to survive [73]. Melanoma is therefore, frequently seen on areas of the body that are exposed to the sun only intermittently (ie, the lower legs in women and the back in men) [18]. It is more common in people with indoor occupations and those who receive intermittent UVR exposure on vacations and weekends. Mutations in the proto-oncogene *BRAF* have been found in high frequency in both melanomas and melanocytic nevi, predominantly in melanomas from intermittently sun-exposed areas (and much less frequently in melanomas from unexposed areas or chronically UV-exposed areas), suggesting that this type of mutation is UV-induced [70]. Artificial UV exposure from tanning beds is believed to be a potential contributor to the increase in incidence of melanoma in young women. Exposure to excess UVR is the most important modifiable risk factor for melanoma [18].

8.4.4 Photoaging

Skin aging occurs from both intrinsic (chronologic) and extrinsic (environmental) factors. Intrinsic aging is usually under the control of genetic and hormonal factors. Extrinsic aging results from DNA damaging agents such as UVR, radical oxygen species (ROS), cigarette smoking, and many chemotherapeutic drugs, notably cysplatin [93]. By far, the greatest source of extrinsic aging is accumulated and unprotected UVR exposure [32, 86]. Skin ages even in sun-protected areas, but much more slowly. The dermis becomes thin, collagen content diminishes by about 1% per year throughout adult life and becomes less elastic, and fibroblasts in the dermis decrease in number, leading to reduced collagen synthesis [95].

Due to its ability to penetrate deeper into the dermis and damage connective tissue, UVA in particular is thought to play an important role in the dermal changes of photoaging. [73, 86]. UVA leads to ROS production, which damages DNA, proteins, and lipids. ROS also trigger cytokine cascades, leading to alteration of the structural components of the skin and photoaging [63]. UV radiation directly activates cell surface receptors, initiating intracellular signals that eventually lead to the increased transcription of matrix metalloproteinases (MMP) and decreased expression of the procollagen I and III genes which leads to collagen degradation and down-regulation of collagen synthesis [28, 29]. Collagen gives the skin its strength and elasticity; loss of collagen fibers thus leads to decreased elasticity, increased fragility, and decreased capacity for wound healing [97]. Keratinocytes and fibroblasts from sun-exposed areas have reduced proliferative ability compared to those from sun-protected sites [32]. Immunosuppression is also evident in aging skin. The density of Langerhans cells (LCs) and T lymphocytes in the skin decreases greatly [34, 85]. LCs have reduced ability to migrate from the epidermis in response to cytokines [8] and T cells become less responsive to specific antigens [47, 60]. These changes lead to an increased susceptibility to photocarcinogenesis and chronic skin infections [17].

Histologically, the photodamaged epidermis demonstrates variability in thickness, with areas of atrophy and acanthosis (hypertrophy) and a variation in the degree of pigmentation. The dermis displays loss of mature collagen, reduction in the density of anchoring fibrils affecting epidermal adhesion to the dermis, and tangled clumps of elastin. In contrast to the mostly hypocellular sun-protected skin, photodamaged skin displays an abundance of inflammatory cells. It also contains a higher number of fibroblasts, which display an irregular stellate shape. [32, 96]. Clinically, photoaged skin demonstrates wrinkling, irregular pigmentation (lentigines, freckling, melasma, dyspigmentation), dryness, roughness, thinning, solar keratosis, telangectasias, and a variety of premalignant lesions, such as actinic keratoses [53, 86, 98].

8.5 Photoprotection

Although the synthesis of melanin protects epidermal DNA by various mechanisms, including formation of melanin caps over basal layer nuclei and scavenging of ROS [1], the damaging effects of acute and chronic UVR exposure such as sunburn, photoaging, and photocarcinogenesis, make it necessary to advocate extrinsic photoprotective measures. Photoprotection includes avoiding sun exposure during peak hours (10 am–4 pm), use of protective clothing (including hat and sunglasses), and use of topical sunscreens [73]. It is important to note that clouds absorb only 70% of the sunrays [53], hence sun protection remains essential even on cloudy days.

The amount of protection afforded by a sunscreen is measured by its sun protection factor (SPF). SPF is measured using the minimal erythema dose (MED), which is the lowest dose of UVR required to produce minimal erythema, a faint pink response on the skin best appreciated 16–24 h after UV exposure [40] (Fig. 8.4). SPF is calculated as the amount of time required to produce minimal erythema (i.e. MED) in sunscreen-protected skin over the MED in unprotected skin. This means, that a sunscreen with SPF of 10 allows a person who normally burns in 20 minutes to be exposed for as long as 200 min before burning occurs [53, 61]. Because SPF measures cutaneous erythema, it is predominantly a reflection of the biologic effect of UVB. There is no direct correlation between SPF and UVA protection [59]. The UVA protection factor is currently not labeled on sunscreens, but all sunscreens labeled as "broad spectrum" protect against both UVA and UVB and are produced by combining filters with varying UV absorption spectra. Only sunscreens with an SPF of at least 15 can be designated as sunscreens that decrease the risk of skin cancer and early skin aging caused by UVR exposure. "Water resistant" and "very water resistant" sunscreens claim to maintain the SPF after 40–80 minutes of sweating or swimming [91].

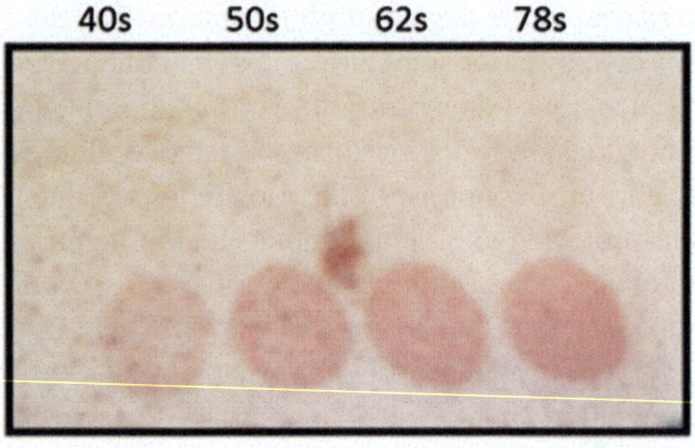

Fig. 8.4 Minimal erythema dose (MED). Test sites are exposed to UVR at increasing time intervals. The exposed areas are examined after 24-h. The site that shows the first visible erythema reaction (i.e. shortest time duration to produce full circle appreciable erythema) represents the MED

Sunscreens can be organic (formerly known as chemical sunscreens) or inorganic (formerly known as physical sunscreens) (Table 8.2). Organic sunscreens contain aromatic compounds that absorb UVB and UVA and re-emit the energy as insignificant quantities of heat [76]. UBV absorbing organic compounds include PAVA derivatives, cinnamates, and salicylates, among others. UVA absorbing compounds include avobenzone and meradimate. UVB and UVA absorbing compounds include benzophenones, ecamsule, drometrizole, bemotrizinol, and bisoctrizole. Inorganic sunscreens contain mineral compounds such as zinc oxide and titanium dioxide, which reflect both UVA and UVB. In contrast to some organic sunscreens, inorganic sunscreens are more stable and have a lower irritating and sensitizing potential. In the past, inorganic sunscreens would form a thick white layer on the skin. To overcome this cosmetically unacceptable result, nanotechnology has been used to produce titanium dioxide and zinc oxide nanoparticles, which when applied form a thin transparent layer on the skin without compromising the level of UVA and UVB protection [92]. Safety regarding the use of nanoparticles is an ongoing subject of discussion. Although there is concern for systemic absorption and toxicity [36, 81], a review conducted by the Australian Therapeutic Goods Administration in 2013 [4], demonstrated penetration is limited to the stratum corneum and that their use is unlikely to cause any harm.

Table 8.2 Sunscreens

Sunscreen	Range of protection
Organic	
PABA derivatives	
PABA (para-aminobenzoic acid)	UVB
Padimate O (octyl dimethyl PABA)	UVB
Cinnamates	
Octinoxate (octyl methoxycinnamate)	UVB
Cinoxate	UVB
Salicylates	
Octisalate (octyl salicylate)	UVB
Homosalate	UVB
Trolamine salicylate	UVB
Benzophenones	
Oxybenzone (benzophenone-3)	UVB, UVA2
Sulisobenzone (benzophenone-4)	UVB, UVA2
Dioxybenzone (benzophenone-8)	UVB, UVA2
Others	
Octocrylene	UVB
Ensulizole (phenylbenzimidazole sulfonic acid)	UVB
Avobenzone (butyl methoxydibezoyl methane, Parsol 1789)	UVA1
Ecamsule (terephthalylidene dicamphor sulfonic acid, Mexoryl SX)	UVB, UVA2
Drometrizol trisiloxane (Mexoryl XL)	UVB, UVA2
Meradimate (menthyl anthranilate)	UVA2
Bemotrizinol (bis-ethylhexyloxyphenol methoxyphenol triazine, Tinosorb M)	UVB, UVA2
Bisoctrizole (methylene bisbenzotriazolyl tetramethylbutylphenol, Tinosorb M)	UVB, UVA2
Inorganic	
Titanium dioxide	UVB, UVA2, UVA1
Zinc oxide	UVB, UVA2, UVA1

Adapted with the permission from: Sunscreens: An Update, The Medical Letter 2008; 50:70

In general, sunscreens have an excellent safety profile. However, some compounds in sunscreens have been reported to cause allergic and irritant contact dermatitis, phototoxic and photoallergic reactions, contact urticaria, and rare anaphylactic reactions [5, 39, 41, 78]. Agents such as PABA, amyl-dimethyl-PABA, or benzophenone-10, are now rarely used in sunscreens. Top allergens in sunscreens are benzophenone-3, DL-alpha-tocopherol, and fragrances.

Sunscreens come in lotion and cream formulations (oil-in-water emulsions) or liquid, spray, and gel formulations (ethanol/oil-based). Lotions are thinner and less greasy than creams [83]. Ethanol based products are fast drying and leave a cooling sensation on the skin but they may be irritating. Sunscreens should be applied evenly and in sufficiatent amounts to cover all sun exposed skin. It should be applied 15–30 min before sun exposure and reapplied at least every 2 h as and after water exposure.

Sunscreen is recommended for everyone regardless of skin phenotype. They are especially useful for individuals with light skin (phototypes I, II, and III). Wearing clothes is also important for sun protection. Ultraviolet protection factor (UPF) is an in vitro measurement of the amount of UV that penetrates a fabric, resulting in cutaneous erythema [38].

8.6 Phototesting and Phototherapy

Phototesting is the process of evaluating a patient's sensitivity to specific wavelengths of UV and visible light to help guide light treatment and for the diagnostic work up of suspected photodermatosis. Phototesting includes determining the MED, reproducing photodermatosis using a provocation phototest, and photopatch testing to identify photoallergy [27].

As previously mentioned, the minimal erythema dose (MED) is a way to measure an individual's sensitivity to UVR and is used to determine SPF. The MED can also be used as a guide to quantitatively determine the appropriate dosage of UV radiation to administer in phototherapy [40]. MED is measured by exposing adjacent areas of skin, preferably sun-protected areas, to increasing doses of UVR. The areas are then graded based on their degree of erythema, either visually or with a chromameter. The erythematous skin that was exposed to UVR for the shortest duration is the MED (Fig. 8.4) [11, 40]. MED is affected by the type of lamp used, the distance of the skin from the lamp, and previous exposure of the skin to UVR [40]. MED is frequently used for UVB phototherapy while MPD (minimal phototoxic dose) is used for PUVA therapy [43]. Unlike MED, MPD requires oral intake of a photosensitizing agent (8-methoxypsoralen) prior to phototestitng. However, because of the relative complexity of performing such tests an SPT based protocol is more commonly used to deliver therapy. Provocative phototesting is used to reproduce photodermatoses. UVA and/or UVB are the most commonly used light sources. Repeat exposure is recommended in order to provoke certain photodermatoses, such as PLE (polymorphous light eruption) [79]. Photopatch testing is used to diagnose photoallergy (photocontact dermatitis), which should be suspected in any patient with history of dermatitis after sun exposure or an eczematous eruption predominantly affecting light-exposed sites [15]. Sunscreen agents are currently the most common photoallergens [16].

The therapeutic benefits of sunlight are well known. However, because of its varying intensity and availability, it has not been used as a standard light source for phototherapy, with the exception perhaps of the recent interest in daylight photodynamic therapy (PDT) in Europe and other continents. Artificial light sources were developed to overcome the disadvantages and limitations of natural sunlight [61] and are currently used to treat many photoresponsive skin diseases. They can simulate solar radiation or isolate certain wavelengths of the optical radiation spectrum. Most medical light devices generate radiation by converting electrical energy to light energy. Optical filters and specific chromophores are used to isolate certain wavelengths. Mirrors, lenses, and fibers are then used to direct the light to the specified target [58]. Medical light sources include the arc lamp, excimer laser or lamp, fluorescent lamp, light emitting diodes (LEDs), and lasers.

8.6.1 Arc Lamps

Arc lamps, also known as gas discharge lamps, are composed of two electrodes with plasma (the arc) between them; these are sealed within a transparent envelope with gas (eg, mercury or xenon). When high voltage is applied, electrons become excited and light is emitted when they return to the ground state [58]. Different gases and pressures lead to different spectral output. Xenon lamps are used as solar simulators. The arc lamp was the first effective artificial light source.

8.6.2 Excimer

The excimer is a complex of excited gases, which upon decomposition, give off excess energy in the form of UVR. The excimer exists both as a lamp

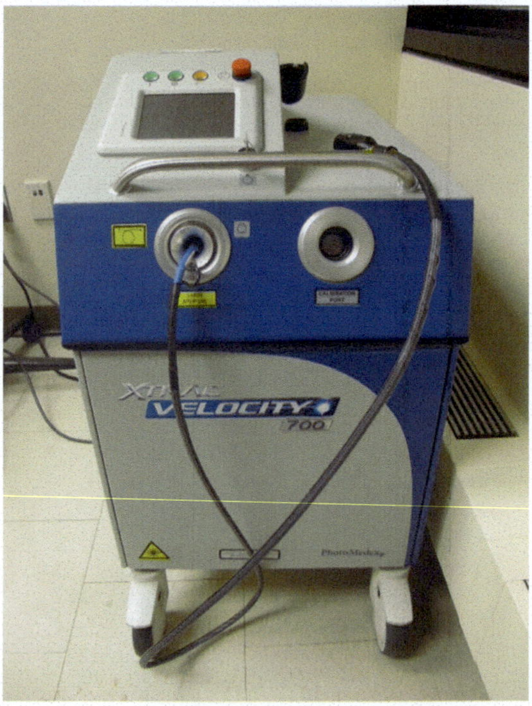

Fig. 8.5 Excimer laser

and a laser. The lamp is a polychromatic (wavelengths from 306 to 310 nm), nontargeted (incoherent) light used to treat a range of body surface areas. The laser, on the other hand, is a monochromatic (308 nm), intermittent (pulsing) light, which can deliver targeted therapy (coherent) (Fig. 8.5). It is commonly used for the treatment of psoriasis, atopic dermatitis, and vitiligo [68].

8.6.3 Fluorescent Lamps

Fluorescent lamps are cylindrical glass tubes coated with phosphors and containing mercury. When current is applied to the ends of the tubes, the mercury is vaporized to a higher state and radiation is produced when the mercury falls to its ground state. The phosphors coating the tube act as a fluorophores (i.e. chromophores for fluorescence), absorbing the light and then re-emitting it at longer wavelengths [58]. Different phosphors lead to UVA, UVB, or visible light. Fluorescent lamps can be used for both treatment and diagnosis of dermatologic conditions. In fact, they are the most frequently used sources of therapeutic UVR in dermatology. They can be used for full body treatment or more targeted treatment such as for hands and feet. The Wood's lamp, commonly used for diagnosis of vitiligo, fungal infections, and erythrasma, emits UVA, which is absorbed by skin fluorophores (collagen, elastin, and porphyrins) and then re-emitted at a longer wavelength as visible light [7].

8.6.4 Light-Emitting Diodes (LEDs)

LEDs are semiconductors that convert electrical current into narrow band light in wavelengths ranging from 274 to 1300 nm. LEDs can deliver the same wavelengths as lasers but at lower energy output. Therefore, LEDs provide a more gentle delivery of light and do not carry the same risk of tissue damage as lasers do [6]. Because LEDs can be made into panels, they can cover greater body surface areas compared to lasers [24], resulting in faster treatment times. LEDs have been used safely in the treatment of neoplasms in photodynamic therapy (PDT), acne, cosmetic rejuvenation, and other indications [6].

8.6.5 Lasers

Laser is an acronym for light amplification by stimulated emission of radiation. In lasers, the incident and emitted photons are of the same wavelength, phase, and direction, this characteristic gives lasers their monochromatic (i.e. single wavelength) and coherent spectral output [58]. The wavelength is determined by a lasing medium (e.g. solid, liquid, gas) in the optical cavity of the laser through which the light passes. It is selected based on the depth and absorption characteristics of the target chromophore. Hemoglobin, melanin, artificial pigment (tattoos), and collagen are some of the chromophores targeted by lasers.

Once laser energy is absorbed in the skin, photothermal, photochemical, or photomechanical effects are possible. Photothermal and photomechanical reactions are the most commonly observed effects in current laser practice. Photothermal effects occur when a chromophore absorbs a specific wavelength and the conversion of absorbed energy into heat leads to destruction of the target (chromophore). Rapid thermal expansion can lead to acoustic waves and subsequent photomechanical destruction of the absorbing tissue [84].

Currently, lasers work on the principle of selective photodermolysis (i.e. selective thermal damage), in which a wavelength is chosen that will be preferentially absorbed by the target tissue (chromophore) and cause its destruction with minimal thermal damage to the surrounding tissue. To limit the amount of thermal energy deposited within the skin, the exposure duration of tissue to light (pulse duration) must be adequate and is chosen based on the size of the target; smaller targets require shorter pulse [58]. Lastly, the energy density (fluence, measured in joules per square centimeter) must be sufficient to achieve destruction within the allotted time [3]. Lasers have been used in the treatment of benign vascular and pigmented birthmarks, hypertrophic scars and keloids, removal of facial or body hair, tattoos, and rhytides.

Some of the most common light treatments currently used in dermatology include narrow band UVB, UVA1, PUVA, photodynamic therapy, and extracorporeal photopheresis.

8.6.6 Narrowband UVB

Narrowband (NB) UVB (311–312 nm) was developed for the treatment of psoriasis. It was shown to be more effective than broadband UVB [69, 89] and allowed for a lower dose of UV to be used. NBUVB is also effective for treatment of atopic dermatitis, early stage mycosis fungoides, vitiligo, and pityriasis rosea [32]. NBUVB can be used in combination with topical agents such as calcipotriol, or oral agents such as retinoids, to augment efficacy and promote faster resolution of skin disease [20, 32]. NBUVB is safe enough to give to children and women during pregnancy [32]. The most common side effect from UVB exposure is an acute phototoxic reaction manifesting as erythema [7]. Other side effects may include conjunctivitis and keratitis (if adequate eye protection is not used during treatment) [43] and an increased long term risk of skin cancer [32] (Fig. 8.6).

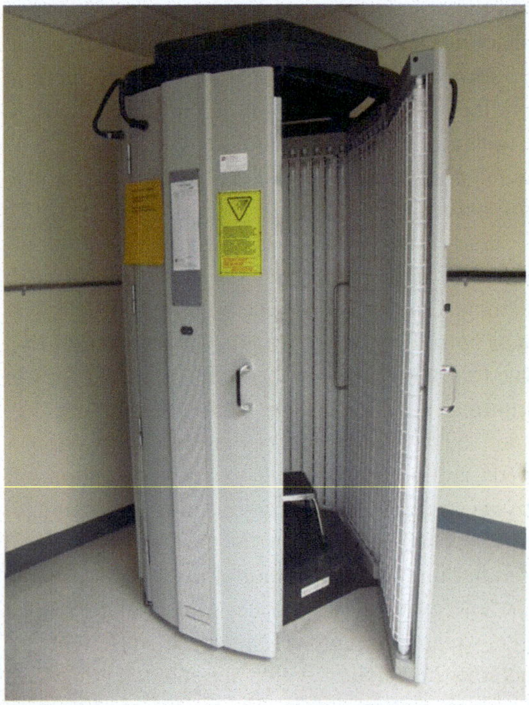

Fig. 8.6 Narrow band UVB light booth

8.6.7 UVA1

By penetrating deep into the dermis, UVA1 phototherapy has led to therapeutic responses without the usual side effects caused by less penetrating UVB wavelengths and UVB-like wavelengths in the UVA-2 region [44]. UVA1 has been shown to induce apoptosis of skin-infiltrating T-helper cells [64] and their depletion from affected skin, eventually leading to clearing of atopic dermatitis [35, 49]. It has also been shown to be an effective treatment for sclerosis by inducing collagenase I expression in a dose dependent manner [50, 55].

8.6.8 PUVA

PUVA, also known as photochemotherapy, is the combination of psoralens (P) plus UVA light. Psoralens are compounds found in plants that when taken orally or applied topically, absorb light and produce photochemical reactions in skin cells resulting in a therapeutic effect [56]. Some of the skin disorders treated with PUVA include psoriasis, dermatitis, vitiligo, polymorphic light eruption, and early stage cutaneous T-cell lymphoma. Immediate side effects include burning, itching, and nausea. Cumulative high-dose exposure to PUVA causes photoaging and increases the risk for skin cancer, in particular squamous cell carcinoma and possibly, melanoma [43, 101].

8.6.9 PDT

Photodynamic therapy (PDT) uses photosensitizing agents to amplify the effects of visible light or lasers. A photosensitizer agent such as aminolevulinic acid is applied to the skin, where it accumulates in the target cells. These cells absorb light and along with oxygen, lead to formation of reactive oxygen species and selective cell apoptosis [82]. PDT is effective in treating neoplasms such as actinic keratosis and superficial nonmelanoma skin cancers (i.e, superficial basal carcinoma and Bowen's disease). Topical PDT can also be used for other non-neoplastic indications such as psoriasis, localized scleroderma, acne, and skin rejuvenation [42, 45, 87]. Most common side effects include phototoxic reactions and pain. The pain is often referred to as smarting reaction and it may require analgesia for control [101].

8.6.10 Extracorporeal Photopheresis

Extracorporeal photopheresis (ECP), also known as extracorporeal photochemotherapy, refers to a type of systemic light treatment, in which leukocytes are separated from the patient's blood, combined with methoxypsoralen and irradiated with UVA (PUVA). The treated white cells are then re-infused into the patient. ECP is a first-line treatment for Sézary syndrome (leukemic CTCL). Other indications include graft-versus host disease and systemic scleroderma, among others [48].

8.7 Summary

UVR is the area of the electromagnetic spectrum with the greatest biological impact on cutaneous health and disease. UVA penetrates deeply into the dermis while UVB is absorbed in the epidermis. Acute changes after UVR exposure include sunburn, tanning, epidermal hyperplasia, synthesis of vitamin D, and pro-inflammatory as well as immunosuppressive changes. Chronic changes result mostly in photocarcinogenesis and photoaging. Photoprotection is therefore important to protect against the harmful effects of UVR exposure. Artificial light sources can be used for treatment of several skin conditions such as psoriasis, atopic dermatitis, CTCL, vitiligo, scleroderma, acne, and even facial rejuvenation.

References

1. Agar N, Young AR (2005) Melanogenesis: a photoprotective response to DNA damage? Mutat Res 571(1–2):121–132
2. Anderson RR, Parrish JA (1981) The optics of human skin. J Invest Dermatol 77:13–19
3. Anderson RR, Parrish JA (1983) Selective photothermolysis: precise microsurgery by selective absorption of pulsed radiation. Science 220:524–527
4. Australian Government (2013) Department of health and ageing. Therapeutic Goods Administartion. A review of the scientific literature on the safety of nanoparticulate titanium dioxide or zinc oxide in sunscreens. https://www.tga.gov.au/sites/default/files/sunscreens-nanoparticles-review-2013.pdf
5. Avenel-Audran M, Dutartre H, Goossens A et al (2010) Octocrylene, an emerging photoallergen. Arch Dermatol 146:753
6. Barolet D (2008) Light-emitting diodes (LEDs) in dermatology. Semin Cutan Med Surg 27(4):227–238. doi:10.1016/j.sder.2008.08.003
7. Baron ED, Suggs AK (2014) Introduction to photobiology. Dermatol Clin 32(3):255–266. doi:10.1016/j.det.2014.03.002
8. Bhushan M, Cumberbatch M, Dearman RJ et al (2002) Tumour necrosis factor-alphainduced migration of human Langerhans cells: the influence of aging. Br J Dermatol 146:32–40
9. Bikle DD, Nemanic MK, Gee E et al (1986a) 1,25-Dihydroxyvitamin D3 production by human keratinocytes kinetics and regulation. J Clin Invest 78:557–566
10. Bikle DD, Nemanic MK, Whitney JO et al (1986b) Neonatal human foreskin keratinocytes produce 1,25-dihydroxyvitamin D3. Biochemistry 25:1545–1548
11. Bodekaer M, Philipsen PA, Karlsmark T et al (2013) Good agreement between minimal erythema dose test reactions and objective measurements: an in vivo study of human skin. Photodermatol Photoimmunol Photomed 29(4):190–195
12. Bolognia JL, Schaffer JV, Duncan KO et al (eds) (2014) Dermatology essentials. Saunders/Elsevier, Oxford
13. Brenner M, Hearing VJ (2008) The protective role of melanin against UV damage in human skin. Photochem Photobiol 84(3):539–549
14. Bruls WAG, Slaper H, van der Leun JC et al (1984) Transmission of human epidermis and stratum corneum as a function of thickness in the ultraviolet and visible wavelegths. Photochem Photbiol 40:485–495
15. Bruynzeel DP, Ferguson J, Andersen K et al (2004) Photopatch testing: a consensus methodology for Europe. J Eur Acad Dermatol Venereol 18(6):679–682
16. Bryden AM, Moseley H, Ibbotson SH et al (2006) Photopatch testing of 1155 patients: results of

the UK multicentre photopatch study group. Br J Dermatol 155:737–747
17. Cerimele D, Celleno L, Serri F et al (1990) Physiological changes in ageing skin. Br J Dermatol 122(Suppl 35):13–20
18. Chen S, Geller AC, Tsao H (2013) Update on the epidemiology of melanoma. Curr Dermatol Rep 2(1):24–34
19. Courdavault S, Baudouin C, Charveron M et al (2004) Larger yield of cyclobutane dimer than 8-oxo-7, 8-dihydroguanine in the DNA of UVA-irradiated human skin cells. Mutat Res 556(1–2):135–142
20. Coven TR, Burack LH, Gilleaudeau R et al (1997) Narrowband UV-B produces superior clinical and histopathological resolution of moderate-to-severe psoriasis in patients compared with broadband UV-B. Arch Dermatol 133(12):1514–1522
21. Darr D, Fridovich I (1994) Free radicals in cutaneous biology. J Invest Dermatol 102:671–675
22. De Fabo EC, Noonan FP (1983) Mechanism of immune suppression by ultraviolet irradiation in vivo. I. Evidence for the existence of a unique photoreceptor in skin and its role in photoimmunology. J Exp Med 158(1):84–98
23. de Gruijl FR, Sterenborg HJ, Forbes PD et al (1993) Wavelength dependence of skin cancer induction by ultraviolet irradiation of albino hairless mice. Cancer Res 53:53–60
24. Dierickx C, Anderson RR (2002) Visible light treatment of photoaging. Dermatol Ther 18(3):191–208
25. Dupont E, Gomez J et al (2013) Beyond UV radiation: a skin under challenge. Int J Cosmet Sci 35(3):224–232
26. El Ghissassi E, Baan R, Straif K et al (2009) A review of human carcinogens – part D: radiation. Lancet Oncol 10:751–752
27. Faurschou A, Wulf HC (2015) Photodermatoses. In: European dermatology guidelines. European dermatology forum. http://euroderm.org. Accessed 4 Jul 2016
28. Fisher GJ, Datta SC, Talwar HS et al (1996) Molecular basis of sun-induced premature skin ageing and retinoid antagonism. Nature 379(6563):335–339
29. Fisher GJ, Kang S, Varani J et al (2002) Mechanisms of photoaging and chronological skin aging. Arch Dermatol 138(11):1462–1470
30. Fu GK, Lin D, Zhang MY et al (1997) Cloning of human 25-hydroxyvitamin D-1 alpha-hydroxylase and mutations causing vitamin D-dependent rickets type 1. Mol Endocrinol 11:1961–1970
31. Garmyn M, Yarosh DB (2007) The molecular and genetic effects of ultraviolet radiation exposure on skin cells. In: Lim H, Honigsmann H, Hawk JL (eds) Photodermatology. Informa Healthcare, New York, pp 41–54
32. Gawkrodger DJ, Ardern-Jones MR (2012) Dermatology: an illustrated colour text. Churchill Livingstone/Elsevier, Edinburgh
33. Gilchrest B, Eller MS, Geller AC et al (1999) The pathogenesis of melanoma induced by ultraviolet radiation. N Engl J Med 340(17):1341–1348
34. Gilchrest BA, Murphy GF, Soter NA (1982) Effect of chronologic aging and ultraviolet irradiation on Langerhans cells in human epidermis. J Invest Dermatol 79:85–88
35. Grewe M, Bruijnzeel-Koomen CA, Schöpf E et al (1998) A role for Th1 and Th2 cells in the immunopathogenesis of atopic dermatitis. Immunol Today 19(8):359–361
36. Hackenberg S, Kleinsasser N (2012) Dermal toxicity of ZnO nanoparticles: a worrying feature of sunscreen? Nanomedicine 7:461
37. Hamzavi I, Lui H (2005) Using light in dermatology: an update on lasers, ultraviolet phototherapy and photodynamic therapy. Dermatol Clin 23(2):199–207
38. Hatch KL, Osterwalder U (2006) Garments as solar ultraviolet radiation screening materials. Dermatol Clin 24(1):85–100
39. Haylett AK, Chiang YZ, Nie Z et al (2014) Sunscreen photopatch testing: a series of 157 children. Br J Dermatol 171:370
40. Heckman C, Chandler R, Kloss JD et al (2013) Minimal erythema dose (MED) testing. J Vis Exp. doi:10.3791/50175
41. Heurung AR, Raju SI, Warshaw EM (2014) Adverse reactions to sunscreen agents: epidemiology, responsible irritants and allergens, clinical characteristics, and management. Dermatitis 25:289
42. Ibbotson SH (2002) Topical 5-aminolaevulinic acid photodynamic therapy for the treatment of skin conditions other than non-melanoma skin cancer. Br J Dermatol 146(2):178–188
43. Iordanou E, Berneburg M (2010) Phototherapy and photochemotherapy. J Dtsch Dermatol Ges 8(7):533–541
44. Jekler J, Larkö O (1990) Combined UVA-UVB versus UVB phototherapy for atopic dermatitis: a paired-comparison study. J Am Acad Dermatol 22(1):49–53
45. Karrer S, Abels C, Landthaler M et al (2000) Topical photodynamic therapy for localized scleroderma. Acta Derm Venereol 80(1):26–27
46. Kift R, Berry JL, Vail A et al (2013) Lifestyle factors including less cutaneous sun exposure contribute to starkly lower vitamin D levels in UK South Asians compared with the white population. Br J Dermatol 169(6):1272–1278
47. Kligman AM (1979) Perspectives and problems in cutaneous gerontology. J Invest Dermatol 73:39e46
48. Knobler R, Berlin G, Calzavara-Pinton P et al (2014) Guidelines on the use of extracorporeal photopheresis. JEADV 28(Suppl 1):1–37
49. Krutann J, Czech W, Diepgen T et al (1992) High-dose UVA1 therapy in the treatment of patients with atopic dermatitis. J Am Acad Dermatol 26:225–230
50. Krutman J (1999) Therapeutic photomedicine: phototherapy. In: Freedberg IM, Az E, Wolff H et al

(eds) Fitzpatrick's dermatology in general medicine. McGraw-Hill, New York
51. Krutmann J, Schroeder P (2009) Role of mitochondria in photoaging of human skin: the defective powerhouse model. J Investig Dermatol Symp Proc 14:44–49
52. Kvam E, Tyrell RM (1997) Induction of oxidative DNA base damage in human skin cells by UV and near visible radiation. Carcinogenesis 18:2379–2384
53. Lanigan SW, Zaidi Z (2010) Dermatology in clinical practice. Springer, New York
54. Lehmann B, Genehr T, Knuschke P et al (2001) UVB-induced conversion of 7- dehydrocholesterol to 1alpha, 25-dihydroxyvitamin D3 in an in vitro human skin equivalent model. J Invest Dermatol 117:1179–1185
55. LeRoy EC (1974) Increased collagen synthesis by scleroderma skin fibroblasts in vitro: a possible defect in the regulation or activation of the scleroderma fibroblast. J Clin Invest 54(4):880–889
56. Ling TC, Clayton TH, Crawley J et al (2016) British Association of Dermatologists and British Photodermatology Group guidelines for the safe and effective use of psoralen-ultraviolet a therapy 2015. Br J Dermatol 174(1):24–55. doi:10.1111/bjd.14317
57. Longo D, Fauci AS, Kasper DL et al (eds) (2012) Photosensitivity and other reactions to light. Harrison's principles of internal medicine. McGraw-Hill, New York
58. Lui H, Anderson RR (2007) Radiation sources and interaction with the skin. In: Lim H, Honigsmann H, Hawk JL (eds) Photodermatology. Informa Healthcare, New York, pp 29–40
59. Lim H, Honigsmann H (2007) Photoprotection. In: Lim H, Honigsmann H, Hawk JL (eds) Photodermatology. Informa Healthcare, New York, pp 267–278
60. Makinodan T (1980) Immunodeficiencies of ageing. In: Doria G, Eshkol A (eds) The immune system: functions and therapy of dysfunction. Academic Press, New York
61. Marks JG, Miller JJ (2013) Lookingbill and Marks' principles of dermatology. Saunders Elsevier, Philadelphia
62. Matsumoto K, Azuma Y, Kiyoki M et al (1991) Involvement of endogenously produced 1,25-dihydroxyvitamin D-3 in the growth and differentiation of human keratinocytes. Biochim Biophys Acta 1092:311–318
63. McCullough J, Kelly KM (2006) Prevention and treatment of skin aging. Ann N Y Acad Sci 1067:323–331
64. Morita A, Werfel T, Stege H et al (1997) Evidence that singlet oxygen-induced human T helper cell apoptosis is the basic mechanism of ultraviolet-A radiation phototherapy. J Exp Med 186(10):1763–1768
65. Moulin G, Thomas L, Vigneau M et al (1994) A case of unilateral elastosis with cysts and comedones. Favre-Racouchot syndrome. Ann Dermatol Venereol 121:721–723
66. Murphy G, Young AR, Wulf HC et al (2001) The molecular determinants of sunburn formation. Exp Dermatol 10(3):155–160
67. Norval M, Gibbs NK, Gilmour J (1995) The role of urocanic acid in UV-induced immunosuppression: recent advances (1992–1994). Photochem Photobiol 62(2):209–217
68. Park K, Liao W, Murase JE (2012) A review of monochromatic excimer light in vitiligo. Br J Dermatol 167(3):468–478
69. Parrish JA, Jaenicke KF (1981) Action spectrum for phototherapy of psoriasis. J Invest Dermatol 76:359–362
70. Pleasance ED, Cheetham RK, Stephens PJ et al (2010) A comprehensive catalogue of somatic mutations from a human cancer genome. Nature 463:191–196
71. Poon TS, Barnetson RS, Halliday GM (2005) Sunlight-induced immunosuppression in humans is initially because of UVB, then UVA, followed by interactive effects. J Invest Dermatol 125(4):840–846
72. Rabe J, Mamelak AJ, McElgunn PJ et al (2006) Photoaging: mechanisms and repair. J Am Acad Dermatol 55(1):1–19
73. Runger MT (2012) Ultraviolet light. In: Bolognia JL, Jorizzo JL, Schaffer JV (eds) Dermatology. Elsevier Saunders, Philadelphia, pp 1455–1465
74. Rünger TM (2008) C to T transition mutations are not solely UVB-signature mutations, because they are also generated by UVA. J Invest Dermatol 128:2138–2140
75. Rünger TM, Kappes UP (2008) Mechanisms of mutation formation with long-wave ultraviolet light (UVA). Photodermatol Photoimmunol Photomed 24:2–10
76. Sambandan DR, Ratner D (2011) Sunscreens: an overview and update. J Am Acad Dermatol 64:748
77. Schaefer H, Moyal D et al (1998) Recent advances in sun protection. Semin Cutan Med Surg 17(4):266–275
78. Schauder S, Ippen H (1997) Contact and photocontact sensitivity to sunscreens. Review of a 15-year experience and of the literature. Contact Dermatitis 37:221
79. Schornagel IJ, Knol EF, van Weelden H et al (2005) Diagnostic phototesting in polymorphous light eruption: the optimal number of irradiations. Br J Dermatol 153:1234–1236
80. Schwarz T, Halliday GM (2007) Photoimmunology. In: Lim H, Honigsmann H, Hawk JL (eds) Photodermatology. Informa Healthcare, New York, pp 55–74
81. Skocaj M, Filipic M, Petkovic J, Novak S (2011) Titanium dioxide in our everyday life; is it safe? Radiol Oncol 45:227
82. Szeimies RM, Karrer S, Abels C et al (2001) Photodynamic therapy in dermatology. In:

Krutmann J, Honigsmann H, Elmets CA et al (eds) Dermatological phototherapy and photodiagnostic methods. Springer, Berlin
83. Tanner PR (2006) Sunscreen product formulation. Dermatol Clin 24(1):53–62
84. Tanzi EL, Lupton JR, Alster TS (2003) Lasers in dermatology: four decades of progress. J Am Acad Dermatol 49(1):1–31
85. Thiers BH, Maize JC, Spicer SS et al (1984) The effect of aging and chronic sun exposure on human Langerhans cell populations. J Invest Dermatol 82:223–226
86. Tobin DJ (2016) Introduction to skin aging. J Tissue Viability. doi:10.1016/j.jtv.2016.03.002
87. Touma D, Yaar M, Whitehead S et al (2004) A trial of short incubation, broad-area photodynamic therapy for facial actinic keratoses and diffuse photodamage. Arch Dermatol 140(1):33–40
88. Urbach F (1999) The cumulative effects of ultraviolet radiation on the skin: photocarcinogenesis. In: Hawk J (ed) Photodermatology. Arnold Publishers, London
89. Walters IB, Burack LH, Coven TR et al (1999) Suberythemogenic narrow-band UVB is markedly more effective than conventional UVB in treatment of psoriasis vulgaris. J Am Acad Dermatol 40:893–900
90. Walterscheid JP, Ullrich SE, Nghiem DX (2002) Platelet-activating factor, a molecular sensor for cellular damage, activates systemic immune suppression. J Exp Med 195(2):171–179
91. Wang SQ, Lim HW (2011) Current status of the sunscreen regulation in the United States: 2011 Food and Drug Administration's final rule on labeling and effectiveness testing. J Am Acad Dermatol 65(4):863–869. doi:10.1016/j.jaad.2011.07.025
92. Wang SQ, Tooley IR (2011) Photoprotection in the era of nanotechnology. Semin Cutan Med Surg 30:210
93. Wang X, Wong SC, Pan J et al (1998) Evidence of cisplatin-induced senescent-like growth arrest in nasopharyngeal carcinoma cells. Cancer Res 58(22):5019–5022
94. Weedon D (2010) Weedon's skin pathology. Elsevier, Philadelphia
95. Weller RP, Hunter JA (2008) Clinical dermatology. Blackwell Publisher, Malden
96. Wlaschek M, Tantcheva-Poór I, Naderi L (2001) Solar UV irradiation and dermal photoaging. J Photochem Photobiol B 63(1–3):41–51
97. Yaar M (2007) The chronic effects of ultraviolet radiation on the skin: photoaging. In: Lim H, Honigsmann H, Hawk JL (eds) Photodermatology. Informa Healthcare, New York, pp 91–106
98. Yaar M, Gilchrest BA (2001) Skin aging: postulated mechanisms and consequent changes in structure and function. Clin Geriatr Med 17(4):617–630
99. Young AR, Chadwick CA, Harrison GI et al (1998) The similarity of action spectra for thymine dimers in human epidermis and erythema suggests that DNA is the chromophore for erythema. J Invest Dermatol 111:982–988
100. Young AR, Wikonkal NM (2007) The chronic effects of ultraviolet radiation on the skin: photocarcinogenesis. In: Lim H, Honigsmann H, Hawk JL (eds) Photodermatology. Informa Healthcare, New York, pp 107–117
101. Zanolli M (2003) The modern paradigm of phototherapy. Clin Dermatol 21(5):398–406

Ultraviolet A-1 in Dermatological Diseases

9

Torello M. Lotti and Serena Gianfaldoni

Abstract

The exposure to ultraviolet radiations and visible light, or phototherapy, is a well-known therapeutic tool available for the treatment of many dermatological disorders. The continuos medical and technological progresses, of the last 50 years, have involved the field of phototherapy, which evolved from UVA and PUVA in its various forms, to the development of narrow-band UVB (NB-UVB) and NB-UVB micro-focused phototherapies. Further advances in technology have now permitted the introduction of a new device emitting UVA-1 radiations.

Keywords

Phototherapy • UVA-1 • Inflammation • Autoimmunity • Sclerosis

9.1 Introduction

Phototherapy is a therapeutic option, available for many dermatologic diseases, which consist in the patient's exposure to ultraviolet radiation or visible light. Even though the beneficial effects of ultraviolet radiations (UVR) for the treatment of different cutaneous disorders are well known by the ancient Egyptians, the continuous scientific progress of the last century allowed a deeper knowledge of the bio-molecular effects of radiation in the tissue, optimizing their use for different purpose. The significant technological advances in the last decades have allowed the development of new phototherapeutic devices and methods of irradiation. Now a number of different devices are available, each having variable properties in respect of the wavelength (Table 9.1) and modality of emission of radiation providing better dermatological treatments.

T.M. Lotti (✉)
Professor and Chair of Dermatology, University G. Marconi of Rome, Rome, Italy
e-mail: professor@torellolotti.it

S. Gianfaldoni
Dermatologist and Venereologist, Pisa, Italy

Centro Studi per la Ricerca Multidisciplinare Rigenerativa, University G. Marconi of Rome, Rome, Italy
e-mail: serena.gianfaldoni@gmail.com

S. Ahmad (ed.), *Ultraviolet Light in Human Health, Diseases and Environment*, Advances in Experimental Medicine and Biology 996, https://doi.org/10.1007/978-3-319-56017-5_9

Table 9.1 Different wavelengths of ultraviolet radiation used for phototherapy

Type of radiation	Wavelength (nm)
UVA	320–400
UVA-1	340–400
UVA-2	320–340
UVB	290–320
UVB narrow band	311

9.2 UVA Therapy

The first phototherapeutic device, introduced for the treatment of different dermatological conditions, was Ultraviolet-A light (broadband UVA) with a wavelength of 320–400 nm. Even if its efficacy in the treatment of different cutaneous diseases, such as psoriasis, was well known, the subsequent introduction of PUVA [1], a particular form photochemical treatment, made broadband UVA less useful for dermatologic purposes.

PUVA therapy is characterized by the oral intake of psoralens, also known as furocoumarins, followed by the exposure to photoactivating UVA radiations. Psoralens are natural photosensitizing compounds, which occur in parsnips and other plant species. Although different types of psoralens, both natural and synthetic, are available for photo-chemotherapy, 8-methoxypsoralen (MPO) and 4,5′,8-trimethylpsoralen (TMP), two psoralen analogs, are the most commonly used. One of the most important PUVA mechanisms of action on biological systems is the attachment of MPO molecules to DNA causing inter-strand crosslink, leading to possible cell death if not repaired effectively in time. A part of a stimulation of melanogenesis, PUVA therapy inhibits the basal cell division and immune system, mainly by the DNA damage of irradiated cells.

Actually, the most important clinical indications for PUVA are the treatment of vitiligo, psoriasis and mycosis fungoides. The therapeutic protocol varies for different clinical conditions, but, usually, it consists in 2–3 sessions for week, with a progressive increasing radiation's dose on the base of patient's clinical response. The psoralen's toxicity, mainly represented by gastric and ocular damage, limits the range of candidates for PUVA therapy only to adult without controindications. The treatment is not fully safe and side effects have been recorded [1, 2]. The acute side effects are mainly represented by erythema, pruritus and xerosis. Less commonly phototoxic reactions has been described too. More rarely, patients developed burns, due to an incorrect irradiation doses or to sun-bathed after taking psoralen. Chronic side effects include chronic actinic damage and a increased risk of cutaneous carcinogenesis (melanoma and non melanoma skin cancer). More rarely, hypertrichosis has been reported as side effect.

9.2.1 Topical PUVA Treatment

Topical PUVA therapy is a valid therapeutic option to the systemic one, especially for the treatment of localized dermatological diseases. The treatment consists in the application of a psoralen compound (e.g. 0.1–0.01% $8Me_3Pso$ in hydrophilic petrolatum or ethanol), onto the cutaneous lesions, followed by the exposure to UVA light. In general the therapeutic protocol consist in 1 to 3 sessions a week, accordingly to the clinical conditions of patients. During each session, the UVA dose is progressively increased until a mild erythematous reaction develops. Such as PUVA therapy, the topical one is characterized by well described short- and long-time side effects, due to ultraviolet radiations [29].

9.2.2 PUVASOL Therapy

PUVASOL therapy is a different phototherapeutic option [3], especially useful for patients, who cannot refer to hospital for the conventional treatment. It consists of psoralen (8-methoxypsoralen) intake and sunlight exposure, which can be easily performed at patients' home. On the other hand, the lack of a medical control during the therapeutic sessions, makes PUVASOL therapy less safe. Severe reaction, such as erythema, pigmentation, blistering, burning and ocular side effects, are well-described.

9.2.3 Bath-PUVA Therapy

A different variant of PUVA therapy is the bath-PUVA, consisting in a 10 minutes-bath in a solution containing 0.0002% Me_3Pso, and in the successively UVA irradiation of patients. Bath-PUVA is usually performed once or twice a week for a period of 6 or more weeks, accordingly to the clinical condition. While short-term side effects are commonly represented by itching and erythema, the long-term ones include photo-ageing and carcinogenesis.

9.3 Narrow-Band UVB Therapy

The last few decades has seen the emergence and confirmations of narrow-band UVB (NB-UVB) as a valid tool for the treatment of several dermatological diseases, like psoriasis, atopic dermatitis, vitiligo [4] and prurigo. The treatment consists in the patient's exposure to NB-UVB (311 nm), 2–3 times a week, for a period which remains variable for the different dermatological condition. Typically, the starting irradiation dose (0.1 mJ/cm^2) is followed by 20% increments weekly, depending on the clinical response.

NB-UVB mainly acts by inhibiting the immune activity. Differently by PUVA, the absence of photosensitizing substances makes NB-UVB therapy more safe and versatile, easily performing also in children, pregnant, open air workers and patients affected by liver or kidney failure. NB-UVB therapy is well-tolerated by most of patients. Pruritus, xerosis, erythema, and transient hyperpigmentation are the more common acute side effects. A part of a rare description of keratoacanthoma after NB-UVB, chronic side effects have to be determined. Patients' photo-damage seems to be possible.

More recently, the continuous advances in technology have permitted the development of NB-UVB micro-phototherapy [4], which consists of the irradiation limited to the affected skin areas. Even if the mechanisms of action of micro-phototherapy are the same of the classical phototherapy, it acts in a more precise and safe way. The possibility to focus the radiations on skin lesions, reduces the risks of side effects in the uninvolved safe skin, and allows the operator to use higher dose of energy. This fact leads to shortened duration and less frequent treatment sessions, with an increase in patient's compliance and satisfaction. Micro-phototherapy is advised for the treatment of localized dermatological conditions (e.g. localized vitiligo) and for lesions in special body parts (e.g. scalp, ear, nose), which normally could not be treated with the more conventional devices. Micro-phototherapy could be performed in children, pregnant female and, not less important, in claustrophobic patients.

9.4 UVA-1 Therapy

More recently, dermatology has experienced a growing interest in the use of light lamps emitting UVA1 radiations (340–400 nm). The well-documented immune-modulating effects of UVA-1, make this type of phototherapy useful for the treatment of several cutaneous diseases, like atopic dermatitis, psoriasis, scleroderma, mastocytosis, and other [5].

While UVB radiations (290–320 nm) are taken up by epidermis and upper dermis, UVA-1 (340–400 nm), can reach the mid- and lower dermis, so they are more useful for the treatment of dermatologic conditions which lies in those level. The biological effects of UVA-1 are mainly mediated through the formation of reactive oxygen species (ROS) intermediates [6, 7], during the mitochondrial oxidative phosphorylation, which can damage DNA, lipids, proteins and cellular organelles.

This fact may exert different biochemical effects. On the immune system, UVA-1 inhibits effector T-cells, through the direct inhibition of dendritic cells [8], and through the production of interleukin 10 (IL-10) and the decrease of tumour necrosis factor-alpha (TNF-alpha) [9, 10]. The UVA-1 immunomodulating effects are also the result of the apoptosis of T lymphocytes. The T cell death may occur immediately (within 4 hours) through the depolarization of mitochondrial membranes damaged by ROS, or may be delayed (after 24 hours) through the oxidative damage of DNA [11, 12].

In addition, on lymphocytes and cytokines, UVA-1 phototherapy may regulate the expression of the intercellular adhesion molecule-1 (ICAM-1) and other types of adhesion molecules.

ICAM-1 is expressed by keratinocytes, in particular inflammatory skin conditions, where it plays an important role in the maintenance of the pathological process by mediating leukocyte/keratinocyte adhesion. UVA-1 is able to decrease keratinocytes ICAM-1 expression, interfering with their inflammatory action [13]. Another important mechanism of action of UVA-1 is the apoptosis of mast-cells with a notably reduction in the concentration of their mediators [14]. On the epidermis, UVA-1 may act both on keratinocytes and melanocytes. Like UVB and PUVA, UVA-1 induces acanthosis, thickening of the stratum corneum and melanocytes' stimulation.

UVA-1 also induces effects on fibroblast and collagen fibers. It activates matrix-metalloproteinases (MMPs), through the generation of ROS and the increased production of interleukin-1 (IL-1) and −6 (IL-6). MMPs are physiologic mediators of matrix degradation: for example, MMP-1 induces the degradation of collagen fibers [15]. On the other hand, UVA-1 down-regulates the expression of the transforming growth factor (TGF)-beta/SMAD pathway, which is fundamental in the pathogenesis of sclerotic skin diseases [16, 17].

Currently several different devices emitting UVA-1 radiation, both as laser and light are available. UVA-1 is a phototherapeutic option, widely used in dermatology for the treatment of inflammatory, autoimmune or sclerotic diseases (Table 9.2) [18–27]. The treatment can be performed alone or in association with more conventional therapies. Because the lack of photosensitizing drugs, UVA-1 may be performed on children, pregnant and patients with contraindication to psoralens. Moreover, the possibility to use lasers allows the treatment of claustrophobic patients and subjects affected by localized cutaneous diseases, with sparing of uninvolved cutaneous areas.

The procedure changes on the basis of the dermatological conditions which have to be treated. In general, UVA-1 light is administered 3–5 times per weeks. Patients usually start the therapy at a dose 20–30 J/cm2, that is progressively increased to the full dose. The more innovative use of UVA-1 focused lasers allows the operators to use higher dose of energy in a safer way, leading to shorten duration and less frequent treatment sessions [5].

Table 9.2 Clinical indications for UVA-1 treatment

Atopic dermatitis
Dyshidrotic dermatitis
Psoriasis
Pytiriasis rosea
Prurigo
Urticaria pigmentosa/Mastocytosis
Localized scleroderma (morphea)
Systemic lupus erythematosus
Lichen sclerosus
Mycosis fungoides and other cutaneous T-cell lymphoma
Others: Vitiligo, Graft versus host disease (GVHD), Granuloma annulare, Necrobiosis lipoidica, Cutaneous sarcoidosis, Follicular mucinosis, POEMS syndrome, Scleromyxedema, Hypereosinophilic syndrome, Pityriasis lichenoides

9.5 UVA-1 Side Effects

Early constructed UVA-1 lamps used to generate an intense heat, which limited their use. Now this problem has been solved by the introduction of new devices with cooling and filtering systems that remove wavelength above 530 nm. For this reason UVA-1 treatment is usually well-tolerated by most of patients. Rarely, tanning, erythema, pruritus and phototoxic reactions (eczema, urticaria) have been reported as short-term side effects [6]. The long term UVA-1 side effects are required to be investigated.

9.6 Innovative Laser Alba 355®

Laser Alba 355® is a new laser technology based on UVA-1 spectrum with a wavelength of 355 nm. The device consists of a Neodymium-doped yttrium orthovanadate (Nd:YVO4) crystal which is energetically stimulated by a diode laser, with a wavelength of 808 nm. The stimulation of the active medium produces radiations, with a

Table 9.3 Laser Alba 355® technical features

Laser source	Solid state pumped laser diode (DPSS)
Active material	Neodymium-doped yttrium orthovanadate (Nd:YVO4)
Wavelenght	355 nm
Maximum output	7 W
Beam size	2.5 mm
Beam quality	TEM00
Beam divergence	1.5 mrad
Power stability	<1%
Repetition pulse rate	20–25 kHz
Maximum energy per pulse	0.35 mJ
Pulse width	10–15 ns
Brightness	For 20,000 hours
Cooling system	Air

Table 9.4 Applications and protocols of Laser Alba 355®

Pathology	J/cm^2	J (cumulative)	Weekly sessions number
Atopic dermatitis	20	1000	3
Mycosis fungoides	130	2000	2
Lupus erythematosus	6	250	2
Localized scleroderma	100	1500	2
Lichen sclerosus	100	1000	2
Psoriasis	130	1800	3

wavelenght of 1064 nm, which are impulsed through an acousto-optic crystal, producing ultra-short pulsed light (25 nanosec).

The high pulse rate is sent to the Nd:YVO4 crystal in order to duplicate and triplicate the original wavelength of 1064 nm. This fact allows the production of two more harmonic wavelength delivery, respectively of 532 and 355 nm. The laser beam is then filtered in order to produce only 355 nm wavelength specific beam, which is successively amplified and homogenate. A particular diaphragm allows the operator to design different shaped dimensional figures (Table 9.3) [28].

The use of Laser Alba 355® allows the treatment of selected affected skin areas, avoiding exposure of unaffected skin. Several dermatological conditions, such as atopic dermatitis, psoriasis, vitiligo, alopecia areata, mycosis fungoides, lupus erythematous, granuloma annular, morphoea and lichen sclerosus have been treated by this lamp [5].

Although, the mechanisms of action are the same as of the classical UVA-1-phototherapy, Laser Alba 355® acts in a more precise way. Because of the treatment focused onto the cutaneous lesions, the operator can use higher doses of energy, leading to less durable and frequent treatment sessions. The possibility to regulate the time of emission and the spot diameters, on the base of the clinical features of each patient, contributes to the more efficacy of the device increasing the patient's compliance (Table 9.4).

The treatment with Laser Alba 355® is generally safe and well-tolerated. Acute side effects, like erythema or pruritus, have rarely been described. Long term side effects have yet to be investigated.

9.7 Conclusion

Despite the continuous progress in medical and technological knowledges, and the evolution of pharmacological and surgical treatments, phototherapy is an important therapeutic tool available for many dermatologic diseases. The last years have been characterized by several advances in the field of phototherapy, and recently has seen the confirmation of the UVA-1 therapy. This modality consists in the use of light therapy lamps emitting UVA-1 (340–400 nm). The well-documented immune-modulating, antinflammatory and antisclerotic effects of UVA-1, make this type of phototherapy useful for the treatment of different cutaneous disorders, such as atopic dermatitis, psoriasis, scleroderma, mastocytosis, and other. Today, many UVA-1 phototherapy devices are available. Among these the last frontier of UVA-1 treatment is represented by the Laser Alba 355®, which allows the treatment of selected affected skin areas, avoiding exposure of unaffected skin.

References

1. Gianfaldoni S, Zarrab Z, Lotti T (2014) Phototherapy and Vitiligo re-pigmentation: from PUVA to Micro-focused Phototherapy. J Pigm Disord 1:102
2. Lapolla W et al (2011) A review of phototherapy protocols for psoriasis treatment. J Am Acad Dermatol 64:936–949
3. Singh S, Khandpur S, Sharma VK, Ramam M (2013) Comparison of efficacy and side effect profile of oral PUVA vs. oral PUVA sol in the treatment of vitiligo: a 36 week prospective study. J Eur Acad Dermatol Venereol 27(11):1344–1351
4. Gianfaldoni S, Lotti ZM (2014) Vitiligo repigmentation: what's new? J Dermatol Clin Res 2(3):1023
5. Lotti TM, Hercogova J, D'Erme AM et al (2013) UVA-1in dermatology: clinical studies and observations. In UVA1 in dermatology. Evidence, data, hypotheses. Nuova Prhomos Publishing House, Città di Castello, pp 74–160
6. de Gruijl FR (2002) Photocarcinogenesis: UVA vs. UVB radiation. Skin Pharmacol Appl Ski Physiol 15(5):316–320
7. Tewari A, Grage MM, Harrison GI et al (2013) UVA1 is skin deep: molecular and clinical implications. Photochem Photobiol Sci 12(1):95–103
8. Duthie MS, Kimber I, Dearman RJ et al (2000) Differential effects of UVA1 and UVB radiation on Langerhans cell migration in mice. J Photochem Photobiol B 57(2–3):123–131
9. Skov L, Hansen H, Allen M et al (1998) Contrasting effects of ultraviolet A1 and ultraviolet B exposure on the induction of tumour necrosis factor-alpha in human skin. Br J Dermatol 138(2):216–220
10. Grundmann JU, Böckelmann R, Bonnekoh B et al (2001) UV erythema reducing capacity of mizolastine compared to acetylsalicylic acid or both combined in comparison to indomethacin. Photochem Photobiol 74(4):587–592
11. Godar DE, Lucas AD (2005) Ultraviolet-A1 (340–400 nm)-mediated receptor and cytokine changes of transformed lymphocytes. Photodermatol Photoimmunol Photomed 21(1):23–31
12. Breuckmann F, von Kobyletzki G, Avermaete A et al (2003) Mechanisms of apoptosis: UVA1-induced immediate and UVB-induced delayed apoptosis in human T cells in vitro. J Eur Acad Dermatol Venereol 17(4):418–429
13. Krutmann J, Grewe M (1995) Involvement of cytokines, DNA damage, and reactive oxygen intermediates in ultraviolet radiation-induced modulation of intercellular adhesion molecule-1 expression. Source J Invest Dermatol 105(1 Suppl):67S–70S
14. Mikita N, Kanazawa N, Yoshimasu T et al (2009) The protective effects of ultraviolet A1 irradiation on spontaneous lupus erythematosus-like skin lesions in MRL/lpr mice. Clin Dev Immunol 2009:673952. doi:10.1155/2009/673952. Epub 2009 Apr 26
15. Vielhaber G, Grether-Beck S, Koch O et al (2006) Sunscreens with an absorption maximum of > or =360 nm provide optimal protection against UVA1-induced expression of matrix metalloproteinase-1, interleukin-1, and interleukin-6 in human dermal fibroblasts. Photochem Photobiol Sci 5(3):275–282. Epub 2006 Feb 7
16. Gambichler T, Skrygan M, Tomi NS et al (2007) Significant downregulation of transforming growth factor-beta signal transducers in human skin following ultraviolet-A1 irradiation. Br J Dermatol 156(5):951–956. Epub 2007 Mar 23
17. Kreuter A, Hyun J, Skrygan M et al (2006) Ultraviolet A1 phototherapy decreases inhibitory SMAD7 gene expression in localized scleroderma. Arch Dermatol Res 298(6):265–272. Epub 2006 Sep 19
18. Ring J, Alomar A, Bieber T et al (2012) Guidelines for treatment of atopic eczema (atopic dermatitis) part I. J Eur Acad Dermatol Venereol 26(8):1045–1060
19. Letić M (2009) Exposure to sunlight as adjuvant therapy for dyshidrotic eczema. Med Hypotheses 73(2):203–204
20. Zerbinati N et al (2012) A preliminary study to assess the efficacy of a new UVA1 laser for treatment of psoriasis. Photomed Laser Surg 30(10):610–614
21. Beattie PE, Dawe RS, Ferguson J et al (2006) UVA1 phototherapy for genital lichen sclerosus. Clin Exp Dermatol 31(3):343–347
22. Su O, Onsun N, Onay HK et al (2011) Effectiveness of medium-dose ultraviolet A1 phototherapy in localized scleroderma. Int J Dermatol 50(8):1006–1013
23. Gambichler T, Terras S, Kreuter A (2013) Treatment regimens, protocols, dosage, and indications for UVA1 phototherapy: facts and controversies. Clin Dermatol 31(4):438–454
24. Simon JC, Pfieger D, Schöpf E (2000) Recent advances in phototherapy. Eur J Dermatol 10(8):642–645
25. Zandi S, Kalia S, Lui H (2012) UVA1 phototherapy: a concise and practical review. Skin Ther Lett 17(1):1–4
26. Oberholzer PA, Cozzio A, Dummer R et al (2009) Granulomatous slack skin responds to UVA1 phototherapy. Dermatology 219(3):268–271
27. Lotti TM, Hercogova J, D'Erme AM et al (2013) UVA-1in dermatology: clinical studies and observations. In UVA1 in dermatology. Evidence, data, hypotheses. Nuova Prhomos Publishing House, Città di Castello, pp 160–171
28. Lotti TM, Hercogova J, D'Erme AM et al (2013) UVA -1 Light vs UVA –1 Laser emission Devices. In UVA1 in dermatology. Evidence, data, hypotheses. Nuova Prhomos Publishing House, Città di Castello, pp 68–71
29. Patrizi A, Raone B, Ravaioli GM (2015) Management of atopic dermatitis: safety and efficacy of phototherapy. Clin Cosmet Investig Dermatol 8:511–520

Photodermatoses in the Pigmented Skin

10

Vinod Kumar Sharma and Kanika Sahni

Abstract

Skin colour (specifically in relation to its melanin content and composition) has a marked influence on its interaction with ultraviolet light. Eumelanin has mainly photoprotective properties while pheomelanin has the ability to cause formation of reactive oxygen species. This difference is responsible for the difference in incidence and presentation of various idiopathic photodermatoses in dark skinned patients compared to those with lighter skin types. Certain conditions are peculiar to darker skins including pin point popular variant of polymorphous light eruption. These differences are discussed in this chapter while also highlighting the challenges faced in performing phototesting in patients with dark skin.

Keywords

Photodermatoses • Dark skin

10.1 Introduction

Sunlight affects the human body in many different ways including but not limited to a role in vitamin D synthesis, psychological and emotional well being. On the downside, it is also responsible in the causation of many diseases including systemic disorders like connective tissue disorders and many cutaneous disorders including photodermatoses. Sensitivity of human skin to ultraviolet light is influenced by many factors including latitude and altitude, thickness of ozone layer and characteristics of the skin most importantly the skin color or phototype. Dark skin is inherently protected from the effects of ultraviolet light due to the higher melanin content and higher ratio of eumelanin to pheomelanin [1–3].

Photodermatoses are groups of cutaneous disorders caused or exacerbated by exposure to electromagnetic radiation of visible or ultraviolet spectrum [4, 5]. Photodermatoses are broadly grouped into idiopathic acquired photodermatoses, inherited conditions like porphyria and genodermatoses, drug induced photosensitivity and photoaggravated dermatoses. Idiopathic photodermatoses include chronic actinic dermatitis (CAD), polymorphous light eruption (PMLE), actinic prurigo (AP), solar urticaria

V.K. Sharma (✉) • K. Sahni
Department of Dermatology and Venereology, All India Institute of Medical Sciences,
Ansari Nagar, New Delhi 110029, India
e-mail: vksiadvl@gmail.com

S. Ahmad (ed.), *Ultraviolet Light in Human Health, Diseases and Environment*, Advances in Experimental Medicine and Biology 996, https://doi.org/10.1007/978-3-319-56017-5_10

(SU) and hydroa vacciniforme (HV).The pattern and spectrum of photodermatoses in dark skins differs from that in fair skinned populations and is discussed in this chapter.

10.2 Skin Color and Skin Phototypes [6]

Classification of skin color does not have clear cut definitions. It differs between races and geographic latitudes as white or light skin with red hair in Europeans and Americans to dark or black skin in Africans and African Americans with intermediate skin color seen in Hispanics and Asians. Constitutive skin color is determined by the differences in genetic composition as well as differences in geographic regions as the variation with geographic region is much greater for skin color compared to other phenotypic traits e.g. craniometric traits [7, 8]. Genetic composition particularly the melanocortin 1 receptor (MC1R) gene determines whether melanogenesis will follow primarily the eumelanogenesis or pheomelanogenesis pathway [9]. Dark skins show higher concentration of eumelanin which is a more stable molecule and highly photoprotective. When eumelanin absorbs light, it nearly completely converts it into heat and prevents the formation of free radical species and hence lowers the risk of ultraviolet induced carcinogenesis. In contrast lighter skins are richer in pheomelanin which on excitement by ultraviolet light, leads to formation of reactive oxygen species which cause phototoxic reactions and DNA damage thus contributing to sunburn and carcinogenesis respectively. The distribution and size of melanosomes also contributes to skin color as larger, more numerous, more neutral pH melanosomes which are more evenly distributed throughout the epidermis contribute to a darker skin phenotype [10–21]. The risk of skin cancers in dark skins is nearly 70 fold less than in light skins as melanin acts as a very effective sun protectant. Skin pigmentation also influences the cutaneous micro flora as microbial colonization is much greater in albinos compared to normally pigmented individuals. Higher colony forming units were in turn associated with a greater degree of sun damage [22–25].

Skin phototypes were initially described by Fitzpatrick who classified individuals into six different skin phototypes depending on the burning and tanning response of the skin to mid day sun exposure with Fitzpatrick skin type I being that which always burns and never tans and skin type VI being that which never burns, always tans. However, this scale was not very accurate for use in Asian skins as the responses differ compared to lighter skinned populations. Recently a simplified color bar tool has been developed for self assessment of skin type by patient based on colour of upper inner arm skin [26–31].

The differences in the pigmentation system in light and dark skins are summarized in Table 10.1.

10.3 Prevalence Data on Photodermatoses in General Population

Community-based studies which throw light on the prevalence of photodermatoses in the general population are few. One such questionnaire-based study to assess self reporting of photodermatoses, performed on 4899 patients in four different regions of Yunnan province of China, found that the incidence of PMLE was 0.65% and that of CAD was 0.18%. In addition, higher altitude was found to correlate positively with the prevalence of PMLE (but not CAD) [32]. Another population-based study from Scotland found CAD to be the commonest photodermatosis with the prevalence of various photodermatoses per 100,000 population as follows: CAD (16.5), porphyrias cutanea tarda (7.6), solar urticaria (3.9), actinic prurigo (3.3), erythropoeitic protoporphyrias (2.3) and hydroa vacciniforme (0.47) [33]. Photodermatoses accounted for 22.9% of patients in a study from Ethiopia, amongst which PMLE was the commonest followed by hyperpigmentation [34].

Among dermatologic clinic based prevalence data, photodermatoses accounted for 12.3% of all dermatology outpatients in a study from Michigan, compared to only 0.4% of the outpa-

Table 10.1 Comparative biology of the pigmentary system in dark and light coloured skin

Characteristic	Dark skin	Light skin
Melanin content	Higher total melanin	Lower total melanin
Eumelanin: pheomelanin ratio	Higher	Lower
Distribution and size of melanosomes	Discretely located larger melanosomes (800nm) throughout the epidermis	Smaller melanosomes (400 nm) in clusters mostly in basal and suprabasal location
Resistance of melanosomes to degradation by lysosomal enzymes	Higher	Lower
Transfer of melanosomes	Melanosomes transferred to keratinocytes individuallly	Melanosomes transferred to keratinocytes as complexes
pH inside melanosomes	Neutral	Acidic
Tyrosinase activity	Lower activity of tyrosinase	Higher activity of tyrosinase
Dopaquinone cyclization	Higher rate	Lower rate
CD-quinone cyclization	Lower rate	Higher rate
Apoptotic cell removal	Higher and more efficient	Less efficient
Result of UV exposure	More pronounced tanning response	Usually more prominent burning response
Penetration of UVL to upper dermis	Significantly lower compared to lighter skin	Significantly higher compared to dark skin
DNA damage after UVL exposure	Less	Higher

tients in a study from Lagos, Nigeria [35, 36]. Two studies performed in Singapore 9 years apart found an incidence of idiopathic photodermatoses from 0.014 to 0.059% [37, 38].

To date, there aren't enough studies on photodermatoses in dark skins to be able to delineate any differences in the pattern and incidence of photodermatoses attributed to skin colour or phototype.

10.4 Relative Frequency of Photodermatoses

The studies currently available in the literature on photodermatoses in dark versus light skin types are summarized in Table 10.2. In general, photodermatoses have accounted for between 12.3 and 22.9% of total dermatologic out patients in different studies and in darker skins, PMLE is the most common idiopathic acquired photodermatosis followed by CAD while in lighter skins, most commonly seen photodermatoses are PMLE (though less common than dark skins) and actinic prurigo [34, 37–42]. In our experience on Indian patients, PMLE accounts for more than half of all photodermatoses followed by CAD, collagen vascular disease and photoaggravated atopic dermatitis [41].

10.5 Polymorphous Light Eruption

PMLE accounts for the majority of patients with idiopathic photodermatoses in both light and dark skinned patients. PMLE presents in a polymorphous fashion with many morphologic variants including papular, papulovesicular, plaque, eczematous, erythema multiforme-like, prurigo-like, and urticarial lesions.

As the name suggests, there are many clinical variants including papular, papulovesicular (Fig. 10.1), plaque-like, vesiculobullous, insect bite-like, erythema multiforme-like, prurigo-like, erythematous-edematous and urticarial lesions (Fig. 10.2). Traditionally it is reported to affect females slightly more often than males and highest incidence is in 2nd and 3rd decades of life. Photoexposed sites like neck, dorsae of forearms and hands are most commonly involved [45, 46] (Fig. 10.3).

Table 10.2 Relative frequency of photodermatoses in various studies

Author/year	City, Country	Inclusion criteria	Race/ phototype	No. of patients	Idiopathic (%)	PMLE (%)	CAD (%)	AP (%)	SU (%)	HV (%)	Phototoxicity (%)	Drug induced (%)	Photo allergic CD (%)	Porphyria (%)	Photo Aggravated dermatoses other than CTD (%)	CTD (%)	Others (%)
Kerr et al 2007 [35]	Detroit/ USA	Photosensitive disorders only (not photoaggravated)	African-American	135	80.7	67.4	11.1	–	2.2	0	13.3	NR	0.7	0.7	NR	NR	4.4
			Caucasian	110	56.2	41.1	7.1	–	8	0	10.7	NR	2.6	21.4	NR	NR	7.27
Wong et al. 2005 [37]	Singapore	Phototested	NR	141	49	25	14	4	6	0	NR	13	4	NR	23	NR	–
Khoo et al. 1996 [38]	Singapore	Phototested	NR	152	27	13	5	4	5	0	NR	11	3	NR	32	1.7	–
Stratigos et al. 2003 [43]	Greece			310	47	30.6	4.83	0.96	8.38	0.3	NR	4.5	5.2	4.5	27	3.2	8.1
Crouch et al. 2003 [44]	Australia			397	53	29.72	9.57	4.53	9.82	0	0.5	6.8	1.7	0	29.4	2.51	19.8
Olumide et al. 1987 [36]	Nigeria			64	3.1	3.1	–	–	–	–	–	–	–	–	–	–	96.9
	New York	Phototested	NR	203	47	26	17	0	4	0	NR	7	8	NR	NR	NR	NR
Wadhwani, et al. [41]	Delhi, India		Fitzpatrick IV–V	362	73.5	59.7	13.8	0	0	0	0	2.49	–	0.2	13.81	7.7	2.2
Nakamura et al. [42]	Michigan, USA	Photosensitive disorders only (not photoaggravated)	African-American	138		86.2	2.9	0	0.7	0	–	0.7	0	0	NR	NR	9.4
			Caucasian	63		54	1.6	0	1.6	0	–	15.9	1.6	7.9	NR	NR	17.4

PMLE Polymorphous light eruption, *CAD* Chronic actinic dermatitis, *AP* actinic prurigo. *SU* solar urticaria, *HV* hydroa vacciniforme, *CD* contact dermatitis, *CTD* connective tissue disease, *NR* not recorded

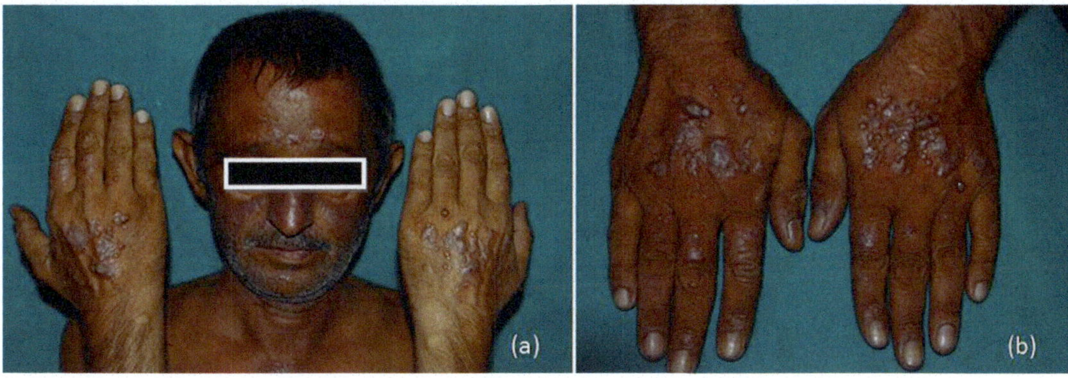

Fig. 10.1 28- year old male (skin type IV) with papulovesicular polymorphous light eruption (PMLE) over cheeks and external ear

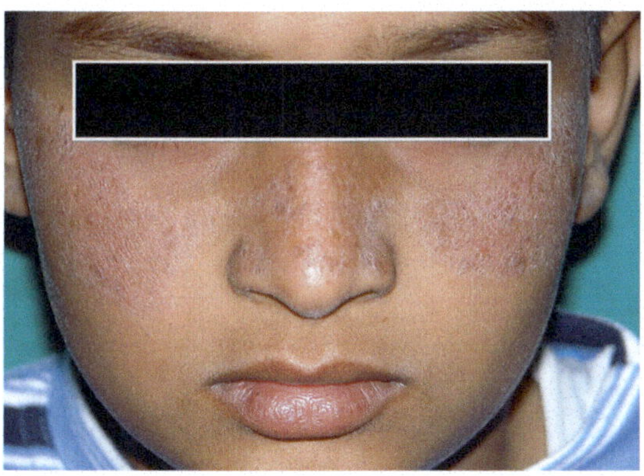

Fig. 10.2 10-year old boy (skin type IV) with erythematous scaly plaques in malar area of face with family history of similar lesions (familial PMLE)

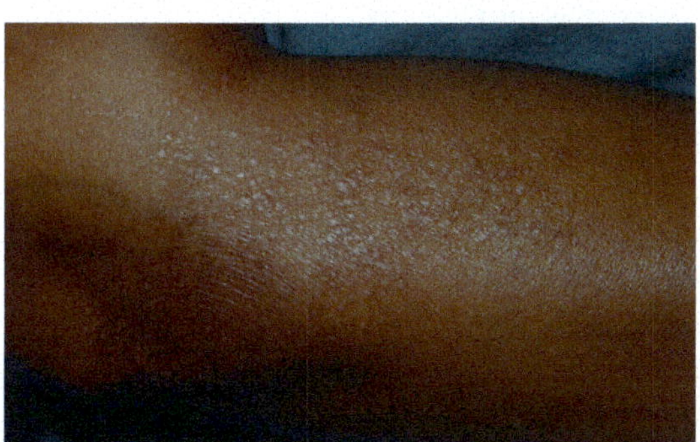

Fig. 10.3 30 year old male (skin type V) with polymorphous light eruption (PMLE). Grouped pinhead-sized shiny papules over the forearm

Peculiarly, darker skin types IV-VI are often found to demonstrate a unique pin-point papular variant. Presentation is in the form of multiple pin point reddish to skin colored papules over the sun exposed areas. An Indian study first reportedthe presentation with tiny lichenoid grouped papules by the name of summertime actinic lichenoid eruption (SALE) [47]. This is possibly the same entity as the Japanese reported entity of micropapular light eruption [48] or the more recently termed pinpoint papular variant of PMLE described in African Americans [49]. In an Asian study, this variant has been found to account for nearly 30% of all PMLE and in our experience too, nearly one third of PMLE patients have such a morphology [41]. An Asian study found this variant to account for nearly one third of cases [50]. We have also conducted a clinicopathologic study and defined an entity called photosensitive spongiotic/ lichenoid eruption of micropapules and plaques (PSLEMP) or photosensitive spongiotic/ lichenoid eruption (PSLE) which has lesions of varied clinical morphologies but demonstrating spongiotic or lichenoid tissue reaction on histopathology [51]. Other authors have also demonstrated both spongiosis and lichenoid tissue reaction in addition to other histologic changes including RBC extravasation and parakeratosis [52].

Another name to this entity has been given by an Italian group as benign summer light eruption (BSLE) which is characterized by preponderance of women, shorter latency, lack of involvement of the face and absence of relapse during summer [53]. PMLE is uncommonly reported even from Hispanic populations [54]. Seasonal variation is common in India with an exacerbation noted in March and September [55].

10.6 Chronic Actinic Dermatitis

Chronic actinic dermatitis (CAD) occurs due to a delayed type hypersensitivity to an altered cutaneous antigen and some authors have also included in this spectrum, a CAD-like photosensitive eruption secondary to airborne allergens. It is histologically characterized by changes of chronic dermatitis occasionally associated with mitotic figures and lymphocytes with large hyperchromatic nuclei. It is clinically characterized by a persistent eczematous eruption affecting predominantly the photoexposed skin with histopathology of chronic eczema with or without lymphoma like changes with reduced MED to UVA, UVB and/or visible light [56–59]. The spectrum of radiation which is responsible for the photosensitivity in CAD has been found to be both UVA and UVB in nearly 2/3rds of patients whereas it is only UVA in around 1/4th of patients. It is most commonly a disease of elderly with mean age of onset in a study on Caucasians being 62.7 years [60].

A recent retrospective study on environmental dermatoses in native Ladakhis, lowlanders and tourists conducted in the high altitude region of Ladakh, India, found that majority of environment induced dermatoses were induced by ultraviolet light including sun burn, melasma, PMLE, CAD and actinic cheilitis. Among these, melasma, CAD and actinic cheilitis occurred much more commonly in native Ladakhis compared to lowlanders and tourists [61].

CAD-like picture may occur as a response to allergic contact dermatitis as patch and photopatch tests have been found positive to a number of allergens (including sesquiterpene lactone mix, fragrance compounds, colophony and rubber chemicals) in patients with a clinical picture of chronic actinic dermatitis [62]. In darker skinned patients of CAD, positivity to parthenium and PPD was commonly seen on patch testing [59, 63].

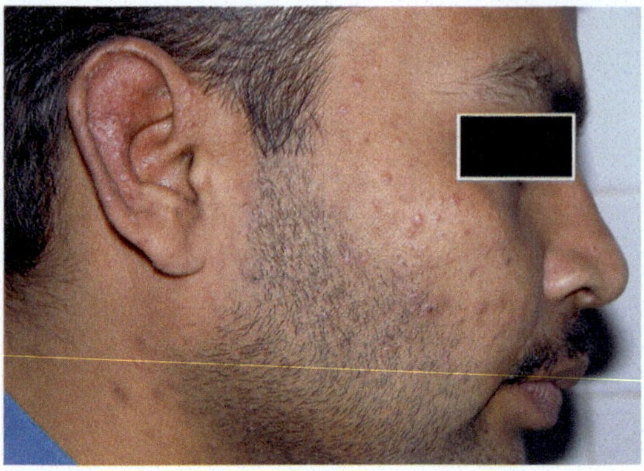

Fig. 10.4 45 year old Indian farmer (skin type V) with chronic actinic dermatitis. (**a**) Lichenified plaques on face with (**b**) prurigo-like lesions on the dorsae of hands

Conversely, when studying patients with contact dermatitis to hair dye a CAD-like clinical presentation was one of the clinical patterns seen in 2 out of the 80 patients [64]. In a cohort of Parthenium dermatitis patients from India, one of the authors has found progression from a classical airborne contact dermatitis pattern to a mixed and CAD like pattern of clinical involvement in more than half of the patients over a mean follow up of 4.2 years. Additionally, photopatch testing was found positive to parthenium in 6 out of 19 patients tested [65]. A recent review on chronic actinic dermatitis also mentions that patients with CAD in the United Kingdom also often demonstrate positive patch testing to relevant Compositae, possibly resulting from exposure to such plants during gardening [66].

In our experience, chronic actinic dermatitis is 2nd most common idiopathic photodermatosis following only PMLE. Our patients have a relatively earlier onset at a mean age of around 44 years and most commonly present with lichenified plaques on photoexposed sites, with prurigo-like lesions occurring commonly in around 20% of cases. (Figs. 10.2 and 10.4) Though studies in West have reported a relatively good prognosis for CAD patients with spontaneous resolution occurring in almost half of the patients over 15 years of follow up, we find CAD to be a chronic and persistent dermatosis [67].

Another study has looked into the effect of skin type on long term prognosis with patients with skin types I and II having higher likelihood of resolution or improvement while patients with photoype IV reported no change [68].

10.7 Actinic Prurigo

Actinic prurigo is a chronic photodermatosis which has many racial and ethnic predispositions, eg. Latin American mesitzos, North American Indians/Alaskan natives and ethnic Chinese. HLA subtypes that predispose to AP include HLA-A24 and HLA-CW4 [37, 69–71]. In the study from Singapore, later age of onset and lack of association with actinic cheilitis may point to an prurigo like variant of CAD. In our experience, actinic prurigo is rare in Indians but we commonly encounter actinic prurigo-like lesions in patients with CAD. Adult onset actinic prurigo (mean age: 36.86 years) in type IV and V skin has been described from Thailand [72]. We have recently reported a case of a 28-year old Indian woman with allergic contact dermatitis to Parthenium presenting as actinic prurigo [73].

10.8 Solar Urticaria

Solar urticaria is more commonly reported in fair skinned populations and we did not encounter any patient of solar urticaria in our cohort. In data from Singapore, patients had a higher age of onset, were more often male and reacted positively to visible light in majority of cases [37]. This was in contrast to data from Europe and Americas [74]. In another cohort of patients with urticaria recruited from our center, 3 out of 515 patients were found to have solar urticaria [75].

10.9 Hydroavacciniforme

Hydroa vacciniforme did not feature in the three large case series from Singapore, India and Nigeria. A single report of a patient from South Africa presenting with crusts and vacciniform scars in sun-exposed areas was possibly hydroa vacciniforme [76]. Recently, there have been two reports in dark-skinned children, one from Morocco and the other in a Malay child from Singapore where clinical features were not different from that described in the White population [77, 78].

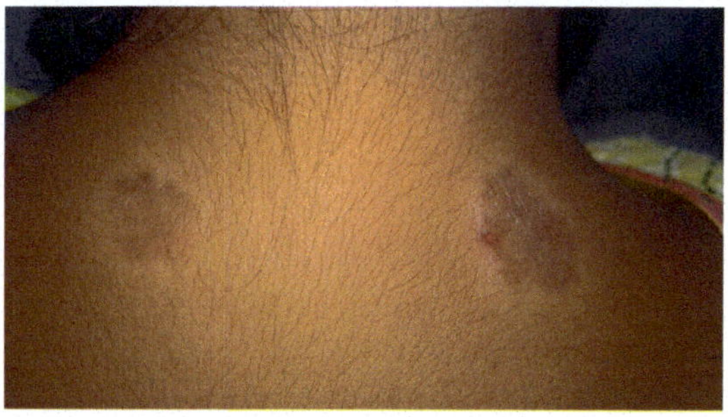

Fig. 10.5 30 year old lady (skin type V) with actinic lichen planus. Well defined lichenoid plaques with peripheral rim of hypopigmentation over the upper back

10.10 Actinic Lichen Planus

Actinic lichen planus has been almost exclusively reported in dark skinned young patients from tropical areas including Middle East, Africa and India [79, 80]. It however is not reported from West Africa [81].The classical presentation is in the form of annular hyperpigmented macules with a halo of hypopigmentation occurring over sun exposed sites, classically over forehead. (Fig. 10.5) We have observed it to occur in 8 out of our 364 patients of photodermatoses whereas among a cohort of lichen planus cases from our institute, actinic lichen planus accounted for around 20% of all cases [41, 82].

10.11 Phytophotodermatitis

Phytophotodermatitis is a UV-induced phototoxic reaction to plants containing allergens most commonly psoralens followed by exposure to UVA light. It is reported mainly from the tropics in the form of sun burn like picture presenting as blisters following sun exposure [83]. In darker patients, the only sign of phytophotodermatitis may be the residual hyperpigmentation without any previous signs or symptoms of sunburn.

10.12 Photosensitive Nutritional Dermatoses

A nutritional disorder presenting with dermatitic or hyperpigmented lesions in a photodistributed location is pellagra secondary to deficiency of niacin. Though it is mainly of historic importance when it occurred in 3rd world countries where millets were the principal sereal in the diet, we have not observed any case in our series [84, 85]. In the current age, it may occur in patients with immunosuppression or those on antituberculous therapy. Was commonly observed in third world countries with high incidence of malnutrition particularly where millet or maize was the principal cereal in the diet. Both systemic and occupational exposure to pyridoxine (vitamin B6) has been known to cause photosensitivity [86, 87].

10.13 Drug-Induced Photosensitivity

Certain topically or systemically administered drugs or chemicals are known to enhance the photosensitivity of cellular components. These include antibiotics (esp fluoroquinolones), non steroidal anti inflammatory drugs and retinoids and more recently antihypertensives like thiazides and diltiazem. This may manifest in the form of photosensitive drug eruption or photodistributed hyperpigmentation [88–90]. Implicated drugs in our series were NSAIDs, doxycycline, antihypertensives and antibiotics [53]. Caucasians are possibly more prone to develop phototoxic drug reactions compared to African Americans probably because of the protection offered by melanin to darker skins [42].

10.14 HIV and Photosensitivity [91]

Human immunodeficiency virus infected patients demonstrate photosensitivity disorders in a myriad of manifestations. These most commonly include photosensitive lichenoid eruptions and erythroderma. Other manifestations of photosensitivity in these patients include porphyrias, chronic actinic dermatitis and photosensitive granuloma annulare.

Photosensitivity has been reported in around 5% of HIV positive patients. Photodistributed dermatoses including lichenoid disorders and hyperpigmentation occur more commonly among African Americans with low CD4 counts of <50 cells/mL and patients receiving ART especially saquinavir. In fact even on adjusting for CD4 counts and HAART, ethnicity continued to remain a risk factor for photosensitivity [91, 92]. There has been a recently reported and unexplained higher incidence of photodermatitis in HIV positive tribals from Bastar in India [93].

HIV photodermatitis presenting with widespread vitiligo-like depigmentation has been reported in a 60-year old man with AIDS and a CD4 count of who developed photodistributed, depigmented macules and patches with a hyperpigmented border on the photoexposed areas of the dorsal forearms, posterior neck and "V" area of the chest. Though the skin phototype is not mentioned, it appears to be a dark skinned patient from the provided photographs [94].

10.15 Phototesting in the Dark Skins

Traditionally the minimal erythema dose has been defined variously for different phototypes by Fitzpatrick. Phototesting has been carried out in a number of studies in both light and dark skinned patients with photodermatoses. Amongst darker skins, in Korean patients, MED to ultraviolet B to correlate only weakly with skin phototype with only minor increases between different phototypes from type II to type V [95]. Similar results were seen in another study by the same group of investigators. Median MED to NBUVB was found to be 750, 950 and 1075 mJ/cm [2] in Korean patients with skin phototypes III, IV and V respectively [96]. A similar study in patients with skin types III and IV from Bahrain reported a mean MED value of 112.22 ± 32.53 mJ/cm [2] [97]. Phototesting guidelines for skin phototypes IV to VI are not yet in place [98, 99]. In an Indian study even after very high irradiation for up to 45 min MED to UVA could not be elicited in any of the 100 normal individuals studied [100].

In our experience on phototesting patients of CAD, only about half of them have detectable MED to UVA. Other Indian studies have also been able to elicit an MED to UVA in majority of patients with photodermatoses in comparison to controls [41, 59, 101]. Similar data exists from Singapore and MEDs were elicitable in most patients with photodermatoses [37]. We have also studied 101 Indian patients with dermatitis on photoexposed sites and performed patch and photopatch testing and found positivity most commonly to Parthenium hysterophorus which was positive in three (4%) photo-patch and 52 (52%) patch tests. This was followed by positive photo-patch test to other allergens including fragrance mix, balsam of Peru, thiuram mix, Compositae mix and promethazine hydrochloride [102].

10.16 Conclusion

This chapter reviews the differences in composition of dark and light skins and their interaction with ultraviolet light and proposes likely pathogenic mechanisms responsible for the difference in spectrum of photodermatoses in dark and light skinned populations. It also specifically highlights the conditions more commonly or exclusively seen in dark skinned populations like pin point popular type of PMLE and prurigo like lesions in CAD. We believe lichenoid photosensitivity disorders are a specific subtype of photosensitivity dermatoses particularly seen in dark skins and may be included in future classifications.

References

1. LoPiccolo MC, Lim HW (2010) Vitamin D in health and disease. Photodermatol Photoimmunol Photomed 26:224–229
2. Moan J, Juzeniene A (2010) Solar radiation and human health. J Photochem Photobiol B 101:109–110
3. Hawk JLM (1999) Photodermatology. Arnold, London
4. Kim JJ, Lim HW (1999) Evaluation of the photosensitive patient. Semin Cutan Med Surg 18:253–256
5. Millard TP, Hawk JL (2002) Photosensitivity disorders: cause, effect and management. Am J Clin Dermatol 3:239–246
6. Relethford JH (2002) Apportionment of global human genetic diversity based on craniometrics and skin colour. Am J Phys Anthropol 118:393–398
7. Relethford JH (1997) Hemispheric difference in human skin color. Am J Phys Anthropol 104:449–457
8. Jablonski NG, Chaplin G (2003) The evolution of human skin coloration. J Hum Evol 39:57–106
9. Sturm RA, Box NF, Ramsay M (1998) Human pigmentation genetics: the difference is only skin deep. BioEssays 20(9):712–721
10. Ito S, Wakamatsu K (2003) Quantitative analysis of eumelanin and pheomelanin in humans, mice, and other animals: a comparative review. Pigment Cell Res 16:523–531
11. Szabo G, Gerald AB, Pathak MA, Fitzpatrick TB (1969) Racial differences in the fate of melanosomes in human epidermis. Nature 222:1081–1082
12. Brenner M, Hearing VJ (2008) The protective role of melanin against UV damage in human skin. Photochem Photobiol 84:539–549
13. Parra EJ 2007 Human pigmentation variation: evolution, genetic basis, and implications for public health. Am J Phys Anthropol Suppl 45:85–105
14. Chedekel MR, Smith SK, Post PW, Vessell DL (1978) Photodestruction of pheomelanin: role of oxygen. Proc Natl Acad Sci U S A 75:5395–5399
15. Sarna T, Menon IA, Sealy RC (1984) Photoinduced oxygen consumption in melanin systems-II. Action spectra and quantum yields for pheomelanins. Photochem Photobiol 39:805–809
16. Ye T, Hong L, Garguilo J, Pawlak A, Edwards GS, Nemanich RJ et al (2006) Photoionization thresholds of melanins obtained from free electron laser-photoelectron emission microscopy, femtosecond transient absorption spectroscopy and electron paramagnetic resonance measurements of oxygen photoconsumption. Photochem Photobiol 82:733–737
17. Kadekaro AL, Kavanagh RJ, Wakamatsu K, Ito S, Pipitone MA, Abdel-Malek ZA (2003) Cutaneous photobiology. The melanocyte vs. the sun: who will win the final round? Pigment Cell Res 16:434–447
18. Krutmann J (2000) Ultraviolet A radiation-induced biological effects in human skin: relevance for photoaging and photodermatosis. J Dermatol Sci 23(Suppl 1):S22–S26
19. Dalle Carbonare M, Pathak MA (1992) Skin photosensitizing agents and the role of reactive oxygen species in photoaging. J Photochem Photobiol B 14:105–124
20. Halliday GM, Byrne SN, Damian DL (2011) Ultraviolet A radiation: its role in immunosuppression and carcinogenesis. Semin Cutan Med Surg 30:214–221

21. Kollias N, Sayre RM, Zeise L, Chedekel MR (1991) Photoprotection by melanin. J Photochem Photobiol B 9:135–160
22. Fitzpatrick TB (1975) Soleil et peau. J de Medecine Esthetique 2:33–34
23. Sachdeva S (2009) Fitzpatrick skin typing: applications in dermatology. Indian J Dermatol Venereol Leprol 75:93–96
24. Fitzpatrick TB (1988) The validity and practicality of sun-reactive skin types I through VI. Arch Dermatol 124:869–871
25. Chung JH, Koh WS, Youn JI (1994) Relevance of skin phototyping to a Korean population. Clin Exp Dermatol 19:476–478
26. Stanford DG, Georgouras KE, Sullivan EA, Greenoak GE (1996) Skin phototyping in Asian Australians. Aust J Dermatol 37(Suppl. 1):S36–S38
27. Ho BK, Robinson JK (2015) Color bar tool for skin type self-identification: a cross-sectional study. J Am Acad Dermatol 73(2):312–3.e1
28. Halder RM, Bang KM (1988) Skin cancer in blacks in the United States. Dermatol Clin 6:397–405
29. Gloster HM Jr, Neal K (2006) Skin cancer in skin of color. J Am Acad Dermatol 55:741–760. 761–744
30. Kaidbey KH, Agin PP, Sayre RM, Kligman AM (1979) Photoprotection by melanin-a comparison of black and Caucasian skin. J Am Acad Dermatol 27:787–788
31. Kiprono SK, Masenga JE, Chaula BM, Naafs B (2012) Skin flora: differences between people affected by Albinism and those with normally pigmented skin in Northern Tanzania – cross sectional study. BMC Dermatol 12:12
32. Deng D, Hang Y, Chen H, Li H (2006) Prevalence of photodermatosis in four regions at different altitudes in Yunnan province. China J Dermatol 33:537–540
33. Dawe RS (2009) Prevalences of chronic photodermatoses in Scotland. Photodermatol Photoimmunol Photomed 25:59–60
34. Shibeshi D (2000) Pattern of skin diseases at the university teaching hospital, Addis Ababa. Ethiopia Int J Dermatol 39:822–825
35. Kerr HA, Lim HW (2007) Photodermatoses in African Americans: a retrospective analysis of 135 patients over a 7-year period. J Am Acad Dermatol 57:638–643
36. Olumide YM (1987) Photodermatoses in Lagos. Int J Dermatol 26:295–299
37. Wong SN, Khoo LS (2005) Analysis of photodermatoses seen in a predominantly Asian population at a photodermatology clinic in Singapore. Photodermatol Photoimmunol Photomed 21:40–44
38. Khoo SW, Tay YK, Tham SN (1996) Photodermatoses in a Singapore skin referral centre. Clin Exp Dermatol 21:263–268
39. Adams PC, Reboussin DM, Barton JC, McLaren CE, Eckfeldt JH, McLaren GD et al (2005) Hemochromatosis and iron-overload screening in a racially diverse population. N Engl J Med 352:1769–1778
40. Meola T, Sanchez M, Lim HW, Buchness MR, Soter NA (1997) Chronic actinic dermatitis associated with human immunodeficiency virus infection. Br J Dermatol 137:431–436
41. Wadhwani AR, Sharma VK, Ramam M, Khaitan BK (2013) A clinical study of the spectrum of photodermatoses in dark-skinned populations. Clin Exp Dermatol 38(8):823–829
42. Nakamura M, Henderson M, Jacobsen G, Lim HW (2014) Comparison of photodermatoses in African-Americans and Caucasians: a follow-up study. Photodermatol Photoimmunol Photomed 30(5):231–236
43. Stratigos AJ, Antoniou C, Papathanakou E, Daboudi M, Tranaka K, Tsara K, Katsambas AD (2003) Spectrum of idiopathic photodermatoses in a Mediterranean country. Int J Dermatol 42:449–454
44. Crouch RB, Foley PA, Baker CS (2003) Analysis of patients with suspected photosensitivity referred for investigation to an Australian photodermatology clinic. J Am Acad Dermatol 48:714–720
45. Tutrone WD, Spann CT, Scheinfeld N, Deleo VA (2003) Polymorphic light eruption. Dermatol Ther 16:28–39
46. Hawk JLM, Norris PG (1999) Abnormal responses to ultraviolet radiation: idiopathic. In: Fitzpatrick's dermatology in general medicine, 5th edn. McGraw-Hill, New York, pp 1573–1589
47. Bedi TR (1978) Summertime actinic lichenoid eruption. Dermatologica 157:115–125
48. Horio T, Danno K, Furukawa F et al (1986) Micropapular light eruption. Nippon Hifuka Gakkai Zashi 96:519–522
49. Kontos AP, Cusack CA, Chaffins M, Lim HW (2002) Polymorphous light eruption in African Americans: pinpoint papular variant. Photodermatol Photoimmunol Photomed 18:303–306
50. Chiam LY, Chong WS (2009) Pinpoint papular polymorphous light eruption in Asian skin: a variant in darker-skinned individuals. Photodermatol Photoimmunol Photomed 25:71–74
51. Shah N (2007) A study of clinicopathological features of photosensitive lichenoid eruption. MD Dermatology thesis, All India Institute of Medical Sciences, New Delhi, November 2007
52. Bansal I, Kerr H, Janiga JJ, Qureshi HS, Chaffins M, Lim HW, Ormsby A (2006) Pinpoint papular variant of polymorphous light eruption: clinical andpathological correlation. J Eur Acad Dermatol Venereol 20:406–410
53. Guarrera M, Cardo P, Rebora AE, Schena D, Calzavara-Pinton P, Venturini M et al (2011) Polymorphous light eruption and benign summer light eruption in Italy. Photodermatol Photoimmunol Photomed 27:35–39
54. Lew R, Jacob J (2014) Polymorphous light eruption: a common skin disease uncommonly recognized in the Hispanic population. Oxf Med Case Rep 2014(8):145–147

55. Sharma L, Basnet A (2008) A clinicoepidemiological study of polymorphic light eruption. Indian J Dermatol Venereol Leprol 74:15–17
56. Hawk JL (2004) Chronic actinic dermatitis. Photodermatol Photoimmunol Photomed 20:312–314
57. Sidiropoulos M, Deonizio J, Martinez-Escala ME, Gerami P, Guitart J (2014) Chronic actinic dermatitis/actinic reticuloid: a clinicopathologic and immunohistochemical analysis of 37 cases. Am J Dermatopathol 36(11):875–881
58. Hawk JL, Magnus IA (1979) Chronic actinic dermatitis- an idiopathic photosensitivity syndrome including actinic reticuloid and photosensitive eczema [proceedings]. Br J Dermatol 101(Suppl 17):24
59. Somani VK (2005) Chronic actinic dermatitis–a study of clinical features. Indian J Dermatol Venereol Leprol 71:409–413
60. Lim HW, Morison WL, Kamide R, Buchness MR, Harris R, Soter NA (1994) Chronic actinic dermatitis. An analysis of 51 patients evaluated in the United States and Japan. Arch Dermatol 130:1284–1289
61. Singh G, Chatterjee M, Grewal R, Verma R (2013) Incidence and care of environmental dermatoses in the high-altitude region of Ladakh, India. Indian J Dermatol 58(2):107–112
62. Menagé H, Ross JS, Norris PG, Hawk JL, White IR (1995) Contact and photocontact sensitization in chronic actinic dermatitis: sesquiterpene lactone mix is an important allergen. Br J Dermatol 132:543–547
63. Kar HK, Langar S, Arora TC, Sharma P, Raina A, Bhardwaj M (2009) Occurrence of plant sensitivity among patients of photodermatoses: a control-matched study of 156 cases from New Delhi. Indian J Dermatol Venereol Leprol 75:483–487
64. Gupta M, Mahajan VK, Mehta KS, Chauhan PS (2015) Hair dye dermatitis and p-phenylenediamine contact sensitivity: a preliminary report. Indian Dermatol Online J 6(4):241–246
65. Sharma VK, Sethuraman G, Bhat R (2005) Evolution of clinical pattern of parthenium dermatitis: a study of 74 cases. Contact Dermatitis 53:84–88
66. Paek SY, Lim HW (2014) Chronic actinic dermatitis. Dermatol Clin 32(3):355–361
67. Dawe RS, Crombie IK, Ferguson J (2000) The natural history of chronic actinic dermatitis. Arch Dermatol 136(10):1215–1220
68. Wolverton JE, Soter NA, Cohen DE (2014) The natural history of chronic actinic dermatitis: an analysis at a single institution in the United States. Dermatitis 25(1):27–31
69. Hojyo-Tomoka MT, Vega-Memije ME, Cortes-Franco R, Domínguez-Soto L (2003) Diagnosis and treatment of actinic prurigo. Dermatol Ther 16:40–44
70. Kryatova MS, Okoye GA (2016) Dermatology in the North American Indian/Alaska native population. Int J Dermatol 55(2):125–134
71. Suárez A, Valbuena MC, Rey M, de Porras Quintana L (2006) Association of HLA subtype DRB10407 in Colombian patients with actinic prurigo. Photodermatol Photoimmunol Photomed 22:55–58
72. Akaraphanth R, Sindhavananda J, Gritiyarangsan P (2007) Adult-onset actinic prurigo in Thailand. Photodermatol Photoimmunol Photomed 23:234–237
73. Singh S, Khandpur S, Sharma VK (2015) Allergic contact dermatitis to Parthenium Hysterophorus mimicking actinic prurigo. Indian J Dermatol Venereol Leprol 81(1):82–84
74. Ryckaert S, Roelandts R (1998) Solar urticaria – a report of 25 casesand difficulties in phototesting. Arch Dermatol 134:71–74
75. Sharma VK, Kumar U (2008) Urticaria: AIIMS approach and experience. In: Sharma VK (ed) Handbook of Urticaria, 1st edn. Massey Art Press, New Delhi, pp 82–87
76. Jacyk WK, Moosa Y (2010) Crusts and vacciniform scars on sun-exposed skin. Clin Exp Dermatol 35:97–98
77. Awatef K, Zahra MF (2016) Hydroa vacciniforme on a dark skin with mucosal involvement. Pan Afr Med J 23:71
78. Chee JN, Koh MJ, Liew HM (2015) Progressive scarring facial lesions in a boy. Clin Case Rep 4(2):120–122
79. (1962) Katzenellenbogen: Lichen planus actinicus (Lichen planus in subtropical countries). Dermatologica 124:10
80. Dostrovsky A, Sagher F (1949) Lichen planus in subtropical countries. Arch Dermatol Syphilol 59:308
81. Alabi GO, Akinsanya JB (1981) Lichen planus in tropical Africa. Trop Geogr Med 33:143–147
82. Kachhawa D, Kachhawa V, Kalla G, Gupta LP (1995) A clinico-aetiological profile of 375 cases of lichen planus. Indian J Dermatol Venereol Leprol 61:276–279
83. Domínguez-Soto L, Hojyo-Tomoka MT, Vega-Memije E, Cortés-Franco R, Waxtein L, Guevara E (1999) Photodermatoses in tropical countries. Clin Dermatol 17:237–243
84. Karthikeyan K, Thappa DM (2002) Pellagra and skin. Int J Dermatol 41:476–481
85. Wan P, Moat S, Anstey A (2011) Pellagra: a review with emphasis on photosensitivity. Br J Dermatol 164(6):1188–1200
86. Bajaj AK, Rastogi S, Misra A, Misra K, Bajaj S (2001) Occupational and systemic contact dermatitis with photosensitivity due to vitamin B6. Contact Dermatitis 44(3):184
87. Morimoto K, Kawada A, Hiruma M, Ishibashi A (1996) Photosensitivity from pyridoxine hydrochloride (vitamin B6). J Am Acad Dermatol 35(2 Pt 2):304–305
88. Kubo Y, Fukumoto D, Ishigami T, Hida Y, Arase S (2010) Diltiazem-associated photodistributed hyperpigmentation: report of two Japanese cases and published work review. J Dermatol 37:807–811

89. Masuoka E, Bito T, Shimizu H, Nishigori C (2011) Dysfunction of melanocytes in photoleukomelanoderma following photosensitivity caused by hydrochlorothiazide. Photodermatol Photoimmunol Photomed 27:328–330
90. Desai N, Alexis AF, DeLeo VA (2010) Facial hyperpigmentation caused by diltiazem hydrochloride. Cutis 86:82–84
91. Bilu D, Mamelak AJ, Nguyen RH, Queiroz PC, Kowalski J, Morison WL et al (2004) Clinical and epidemiologic characterization of photosensitivity in HIV-positive individuals. Photodermatol Photoimmunol Photomed 20(4):175–183
92. Gregory N, Deleo VA (1994) Clinical manifestations of photosensitivityin patients with human immunodeficiency virus infection. Arch Dermatol 130:630–633
93. Singh H, Singh P, Tiwari P, Dey V, Dulhani N, Singh A (2009) Dermatological manifestations in HIV-infected patients at a tertiary care hospital in a tribal (Bastar) region of Chhattisgarh, India. Indian J Dermatol 54:338–341
94. Philips RC, Motaparthi K, Krishnan B, Hsu S (2012) HIV photodermatitis presenting with widespread vitiligo-like depigmentation. Dermatol Online J 18(1):6
95. Youn JI, Oh JK, Kim BK, Suh DH, Chung JH, Oh SJ, Kim JJ, Kang SH (1997) Relationship between skin phototype and MED in Korean, brown skin. Photodermatol Photoimmunol Photomed 13:208–211
96. Youn JI, Park JY, Jo SJ, Rim JH, Choe YB (2003) Assessment of the usefulness of skin phototype and skin color as the parameter of cutaneous narrow band UVB sensitivity in psoriasis patients. Photodermatol Photoimmunol Photomed 19(5):261–264
97. Venkataram MN, Haitham AA (2003) Correlating skin phototype and minimum erythema dose in Arab skin. Int J Dermatol 42:191–192
98. Farr PM, Dawe RS (2007) Phototesting. In: Lim HW, Hoenigsman H, Hawk JLM (eds) Photodermatology. Informa Health Care, New York, pp 433–440
99. Faurschou A, Wulf HC. European dermatology guideline for the photodermatoses. Phototesting. [Online]. Available from: URL: http://www.euroderm.org/content/guidelines.htm
100. Mehta RV, Shenoi SD, Balachandran C, Pai S (2004) Minimal erythema response (MED) to solar simulated irradiation in normal Indian skin. Indian J Dermatol Venereol Leprol 70:277–279
101. Bejoy P, Srinivas CR, Shenoi SD (1998) Phototesting in the idiopathic photodermatoses among Indians. Indian J Dermatol 43:1–3
102. Sharma VK, Bhari N, Wadhwani AR, Bhatia R (2016) Photo-patch and patch tests in patients with dermatitis over the photo-exposed areas: a study of 101 cases from a tertiary care centre in India. Australas J Dermatol 10

11 Psycho-Neuro-Endocrine-Immunology: A Psychobiological Concept

Katlein França and Torello M. Lotti

Abstract

Psycho-Neuro-Endocrine-Immunology (P.N.E.I.) is a scientific field of study that investigates the link between bidirectional communications among the nervous system, the endocrine system, and the immune system and the correlations of this cross-talk with physical health. The P.N.E.I. innovative medical approach represents a paradigm shift from a strictly biomedical view of health and disease taken as hermetically sealed compartments to a more interdisciplinary one. The key element of P.N.E.I. approach is represented by the concept of bidirectional cross-talk between the psychoneuroendocrine and immune systems. The Low Dose Medicine is one of the most promising approaches able to allow the researchers to design innovative therapeutic strategies for the treatment of skin diseases based on the rebalance of the immune response.

Keywords

Psycho-Neuro-Endocrine- Immunology • Low dose medicine • Homeostatic equilibrium • Psychodermatology

K. França
Centro Studi per la Ricerca Multidisciplinare e Rigenerativa, Universita Degli Studi "G. Marconi", Rome, Italy

Institute for Bioethics & Health Policy; Department of Dermatology & Cutaneous Surgery; Department of Psychiatry & Behavioral Sciences, University of Miami Miller School of Medicine, Miami, FL, USA

T.M. Lotti (✉)
Centro Studi per la Ricerca Multidisciplinare e Rigenerativa, Universita Degli Studi "G. Marconi", Rome, Italy
e-mail: professor@torellolotti.it

S. Ahmad (ed.), *Ultraviolet Light in Human Health, Diseases and Environment*, Advances in Experimental Medicine and Biology 996, https://doi.org/10.1007/978-3-319-56017-5_11

11.1 Introduction

In the field of experimental psychology applied to medicine, the second half of the 80′ is characterized by the theorization and divulgation, due to Dr. Robert Ader and colleagues of the principles of Psycho-Neuro-Endocrine-Immunology (P.N.E.I.). This is a scientific field of study that investigates the link between bidirectional communications among the nervous system, the endocrine system, and the immune system and the correlations of this cross-talk with physical health.

The sentence "For this is the great error of our day that the physicians separate the soul from the body" (Hippocrates, sixth century B.C.) clearly represents the *primum movens* of Dr. Ader's studies against the traditional scotomized medical view which is described, for example, by the assertion that the immune system is autonomous, with its self-regulatory and functions separate and independent from the rest of the body.

Dr. Ader's initial research (in the 1970s) on the conditioning of the immune system by psychosocial factors become a cornerstone for studies that described the vast communications network among immune cells, hormones and neurotransmitters; Ader's early observations were also confirmed by Ader himself and other researchers at Harvard University during the 1980s.

Importantly, Dr. Ader's work devoted to the postulation and the development of the new science of P.N.E.I., these old views become less legitimate and, nowadays, the P.N.E.I. concepts guide the scientific community to a unified vision of the biological functions of the body [1–4]. The P.N.E.I. innovative medical approach represents a paradigm shift from a strictly biomedical view of health and disease taken as hermetically sealed compartments to a more interdisciplinary one. After years of ostracism and diffidence, mind-body interactions are now well recognized, deeply studied in the medical literature and taught at most important medical schools.

In 1983 Dr. Ader wrote: "Converging data from a variety of disciplines suggest that the immune system is integrated with other physiological systems and, like all such systems operating in the interests of homeostasis, is sensitive to regulation or modulation by the brain; thus, the immune system stands as a potential mediator of a variety of psychophysiologic effects" [5]. The concept of cross-talk between P.N.E.I. system components and the pivotal role of immune system clearly appear in this sentence.

More than 30 years have passed since Ader's pioneering observations and the P.N.E.I. concept is now well established and accepted, despite of the initial resistances. An example of modern full integrated medical approach to reduce (e.g. in healthcare workers) job-related distress symptoms and adrenocortical activity is represented by Psycho-Neuro-Endocrino-Immunology-based meditation (PNEIMED) an innovative approach that combines the teaching of philosophy and meditation practice of with biomedical analysis from a systemic and integrative perspective [6]. Moreover, from the biochemical point of view, the advances in the fields of molecular biology and physiopathology identified hormones, neuropeptides, cytokines, and growth factors as the signaling molecules involved in both physiological and pathological biological processes, in clear accordance with the principles of P.N.E.I.

11.2 Bidirectional P.N.E.I. Cross-Talk in Dermatology: The Gut-Brain-Skin Axis

The key element of P.N.E.I. approach is represented by the concept of bidirectional cross-talk [7] between the psychoneuroendocrine and immune systems. (Fig. 11.1).

The psychoneuroendocrine system can influence the immune response and, therefore, the capacity of the organism to react against diseases; conversely, the immune system can influence the neuroendocrine functions of the whole body. Such cross-talk among systems is carefully trimmed by feedback loops that simultaneously act in order maintain the homeostatic equilibrium.

This complex interplay is mediated by a wide network of cytokines, hormones, growth factors,

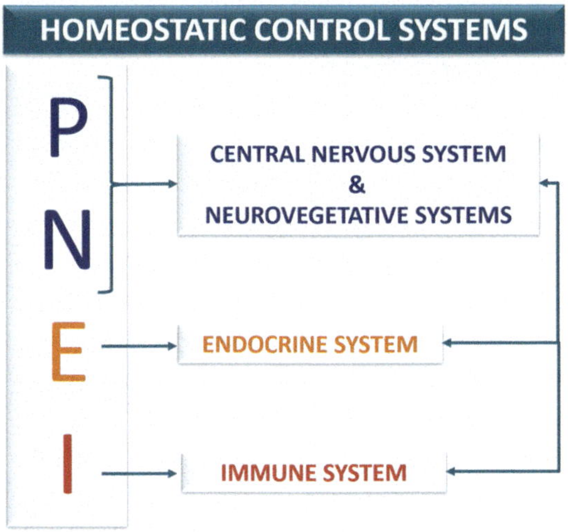

Fig. 11.1 Homeostatic Control Systems and the role of PNEI

neuropeptides and other intermediate molecules collectively named signaling (or messenger) molecules which are the "ABC", the fundamental language of the physiological cross-talk which efficiently regulates cellular responses to both endogenous and exogenous stimuli.

The state of health or disease of a whole body can be depicted by the fluctuations of signaling molecules circulating levels: if the fluctuations are outside the homeostatic range (upper or lower than the physiological limits) we consider this status as a pathologic one.

Gut and skin roles and relations with other organs and tissues are paradigmatic examples of the P.N.E.I. logic. Gut and skin are crucial contact organs through which the mammalian body communicates with the environment. They show some important characteristics in common: they are richly vascularized and innervated and they are also heavily colonized by specific microbial strains [8, 9]. Gut and skin can be considered as complex immune and neuroendocrine organs integrated into the whole immune-endocrine systems and their correct functioning is crucial in order to guarantee the homeostasis and, consequently, the survival of the entire organism [10].

All the P.N.E.I. axes, such as the Gut-Brain Axis and the Gut-Skin Axis, are multi-level networks; they are continuously physiologically modulated by the cellular signaling exchange driven by cytokines, neuro-peptides, neurohormones and other messenger molecules. In physiological conditions, this continuous cross-talk maintains the P.N.E.I. homeostasis of the axes. Recently, the concept of Gut-Brain-Skin Axis has been discussed by Petra Arck and colleagues [11]. In 2009 the researchers observed for the first the connection between the well-known Gut-Brain Axis and Gut-Skin Axis and, through experimental data (*in vivo* mice model), described the effectiveness of a probiotic-based treatment for the reduction of stress-induced neurogenic skin inflammation and hair growth inhibition. These evidences are in line with the observations of John H. Stokes and Donald M. Pillsbury who firstly theorized the gut-brain-skin unifying vision in 1930 [12] (clear example of P.N.E.I. approach application *ante litteram*).

Arck and colleagues validated the unifying model Gut-Brain-Skin Axis in order to highlight the idea that beneficial effects on skin homoeostasis and skin inflammation can be achieved by the assumption of the right kind of probiotics.

The complexity of the Gut-Brain-Skin Axis induces a deep reflection on its regulation, with particular emphasis to the role of the signaling molecule involved in this network, their imbalance at skin level is linked with the majority of inflammation-related and autoimmune skin diseases. (Fig. 11.2).

The Gut-Brain-Skin Axis is a P.N.E.I. microcosm that acts as a homeostatic controller not only of its own systems but the whole organism. Both the intestinal mucosa and the skin have in fact nervous competence (are able to secrete neuropeptides and neurohormones), endocrine (are able to secrete hormones), immune (are able to secrete cytokines) and they are in intimate connection with other organs, systems and apparatuses. By virtue of these interactions it appears evident that the presence of a state of physiological inflammation represents a normal phenomenon both in the intestine and at skin level. The intestinal mucosa and skin are constantly exposed to a heavy antigenic charge mainly represented by bacterial flora. The tolerance of the microbiota is the key physiological inflammation.

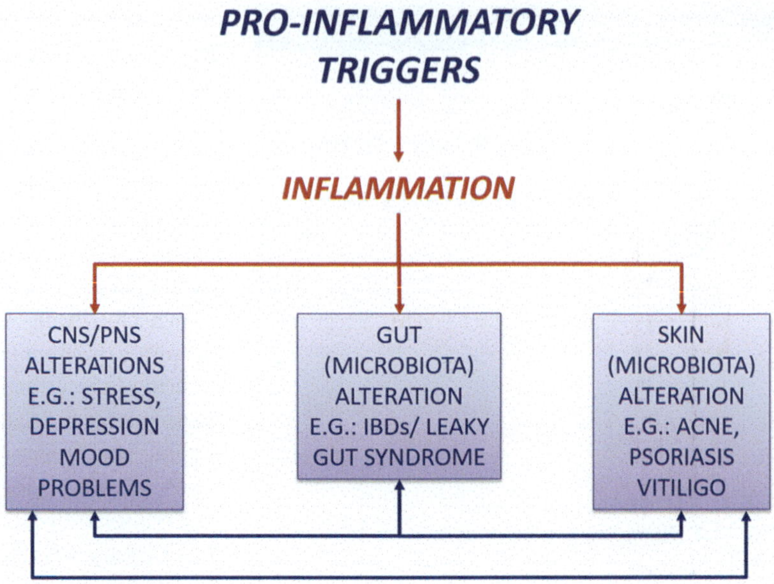

Fig. 11.2 Pro Inflammatory Triggers and the Inflammation process

These P.N.E.I. concepts also offered the opportunity and the tools to study the inflammatory phenomenon in all its complexity and to identify the homeostatic mechanisms governing all stages of the inflammatory phenomenon, from the onset to its resolution.

From a P.N.E.I. point of view, inflammation is such an essential physiological process homeostatically controlled in order to trigger it, develop it and turn it off.

The healthy status of an organism coincides with the condition of homeostasis, in which the vital parameters (pH, temperature, glycaemia, and oxygen's partial pressure) are maintained within a precise and defined range and whose deviation up or down is identified with the pathological state. Inflammation is fully embedded in the physiological functions in homeostatic control. There is thus a level of inflammation, falling within the parameters of "normality", defined physiological inflammation [13].

In the intestine, physiological controlled inflammation is necessary for immunological function, as regulatory immune cells are triggered by intestinal microbiota and food constituents in order to regulate pro-inflammatory pathways and maintain the correct immunocompetence.

Phlogogenic events such as epithelial barrier disruptions, sudden changes of microbiota composition, altered immune balance and, finally, homeostatic balance disruption can contribute to disease onset. Physiological inflammation is overcome by a low-intensity chronic inflammatory condition named Low Grade Chronic Inflammation (L.G.C.I.) [14–16].

L.G.C.I. and quali/quantitative alterations of the microbiota may contribute to the onset of local diseases characterized by alterations in the permeability of the intestinal mucosa [IBD (Inflammatory Bowel Disease), IBS (Irritable Bowel Syndrome), gluten sensitivity, leaky gut syndrome].

Also systemic diseases such as autistic spectrum disorders, the Anxious-Depressive syndromes, Alzheimer's disease, type II diabetes, obesity, psoriasis, rheumatoid arthritis, BPCO (Bronco-Pulmonary-Chronic-Obstructive pulmonary disease) or the RRI (Recurrent Respiratory Infections) are linked with inflammatory conditions and P.N.E.I. homeostasis alterations. Interestingly also skin microbial changes and loss of physiological immunocompetence are related with some local and systemic diseases such as acne vulgaris, vitiligo and atopic dermatitis [9, 12, 17].

11.3 Alteration of P.N.E.I. Homeostasis, Inflammation and Dermatologic Diseases

Focusing the attention on skin compartment, it is important to remember that skin defense system is composed of three main levels: the skin mechanical barrier, the innate immunity, and the acquired immunity [18, 19]; these levels have specific roles in order to react against external and internal inflammatory triggers. An example of the intercellular cross-talk at cutaneous level is the complex of signaling pathways that regulate the functional interactions between keratinocytes and melanocytes, fundamental for the skin pigmentation. Keratinocytes produce growth factors and other signaling molecules, which can drive melanocytes' migration, differentiation and melanin synthesis. Keratinocytes-melanocytes cross-talk represent a small P.N.E.I. network at epidermal level: the psycho-neuro component is guarantee by the embryologic origin of melanocytes which derive from the same embryonic layer that origins some neuronal cell lines, the neural crest [20]. The intercellular cross-talk between keratinocytes and melanocytes is homeostatically regulated condition by growth factors and cytokines of endocrine origin. The immune function is linked with represented by the involvement of melanocytes in the anti-oxidative stress protective mechanisms mediated by keratinocyte-derived b-FGF (basic-Fibroblast Growth Factor). (Fig. 11.3).

These observations highlight the pivotal role of P.N.E.I. homeostatic mechanisms in the maintenance of healthy skin conditions.

An alteration of skin structure (due to infection or mechanical/chemical injuries) and/or the loss of immune skin homeostasis contributes to the pathogenesis of inflammatory skin diseases that are characterized by the breakdown of the homeostatic cross-talk; the role played by the Immune System in the context of the P.N.E.I. network within the "skin system" is crucial for the maintenance of the physiological inflammation.

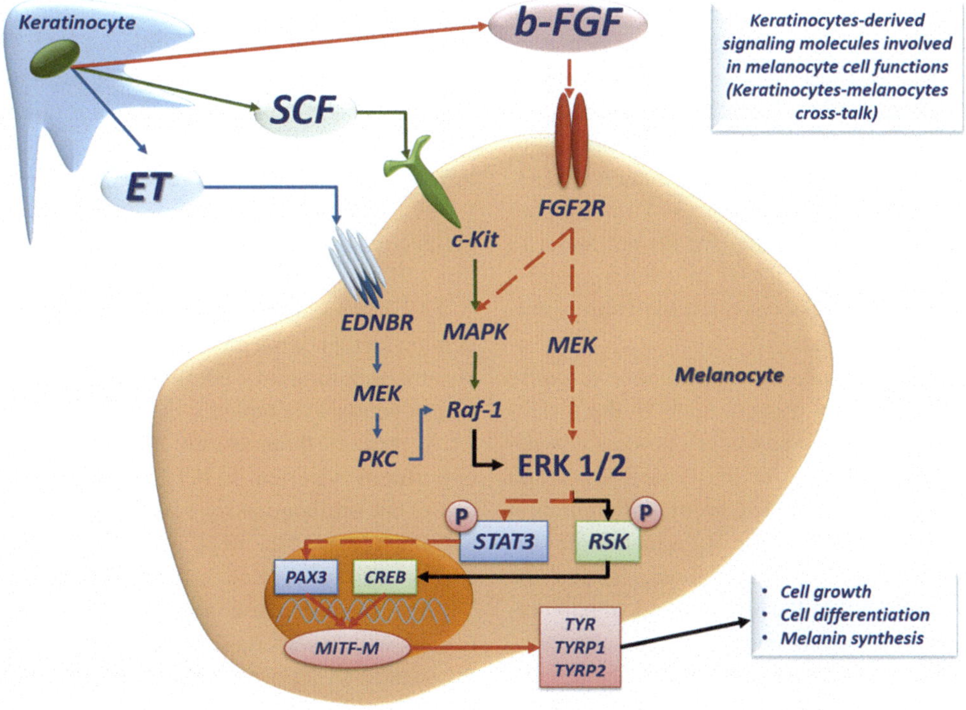

Fig. 11.3 Keratinocytes-Melanocytes cross-talk

The physiological inflammatory process is supported by a panel of Th1-related cytokines, which comprise IL-1, TNF-α and IL-6 that exert their role in a precise chronobiology. Within 72–96 h after the proinflammatory stimulus, the response is managed by a sequence of cytokines activation: and deactivation: IL-1 and TNF-α (primary inflammatory cytokines) induce the production of adhesive molecules, chemokines, growth factors and lipid mediators such as prostaglandins and nitric oxide (NO). These mediators stimulate leukocyte recruitment at the site of inflammation by amplifying the innate immune mechanisms. Then, IL-6 acts as secondary mediator, responsible for maintaining the inflammatory response, and stimulates the production of acute phase proteins in the liver. This chronobiology reflects the temporal scan triggering mechanisms and maintenance of the acute inflammatory phenomenon, which is followed by the progressive decrease in the levels of IL-1 TNF-α and IL-6 and increased levels of IL-10, the most important Th2 anti-inflammatory cytokine, typical of the phase the phenomenon of inflammation resolution [21, 22].

In presence of Low Grade Chronic Inflammation the two phases of inflammation maintenance and resolution coexist. The inflammation is continuously enhanced without an effective *restitutio ad integrum*; the phases of sequential release of cytokines are altered, IL-1, TNF-α and IL-6 levels are about 3–4 times higher than baseline. Contextually, we do not assist to the up-regulation of IL-10 anti-inflammatory. Inflammation persists over time, like a fire smoldering under the ashes.

The persistence of an altered immune response to pro-inflammatory triggers leads to the instauration of a chronic inflammatory process characterized by the absence of the typical signs and symptoms, the Low Grade Chronic Inflammation (LGCI). A relevant number of dermatologic diseases include within the etiologic factors the presence of a shift of the immunological balance, which reflects an imbalance between the cytokines expressed by Th1/Th17 and T_{reg}/Th2 lymphocyte subpopulations [23, 24].

The so-called "Th1/Th2 shift" paradigm is supported by the evidence that Th1 cytokines hyper-production is strictly linked with inflammatory and autoimmune skin diseases such as psoriasis, vitiligo and alopecia areata.

An example of the complexity of the skin P.N.E.I. cross-talk is given by the deep analysis of the inflammatory mechanism at skin level. In 1999 Caroline Robert and Thomas S. Kupper published on *The New England of Medicine* [25] an exhaustive review on immune imbalance related to inflammatory skin diseases. The authors highlighted the fundamental role of T cell-mediated immune surveillance in both physiological and pathological skin conditions pointing out the central role of a class of memory T cells characterized by the presence of the Cutaneous Lymphocyte Antigen (CLA) on their surface and responsible for skin-homing T cell.

CLA-positive T-cells are generated in lymph nodes draining skin and recruited back to the skin during inflammation. The presence of LGCI is a potent trigger for CLA^+ T-cells and their continuous activation is linked with the inappropriate immune surveillance, which characterizes for example psoriasis, allergic contact dermatitis and atopic dermatitis. Also in vitiligo CLA^+ T-cells contribute to the massive death of melanocytes driving the skin-homing (mainly near disappearing melanocytes) of $CD8^+$ T cells at perilesional level. The increased *in situ* presence of a CLA^+/$CD8^+$ T cells is responsible of the destruction of melanocytes with consequent skin depigmentation [26].

LGCI is one of the most important etiopathogenetic factors of the most dramatic dermatologic chronic inflammatory autoimmune diseases and consequently a therapeutic target.

At present, there are no classical therapeutic opportunities to treat LGCI because the chronic use of anti-inflammatory active principles studied for the management of acute phenomena shows an unfavorable efficacy/adverse effects balance; in particular chronic NSAID use is connected with an increased incidence of chronic diseases such as heart failure and hypertension [27].

Since the 1990s, anti-cytokine therapy was proposed and tested for the treatment of inflammatory and autoimmune diseases mainly counteracting the expression of Th1 proinflammatory cytokines such IL-1 and TNF-α. Moreover, the therapeutic use of Th2 cytokines (e.g. IL-10) and specific antibodies was applied for alopecia areata, psoriasis and atopic dermatitis treatment.

However, side effects due to high dosages normally used for these molecules have slow down the development of possible new drugs [28]. The most important and limiting pitfalls connected with the use of high dosage cytokines and other signal molecules are:

- The need of high doses of active molecules in order to reach the therapeutic goal
- The low compliance of systemic administration performed by injective routes.

An innovative approach for the treatment of LGCI based on new therapeutic tools and concepts is need. The Low Dose Medicine (LDM) fulfills these specifications.

11.4 Low Dose Medicine (LDM): Theoretical, Physiological and Biochemical Basis

The Low Dose Medicine (LDM) is an innovative therapeutic approach based on the most advanced knowledge in Molecular Biology, P.N.E.I. and research results in the field of low dose pharmacology.

LDM has deep roots within the fundamental P.N.E.I. principles resumed in the centrality of the human being as a whole mind-body entity. Each patient is considered as a unique identity; this assumption guides the study of a specific therapeutic approach for a particular disease.

The primary outcome of the LDM approach is the restoration/preservation of the homeostatic equilibrium; the oral administration of the appropriate biological signaling molecules, which are selected after identification of the altered P.N.E.I. networks, is the therapeutic tool that allows reaching the expected outcome.

The use of biological molecules which control and drive the intercellular cross-talk in order to restore the physiological homeostasis is the innovative core of LDM. The main characterizing points of LDM approach are:

- Oral administration of signaling molecules
- Systemic and synergistic activity of the orally administered molecules
- Accurate modulating action of specific signaling pathways exerted by the orally administered molecules.

The most representative aspect of LDM is the significative efficacy of orally administered low-dose signaling molecules. From a biochemical point of view, cytokines, hormones, neuropeptides and growth factors are oligo-peptides and small protein sequences. Oligo-peptides and small fragments of proteins reach the intestinal tract and here exert their biological actions [29].

The proposed action mechanism for orally delivered signaling molecules involves the intestinal M cells which act as a carrier of signaling to T cells in Peyer's patches lymph nodes [30]. The interaction between delivered signaling molecules and M cell is the key event that underlay the effectiveness of this administration route [31–33].

The critical issue of oral administration route is the low bioavailability (within 1 and 2%) of signaling molecules and oligo-peptides in general [34]; to overcome this critical pitfall an effective drug delivery system is needed.

The SKA (Sequential Kinetic Activation) technology, codified and standardized by GUNA S.p.a. -Italy-, makes the low doses of signaling molecules able to be active even below the minimum dose classically considered as effective. SKA technology also permits low dose molecules to be, as effective as classic recombinant peptides administered at higher concentrations overcoming the high dose-related adverse effects.

The action mechanism of SKA low dose signaling molecules takes advantage of the activation of some units of cellular (or plasmatic) receptors by virtue of their low concentration. Low dose SKA signaling molecules are adminis-

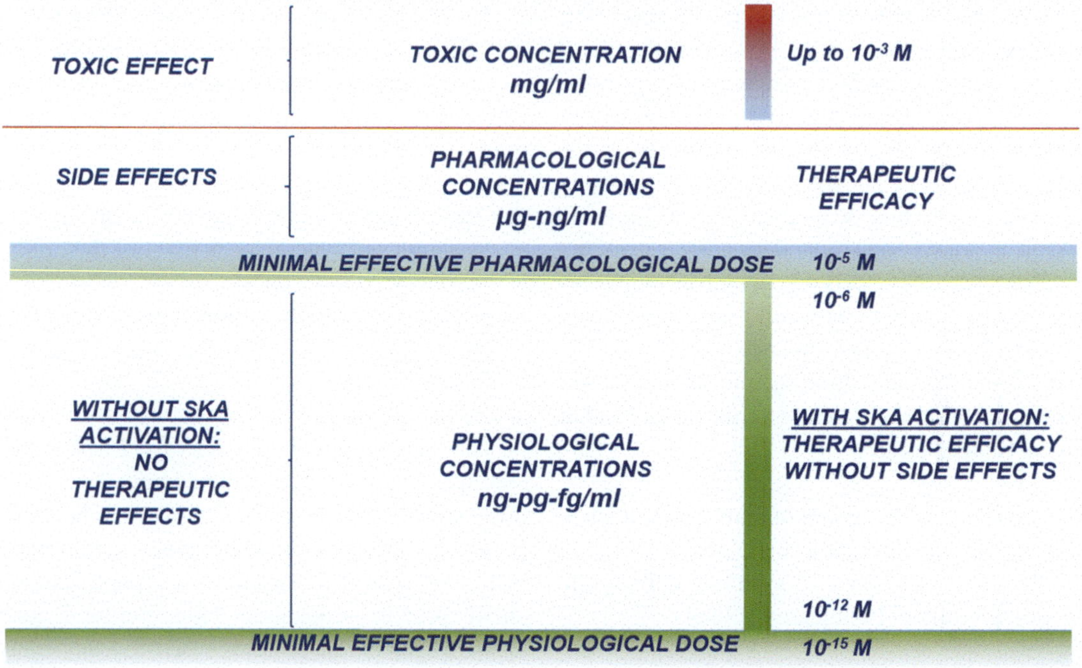

Fig. 11.4 SKA activation and therapeutic efficacy

tered in agreement with their physiological working levels [between 10^{-6} (micromolar) for hormones [35] and 10^{-12} (picomolar) for other messenger molecules [36]] (Fig. 11.4).

A demonstration of the effectiveness of low doses of signaling molecules is possible referring to the effects demonstrable in accordance with Arndt-Schultz experiments [37, 38]. W. H. Hauss observed ad explained the effects of Arndt-Schultz's Law on "mesenchymal non-specific reaction" [39]. Recently, the research of Edward J. Calabrese on the phenomenon called "hormesis", that means "different behavior of the same substance at different doses" [37, 40] further elucidated Arndt-Schultz observations. Furthermore, the pharmacological importance of low doses is recognized from both regulatory bureaus and industries in the pharmacological field [41–43].

The biological response due to low doses of signaling molecules is also linked with the characteristic of the specific ligand/receptor binding. Receptors for both class-I and class-II helical signaling molecules [44, 45] undergo ligand-induced receptor homo- or hetero-oligomerization laws [46], which can explain the dose-dependent mixed agonist/antagonist activity of some cytokines and oligo-peptides [46, 47], a characteristic linked with low dose-response [41].

The peculiar ligand/receptor interactions exerted by low dose SKA molecules induces the activation and fine regulation of a great number of intercellular signaling pathways, contributing to the restoration and/or protection of the biological function of the entire P.N.E.I. network. Low dose SKA molecules are able to activate (or reactivate) the P.N.E.I. self-tuning intra- and intercellular pathways representing the innovative and highly effective tools of LDM [48–56].

11.5 Low Dose Medicine and Skin Diseases: Preclinical Studies

As previously described, the skin diseases etiology is complex and the alteration of both innate and adaptive immune responses occupies a relevant role in both diseases' onset and maintenance.

From a biological point of view, the importance of LGCI in both psoriasis onset and progression and the efficacy of Low Dose SKA molecules in the reduction of its negative impact were recently evaluated by Barygina V and colleagues.

A panel of *in vitro* experiments was performed on fibroblasts obtained from lesional skin of psoriatic patients [54] evaluating the oxidative stress level as marker of an inflammatory condition.

Extracellular Reactive Oxygen Species (ROS), over-expressed by fibroblasts, exert a pro-inflammatory action in psoriatic lesional skin; the effectiveness of low doses SKA interleukin-4; 10, basic-Fibroblast Growth factor and β-Endorphin (IL-4, IL-10, bFGF, and b-End – 10 fg/ml) in reduction of ROS production by lesional fibroblasts highlight one of the possible LDM medicines' action mechanism against LGCI, crucial etiologic component of psoriasis onset and progression.

Barygina V. and colleagues also designed and performed a basic preclinical *in vitro* study [55] in order to evaluate the effects of low dose SKA IL-4, IL-10, b-FGF, and β-End (10 fg/ml) in the modulation of intra- and extra-cellular oxidative stress and on the proliferation of human perilesional keratinocytes (PL) from the skin of Vitiligo patients (in vitro study on cells obtained from lesion skin biopsies). Vitiligo, a highly psychologically disabling skin disorder characterized by a progressive depigmentation, is another example of a dermatologic disease characterized by the presence of LGCI and related excessive oxidative stress.

Obtained results showed that low dose SKA IL-4, IL-10, and b-FGF are effective significantly reducing the intra-cellular oxidative stress rates. Furthermore, low dose SKA IL-4 and b-FGF are also able to reduce the extra-cellular oxidative stress.

Low dose SKA IL-10, b-FGF, and β-End induce a significative increase of keratinocytes viability compared to untreated perilesional cells. IL-4, IL-10, β-End, and b-FGF show positive effect on both redox mechanism effectiveness and cell viability without interferes with keratinocytes cell cycle.

11.6 Low Dose Medicine and Skin Diseases: Clinical Results

In 2014, the first study conducted on a dermatologic disease (psoriasis vulgaris) in order to test the LDM approach with the oral administration of low dose SKA activated cytokines was published. Roberti ML. ad colleagues designed and performed a multicenter double-blind placebo-controlled clinical study [52] in order to test the efficacy of low dose SKA interleukin-4, -10 and -11 (IL-4; IL-10; IL-11 at the concentration of 10 fg/ml) for the therapy of psoriasis vulgaris.

The main outcomes chosen for the evaluation of the treatment with low dose SKA interleukins were

- Presence and extension of psoriatic plaques evaluated in agreement with PASI (Psoriasis Area Severity Index) scale.
- Improvement of the quality of life parameters evaluated in agreement with DLQI (Dermatology Life Quality Index) rating scales.

The results revealed the efficacy (and safety) of oral administered low dose SKA interleukins in the reduction of both evaluated scores. The study also highlighted the long-lasting efficacy of the proposed treatment opening the opportunity to formulate a treatment protocol for psoriasis and other dermatologic chronic diseases characterized by an immune imbalance with the presence of a LGCI status.

In 2015, another interesting study in the field of LDM applied to the treatment of psoriasis was published by Lotti and colleagues [53]. The results of a spontaneous retrospective observational clinical study were collected and evaluated.

The clinical outcomes of the most up-to-date therapeutic approach for psoriasis treatment based on UV-A-1 phototherapy combined with low dose SKA cytokine therapy were evaluated. And revealed that the combination of UV-A-1 phototherapy with laser *plus* low dose SKA interleukin-4 and -10 and low dose SKA antibodies anti IL-1α/β is more effective than UV-A-1 phototherapy alone and also equally safe. The combination of phototherapy and LDM represent an innovative strategy for the treatment of inflammatory skin diseases such as psoriasis vulgaris.

Lotti T. and colleagues also performed a retrospective spontaneous clinical study comparing the effectiveness of current Vitiligo treatments with LDM therapy [56].

In this study, some groups of patients treated in accordance with standard and experimental therapeutic protocols were evaluated; two groups, treated respectively with orally administered low dose SKA IL-4, IL-10, Anti-IL-1 antibodies, and low dose SKA basic- Fibroblast Growth Factor (b-FGF) were evaluated and compared with other groups of patients who received topical treatments with a cortisone cream (alone or in combined associations with both groups of low dose SKA molecules) and phototherapy (narrow-band UV-B radiation) alone or in combined associations with the low dose SKA molecules. Two groups of subjects treated with natural sunlight exposure and systemic oral intake of *G. biloba* extract were evaluated as control groups.

An inclusion criteria applicable for all the current vitiligo treatments is that the skin surface presenting vitiliginous lesion not exceeding the 15% of the total skin surface. The study highlighted that the low dose SKA treatment effectively reduces the depigmented skin areas and brakes the spreading of the vitiliginous lesions, in particular when co-administered with UV-B phototherapy with a significative reduction of the depigmented areas. The effectiveness of the association of low dose SKA treatments with the topical UV-B treatment opens news scenarios for the combined use phototherapy and LDM.

11.7 Conclusions

Many dermatologic diseases have a complex pathogenesis; the inflammatory phenomena is one of the most important etiological component, it is driven by the imbalance between Th1/Th17 and Th2/T_{reg} responses and induces a profound alteration in immune response homeostasis. The consequent disruption of P.N.E.I. equilibrium has not only local but also systemic negative outcomes that compromise the whole body health conditions.

An effective therapeutic action exerting a rebalance action of the immune inflammatory response, not adequately managed with currently available therapies, is needed. Today, the Low Dose Medicine is one of the most promising approaches able to allow the researchers to design innovative therapeutic strategies for the treatment of skin diseases based on the rebalance of the immune response.

The availability of Low Dose SKA signaling molecule is the cardinal point of LDM because the effective and safe oral administration of low dose SKA signaling molecules represents the innovative core of the entire strategy for the treatment of dermatological diseases characterized by an immune imbalance and LGCI such as in psoriasis vulgaris and vitiligo.

Preclinical and clinic results confirm the effectiveness of LDM approach and give to the physician the therapeutic tool and theoretic basis for a fine tuning of the immune system in order to restore its homeostatic equilibrium in accordance with P.N.E.I. principles.

References

1. Commins SP, Borish L, Steinke JW (2010) Immunologic messenger molecules: cytokines, interferons, and chemokines. J Allergy Clin Immunol 125(2 suppl 2):S53–S72
2. Bacchus W, Aubel D, Fussenegger M (2013) Biomedically relevant circuit-design strategies in mammalian synthetic biology. Mol Syst Biol 9:691
3. Ader R, Cohen N (1993) Psychoneuroimmunology: conditioning and stress. Annu Rev Psychol 44:53–85
4. Ader R, Cohen N, Felten D (1995) Psychoneuroimmunology: interactions between the nervous system and the immune system. Lancet 345(8942):99–103
5. Ader R (1983) Developmental psychoneuroimmunology. Dev Psychobiol 16(4):251–267. (1 983)
6. Bottaccioli F, Carosella A, Cardone R, Mambelli M, Cemin M, D'Errico MM, Ponzio E et al (2014) Brief training of psychoneuroendocrinoimmunology-based meditation (PNEIMED) reduces stress symptom ratings and improves control on salivary cortisol secretion under basal and stimulated conditions. Explore (NY) 10(3):170–179

7. Haroon E, Raison CL, Miller AH (2012) Psychoneuroimmunology meets neuropsychopharmacology: translational implications of the impact of inflammation on behavior. Neuropsychopharmacology 37(1):137–162
8. Tojo R, Suárez A, Clemente MG (2014) De los Reyes-Gavilán CG, Margolles A, Gueimonde M, Ruas-Madiedo P. Intestinal microbiota in health and disease: role of bifidobacteria in gut homeostasis. World J Gastroenterol 20(41):15163–15176
9. Myles IA, Williams KW, Reckhow JD, Jammeh ML, Pincus NB, Sastalla I, Saleem D et al (2016) Transplantation of human skin microbiota in models of atopic dermatitis. JCI Insight 1(10):pii: e86955
10. O'Neill CA, Monteleone G, McLaughlin JT, Paus R (2016) The gut-skin axis in health and disease: a paradigm with therapeutic implications. BioEssays 38:1167–1176
11. Arck P, Handjiski B, Hagen E, Pincus M, Bruenahl C, Bienenstock J, Paus R (2010) Is there a 'gut-brain-skin axis'? Exp Dermatol 19(5):401–405
12. Bowe WP, Logan AC (2011) Acne vulgaris, probiotics and the gut-brain-skin axis – back to the future? Gut Pathog 3(1):1
13. Kotas ME, Medzhitov R (2015) Homeostasis, inflammation, and disease susceptibility. Cell 160(5):816–827
14. Hueston CM, Deak T (2014) The inflamed axis: the interaction between stress, hormones, and the expression of inflammatory-related genes within key structures comprising the hypothalamic-pituitaryadrenal axis. Physiol Behav 124:77–91
15. Khan MJ, Gerasimidis K, Edwards CA, Shaikh MG (2016) Role of gut microbiota in the aetiology of obesity: proposed mechanisms and review of the literature. J Obes 2016:7353642
16. Mancuso P (2016) The role of adipokines in chronic inflammation. Immunotargets Ther 5:47–56
17. Ganju P, Nagpal S, Mohammed MH, Nishal Kumar P, Pandey R, Natarajan VT, Mande SS et al (2016) Microbial community profiling shows dysbiosis in the lesional skin of Vitiligo subjects. Sci Rep 6:18761
18. Turvey SE, Broide DH (2010) Innate immunity. J Allergy Clin Immunol 125(2 Suppl 2):S24–S32
19. Dainichi T, Hanakawa S, Kabashima K (2014) Classification of inflammatory skin diseases: a proposal based on the disorders of the three-layered defense systems, barrier, innate immunity and acquired immunity. J Dermatol Sci 76(2):81–89
20. Rawles ME (1947) Origin of pigment cells from the neural crest in the mouseembryo. Physiol Zool 20:248–266
21. Hussain T, Tan B, Yin Y, Blachier F, Tossou MC, Rahu N (2016) Oxidative stress and inflammation: what polyphenols can do for us? Oxidative Med Cell Longev 2016:7432797
22. Trifunović J, Miller L, Debeljak Ž, Horvat V (2015) Pathologic patterns of interleukin 10 expression--a review. Biochem Med (Zagreb) 25(1):36–48
23. Deng Y, Chang C, Lu Q (2016) The inflammatory response in psoriasis: a comprehensive review. Clin Rev Allergy Immunol 50(3):377–389
24. Zhen Y, Yao L, Zhong S, Song Y, Cui Y, Li S (2016) Enhanced Th1 and Th17 responses in peripheral blood in active non-segmental vitiligo. Arch Dermatol Res 308:703–710
25. Robert C, Kupper TS (1999) Inflammatory skin diseases, T cells, and immune surveillance. N Engl J Med 341(24):1817–1828
26. van den Wijngaard R, Wankowicz-Kalinska A, Le Poole C, Tigges B, Westerhof W, Das P (2000) Local immune response in skin of generalized vitiligo patients. Destruction of melanocytes is associated with the prominent presence of CLA+ T cells at the perilesional site. Lab Investig 80(8):1299–1309
27. Marcum ZA, Hanlon JT (2010) Recognizing the risks of chronic nonsteroidal anti-inflammatory drug use in older adults. Ann Longterm Care 18(9):24–27
28. Dinarello CA (2003) Anti-cytokine therapeutics and infections. Vaccine 21(Suppl 2):S24–S34
29. Tessaro I, Modina SC, Franciosi F, Sivelli G, Terzaghi L, Lodde V et al (2015) Effect of oral administration of low-dose follicle stimulating hormone on hyperandrogenized mice as a model of polycystic ovary syndrome. J Ovarian Res 8(1):64
30. Yun Y, Cho YW, Park K (2013) Nanoparticles for oral delivery: targeted nanoparticles with peptidic ligands for oral protein delivery. Adv Drug Deliv Rev 65(6):822–832
31. Burnett AF, Biju PG, Lui H, Hauer-Jensen M (2013) Oral interleukin 11 as a countermeasure to lethal total-body irradiation in a murine model. Radiat Res 180(6):595–602
32. Hanson ML, Hixon JA, Li W, Felber BK, Anver MR, Stewart CA et al (2014) Delivery of IL-27 recombinant bacteria attenuates immune colitis in mice. Gastroenterology 146(1):210–221
33. Forster K, Goethel A, Chan CW, Zanello G, Streutker C, Croitoru K (2012) An oral CD3-specific antibody suppresses T-cell-induced colitis and alters cytokine responses to T-cell activation in mice. Gastroenterology 143(5):1298–1307
34. Renukuntla J, Vadlapudi AD, Patel A, Boddu SH, Mitra AK (2013) Approaches for enhancing oral bioavailability of peptides and proteins. Int J Pharm 447(1–2):75–93
35. Vandenberg LN, Colborn T, Hayes TB, Heindel JJ, Jacobs DR Jr, Lee DH et al (2012) Hormones and endocrine-disrupting chemicals: low-dose effects and nonmonotonic dose responses. Endocr Rev 33(3):378–455
36. Biancotto A, Wank A, Perl S, Cook W, Olnes MJ, Dagur PH et al (2013) Baseline levels and temporal stability of 27 multiplexed serum cytokine concentrations in healthy subjects. PLoS One 8(12):e76091
37. Amendola A, Cerioli NL, Migliore L (2006) Ormesi: la rivoluzione dose-risposta. APAT, Roma

38. Stumpf WE (2006) The dose makes the medicine. Drug Discov Today 11(11/12):550–555
39. Hauss WH, Hulsing GJ, Gerlach U (1968) Die unspezifische Mesenchymreaktion Zur Pathogenese der reaktiven Mesenchymerkrankungen. Georg Thieme Verlag, Stuttgart, pp 29–35
40. Mattson MP, Calabrese EJ (2010) Hormesis a revolution in biology, toxicology and medicine. Springer, New York/Dordrecht/Heidelberg/London
41. VV AA (1994) Guideline for industry dose-response information to support drug registration. ICH. London
42. Olson HM, Kadyszewski E, Beierschmitt W (2001) Hormesis – a pharmaceutical industry perspective. Crit Rev Toxicol 31(4–5):659–661
43. Maynard KI (2011) Hormesis pervasiveness and its potential implications for pharmaceutical research and development. Dose-Response 9:377–386
44. Krause CD, Pestka S (2005) Evolution of the class 2 cytokines and receptors, and discovery of new friends and relatives. Pharmacol Ther 106(3):299–346
45. Huising MO, Kruiswijk CP, Flik G (2006) Phylogeny and evolution of class-I helical cytokines. J Endocrinol 189(1):1–25
46. Wells JA (1996) Binding in the growth hormone receptor complex. Proc Natl Acad Sci U S A 93(1):1–6
47. Scheller J, Chalaris A, Schmidt-Arras D, Rose-John S (2011) The pro- and anti-inflammatory properties of the cytokine interleukin-6. Biochim Biophys Acta 1813(5):878–888
48. Gariboldi S, Palazzo M, Zanobbio L, Dusio GF, Mauro V, Solimene U et al (2009) Low dose oral administration of cytokines for treatment of allergic asthma. Pulm Pharmacol Ther 22:497–510
49. D'Amico L, Ruffini E, Ferracini R, Roato I (2012) Low dose of IL-12 stimulates T cell response in cultures of PBMCs derived from non-small cell lung cancer patients. J Cancer Ther 3:337–342
50. Cardani D, Dusio GF, Luchini P, Sciarabba M, Solimene U, Rumio C (2013) Oral Administration of Interleukin-10 and anti-IL-1 antibody ameliorates experimental intestinal inflammation. Gastroenterology Research 6:124–133
51. Radice E, Miranda V, Bellone G (2014) Low-doses of sequential-kinetic-activated interferon-gamma enhance the ex vivo cytotoxicity of peripheral blood natural killer cells from patients with early-stage colorectal cancer. A preliminary study intern. Immunopharmacology 19:66–73
52. Roberti ML, Ricottini L, Capponi A, Sclauzero E, Vicenti P, Fiorentini E et al (2014) Immunomodulating treatment with low dose interleukin-4, interleukin-10 and interleukin-11 in psoriasis vulgaris. J Biol Regul Homeost Agents 28:133–139
53. Lotti T (2015) Successful combination treatment for psoriasis with phototherapy and low-dose cytokines: a spontaneous, retrospective observational clinical study. Hautarzt 66(11):849–854
54. Barygina V, Becatti M, Lotti T, Taddei N, Fiorillo C (2016) Low dose cytokines reduce oxidative stress in primary lesional fibroblasts obtained from psoriatic patients. J Dermatol Sci pii:S0923-1811(16)30119-0
55. Barygina V, Becatti M, Lotti T, Moretti S, Taddei N, Fiorillo C (2015) Treatment with low-dose cytokines reduces oxidative-mediated injury in perilesional keratinocytes from vitiligo skin. J Dermatol Sci 79(2):163–170
56. Lotti T, Hercogova J, Wollina U, Chokoeva AA, Zarrab Z, Gianfaldoni S et al (2015) Vitiligo: successful combination treatment based on oral low dose cytokines and different topical treatments. J Biol Regul Homeost Agents 29(1 Suppl):53–58

Part IV

UV Light Benefits to Man

Ultraviolet B Radiation: The Vitamin D Connection

12

Michael F. Holick

Abstract

Vitamin D is known as the sunshine vitamin. During exposure to sunlight the skin transforms 7-dehydrocholesterol into vitamin D_3. Throughout evolution vitamin D_3 has played a pivotal role in the evolution of vertebrates. Vitamin D is not only critically important for bone health but has a multitude of other biologic functions to help reduce chronic illnesses. This Chapter reviews how vitamin D is produced in the skin, factors that affect its production and a prospective on how to obtain vitamin D from sensible sun exposure.

Keywords

Sunlight • Vitamin D • Rickets • UVB radiation • Vitamin D deficiency

12.1 Prehistoric and Historic Perspectives

As life evolved in the ocean it took advantage of the sun's energy to generate a variety of photochemical compounds essential for the evolution of life on earth. Photosynthesis resulted in the generation of carbohydrates to provide energy source for these early life forms. In addition to the production of carbohydrates during exposure to visible radiation early life forms were also producing vitamin D as a result of being exposed to solar ultraviolet radiation. One of the early phytoplankton species, Emiliania huxlei (a coccolithophore which has calcium carbonate containing exoskeleton) which has existed unchanged in the Sargasso Sea (the Atlantic Ocean) for more than 250 million years was found to have a large quantity of the vitamin D_2 precursor ergosterol. When exposed to ultraviolet radiation that it was converted ultimately to vitamin D_2 [1]. Thus the photosynthesis of vitamin D has been occurring throughout evolution in organisms exposed to sunlight. Although the function of ergosterol and vitamin D2 are unknown in these primitive organisms it has been suggested that one of the functions was to act as a natural sunscreen to efficiently absorb

M.F. Holick (✉)
Section Endocrinology, Nutrition and Diabetes, Department of Medicine, Physiology and Biophysics, Boston University Medical Center, Boston, MA 02118, USA
e-mail: mfholick@bu.edu

S. Ahmad (ed.), *Ultraviolet Light in Human Health, Diseases and Environment*, Advances in Experimental Medicine and Biology 996, https://doi.org/10.1007/978-3-319-56017-5_12

ultraviolet B (UVB; 290–320 nm) radiation that was potentially damaging to UV sensitive macromolecules including DNA. Recently it was observed that vitamin D was most effective in prolonging the lifespan of the round worm Caenorhabditis elegans by improving protein homeostasis and slowing the aging process [2].

It was also speculated that early vertebrates including the dinosaurs required vitamin D to maintain a healthy skeletal structure. Their quick demise could've been caused in part by their inability to produce any vitamin D as a result of the asteroid strikingly earth and releasing into the atmosphere from its impact and fires so much debris that it prevented any vitamin D producing radiation from reaching the earth surface. Curiously nocturnal rodents adapted to their sunless environment and did not require vitamin D to survive. Ultimately it was these nocturnal rodents that gave rise to the evolution of hominids and humans [1].

Most vertebrates exposed to sunlight are able to produce vitamin D in their skin with the exception of cats that apparently obtained their vitamin D from their dietary sources. Captive vertebrates including amphibians, reptiles and nonhuman primates are at extremely high risk for vitamin D deficiency and metabolic bone disease due to lack of direct sun exposure [3, 4]. The vitamin D producing ultraviolet radiation is absorbed by glass and thus these animals require dietary vitamin D or exposure to artificial ultraviolet radiation to maintain a healthy skeleton and normal calcium metabolism [5].

For humans the lack of sun exposure as a result of the industrial revolution resulted in a devastating bone disease commonly known as rickets [6, 7]. This disease was recognized in the mid 1600s that caused severe growth retardation and skeletal deformities especially of the lower legs. The first insight for the role of sunlight for bone health was made by Sniadecki in 1822 when he reported that from his experience as a physician he observed that rickets was common in children living in Warsaw Poland and he very seldom observed rickets in children living in the rural areas outside of this industrialized city. He made the association that it was lack of sunlight in the dark and poorly lit streets in Warsaw that was responsible for this devastating skeletal disorder [6–8]. 70 years later Palm reported that from his clinical experience in London and contacts with his colleagues who were in India and China that rickets was extremely common in London and yet his colleagues reported back to him that rickets was uncommon in children living in squalor in India and China. He concluded the only common denominator that could explain this dramatic difference was the fact that the pall of smoke from coal burning in the atmosphere and buildings built in close proximity prevented any sunlight from reaching children who were outside in London. He advocated sunbathing to treat and prevent rickets [7, 9].

It was incomprehensible to the medical community at the time to believe that exposure of the skin to sunlight could have any beneficial effect of the skeleton and the observations by Sniadecki and Palm [7] were dismissed. It wasn't until 1919 when Huldschinsky reported that exposure of children to radiation from a mercury arc lamp was effective in improving mineralization of the skeleton based on the analysis of x-rays before and after the exposure for several weeks. He also reported that exposure of one arm to the ultraviolet radiation was effective in improving the mineralization of the bones in the unexposed arm as it did in the exposed arm. He concluded that as a result of the ultraviolet radiation exposure something was produced in the skin that circulated in the body to have an effect on the skeleton in the arm not exposed to the mercury arc lamp [7, 10, 11].

Two years later Hess and Unger exposed children with rickets to sunlight on the roof of their hospital in New York City and reported significant improvement in their rickets [12]. Thus it was established that exposure to ultraviolet radiation and sunlight were effective in treating and preventing rickets.

However it was also perplexing that rickets in dogs and children could be treated effectively with cod liver oil. Originally it was considered that the vitamin A in cod liver oil had antirachitic activity. However when cod liver oil was heated and exposed to oxygen which destroyed vitamin A activity the antirachitic activity remained intact. This resulted in McCollum calling this new antirachitic factor vitamin D [7, 10, 13, 14].

By the turn of the twentieth century more than 90% of children living in the Netherlands and North Eastern United States had evidence of rickets [7]. A large campaign was mounted worldwide to find the cause and cure for this crippling metabolic bone disease [13]. A study in rodents demonstrated that cod liver oil was as effective as exposure to UV radiation in treating rickets [7]. Thus it was concluded that the antirachitic factor in cod liver oil could be produced in the skin when exposed to sunlight. As a result of these observations Steenbock and Black [15] and Hess and Weinstock [16] began exposing a variety of foods including cotton seed oil, corn oil and milk to ultraviolet radiation demonstrated this process produced the antirachitic factor and was effective in preventing rickets in rodents.

It was also recognized at the same time that yeast exposed to UV radiation resulted in the production of the antirachitic factor. An analysis of the yeast resulted in the identification of the precursor of vitamin D as ergosterol. As a result ergosterol was added to milk as well as wide variety of other foods and drinks followed by exposure to ultraviolet radiation resulting in them having antirachitic activity [6, 7, 17]. It was initially the irradiation of milk containing ergosterol that was effective in preventing rickets in children. Once it was determined that the ergosterol was the precursor of vitamin D it was irradiated and the irradiated product was added to milk to fortify it with the antirachitic factor, vitamin D. This process quickly eliminated rickets as a significant health problem in countries that fortified their milk with vitamin D [6, 7].

At the same time in the early 1930s departments of Health in the United States and UK also advocated sensible sun exposure for the prevention of rickets [7, 10, 13]. Originally it was assumed that the vitamin D produced in human skin during sun exposure was the same as the vitamin D produced in UV irradiated yeast. However it was observed that the vitamin D produced from irradiated yeast was less effective in its antirachitic activity in chickens when compared to vitamin D obtained from the irradiation of pig skin [7, 13, 17]. It was finally demonstrated by Windaus that the precursor of vitamin D in mammalian skin was from the precursor of cholesterol, 7-dehydrocholesterol not ergosterol. The difference between the 2 provitamin D's was a double-blind between C22 and C23 and a methyl group on C24 [7, 13, 17].

12.2 The Photochemistry and Photobiology of Vitamin D

In the 1940s a major effort was undertaken to understand the photochemistry of vitamin D. It was demonstrated that during exposure to UV radiation 7-dehydrocholesterol underwent a ring opening between carbons 9 and 10 to form previtamin D_3. It was found that previtamin D_3 was thermodynamically unstable and rearranged its three double bonds (triene) to form the thermodynamically stable vitamin D_3. It was also observed that continued exposure of previtamin D_3 to UV radiation produced a variety of photoproducts including lumisterol, tachysterol and toxisterols [7, 18].

In the 1980s studies were conducted to understand the photochemistry of vitamin D in human skin (Fig. 12.1) [19–21]. An evaluation of the action spectrum for vitamin D_3 (efficiency of various wavelengths in producing vitamin D_3) in human skin revealed that the wavelengths most effective for producing previtamin D_3 were around 298 nm and that UVA (wavelengths above 315 nm) was ineffective (Fig. 12.2) [7, 19]. It was also observed that during prolonged exposure to UVB radiation (290–315 nm) that previtamin D_3 photo isomerized to lumisterol and tachysterol [21]. This observation revealed that sun exposure regulates the production of previtamin D_3 and that excess exposure does not result in the production of intoxicating amounts of vitamin D_3 [20]. Originally it was thought that melanin pigmentation not only in decreased risk for developing skin cancer but also prevented excessive vitamin D from being produced in the skin [22]. Although the former is true the latter is not since sunlight itself is responsible for regu-

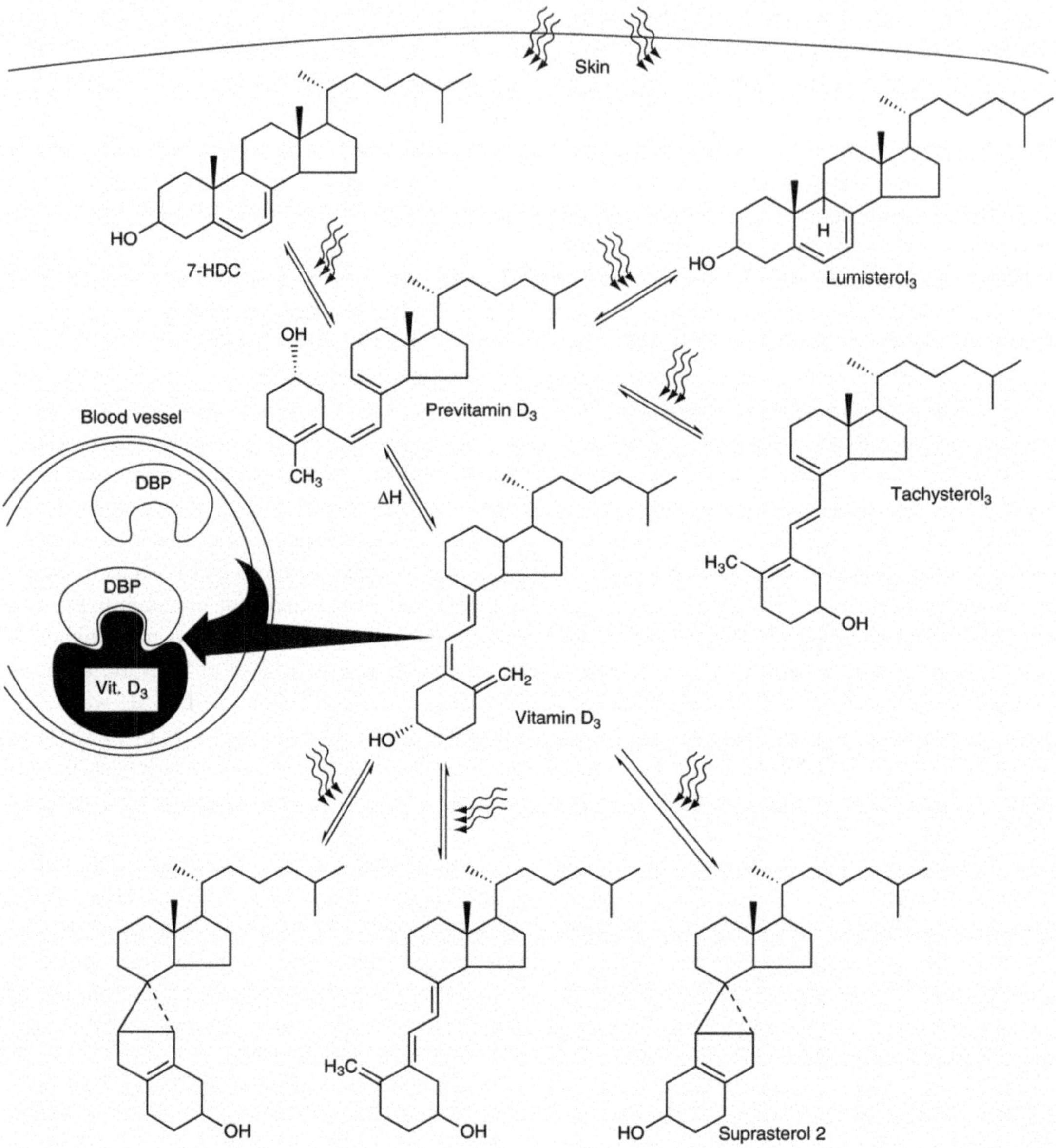

Fig. 12.1 A schematic representation of the photochemical and thermal events that result in the synthesis of vitamin D_3 in the skin, and the photodegradation of previtamin D_3 and vitamin D_3 to biologically inert photoproducts. 7-dehydrocholesterol (7-DHC) in the skin is converted to previtamin D_3 by the action of solar ultraviolet B radiation. Once formed, previtamin D_3 is transformed into vitamin D_3 by a heat-dependent (ΔH) process. Vitamin D_3 exits the skin into the dermal capillary blood system and is bound to a specific vitamin D-binding protein (DBP). When previtamin D_3 and vitamin D_3 are exposed to solar ultraviolet B radiation, they are converted to a variety of photoproducts that have little or no activity on calcium metabolism (Holick, copyright 1995 with permission)

lating the production of vitamin D_3 in the skin. No more than 15% of 7-dehydrocholesterol is converted to previtamin D_3. The continued exposure to sunlight results in the production of lumisterol and tachysterol setting up a photo-equilibrium (Fig. 12.3). Prolonged exposure to UVB radiation will also converted to previtamin D_3 and its photoproducts to other photoproducts known as toxisterols (Fig. 12.4) [7]. These were originally thought to have toxic properties which

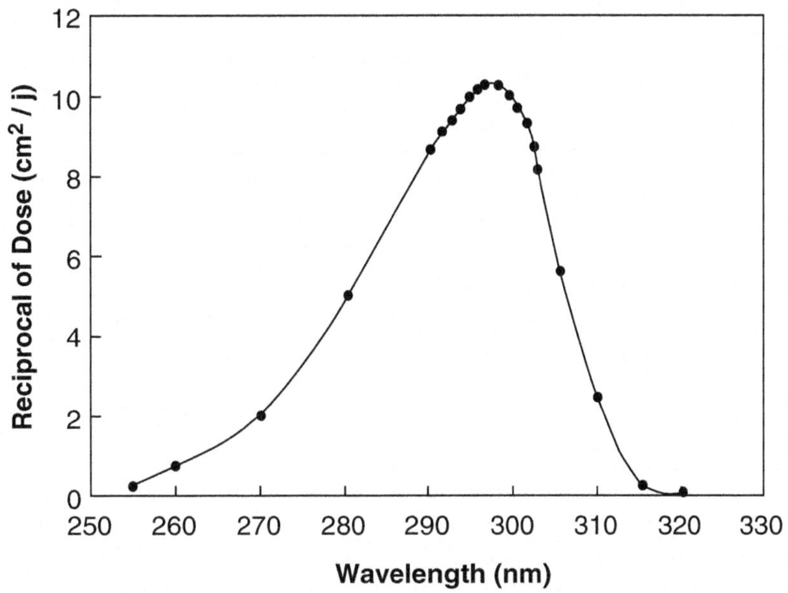

Fig. 12.2 Action spectrum for the conversion of 7-dehydrocholesterol to previtamin D_3 in human skin (Holick, copyright 2007. Reproduced with permission)

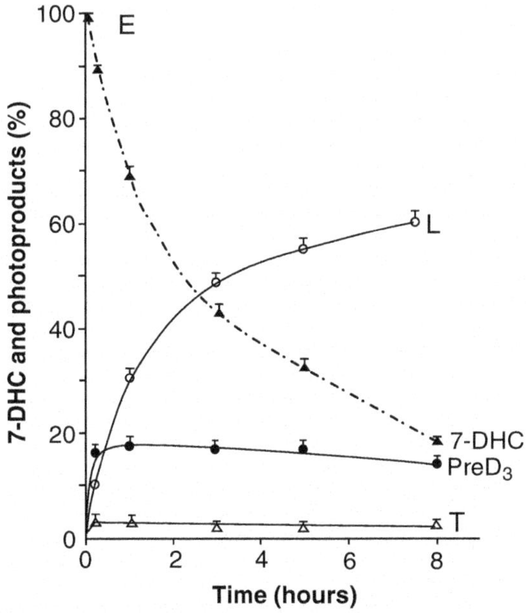

Fig. 12.3 An analysis of the photolysis of 7-dehydrocholesterol (7-DHC) in the basal-cell layer and the appearance of the photoproducts previtamin D_3 (Pre-D_3), lumisterol$_3$ (L), and tachysterol$_3$ (T) with increasing time of exposure to equatorial simulated solar ultra violet radiation. *Bars* above data points show the standard error of the mean of three determinations (Holick, copyright 1981. Reproduced with permission)

is why they were called toxisterols. Vitamin D_3 also will photo isomerized when exposed to ultraviolet B radiation forming suprasterols and 5, 6-trans-vitamin D_3 [7].

The conversion of previtamin D_3 to vitamin D_3 is a temperature-dependent process. At room temperature it takes several days for this process to be completed. Even at body temperature it takes more than a day for most of the previtamin D_3 to be converted to vitamin D_3. However when reptile skin and human skin was exposed to ultraviolet B radiation it was observed that the conversion of previtamin D_3 to vitamin D_3 was approximately tenfold more rapid compared to previtamin D3 isomerizing to vitamin D_3 at the same temperature in an organic solvent [23]. It was quickly determined that this was not due to an enzymatic reaction but rather due to a novel nonenzymatic membrane mediated catalytic mechanism. The 7-dehydrocholesterol being a planar molecule is sandwiched in between the fatty acid hydrocarbon side chain with the 3-hydroxyl oriented to the polar head group of triglyceride in the plasma membrane. When 7-dehydrocholesterol absorbs UVB radiation it undergoes a bond cleavage between carbons 9–10 to form the thermodynamically unstable cis-cis conformer which is maintained within the triglyceride permitting it to rapidly convert to vitamin D_3 and not to its more

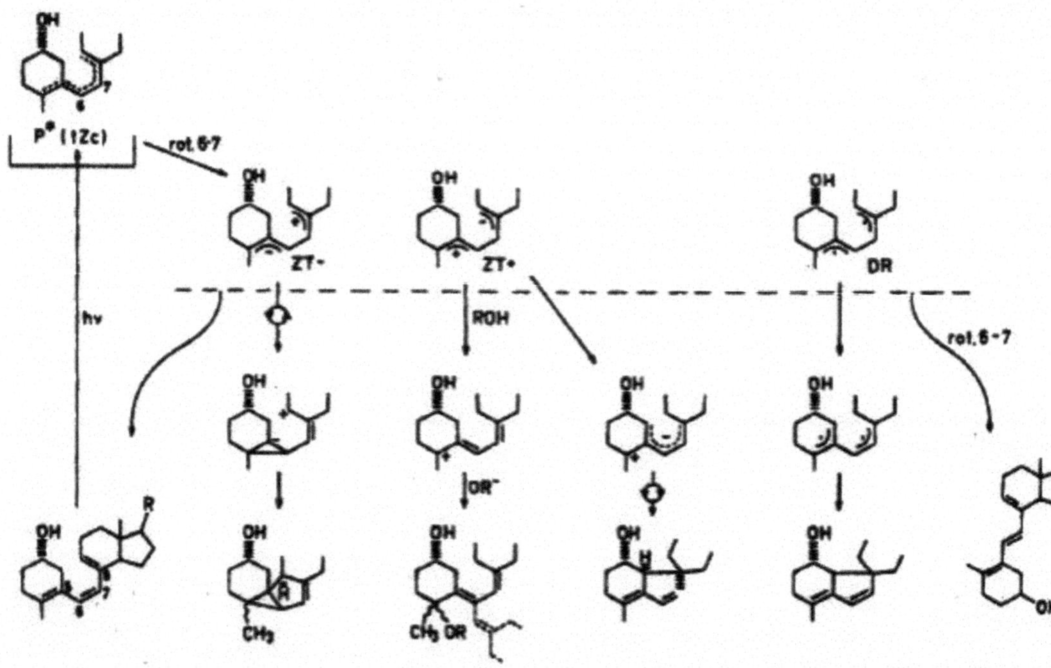

Fig. 12.4 Once previtamin D_3 is formed, it has the ability to rotate around the 6-7 bond. Relaxation via rotation about the 6-7 bond followed by UV irradiation can give rise to a wide variety of toxisterols and tachysterol

thermodynamically stable cis-trans conformer which is not able to isomerize to vitamin D_3 as demonstrated in Fig. 12.5 [7].

12.3 Factors Affecting the Cutaneous Production of Previtamin D3

It had been assumed that one of the driving forces in evolution for dark skinned pigmentation was for the prevention of excessive amounts of vitamin D from being produced that could potentially cause toxicity [22]. However is now recognized that sunlight itself regulates the cutaneous production of vitamin D_3 [7, 20, 24]. Thus melanin pigmentation did not evolve to prevent vitamin D intoxication. However melanin is extremely efficient in absorbing UVB radiation and therefore competes with 7-dehydrocholesterol for solar UVB radiation reducing its conversion to previtamin D_3 [25]. A person with skin type 5 and 6 (never burns always tans) therefore requires a much longer exposure time usually about 5–10 times more compared to a person with skin type 2 (always burns and sometimes tans). This is the explanation for why people of color are at much higher risk for vitamin D deficiency.

Sunscreens mimic melanin and efficiently absorb UVB radiation. A sunscreen with a sun protection factor of 30 applied properly would be expected to absorb approximately 97–98% of the UVB radiation striking the skin and therefore reduces the skin's capacity to produce vitamin D_3 by 97–98% [26].

As stratospheric ozone efficiently absorbs solar UVC (180–290 nm) and a large amount but not all UVB radiation (290–315 nm) is the explanation for why approximately only 1% of the solar UVB radiation ever reaches the earth surface. An increase in the path length by which UVB radiation passes through results in a further decrease in how much UVB radiation reaches the earth surface (Fig. 12.6). This phenomenon explains why time of day, season, latitude, altitude as well as weather conditions have such a dramatic effect on the cutaneous production of vitamin D_3 [7, 27]. The zenith angle of the sun is

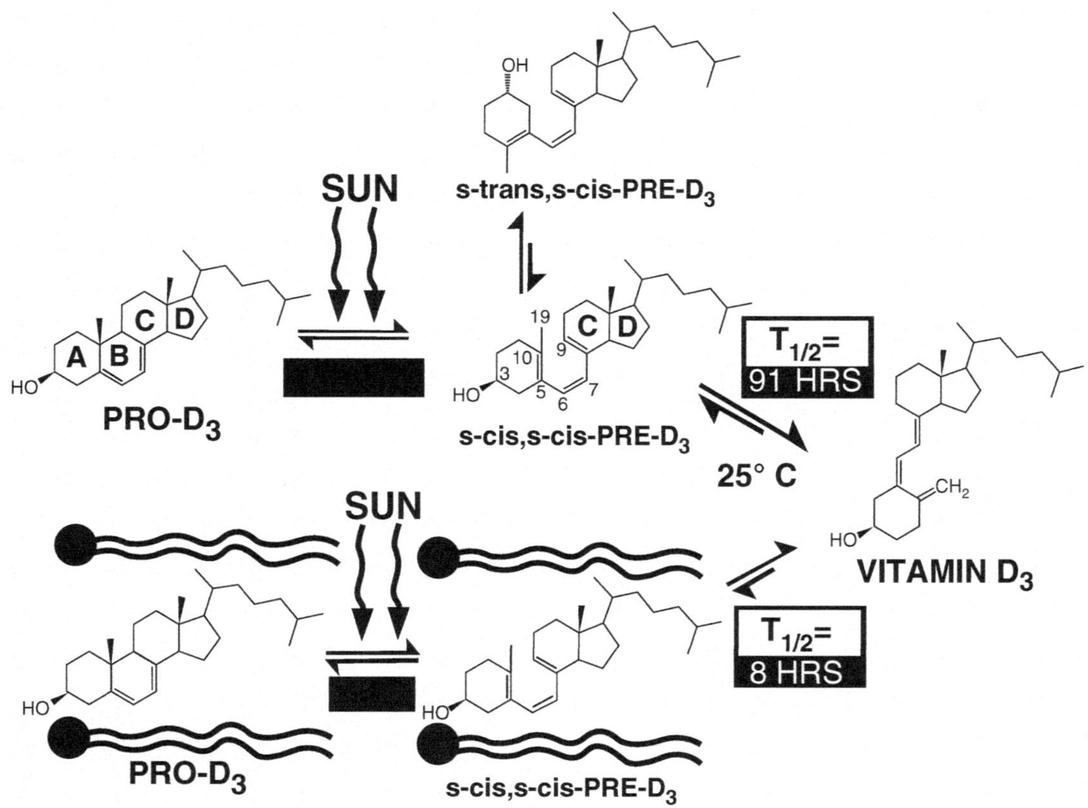

Fig. 12.5 Photolysis of provitamin D_3 (pro-D_3. 7-dehydrocholesterol) into previtamin D_3 (pre-D_3) and its thermal isomerization to vitamin D_3 in hexane and in lizard skin at 25 °C. In hexane pro-D_3 is photolyzed to s-cis,s-cis-pre-D_3. Once formed, this energetically unstable conformation undergoes a conformational change to the s-trans,s-cis-pre-D_3. Only the s-cis,s-cis-pre-D_3 can undergo thermal isomerization to vitamin D_3. The s-cis,s-cis conformer of pre-D_3 is stabilized in the phospholipid bilayer by hydrophilic interactions between the 3ß-hydroxl group and the polar head of the lipids, as well as by the van der Waals interactions between the steroid ring and side-chain structure and the hydrophobic tail of the lipids. These interactions significantly decrease the conversion of the s-cis,s-cis conformer to the s-trans,s-cis conformer, thereby facilitating the thermal isomerization of s-cis,s-cis-pre-D_3 to vitamin D_3 (Holick, copyright 1995. Reproduced with permission)

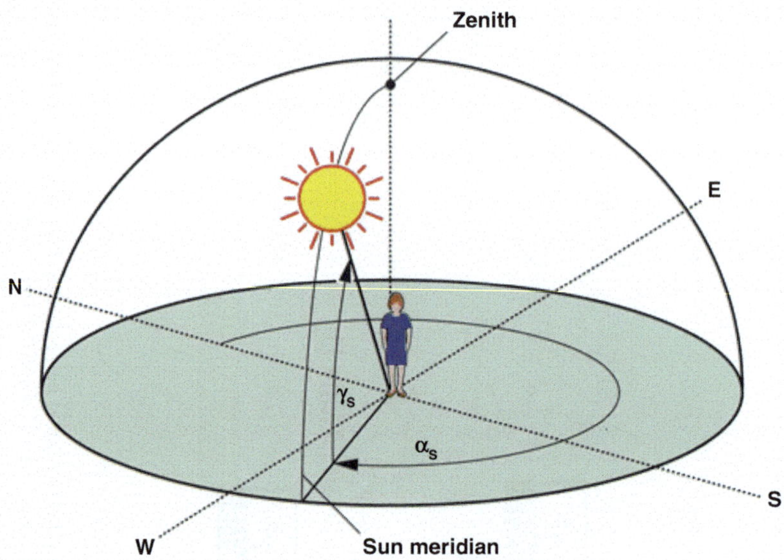

Fig. 12.6 The solar zenith angle is the angle made by the sun's light with respect to the vertical (the sun being directly overhead). This angle is increased at higher latitudes, early morning and late afternoon when the sun is not directly overhead, and during the winter months. As the solar zenith angle increases, the amount of UVB radiation reaching the earth's surface is reduced. Therefore, at higher latitudes, greater distance from the equator, more of the UVB radiation is absorbed by the ozone layer thereby reducing or eliminating the cutaneous production of vitamin D_3 (Holick, copyright 2006. Reproduced with permission)

more oblique in the early morning and late afternoon explaining why very little vitamin D is produced before 9 AM and after 3 PM even in the summertime. The higher latitude for more oblique is the angle of the sun explaining why very little if any vitamin D is produced in the skin during the winter months. In Boston 42° North very little if any vitamin D is produced between November and March. 10° further North in Canada or Europe very little if any vitamin D is produced between October and April (Fig. 12.7) [27]. Clouds as well as air pollution absorb UVB radiation reducing the efficiency of the sun in producing vitamin D3 in the skin. The higher the altitude shorter is the path length and therefore the cutaneous production of vitamin D3 is much more efficient. In Agra, India (169 m) in November at 27° North latitude very little previtamin D3 was produced during sun exposure. An evaluation of previtamin D3 production traveling to base camp of Mt. Everest revealed a gradual increase in the production of previtamin D3 with increasing latitude reaching a maximum of about 400% higher at 5350 m compared to Agra [28].

12.4 The Role of Sunlight and Other Sources of UVB Radiation in Contributing to Vitamin D Status

Once vitamin D_3 is produced in the skin or ingested in the diet and travels to the liver where it is converted to 25-hydroxyvitamin D [25(OH)D] [29, 30]. This is the major circulating form of vitamin D used by doctors to measure a person's vitamin D status. Studies have shown that blood levels of 25(OH)D very with season with a peak blood level occurring at the end of the summer and then nadir at the end of the winter. Hours of sunshine in Denmark was directly associated with blood levels of 25(OH)D [31] (Fig. 12.8). A study of 3.8 million blood samples collected over a two-year period of time in the United States revealed that there was a definite seasonal variation in the circulating blood levels of 25(OH)D [32]. There was also a significant latitudinal effect with blood samples collected in the southern United States having higher circulating

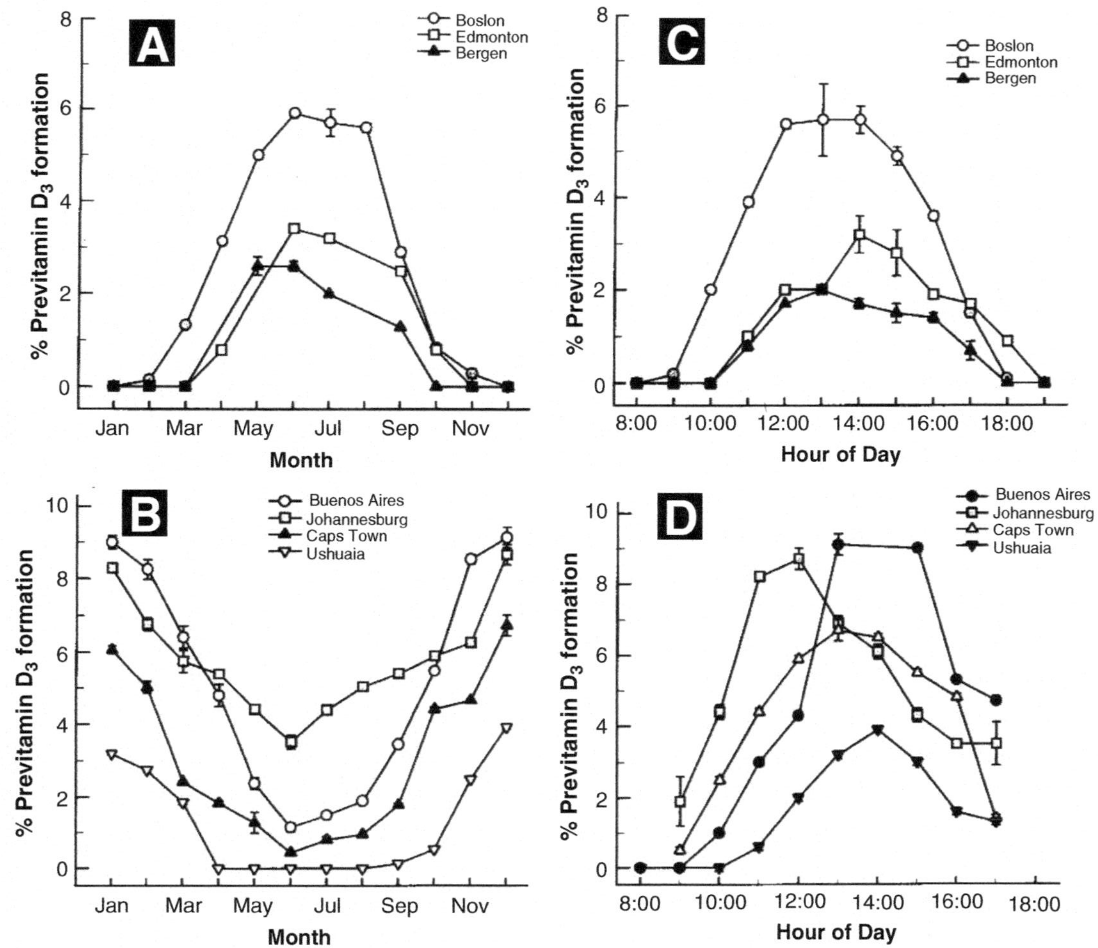

Fig. 12.7 Influence of season, time of day, and latitude on the synthesis of previtamin D_3 in Northern (A and C) and southern hemispheres (B and D). The hour indicated in C and D is the end of the 1-h exposure time (Holick, copyright 1998. Reproduced with permission)

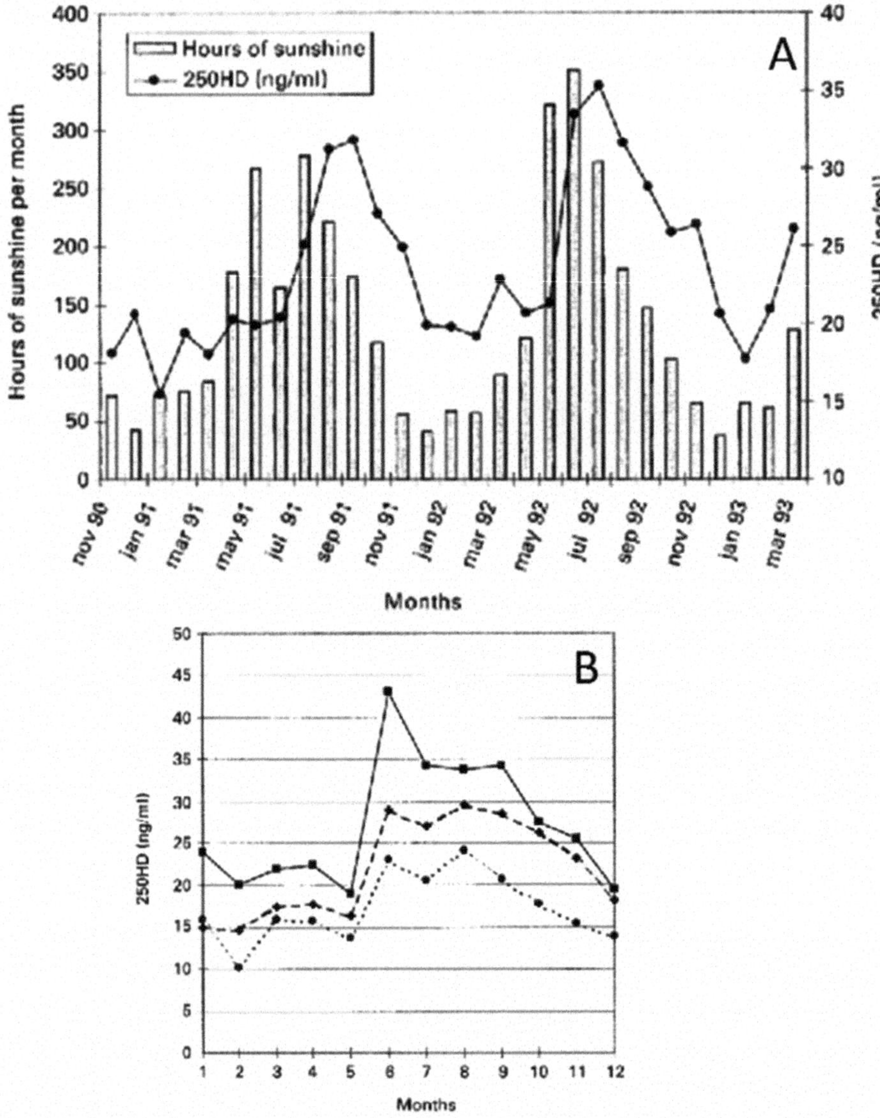

Fig. 12.8 (**a**) Seasonal fluctuation of serum 25(OH)D in healthy perimenopausal Danish women and relationship between hours of sunshine and serum 25(OH)D. (**b**) Seasonal fluctuation of serum 25(OH)D according to frequency of sun exposure. ■, regular sun exposure; ◆, occasional sun exposure; ●, avoiding direct sun exposure (Holick, copyright 2013. Reproduced with permission)

25(OH)D levels at the end of the winter (24 ng/mL) compared to samples collected at the same time from adults living in Northern United States. The study also demonstrated that the contribution of season was very significant. The mean blood levels of 25(OH)D at the end of the winter in northern United States was 21 ng/mL and at the end of the summer rose to 29 ng/mL. This seasonal variation had also a significant physiologic effect on the blood levels of parathyroid hormone (PTH). There was an inverted relationship between PTH and blood levels of 25(OH)D with a 4 week lag time. Thus at the end of the summer PTH levels reached their nadir 4 weeks later and at the end of the winter when 25(OH)D reached its nadir PTH levels were at their highest level 4 weeks later (Fig. 12.9).

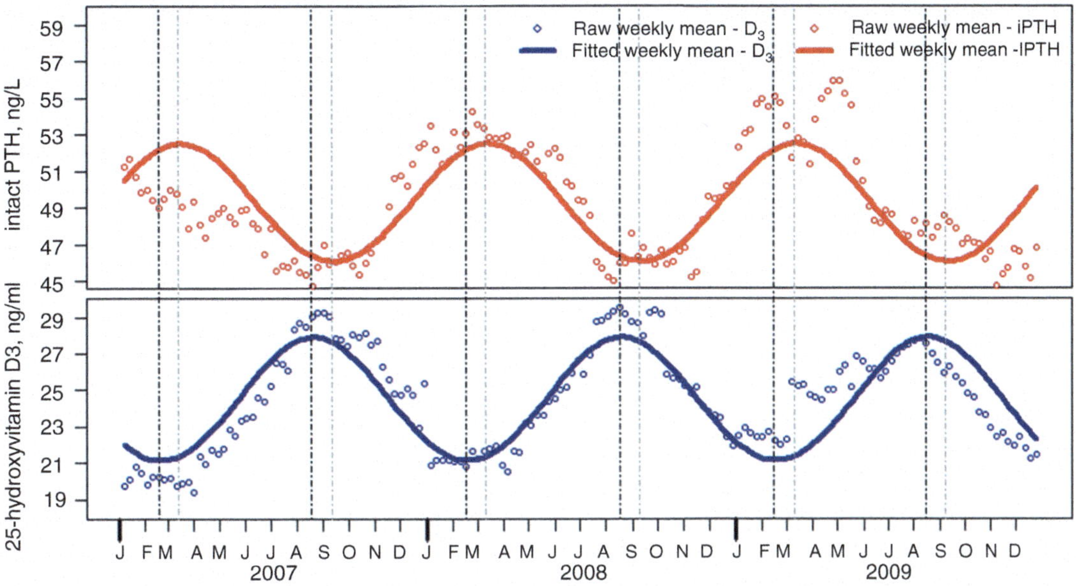

Fig. 12.9 Seasonal Variation of 25-Hydroxyvitamin D3 (bottom panel) and Intact PTH (iPTH) (*top panel*) Weekly Mean Values. The maximum seasonal variation of 25(OH) D3 (peak to trough) was 6.8 ng/mL, reaching its trough in the 8th week (early March) of each year and its peak in the 34th week (early September). Peak iPTH values occurred at week 12 (early April) and trough values at week 37 (late September), a pattern that is roughly reciprocal to that of 25(OH)D3, but lags by 3.5 weeks. Individual points represent the mean of the normalized distribution for each week. The *solid* lines represent the fit. *Dark vertical dashed* lines represent 25-hydroxyvitamin D3 peaks and troughs, and *light vertical dashed* lines represent the iPTH peaks and troughs. To convert 25-hydroxyvitamin D3 from ng/mL to nmol/L, multiply by 2.496 (rounded as 2.5) (Holick, copyright 2015. Reproduced with permission)

Human skin has a large capacity to produce vitamin D. When healthy adults in a bathing suit had their whole bodies exposed to one minimal erythemal dose (slight pinkness to the skin 24 h after the exposure; MED) of UVB radiation in a tanning bed they raised their blood levels of vitamin D to ~20 ng/mL which is equivalent to ingesting approximately 20,000 IUs of vitamin D (Fig. 12.10) [7]. Studies in surgically obtained humans skin have also demonstrated that approximately 250 ng (10 IUs)/in2 of vitamin D is produced when exposed to one MED of UVB radiation [33]. A study in healthy adults who used a tanning bed at least once a week were found to have mean blood levels of 25(OH)D of 48 ng/mL. Healthy adults matched for sex and age at the same time had a mean blood level of 25(OH)D of 18 ng/mL [34].

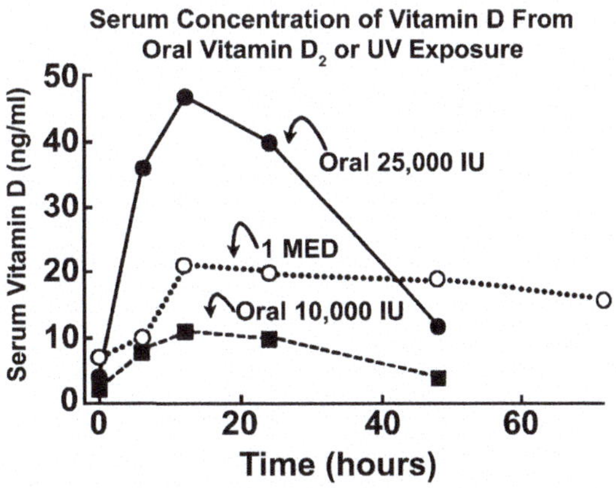

Fig. 12.10 Comparison of serum vitamin D_3 levels after a whole-body exposure (in a bathing suit; bikini for women) to 1 MED (minimal erythemal dose) of simulated sunlight compared with a single oral dose of either 10,000 or 25,000 IU of vitamin D_2 (Holick, copyright 2004. Reproduced with permission)

12.5 Ultraviolet B Induced Extrarenal Synthesis of 1,25-Dihydroxyvitamin D

Once vitamin D is made in the skin or ingested in the diet it travels to the liver to be converted to 25(OH)D. Once formed it reenters the circulation and travels to the kidneys where it is converted to its active form 1,25-dihydroxyvitamin D [1,25(OH)2D] [29, 30]. Patients with chronic kidney disease who are unable to produce 1,25(OH)$_2$D results in a decrease in the efficiency of intestinal calcium absorption leading to a transient decrease in the blood calcium levels. This is immediately recognized by the parathyroid glands resulting in an increase in the production of parathyroid hormone. This causes secondary hyperparathyroidism which results in a metabolic bone disease of the skeleton known as renal osteodystrophy. 1,25(OH)$_2$D$_3$ (calcitriol) and its active analogs have been effectively use for the prevention and treatment of secondary hyperparathyroidism and renal osteodystrophy [29]. However these medications can cause hypercalcemia limiting their use in some patients.

It is recognized that variety of cells and organs not responsible for calcium metabolism also have the capacity to convert 25(OH)D to 1,25(OH)2D including activated monocytes and macrophages and keratinocytes [29, 30]. It had in previously reported that patients with chronic kidney disease who while on dialysis and have no kidney function often have measurable levels of 1,25(OH)$_2$D. Is thought that uremia associated with chronic kidney disease that activate monocytes which have the capacity to produce 1,25(OH)$_2$D [35, 36].

It had been previously reported that keratinocytes have a large capacity to convert 25(OH)D to 1,25(OH)$_2$D [37]. It has also been reported that keratinocytes express not only the 25(OH)D- 1 alpha hydroxylase but also the vitamin D-25 -hydroxylase [38]. In vitro studies reported that vitamin D3 added to culture the skin cells could be converted to 1,25(OH)2D3 [39].

This was the rationale conducting a study in patients with chronic kidney disease on hemodialysis to expose them to UVB radiation to determine if this would be effective not only in raising blood levels of vitamin D$_3$ but also increase the blood levels of 25(OH)D$_3$ and 1,25(OH)$_2$D$_3$. 95 chronic kidney disease patients (mean a 62 years) on hemodialysis were treated with a mean dose of 35,000 IUs of vitamin D$_3$ a week while a group of 14 patients (mean age 51 years) received whole body UVB radiation for 6 months. Skin biopsies were obtained in 3 patients. The group receiving oral vitamin D$_3$ raised their blood levels of 25(OH)D$_3$ by 60% over 18 months compared to an increase of 400% in the group that received UVB radiation for 6 months. In a group of 4 patients who received suberythemal exposures to UVB radiation for up to 10 years also were able to maintain normal circulating levels of 1,25(OH)$_2$D$_3$ (Fig. 12.11). The skin biopsies confirmed that the epidermal cells were expressing the VDR as well as the vitamin D-25-hydroxylase and the 25(OH)D-1- alpha hydroxylase [38].

The 14 patients who received the UVB irradiation for 6 months show an increase in their hematocrit and required less erythropoietin. They demonstrated an increase in maximum oxygen uptake and work load capacity that was associated with decreased lactic acid production. They also demonstrated decreased heart rate and

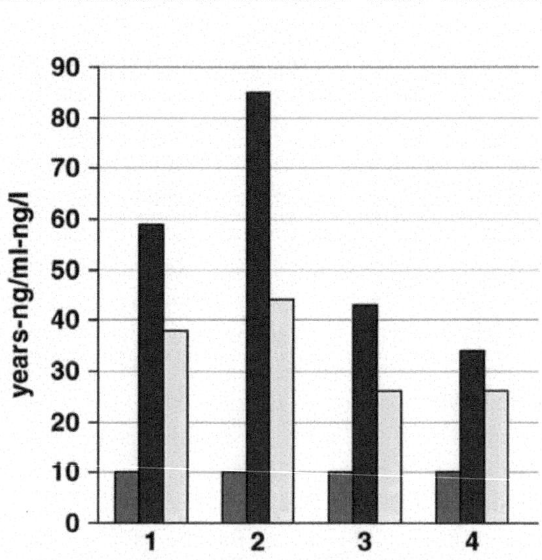

Fig. 12.11 *Vitamin D status of 4 hemodialysis patients over 10 years received suberythemal UVB irradiation one to three times weekly (Vit D_3: gray bars; 25(OH)D_3: black bars; 1,25(OH)$_2D_3$: white bars)* (Holick, copyright 2016. Reproduced with permission)

systolic and diastolic blood pressure with an increase in the R-R interval and beat-to-beat differences [40].

12.6 Sunlight, Skin Cancer and Vitamin D

Sunlight has been and continues to be a major source of vitamin D for children and adults worldwide [7, 41]. The introduction of sunscreens and the worldwide publicity campaign recommending avoidance of all direct sun exposure because of concern for increased risk for skin cancer, has cause a vitamin D deficiency pandemic [42]. Globally 30–40% and 60–80% of children and adults have been reported to be vitamin D deficient based on The Institute of Medicine's definition for maximum bone health and vitamin D insufficiency based on the endocrine Society's definition for maximum bone and muscle health respectively [30, 43–47] (Figs. 12.12 and 12.13).

Most skin cancer is due to excessive exposure to sunlight and the number of sun burning experiences especially as a child and young adult. These cancers, known as melanoma and non-melanoma skin cancer, typically appear the most sun exposed and sun damaged areas including the face, top of the ears and top of the hands [48, 49]. A study in Danish adults who were exposed to high intensity sunlight for 38 h over a 6 days during a vacation in the Canary Islands were able to improve their circulating levels of 25(OH)

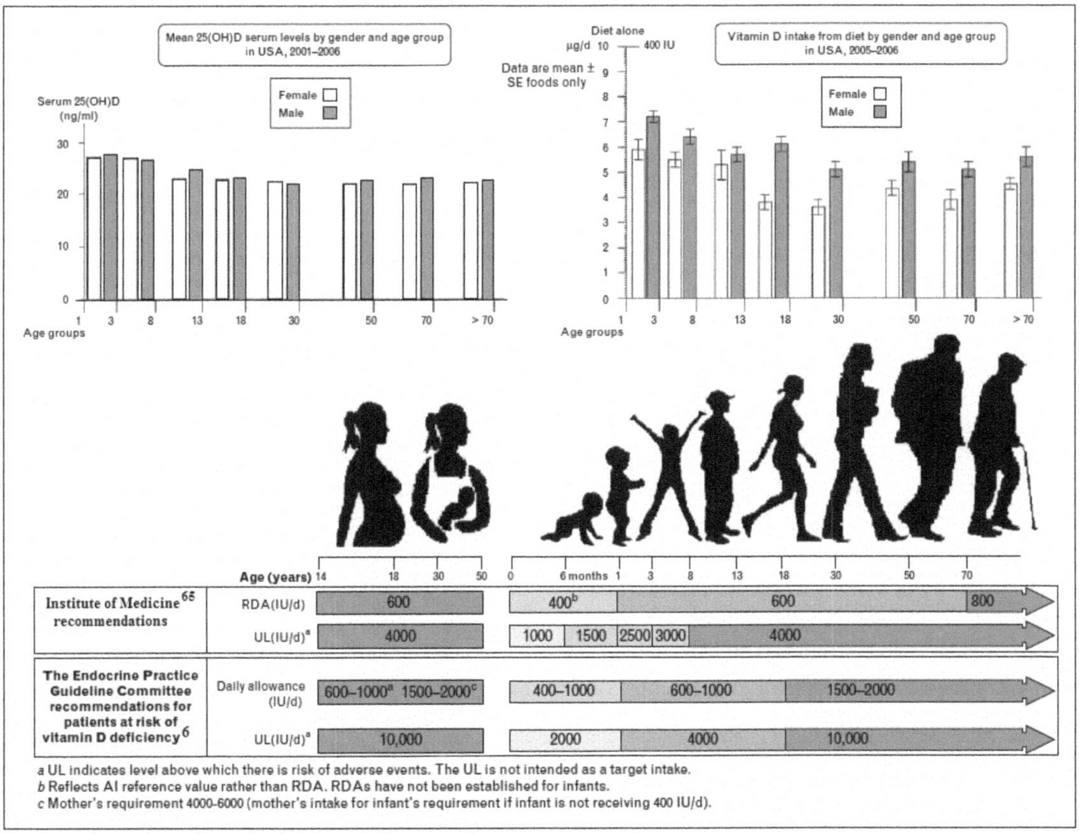

Fig. 12.12 Vitamin D intakes recommended by the Institute of Medicine and the Endocrine Practice Guidelines Committee. 25(OH)D[25-hydroxyvitamin D]; AI [adequate intake]; RDA [recommended dietary allowance]; SE [standard error]; UL [tolerable upper intake level] (Copyright Holick 2013, reproduced with permission)

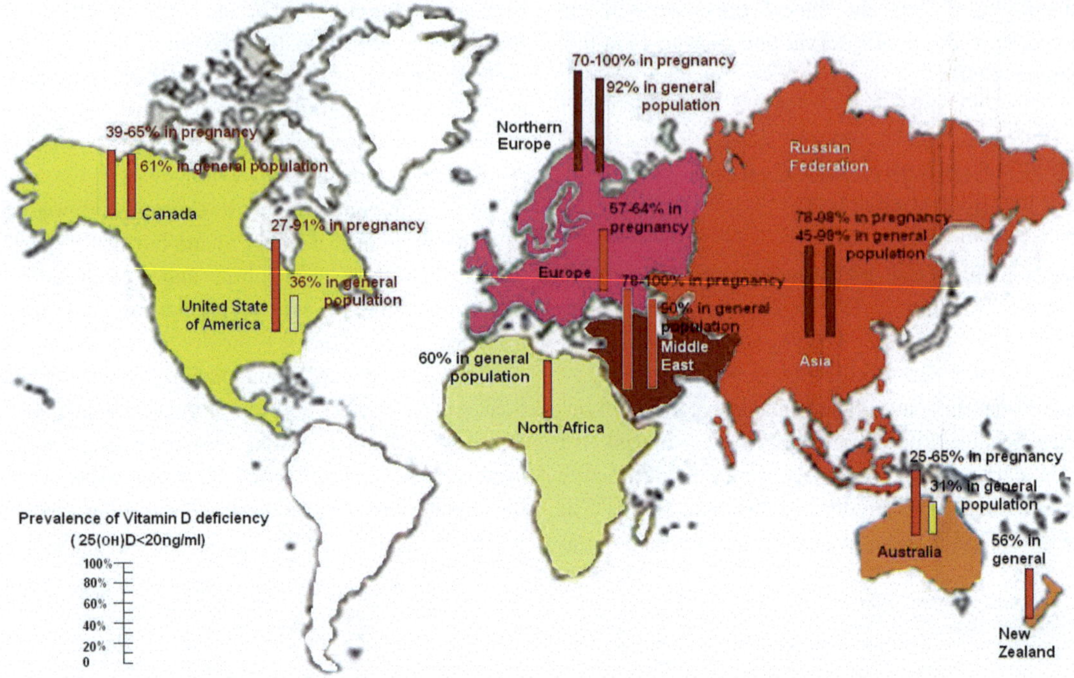

Fig. 12.13 Reported incidence of vitamin D deficiency defined as a 25-hydroxyvitamin D (25[OH]D) level below 20 ng/mL around the globe in pregnant women and the general population. To convert 25(OH)D values to nmol/L, multiply by 2.496 (Copyright Holick 2013, reproduced with permission)

D. However. Peterson et al. [50] also observed a significant and concerning cutaneous DNA damage as measured by increased urinary cyclobutane pyrimidine dimers (CPD), a surrogate for DNA damage. Thus, it was suggested that you could not have your cake and eat it to, i.e. take advantage of the beneficial effect of sun exposure for producing the vital vitamin D_3 without significant DNA damage in the skin. From an evolution perspective this makes little sense since sun-induced synthesis of vitamin D_3 was essential not only for the evolution of none human vertebrates on terra firma but was also essential for the maintenance of skeletal health for hominids including present-day humans [51]. Felton et al. [52]expose healthy British adults with skin type 2 and 4 to an amount of simulated sunlight typical for what would occur during the summer in the UK. Those with skin type 2 received simulated June midday sun light for approximately 13–17 min 6 times weekly for 6 weeks. They reported a 49% increase in circulating levels of 25(OH)D. Initially they observed that this exposure resulted in the formation of CPD and other pyrimidine photoproducts that if unrepaired have been associated with increased risk for nonmelanoma skin cancer. However 24 h after the last exposure skin biopsies and urine revealed significant clearing of the CPD-positive nuclei. This corresponded to undetectable levels of CPD in the urine and no change or accumulation in another marker for DNA damage from baseline, i.e. urinary 8-oxo-2′-deoxyguanosine (8-oxodG), a measure of oxidatively damaged DNA. They compared skin type 2 with type 5, and found that there was more DNA damage done to those with type 2, supporting that our ancestors who migrated further from the equator were at a disadvantage when it comes to UVB skin protection. As has been previously reported, increased skin protecting pigmentation efficiently absorbs UVB radiation and therefore also reduces the number of photons absorbed by 7-dehydrocholesterol, resulting in a decrease in the effectiveness of the sun in producing vitamin D_3,which they also observed by demonstrating a statistically insignificant increase in serum 25(OH)D levels in their Asian subjects. These data support the concept that skin pigment began to devolved as a result of the migration of humans north and south of the equator [51]. A mutation of the melanocortin 1 receptor (MRC1R), which regulates pigmentation in humans and other vertebrates resulted in decreased melanin synthesis resulting in penetration of more of the less intense solar UVB radiation for the production of vitamin D [51, 53, 54]. Asians with skin type 5 demonstrated very little DNA damage from the same amount of simulated sunlight exposure and were unable to make enough vitamin D in their skin to significantly raise their blood level of 25(OH)D [50]. Therefore the degree of skin pigmentation evolved to protect the skin from the damaging effects from excessive sun exposure while at the same time permitting an adequate amount of vitamin D to be produced. This is nicely demonstrated in Maasai herders who have skin type 6 and have circulating levels of 25(OH)D on average of 48 ng/mL [55]. Achieving these levels requires the ingestion of 3000–5000 IUs daily [56].

The most feared form of skin cancer is melanoma. It has been suggested that the major reason to abstain from any direct sun exposure is for the prevention of this deadly cancer [42]. However it is well documented that most melanomas occur on the least sun exposed areas and occupational sun exposure is associated with a reduced risk. The major risk factors are number of sunburns as a child and young adult, being red headed, having a large number of moles on the body and a genetic predisposition for developing it [48].

There are other numerous studies relating vitamin D deficiency with increased risk for many acute and chronic illnesses. These include increased risk for preeclampsia and the need for a cesarean section, autoimmune diseases including Type 1 and 2 diabetes, multiple sclerosis, cardiovascular disease, infectious diseases, neurocognitive dysfunction, deadly cancers including breast and colon cancers (Fig. 12.14) [29, 30]. What is also remarkable are the earlier observations that living at higher latitudes with less vitamin D producing sun exposure was associated with increased risk for mortality, multiple sclerosis, Type 1 diabetes, hypertension and deadly cancers [7, 30, 41].

Besides the cutaneous production of vitamin D, exposure to solar UVB radiation also increases the production of β-endorphin. Exposure to solar UV radiation is also associated with increased production of nitric oxide and carbon monoxide both of which cause vasodilation and can lower blood pressure. It also increases the expression of the proopiomelanocortin (POMC) gene increase in the production of adrenocorticotropin hormone which helps to regulate the immune system [41].

12.7 Conclusion

There needs to be an acknowledgment by health care officials worldwide that sensible sun exposure is important not only for the production of vitamin D but also for overall health and wellbeing. A study of Australian dermatologist in the summer revealed that their use of a sunscreen resulted in 87% of them being vitamin D deficient at the end of the summer [57]. The World Health Organization on its website regarding sunlight and health state that some sunlight exposure is important for the production of vitamin D. However time of day, season, latitude, altitude and skin pigmentation all influence the efficiency of the skin to produce vitamin D during sun exposure. An app dminder.info which is free for the iphone and android formats provides guidance for sensible sun exposure and informs to user not only how much vitamin D they can produce when exposed to sunlight but also provides a warning to wear sun protection after that sensible sun exposure to prevent sunburning. For a wide variety of reasons it is not reasonable to expect that you can obtain an adequate amount of vitamin D from sun exposure unless you are outdoors all the time and exposed to a significant amount of your skin to sunlight such as a lifeguard or a Maasai herder or frequent a tanning salon [7, 34, 55, 58]. Following

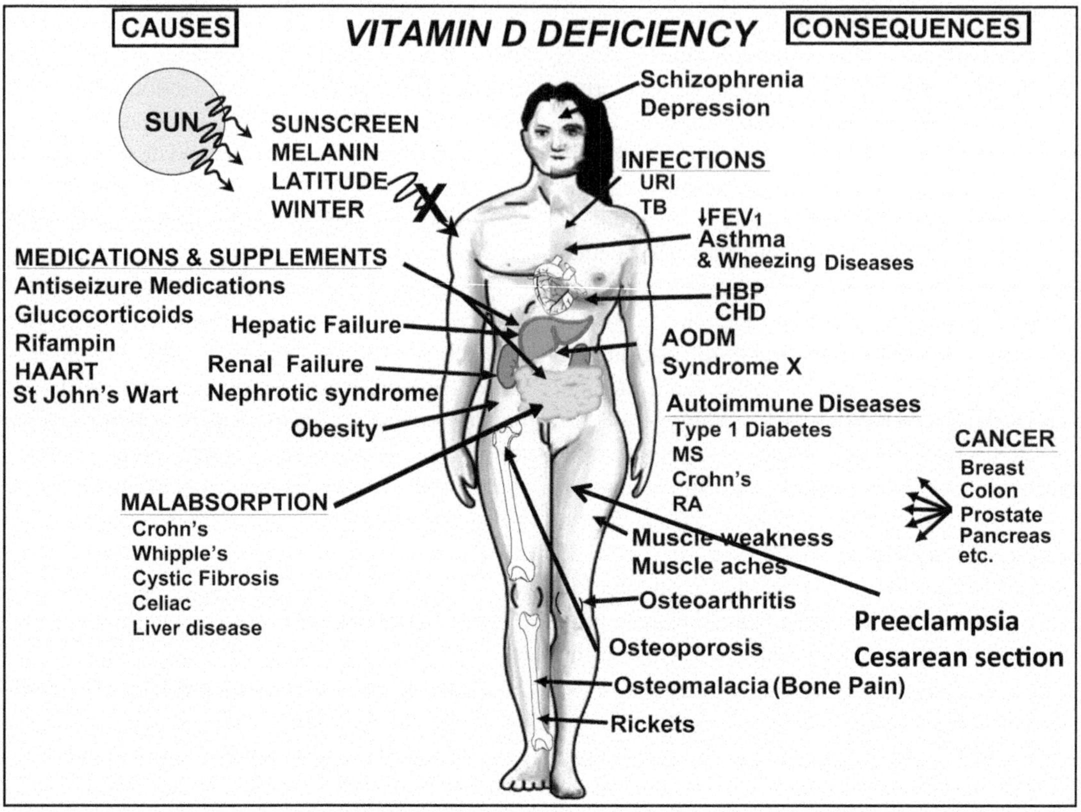

Fig. 12.14 A Schematic representation of the major causes for vitamin D deficiency and potential health consequences (Holick, copyright 2007. Reproduced with permission)

the recommendations of the Endocrine Society will help to achieve blood levels of 25(OH)D in the desired range above 30 ng/mL [45]. The amount recommended are 400–1000 IUs, 600–1000 IUs and 1500–2000 IUs daily for children under 1 year, children 1 year and older and adults respectively. Obese children and adults require 2–3 times more vitamin D to satisfy their requirement. For simplicity I recommend all children can take 1000 IUs daily and teenagers and adults 2000 IUs daily as a supplement. I also recommend that this amount of vitamin D be taken daily throughout the entire year even in the summer. This amount of vitamin D along with any vitamin D available in the diet and sun exposure will not cause vitamin D intoxication [7, 59]. The safe upper level or vitamin D for children is 4000 IUs daily and 10,000 IUs daily for adults as recommended by the Endocrine Society [45].

References

1. Holick MF (2003) Vitamin D: a millennium perspective. J Cell Biochem 88:296–307
2. Mark KA, Dumas KJ, Bhaumik D, Schilling B, Davis S, Oron TR, Sorensen DJ, Lucanic M, Brem RB, Melov S, Ramanathan A, Gibson BW, Lithgow GJ (2016) Vitamin D promotes protein homeostasis and longevity via the stress response pathway genes skn-1, ire-1, and xbp-1. Cell Rep 17:1227–1237
3. Power ML, Oftedal OT, Savage A, Blumer ES, Soto LH, Chen TC, Holick MF (1997) Assessing vitamin D status of callitrichids: baseline data from wild cotton-top tamarins (*Saguinus oedipus*) in Colombia. Zoo Biol 16:39–46. 177
4. Ferguson GW, Gehrmann WH, Peavy B, Painter C, Hartdegen R, Chen TC, Holick MF, Pinder JE (2010) Restoring vitamin D in monitor lizards: exploring the efficacy of dietary and UVB sources. J Herp Med Surg 19:81–88
5. Holick MF (2004) Sunlight and vitamin D for bone health and prevention of autoimmune diseases, cancers, and cardiovascular disease. Am J Clin Nutr 80:1678S–1688S

6. Holick MF (1999) Biologic effects of light: historical and new perspectives. In: Holick MF, Jung EG (eds) Biologic effects of light. Kluwer Academic Publishers, Boston, pp 11–32
7. Wacker M, Holick MF (2013) Sunlight and vitamin D: a global perspective for health. Dermato-Endocrinol 5(1):51–108
8. Molozolowski W (1939) Jedrezej Sniadecki (1786–1883) on the cure of rickets. Nature 143:121
9. Palm TA (1890) The geographic distribution and etiology of rickets. Practitioner 45:270–279. 321–342
10. Rajakumar K, Greenspan SL, Thomas SB, Holick MF (2007) Solar ultraviolet radiation and vitamin D: a historical perspective. Am J Public Health 97:1746–1754
11. Huldschinsky K (1919) Heilung von Rachitis durch Kunstliche Hohensonne. Dtsch Med Wochenschr:90–91
12. Hess AF, Unger LJ (1922) Use of the carbon arc light in the prevention and cure of rickets. JAMA 78(21):1596–1598
13. Holick MF (2006) Resurrection of vitamin D deficiency and rickets. J Clin Invest 116(8):2062–2072
14. McCollum EF, Simmonds N, Becker JE, Shipley PG (1922) Studies on experimental rickets; and experimental demonstration of the existence of a vitamin which promotes calcium deposition. J Biol Chem 53:293–312
15. Steenbock H, Black A (1924) The reduction of growth-promoting and calcifying properties in a ration by exposure to ultraviolet light. J Biol Chem 61:408–422
16. Hess AF, Weinstock M (1924) Antirachitic properties imparted to inert fluids and to green vegetables by ultraviolet irradiation. J Biol Chem 62:301–313
17. Eliot MM, Park EA (1938) Rickets. In: Brennemann's practice of pediatrics, vol 1. W.F. Prior Company Inc, Hagerstown, pp 1–110
18. Havinga E (1973) Vitamin D, example and challenge. Experientia 29:1181–1193. PMID:4758912. http://dx.doi.org/10.1007/BF01935064
19. MacLaughlin JA, Anderson RR, Holick MF (1982) Spectral character of sunlight modulates photosynthesis of previtamin D_3 and its photoisomers in human skin. Science 216:1001–1003
20. Holick MF, MacLaughlin JA, Clark MB, Holick SA, Potts JT Jr, Anderson RR, Blank IH, Parrish JA, Elias P (1980) Photosynthesis of previtamin D_3 in human skin and the physiologic consequences. Science 210:203–205
21. Holick MF, MacLaughlin JA, Dobbelt SH (1981) Regulation of cutaneous previtamin D_3 photosynthesis in man: skin pigment is not an essential regulator. Science 211:590–593
22. Loomis WF (1967) Skin pigment regulation of vitamin D biosynthesis in man. Science 157(3788):501–506
23. Holick MF, Tian XQ, Allen M (1995) Evolutionary importance for the membrane enhancement of the production of vitamin D_3 in the skin of poikilothermic animals. Proc Natl Acad Sci U S A 92:3124–3126
24. Webb AR, de Costa BR, Holick MF (1989) Sunlight regulates the cutaneous production of vitamin D_3 by causing its photodegradation. J Clin Endocrinol Metab 68:882–887
25. Clemens TL, Henderson SL, Adams JS, Holick MF (1982) Increased skin pigment reduces the capacity of skin to synthesis vitamin D_3. Lancet 1(8263):74–76
26. Matsuoka LY, Ide L, Wortsman J, MacLaughlin J, Holick MF (1987) Sunscreens suppress cutaneous vitamin D_3 synthesis. J Clin Endocrinol Metab 64:1165–1168
27. Webb AR, Kline L, Holick MF (1988) Influence of season and latitude on the cutaneous synthesis of vitamin D_3: exposure to winter sunlight in Boston and Edmonton will not promote vitamin D_3 synthesis in human skin. J Clin Endocrinol Metab 67:373–378
28. Holick MF, Chen TC, Sauter ER (2007) Vitamin D and skin physiology: a D-lightful story. J Bone Miner Res 22(S2):V28–V33
29. Holick MF (2007) Vitamin D deficiency. New Engl J Med 357:266–268
30. Hossein-nezhad A, Holick MF (2013) Vitamin D for health: a global perspective. Mayo Clin Proc 88(7):720–755
31. Brot C, Vestergaard P, Kolthoff N, Gram J, Hermann AP, Sorensen OH (2001) Vitamin D status and its adequacy in health Danish perimenopausal women: relationships to dietary intake, sun exposure and serum parathyroid hormone. Br J Nutr 86(1):S97–S103
32. Kroll MH, Bi C, Garber CC et al (2015) Temporal relationship between vitamin D status and parathyroid hormone in the United States. PLoS One 10(3):e 0118108. doi:10.1371/journal.pone.0118108. eCollection 2015
33. Obi-Tabot E, Tian XQ, Chen T, Holick MF (2000) A human skin equivalent model that mimics the photoproduction of vitamin D_3 in human skin. Vitro Cell Develop Biol 36:201–204. 210
34. Tangpricha V, Turner A, Spina C, Decastro S, Chen T, Holick MF (2004) Tanning is associated with optimal vitamin D status (serum 25-hydroxyvitamin D concentration) and higher bone mineral density. Am J Clin Nutr 80:1645–1649
35. Dusso AS, Brown AJ (2005) Slatopolsky. Vitamin D. Am J Physiol Renal Physiol 289:F8–F28
36. Adams JS, Singer FR, Gacad MA, Sharma OP, Hayes MJ, Vouros P, Holick MF (1985) Isolation and structural identification of 1,25-dihydroxyvitamin D_3 produced by cultured alveolar macrophages in sarcoidosis. J Clin Endo Metab 60:960–966. 73
37. Bikle DD, Nemanic MD, Whitney JO, Elias PO (1986) Neonatal human foreskin keratinocytes produce 1,25-dihydroxyvitamin D_3. Biochemistry 25:1545–1548
38. Krause R, Roth HJ, Kaase H, Strange R, Holick MF (2016) Vitamin D status in chronic kidney disease-UVB irradiation is superior to oral supplementation. Anticancer Res 36:1397–1402

39. Lehmann B, Meurer M (2010) Vitamin D metabolism. Dermatol Ther 23:2–12
40. Krause R, Strange R, Kaase H, Holick MF (2016) UV irradiation and pleotropic effects of vitamin D in chronic kidney disease-benefits on cardiovascular comorbidities and quality of life. Anticancer Res 36:1403–1418
41. Holick MF (2016) Biologic effects of sunlight, ultraviolet radiation, visible light, infrared, and vitamin D for health. Anticancer 36:1345–1356
42. Wolpowitz D, Gilchrest BA (2006) The vitamin D questions: how much do you need and how should you get it? J Am Acad Dermatol 54:301–317
43. Daly RM, Gagnon C, Lu ZX et al (2012) Prevalence of vitamin D deficiency and its determinants in Australian adults aged 25 years and older: a national, population-based study. Clin Endocrinol 77:26–35
44. IOM (Institute of Medicine) (2011) Dietary reference intakes for calcium and vitamin D. Committee to review dietary reference intakes for calcium and vitamin D. The National Academies Press Institute of Medicine, Washington, DC
45. Holick MF, Binkley NC, Bischoff-Ferrari HA, Gordon CM, Hanley DA, Heaney RP, Murad MH, Weaver CM (2011) Evaluation, treatment, and prevention of vitamin D deficiency: an Endocrine Society clinical practice guideline. J Clin Endocrinol Metab 96(7):1911–1930
46. Pludowski P, Holick MF, Pilz S, Wagner CL, Hollis BW, Grant WB, Shoenfeld Y, Lerchbaum E, Llewellyn DJ, Kienreich K, Soni M (2013) Vitamin D effects on musculoskeletal health, immunity, cardiovascular disease, cancer, fertility, pregnancy, dementia and mortality – a review of recent evidence. Autoimmunity Rev 12:976–989. 482
47. Nesby-O'Dell S, Scanlon KS, Cogswell ME, Gillespie C, Hollis BW, Looker AC (2002) Hypovitaminosis D prevalence and determinants among African American and white women of reproductive age: third national health and nutrition examination survey, 1988-1994. Am J Clin Nutr 76:187–192
48. Kennedy C, Bajdik CD, Willemze R, de Gruijl FR, Bavinck JN (2003) The influence of painful sunburns and lifetime of sun exposure on the risk of actinic keratoses, seborrheic warts, melanocytic nevi, atypical nevi and skin cancer. J Invest Dermatol 120(6):1087–1093
49. Moan J, Porojnicu AC, Dahlback A, Setlow RB (2008) Addressing the health benefits and risks, involving vitamin D or skin cancer, of increased sun exposure. Proc Natl Acad Sci U S A 105(2):668–673
50. Petersen B, Wulf HC, Triguero-mas M et al (2014) Sun and ski holidays improve vitamin D status, but are associated with high levels of DNA damage. J Invest Dermatol 134:2806–2813
51. Holick MF (2016) Can you have your cake and eat it too? The sunlight D-lema. Br J Dermatol 175:1129–1131
52. Felton SJ, Cooke MS, Kift R et al (2016) Concurrent beneficial (vitamin D production) and hazardous (cutaneous DNA damage) impact of repeated low-level summer sunlight exposures. Br J Dermatol 175:1320–1328
53. Lalueza-Fox C, Römpler H, Caramelli D et al (2007) A melanocortin 1 receptor allele suggests varying pigmentation among Neanderthals. Science 318:1453–1455
54. Greaves M (2014) Was skin cancer a selective force for black pigmentation in early hominin evolution? Proc Biol Sci 281:2013–2955
55. Luxwolda MF, Kuipers RS, Kema IP, Dijck-Brouwer DA, Muskiet FA (2012) Traditionally living populations in East Africa have a mean serum 25-hydroxyvitamin D concentration of 115 nmol/l. Br J Nutr 108(9):1557–1561
56. Heaney RP, Davies KM, Chen TC, Holick MF, Barger-Lux MJ (2003) Human serum 25-hydroxycholecalciferol response to extended oral dosing with cholecalciferol. Am J Clin Nutr 77:204–210
57. Czarnecki D, Meehan CJ, Bruce F (2009) The vitamin D status of Australian dermatologists. Clin Exper Derm 34:621–638
58. Baggerly C, Cuom R, French C, Garland C, Gorham E, Grant W, Heaney R, Holick M (2015) Sunlight and vitamin D: necessary for public health. J Am College Nutr 34(4):359–365. PMID: 26098394. 533
59. Holick MF (2015) Vitamin D is not as toxic as was once thought: a historical and an up-to-date perspective. Mayo Clinic Proc 90(5):561–564. 530

Role of Vitamin D in Rheumatoid Arthritis

13

Mattia Bellan, Pier Paolo Sainaghi, and Mario Pirisi

Abstract

Vitamin D is a fat soluble hormone, with a well described role in bone health and calcium/phosphate metabolism. Recent evidences have related vitamin D to other physiological functions and pathological conditions. Specifically, vitamin D has widely proven activities on immune system and evidences suggest that it may be implicated in the pathogenesis of rheumatoid arthritis (RA). The relationship between vitamin D and RA is complex, also because a deficitary vitamin D status, which is very common in RA patients, can contribute to the increased risk of osteoporosis typical of RA. In this chapter, will be described and discussed the main aspects of the relationship between RA and vitamin D.

Keywords

Vitamin D • Rheumatoid arthritis • Hypovitaminosis D

M. Bellan (✉) • M. Pirisi
Department of Translational Medicine, Università del Piemonte Orientale UPO,
via Solaroli 17, 28100 Novara, Italy

Immunorheumatology Unit, Internal Medicine Division, AOU Maggiore della Carità, Novara, Italy
e-mail: bellanmattia@yahoo.it

P.P. Sainaghi
Immunorheumatology Unit, Internal Medicine Division, AOU Maggiore della Carità, Novara, Italy

List of Abbreviations

$1,25(OH)_2$ vitamin D	1,25-dihydroxyvitamin D
25(OH) vitamin D	25-hydroxyvitamin D
APC	Antigen presenting cells
ARD	Autoimmune rheumatic diseases
BMD	Bone mineral density
BMI	Bone mass index
CD	Cluster of differentiation
C-EBP	CCAAT-enhancer-binding proteins
CRP	C-reactive protein
CTLA-4	Cytotoxic T-Lymphocyte Antigen 4
DC	Dendritic cells
Dkk-1	Dickkopf-related protein 1

S. Ahmad (ed.), *Ultraviolet Light in Human Health, Diseases and Environment*, Advances in Experimental Medicine and Biology 996, https://doi.org/10.1007/978-3-319-56017-5_13

EGFR	Epidermal growth factor receptor
Foxp3	Forkhead box P3
HIV	Human immunodeficiency virus
IBD	Inflammatory bowel diseases
IFN-γ	Interferon γ
IL	Interleukin
IU	International units
JAK	Janus kinase
MAPK	Mitogen-activated protein kinase
MHC	Major histocompatibility complex
MS	Multiple sclerosis
NF-kB	nuclear factor kappa-light-chain-enhancer of activated B cells
OPG	Osteoprotegerin
PTH	Parathyroid hormone
RA	Rheumatoid Arthritis
RANK	Receptor activator of nuclear factor kappa-B
RANKL	Receptor activator of nuclear factor kappa-B ligand
RCT	Randomized clinical trial
RXR	Retinoid X receptor
SLE	Systemic lupus erythematosus
STAT1α	Signal transducer and activator of transcription 1
T1DM	Type 1 diabetes mellitus
TGFα	Transforming growth factor α
Th	T helper
TLR	Toll like receptor
TNFα	Tumor necrosis factor α
T_{reg}	Regulatory T cells
UCTD	Undifferentiated connective tissue disease
UV	Ultraviolet
VDR	Vitamin D receptor
VDRE	Vitamin D Responsive Elements

13.1 Introduction

The beneficial effect of UV light has been widely demonstrated in different settings of human physiology, being vitamin D a crucial mediator of these actions. In the last decades, evidences deriving from many studies have shed a new light on vitamin D physiology, demonstrating that vitamin D activity is much wider, specially associated with human health, than previously realised, leading to the hypothesis that vitamin D could be involved in many pathophysiological mechanisms. In this context, the relation between vitamin D and RA seems particularly relevant. Some aspects of this topic deserve a deeper consideration. In this chapter, we will focus our presentation on two main issues: the prevalence and treatment of hypovitaminosis D in patients affected by RA and the potential pathophysiological and therapeutic role of vitamin D in RA.

13.1.1 Vitamin D Metabolism and "Classical" Function

Here we start with the vitamin D metabolism along with the main "classical" functions of this hormone. Vitamin D is a fat-soluble steroid molecule derived from both dietary intake and endogenous synthesis. However, the greatest part of vitamin D required for human health originates from endogenous synthesis; therefore, cholecalciferol nowadays is more properly considered a hormone rather than a vitamin.

Ultraviolet rays (UVR) photolyse the cutaneous precursor, 7-dehydrocholesterol, present in epidermal keratinocytes and in dermal fibroblasts, into cholecalciferol, which is then hydroxylated to the circulating form of the hormone, 25(OH) vitamin D, by different liver isoforms of a 25-hydroxylase (CYP2C11, CYP2J3, CYP2R1, CYP3A4, CYP27A1, CYP2D25) [1–4]. 25(OH) vitamin D circulates in the blood stream bound to a vitamin D Binding Protein (DBP), and can also be stored in fat tissue [5]. This is the intermediate metabolite usually measured to define the vitamin D status, because it is more stable and has a longer life than the active form [6]. $1,25(OH)_2$ vitamin D, also called calcitriol, is the active form of the hormone, resulting from a further hydroxylation mediated by a 1-α-hydroxylase (CYP27B1) expressed in the cells of the convoluted proximal tubule of the kidney [7]. While the expression of 25-hydroxylase is restricted to the liver, CYP27B1 is expressed by many other tissues, including placenta, endothelium, prostate, monocytes and macrophages, skin, colon and brain [8, 9].

$1,25(OH)_2$ vitamin D exerts its actions on target cells by binding to a nuclear receptor, the so-called vitamin D receptor (VDR). Following the linkage with $1,25(OH)_2$ vitamin D, VDR heterodimerizes with retinoid X receptor (RXR), leading to the formation of a complex which moves from cytoplasm to nucleus, where it modulates both positively and negatively the expression of many downstream genes, linking specific DNA regions (VDRE) included in the promoter region of the target genes [10].

The best known and described role of vitamin D in human health is the regulation of bone and calcium/phosphate metabolism. $1,25(OH)_2$ vitamin D enhances calcium and phosphate absorption in the gastrointestinal tract [11] and tubular calcium resorption in the kidney [12]. Furthermore, $1,25(OH)_2$ vitamin D has a pro-resorptive role on bone, which is the result of osteoclast activation via RANK/RANKL interaction. In fact, $1,25(OH)_2$ vitamin D is one of the strongest inducers of RANKL in osteoblasts [13] and, on the other hand, a suppressor of the synthesis of OPG, which acts as a decoy receptor downregulating this system [14]. Finally, $1,25(OH)_2$ vitamin D downregulates PTH synthesis both indirectly (increasing calcium concentration) and directly (linking a VDRE in the promoter of the PTH gene) [15]. The net effect of vitamin D activity, therefore, leads to calcium and phosphate increase and PTH plasma concentration decrease (Fig. 13.1).

13.1.2 Hypovitaminosis D: Causes and Consequences

A deficitary 25(OH) vitamin D concentration may have many different causes: UVB intensity and time of exposure, life-style habits as well as

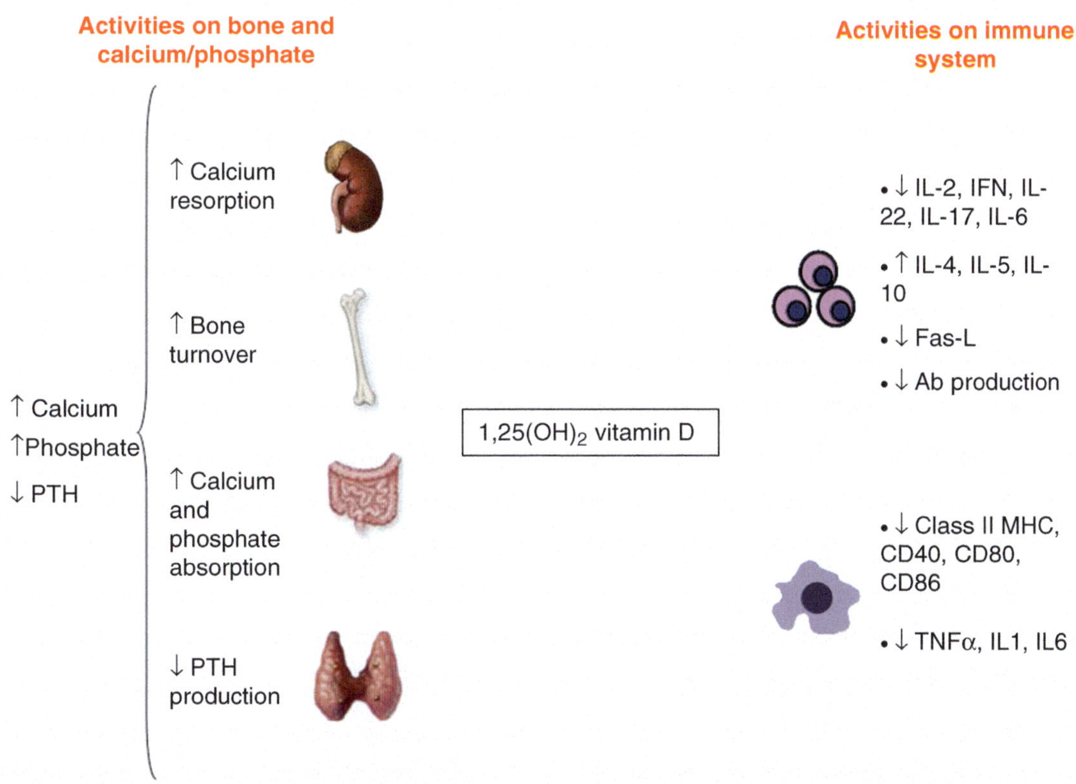

Fig. 13.1 Effects of vitamin D on calcium/phosphate metabolism and on immune system. The net effect of vitamin D leads to an increase of plasma calcium and phosphate concentration and to a decrease of plasma PTH concentration. Vitamin D also acts on T and B cells and on antigen presenting cells, with a complex immunomodulant action

latitude may result in [16–18]. A major determinant of plasma 25 (OH) vitamin D concentrations is the season in which the dosage is performed. Assays carried out on 25 (OH) vitamin D concentrations showed that it is higher in September, following summer in the northern hemisphere, and lower in February [19]. Wearing a sunscreen impairs cholecalciferol; for example, the use of a sun protection factor of 30 reduces vitamin D synthesis by more than 95% [20]. Melanin acts as a natural sunscreen, protecting skin from UVB rays and therefore affecting cholecalciferol production. Dark skinned people require at least three to five times longer exposure to make the same amount of vitamin D as a person with a white skin tone [21]. Furthermore, 7-dehydrocholesterol amount in human epidermis is inversely related to age [22]; thus aging significantly decreases the ability of the skin to produce cholecalciferol [23]. An increased proportion of body fat decreases bioavailability of cholecalciferol [24], due to its preferential accumulation in the adipose tissue [25]. Other authors confirmed an inverse association between BMI and 25(OH) vitamin D concentrations and a higher proportion of hypovitaminosis D in obese patients [26]. Another potential causal factor of hypovitaminosis D is reduced absorption in the gastrointestinal tract. Patients with a fat malabsorption syndrome [27], including those who underwent bariatric surgery [28], are often unable to absorb the fat-soluble vitamin D.

The activity of metabolic enzymes of vitamin D pathways can be influenced by a wide variety of medications, especially anticonvulsants and antiviral agents used for HIV treatment, which enhance CYP24A1, leading to an accelerated catabolism of the active metabolite [29].

Vitamin D deficiency results in abnormalities in calcium, phosphorus and bone metabolism. Specifically, vitamin D deficiency causes a decrease in intestinal dietary calcium and phosphorus absorption; consequently, PTH levels increase in response to the lowering calcium concentrations [30, 31]. Secondary hyperparathyroidism maintains serum calcium in the normal range at the expense of mobilizing calcium from the skeleton and increasing phosphorus wasting in the kidneys. Therefore, the persistence of high PTH levels induces osteoclastic activity, creating local foci of bone weakness and causing a generalized decrease in BMD. Osteopenia and osteoporosis ensue, typically in association with normal serum calcium and low-normal phosphorus concentrations. In fact, secondary hyperparathyroidism is generally effective in maintaining normal calcium concentrations but, inducing phosphaturia, leads to a poor phosphorus status. This results in an inadequate calcium-phosphorus product, causing a mineralization defect in the skeleton. In young children who have little mineral in their skeleton, this defect results in a variety of skeletal deformities classically known as rickets [32], while in adults is associated to osteomalacia. Osteomalacia causes a decrease in BMD and is associated with isolated or generalized aches and pains in bones and muscles [33, 34], since $1{,}25(OH)_2$ vitamin D also regulates muscle cells growth and differentiation [35].

The definition of vitamin D adequacy is still largely controversial and mainly based upon the identification of the plasma 25(OH) vitamin D threshold able to suppress PTH synthesis. Different studies reached different conclusions, but the 75 nmol/l (30 ng/ml) cut-off seems the more reliable [36]. Indeed, the threshold of 75 nmol/l is deemed adequate by many for fracture prevention in the general older population [37]. The 50 nmol/l (20 ng/ml) cut-off limit has been alternatively suggested to differentiate populations at true risk for the effects of vitamin D deficiency [38]. Consequently, the vitamin D status is currently defined deficient for concentration lower than 50 nmol/l (20 ng/ml), insufficient for 50–75 nmol/l (20–30 ng/ml) and adequate for 75–250 nmol/l (30–100 ng/ml) [39], though this classification is not universally accepted [38].

Many cohort studies have been performed worldwide to estimate hypovitaminosis D prevalence, all agreeing in the depiction of a very diffuse health concern [40]. In fact, in accordance with the above definitions, the prevalence of vitamin D deficiency is approximately 36% in otherwise healthy young adults aged 18–29 years, 42% in black women aged 15–49 years, 41% in outpatients aged 49–83 years, up to 57% in general medicine inpatients in the United States. Higher rate of prevalence have been described in Europe (28–100% of healthy and 70–100% of hospitalized adults) [41]. Considering the cut-off of 75 nmol/l, the prevalence is obviously higher, especially in the elderly and among hospitalized patients at higher risk for osteoporosis [42–45].

13.1.3 Vitamin D Status in RA

As previously stated, a first relevant issue dealing with the topic of vitamin D in RA is the prevalence of hypovitaminosis D and its correction in patients affected by this condition. The deleterious effect of hypovitaminosis D on bone can be particularly relevant in this population; in fact, patients affected by inflammatory arthritis, such as RA, are characterized by a decreased BMD, being therefore at higher risk for osteoporosis and fractures [46]. Therefore, the maintainance of a normal vitamin D status is a milestone in the treatment of RA comorbidities. Furthermore, hypovitaminosis D prevalence is very high in patients affected by ARD in general and specifically in RA.

Considering non-supplemented patients affected by ARD, the prevalence of hypovitaminosis D, in a rheumatology outpatient clinic in Northern Italy, has been reported to be as high as 87% [47]. This observation is in line with many other studies [48, 49], agreeing with a 90% prevalence of hypovitaminosis D in rheumatic patients who are not undergoing cholecalciferol supplementation. In this specific context, season, gender and age are confirmed to be main determinants of vitamin D status, since female sex, elderly and measurements performed in spring are additional risk factors for hypovitaminosis D [47, 50, 51]. Whether ARD are independent risk factors for hypovitaminosis is, again, highly debated. In the past, lower plasma 25(OH) vitamin D concentrations have been described in ARD patients with respect to general population [52, 53], though other studies lead to opposite results [54–56].

Similarly to what happens in ARD, a high prevalence of hypovitaminosis D has been observed in RA. Specifically, the prevalences of insufficiency and deficiency have been reported around 85% and 45%, respectively [57, 58]. Interestingly, we have recently reported that ARD patients show an altered vitamin D/PTH ratio, since they had higher plasma PTH for similar vitamin D concentrations; in other words, PTH synthesis seems to be more refractory to plasma vitamin D suppression than in general population, contributing to the developement of a" relative hyperparathyroidism". These results suggest that patients with autoimmune/inflammatory diseases may actually have an impairment of vitamin D metabolism. Different possible explanations are conceivable to explain this finding. Chronic inflammatory processes may reduce parathyroid cells sensitivity to active vitamin D. Alternatively, immune cells might consume $1,25(OH)_2$ vitamin D, at the expense of the amount available to act on bone health [59].

A main issue is how to correct vitamin D status in RA. A guideline dealing with cholecalciferol dietary requirements and supplementation in general population has been released in 2011 by The Endocrine Society Task Force [39]. In case of inadequate vitamin D status, the Task Force suggests the use of 50,000 IU of vitamin D2 or vitamin D3 once a week for 8 weeks to achieve a blood level of 25(OH) vitamin D above 30 ng/ml, followed by a maintenance therapy of 1500–2000 IU/d. However, this recommendation is still largely debated and not universally accepted [60]; the best regimen in the specific subset of RA patients is even less defined.

It has been shown that a high loading dose of 300,000 IU, followed by a maintainance daily dose of 800–1000 IU cholecalciferol [61], could be of advantage, being more effective in inducing PTH suppression along with vitamin D normalization. The potential advantages of this regimen need to be weighted at the light of recent findings [62] suggesting an increase in falls and fractures risk in patients treated with a high cholecalciferol dose (500,000 IU). This last observation has been recently replicated in a randomized clinical trial; although higher monthly doses of vitamin D (60,000 IU) were effective in reaching normal 25(OH) vitamin D plasma concentrations, they had no benefit on lower extremity function and were associated with increased risk of falls compared with 24,000 IU. More specific studies on RA are required to better ponder the potential risks and advantages of high doses regimens in this population. Whatever, the correction of hypovitaminosis D in RA is crucial, since patients affected by inflammatory arthritis are at the higher risk of osteoporosis [63].

Osteoporosis is a clinical condition characterized by a high risk of vertebral and non-vertebral fractures, due to the reduction of BMD. The reasons why osteoporosis occurs in inflammatory arthritis are multiple and not completely understood. The failure of several bone regulatory systems has been claimed to be responsible for this complication of systemic inflammatory diseases even though this issue remains partially unresolved. Patients affected by RA have been reported to be at higher risk of vertebral and non vertebral fractures [64–66]. With respect to the reference population values, female RA patients display lower BMD values at the hip and the spine; the risk of osteoporosis seems higher among patients who are older, postmenopausal, positive for rheumatoid factor, treated with corticosteroids, with longer disease duration and higher burden of disability [67]. In a prospective cohort of 102 RA patients [68] who completed a 5-years follow-up, an annual incidence of vertebral fractures of 3.7/100 patients/year has been reported, higher than in the general population according to other prospective studies [69, 70]. The annual incidence of nonvertebral fractures was also increased. The reasons why patients affected by autoimmune inflammatory diseases are prone to develop osteoporosis are complex. A central role seems to be played by systemic inflammation. In patients with RA the overexpression of several inflammatory cytokines TNF-α, IL-1, IL-6 and IL-17 favors the RANKL-induced osteoclastogenesis [71, 72]. Furthermore TNFα can also induce osteocytes to synthetize sclerostin and Dkk-1, two inhibitors of the Wnt/β catenin pathway, a crucial system for osteoblastic differentiation [73, 74]. In a recent study on RA patients, a OPG/RANKL ratio 5 times lower than that observed in healthy controls has been reported, with an inverse correlation between circulating OPG and the disease activity score DAS28, and a positive correlation between RANKL and CRP. Furthermore, Dkk-1 and sclerostin levels were higher in RA patients than in healthy controls. After 2 months of treatment with tocilizumab (a humanized anti-IL-6 receptor antibody), the OPG/RANKL ratio increased proportionally to clinical improvement and suppression of inflammation; furthermore, sclerostin increased while Dkk-1 decreased with respect to baseline [75]. Similar data have also been obtained with other biologics; in fact, the improvement of inflammation control with infliximab has been associated with a reduction in bone loss [76]. Another major factor involved in the pathogenesis of osteoporosis in rheumatic diseases is the long term use of corticosteroids. It is known that glucocorticoids can induce osteoporosis through different mechanisms: in fact, the use of glucocorticoids reduces the number and the function of osteoblasts and impairs their differentiation and maturation through interference with Wnt/β-catenin signaling [77, 78]. In this context, the apoptosis of osteoblasts and osteocytes is enhanced, the expression of RANKL increased and that of OPG decreased, favoring the activation of osteoclasts. Glucocorticoid treatment is an independent risk factor for bone loss; in a meta-analysis on 2891 steroid users glucocorticoid treatment has been linked dose dependently to bone loss and risk fractures, in particular in the first months of treatment. The risk decreases after stopping therapy. However, doses as low as 5 mg/day have been reported to increase the risk of fractures of approximately 20%; interestingly, higher initial doses are strongly related to the risk of bone loss than higher cumulative doses [79].

A role for vitamin D/PTH system in the pathogenesis of RA-related bone loss can be postulated [80]. Actually, an impairment of vitamin D system has already been claimed as a concausal factor in the pathogenesis of osteoporosis in inflammatory arthritis. In fact, a VDR polymorphism has been linked to bone loss in RA [81]; in particular Rass [82] and colleagues found a lower BMD in RA patients carrying the BB and Bb genotypes of the VDR BsmI polymorphism with respect to carriers of the bb genotype. These results suggest that the B allele may be a marker for increased bone reabsorption and bone loss in RA. The recent observation of a "relative hyperparathyroidism" in ARD patients, could partially explain the alteration of bone metabolism observed during chronic inflammatory conditions.

13.1.4 Vitamin D and Rheumatoid Arthritis

In the last few decades, the discovery that VDR is expressed by many different cell types other than the classical target cells led many authors to explore new putative vitamin D functions. For instance, $1,25(OH)_2$ vitamin D is able to induce epidermal cells differentiation [83], has a crucial role in proliferation and differentiation of the nervous system, affecting neurotrophism, neuroprotection, neurotransmission and neuroplasticity [84], inhibits Renin/Angiotensin/Aldosteron System [85], regulates insulin secretion in vitro and in vivo [86–88]. Furthermore, vitamin D has antiproliferative actions in vitro, inducing p21 and p27 and inhibiting the proliferative signal of TGFα-EGFR, observations that focused the attention of scientists on its potential role in cancer prevention [89].

Probably, the most convincing amount of evidence about new functions of vitamin D has been obtained in the field of immune system and rheumatology. A first clue of vitamin D involvement in this context derived from the isolation of VDR in mononuclear cells [90, 91]. Further studies demonstrated that, in vitro, $1,25(OH)_2$ vitamin D regulates the function of almost all the main actors of immune system, acting on both innate and adaptive immunity. Monocytes and macrophages are among the main targets of vitamin D action. $1,25(OH)_2$ vitamin D affects functional activities of monocytes and macrophages. Tumor cell cytotoxicity, phagocytosis, and mycobactericidal activity of monocytes/macrophages are enhanced by exposure to active vitamin D [92, 93], while monocyte function as an APC is decreased [94], as decreased is the production of crucial proinflammatory cytokines such as IL-6 and TNFα [95]. Furthermore, $1,25(OH)_2$ vitamin D promotes the terminal differentiation of monocytes towards a macrophage phenotype [96] and clearly inhibits, in vitro, the differentiation of murine and human monocytes in DC [97]. $1,25(OH)_2$ vitamin D also impairs DC function as APC, by downregulating MHC II and costimulatory molecules expression [98], and chemotaxis [99], thus affecting adaptive immune system, which is strictly regulated by DC activity.

Interestingly, 25(OH) vitamin D could be directly activated by immune cells. In fact, CYP27B1 gene expression is demonstrable in macrophages [100] and human macrophages obtained from synovial fluid of arthritic patients are able to activate directly 25(OH) vitamin D in significant amounts ex vivo [101]. Similarly, DC are able to express CYP27B1 and to activate directly vitamin D in vitro [102]. These observations are particularly relevant, testifying the presence of a positive autocrine loop in immune cells, able to activate vitamin D locally, where it may play a role in response to bacterial and mycobacterial infections. In fact, in animal models, macrophage expression of CYP27B1 is significantly increased and associated to an upregulation of VDR and other vitamin D responsive genes by infectious diseases [103]. The pathways underlying CYP27B1 induction are now well described: a crucial role is played by the TLR family. Specifically, Lipopolysaccharide would activate TLR2/1 and TLR4 upregulating CYP27B1 expression [104]. Another fundamental trigger for vitamin D autocrine activation is represented by IFN-γ [105]. IFNγ and TLR4 seem to act in a synergistic way activating a complex downstream pathway involving JAK, p38, MAPK and the transcriptional factors NF-Kβ, C-EBP and STAT1α; each of them is essential, because blocking individual pathways is sufficient to block CYP27B1 expression. Interestingly this immune autocrine loop is not downregulated by $1,25(OH)_2$ vitamin D, as observed for CYP27B1 in the kidney [106]. Clinically significant macrophagic vitamin D activation can be observed even in chronic inflammatory conditions; CYP27B1 overexpression in granulomas is the mechanism underlying the development of hypercalcemia in patients affected by tuberculosis and sarcoidosis [107].

The action of 1,25(OH)$_2$ vitamin D is not limited to innate immunity. It acts directly on B cells, inhibiting proliferation and inducing apoptosis of activated B cells; furthermore, 1,25(OH)$_2$ vitamin D inhibits plasma cells and post-switch memory B cells differentiation and significantly reduces immunoglobulin secretion [108]. 1,25(OH)$_2$ vitamin D acts as inhibitor of T cells cytotoxic activity, by suppressing Fas-Ligand expression in activated T cells [109]. 1,25(OH)$_2$ vitamin D has been shown to drive CD4$^+$ differentiation leading to a suppression of Th$_1$ and Th$_{17}$ function towards a more favourable and less inflammatory Th$_2$ or T$_{reg}$ phenotype. In fact 1,25(OH)$_2$ vitamin D reduces the expression of the Th$_1$ associated cytokines IL-2, TNF-α, and IFNγ [110] while the Th$_2$ key cytokines IL-4 and IL-5 are induced [111, 112]. In the last decades, a growing interest has been devoted to the role played by Th$_{17}$ in immune diseases, especially in inflammatory arthritis. This specific subset of T CD4+ cells is able to produce IL-21, IL-6, TNF-α and IL-17, thus playing a pivotal role in inflammation. Th$_{17}$ associated cytokines have also been shown to be inhibited by 1,25(OH)$_2$ vitamin D [113]; furthermore 1,25(OH)$_2$ vitamin D inhibits the differentiation and the maintenance of Th$_{17}$ [114]. In addition to Th$_1$, Th$_2$ and Th$_{17}$ cells, CD4+ T cells can also develop into T$_{reg}$, the main function of which appears to be the maintenance of self-tolerance. 1,25(OH)$_2$ vitamin D favours the development of T$_{reg}$, inducing the expression of CTLA-4 and Foxp3 and inhibiting IL-17, IL-21 and IFNγ expression [115]. Figure 13.1 presents schematically the main actions of vitamin D on the immune system.

Despite this consistent body of evidence obtained by in vitro studies, the role of vitamin D in the development and treatment of autoimmune conditions is still largely debated, mainly because in vivo studies lead to less conclusive findings. However, in the last two decades, an inadequate vitamin D status has been associated with different autoimmune conditions, such as T1DM [116], MS [117], IBD [118], SLE [119], UCTD [120].

Finally, vitamin D has been claimed as a potential actor in the pathogenesis of RA. RA is a very common autoimmune rheumatic condition, with a still unknown pathogenesis and high impact in terms of quality of life and socio-economic costs. Vitamin D status has been widely explored in epidemiologic studies, which raised controversies and lead to inconclusive results. In 2004, Merlino et al. described a lower risk of RA development in patients with a greater vitamin D intake (from food or oral supplements) in a prospective cohort of 29,368 women aged 55–69 years along a 11-years follow-up [121]. This finding was not confirmed in another prospective cohort study [122] and was criticized because vitamin D intake estimation was based on self-questionnaires. In fact, in a study on 79 RA patients and 79 age and sex matched controls, no differences were found with respect to plasma 25(OH) vitamin D concentration measured in blood samples collected 1, 2 and 5 years before onset of the disease [54]. Similar conclusions were reached in a more recent cohort study [123]. A potential clue for vitamin D involvement in RA pathogenesis derives from the observation of a meta-analysis, according to which patients carrying TaqI and FokI VDR polymorphisms are at higher risk for RA development [124]. On the other hand, many studies agreed in correlating vitamin D status with disease activity during RA course [55, 125–127].

In conclusion, the current evidences support a convincing immunomodulatory role for vitamin D in vitro, but further studies are required to better define the relative role that vitamin D could play in development and maintenance of autoimmunity in vivo; a better knowledge of these mechanisms could have important consequences and, maybe, lead to implement our therapeutical instruments. In fact, 1,25(OH)$_2$ vitamin D acts on targets that are not influenced by standard treatment, being particularly promising in addition to drugs commonly used in RA. For example, vitamin D is able to suppress the crucial Th$_{17}$ pathway which is essentially untouched by antiTNF treatment [128]. Anyway, 1,25(OH)$_2$

vitamin D does not only act on inflammation, but directly acts on fibroblasts profile, reducing their erosive potential [129]. The fact that monocytes and macrophages are able to activate vitamin D locally in synovia can have further relevance, since the stimulation of this vitamin D milieu in synovial tissue could be particularly favourable in RA patients, leading to the downregulation of local concentration of key inflammatory cytokines, such as IL-6, TNFα, IL-17 [130], and metalloproteinases production.

In the past, cholecalciferol has been already tested in small RCT as additional treatment in RA with controversial results. While some authors described an improvement in disease activity [131] and in patients pain relief [132], others failed to demonstrate any effect on disease course [133, 134]. The sample size of all these studied has been small, which could have affected the results. Furthermore, it is likely that local concentrations required to exert an immunomodulatory role could be obtained only in the presence of very high plasma 25(OH) vitamin D concentrations. Fairney et al. demonstrated that plasma 25(OH) vitamin D approximately doubles the synovial concentration [135]. Since Jeffery et al. [115] showed an anti-inflammatory effect of 25(OH) vitamin D, in the presence of activating cells such as APCs, at a 50–100 nM concentration, a biological effect in vivo probably requires a plasma 25(OH) vitamin D concentration constantly >100–200 nmol/l, the safety of which is to be tested.

13.2 Conclusion

In summary, though in vitro data are promising, it is not possible to attribute to vitamin D a conclusive role in immune diseases pathogenesis and treatment in vivo yet. Nevertheless, obtaining a normal vitamin D status is paramount in preventing RA-related osteoporosis; therefore, the correction of a deficient vitamin D status should be suggested to each rheumatic patient. Currently, there is no consensus about the best regimen for these patients and further studies are required to better address this important area of medical problems. Furthermore, more research on the postulated role in immune modulation might prove to be more relevant than expected. In the near future, further insights on vitamin D physiology could lead the way to new therapeutic uses of this old, but still promising molecule.

References

1. Rosenheim O, Webster TA (1927) The relation of cholesterol to vitamin D. Biochem J 21:127–129
2. Avioli LV (1969) Absorption and metabolism of vitamin D3 in man. Am J Clin Nutr 22:437–446
3. Ohyama Y, Yamasaki T (2004) Eight cytochrome P450s catalyze vitamin D metabolism. Front Biosci 9:3007–3018
4. DeLuca HF (1969) 25-Hydroxycholecalciferol, the probable metabolically active form of vitamin D. Isolation, identification, and subcellular location. Am J Clin Nutr 22:412–424
5. Daiger SP, Schanfield MS, Cavalli-Sforza LL (1975) Group-specific component (Gc) proteins bind vitamin D and 25-hydroxyvitamin D. Proc Natl Acad Sci U S A 72:2076–2080
6. Zerwekh JE (2008) Blood biomarkers of vitamin D status. Am J Clin Nutr 87:1087–191S
7. Norman AW (1971) Evidence for a new kidney-produced hormone, 1,25-dihydroxycholecalciferol, the proposed biologically active form of vitamin D. Am J Clin Nutr 24:1346–1351
8. Zehnder D, Bland R, Williams MC, McNinch RW, Howie AJ, Stewart PM, Hewison M (2001) Extrarenal expression of 25-hydroxyvitamin d(3)-1 alpha-hydroxylase. J Clin Endocrinol Metab 86:888–894
9. Adams JS, Hewison M (2012) Extrarenal expression of the 25-hydroxyvitamin D-1-hydroxylase. Arch Biochem Biophys 523:95–102
10. Kliewer SA, Umesono K, Mangelsdorf DJ, Evans RM (1992) Retinoid X receptor interacts with nuclear receptors in retinoic acid, thyroid hormone and vitamin D3 signalling. Nature 355:446–449
11. Christakos S, Lieben L, Masuyama R, Carmeliet G (2014) Vitamin D endocrine system and the intestine. Bonekey Rep 3:496
12. Jeon US (2008) Kidney and calcium homeostasis. Electrolyte Blood Press 6:68–76
13. Kim S, Yamazaki M, Zella LA, Meyer MB, Fretz JA, Shevde NK, Pike JW (2007) Multiple enhancer regions located at significant distances upstream of the transcriptional start site mediate RANKL gene expression in response to 1,25-dihydroxyvitamin D3. J Steroid Biochem Mol Biol 103:430–434
14. Khosla S (2001) Minireview: the OPG/RANKL/RANK system. Endocrinology 142:5050–5055

15. Demay MB, Kiernan MS, DeLuca HF, Kronenberg HM (1992) Sequences in the human parathyroid hormone gene that bind the 1,25-dihydroxyvitamin D3 receptor and mediate transcriptional repression in response to 1,25-dihydroxyvitamin D3. Proc Natl Acad Sci U S A 89:8097–8101
16. Zgaga L, Theodoratou E, Farrington SM, Agakov F, Tenesa A, Walker M, Knox S, Wallace AM, Cetnarskyj R, McNeill G, Kyle J, Porteous ME, Dunlop MG, Campbell H (2011) Diet, environmental factors, and lifestyle underlie the high prevalence of vitamin D deficiency in healthy adults in Scotland, and supplementation reduces the proportion that are severely deficient. J Nutr 141:1535–1542
17. Matsuoka LY, Wortsman J, Haddad JG, Hollis BW (1989) In vivo threshold for cutaneous synthesis of vitamin D3. J Lab Clin Med 114:301–305
18. Wacker M, Holick MF (2013) Sunlight and vitamin D: a global perspective for health. Dermatoendocrinol 5:51–108
19. Hyppönen E, Power C (2007) Hypovitaminosis D in British adults at age 45 y: nationwide cohort study of dietary and lifestyle predictors. Am J Clin Nutr 85:860–868
20. Matsuoka LY, Ide L, Wortsman J, MacLaughlin JA, Holick MF (1987) Sunscreens suppress cutaneous vitamin D3 synthesis. J Clin Endocrinol Metab 64:1165–1168
21. Clemens TL, Henderson SL, Adams JS, Holick MF (1982) Increased skin pigment reduces the capacity of skin to synthesise vitamin D3. Lancet 1:74–76
22. MacLaughlin J, Holick MF (1985) Aging decreases the capacity of human skin to produce vitamin D3. J Clin Invest 76:1536–1538
23. Holick MF, Matsuoka LY, Wortsman J (1989) Age, vitamin D, and solar ultraviolet. Lancet 2:1104–1105
24. Snijder MB, van Dam RM, Visser M, Deeg DJ, Dekker JM, Bouter LM, Seidell JC, Lips P (2005) Adiposity in relation to vitamin D status and parathyroid hormone levels: a population-based study in older men and women. J Clin Endocrinol Metab 90:4119–4123
25. Wortsman J, Matsuoka LY, Chen TC, Lu Z, Holick MF (2000) Decreased bioavailability of vitamin D in obesity. Am J Clin Nutr 72:690–693
26. Hyppönen E, Power C (2006) Vitamin D status and glucose homeostasis in the 1958 British birth cohort: the role of obesity. Diabetes Care 29:2244–2246
27. Thompson GR, Lewis B, Booth CC (1966) Absorption of vitamin D3-3H in control subjects and patients with intestinal malabsorption. J Clin Invest 45:94–102
28. Karefylakis C, Näslund I, Edholm D, Sundbom M, Karlsson FA, Rask E (2014) Vitamin D status 10 years after primary gastric bypass: gravely high prevalence of hypovitaminosis D and raised PTH levels. Obes Surg 24:343–348
29. Zhou C, Assem M, Tay JC, Watkins PB, Blumberg B, Schuetz EG, Thummel KE (2006) Steroid and xenobiotic receptor and vitamin D receptor crosstalk mediates CYP24 expression and drug-induced osteomalacia. J Clin Invest 116:1703–1712
30. Holick MF (2007) Vitamin D deficiency. N Engl J Med 357:266–281
31. Heaney RP (2004) Functional indices of vitamin D status and ramifications of vitamin D deficiency. Am J Clin Nutr 80:1706–179S
32. Holick MF (2006) Resurrection of vitamin D deficiency and rickets. J Clin Invest 116:2062–2072
33. Malabanan AO, Turner AK, Holick MF (1998) Severe generalized bone pain and osteoporosis in a premenopausal black female: effect of vitamin D replacement. J Clin Densitometr 1:201–204
34. Plotnikoff GA, Quigley JM (2003) Prevalence of severe hypovitaminosis D in patients with persistent, nonspecific musculoskeletal pain. Mayo Clin Proc 78:1463–1470
35. Boland R (1986) Role of vitamin D in skeletal muscle function. Endocr Rev 7:434–448
36. Chapuy MC, Preziosi P, Maamer M, Arnaud S, Galan P, Hercberg S, Meunier PJ (1997) Prevalence of vitamin D insufficiency in an adult normal population. Osteoporos Int 7:439–443
37. Bischoff-Ferrari H (2009) Vitamin D: what is an adequate vitamin D level and how much supplementation is necessary? Best Pract Res Clin Rheumatol 23:789–795
38. Rosen CJ, Abrams SA, Aloia JF, Brannon PM, Clinton SK, Durazo-Arvizu RA, Gallagher JC, Gallo RL, Jones G, Kovacs CS, Manson JE, Mayne ST, Ross AC, Shapses SA, Taylor CL (2012) IOM committee members respond to Endocrine Society vitamin D guideline. J Clin Endocrinol Metab 97:1146–1152
39. Holick MF, Binkley NC, Bischoff-Ferrari HA, Gordon CM, Hanley DA, Heaney RP, Murad MH, Weaver CM (2011) Endocrine Society. Evaluation, treatment, and prevention of vitamin D deficiency: an Endocrine Society clinical practice guideline. J Clin Endocrinol Metab 96:1911–1930
40. Mithal A, Wahl DA, Bonjour JP, Burckhardt P, Dawson-Hughes B, Eisman JA, El-Hajj Fuleihan G, Josse RG, Lips P, Morales-Torres J (2009) IOF committee of scientific advisors (CSA) nutrition working group. Global vitamin D status and determinants of hypovitaminosis D. Osteoporos Int 20:1807–1820
41. Holick MF (2006) High prevalence of vitamin D inadequacy and implications for health. Mayo Clin Proc 81:353–373
42. Hirani V, Primatesta P (2005) Vitamin D concentrations among people aged 65 years and over living in private households and institutions in England: population survey. Age Ageing 34:485–491
43. Simonelli C, Weiss TW, Morancey J, Swanson L, Chen YT (2005) Prevalence of vitamin D inadequacy in a minimal trauma fracture population. Curr Med Res Opin 21:1069–1074

44. Isaia G, Giorgino R, Rini GB, Bevilacqua M, Maugeri D, Adami S (2003) Prevalence of hypovitaminosis D in elderly women in Italy: clinical consequences and risk factors. Osteoporos Int 14:577–582
45. Perin A, Zanatta E, Pigatto E, Carniello S, Cozzi F (2012) Hypovitaminosis D in an hospitalized old population of Western Friuli. Reumatismo 64:166–171
46. Bultink IE, Vis M, van der Horst-Bruinsma IE, Lems WF (2012) Inflammatory rheumatic disorders and bone. Curr Rheumatol Rep 14:224–230
47. Sainaghi PP, Bellan M, Carda S, Cerutti C, Sola D, Nerviani A, Molinari R, Cisari C, Avanzi GC (2012) Hypovitaminosis D and response to cholecalciferol supplementation in patients with autoimmune and non-autoimmune rheumatic diseases. Rheumatol Int 32:3365–3372
48. Mouyis M, Ostor AJ, Crisp AJ, Ginawi A, Halsall DJ, Shenker N, Poole KE (2008) Hypovitaminosis D among rheumatology outpatients in clinical practice. Rheumatology (Oxford) 47:1348–1351
49. Stoll D, Dudler J, Lamy O, Hans D, So A, Krieg MA, Aubry-Rozier B (2011) High prevalence of hypovitaminosis D in a Swiss rheumatology outpatient population. Swiss Med Wkly 141:13196
50. Reusch J, Ackermann H, Badenhoop K (2009) Cyclic changes of vitamin D and PTH are primarily regulated by solar radiation: 5-year analysis of a German (50 degrees N) population. Horm Metab Res 41:402–407
51. Jacques PF, Felson DT, Tucker KL, Mahnken B, Wilson PW, Rosenberg IH, Rush D (1997) Plasma 25-hydroxyvitamin D and its determinants in an elderly population sample. Am J Clin Nutr 66:929–936
52. Atwa MA, Balata MG, Hussein AM, Abdelrahman NI, Elminshawy HH (2013) Serum 25-hydroxyvitamin D concentration in patients with psoriasis and rheumatoid arthritis and its association with disease activity and serum tumor necrosis factor-alpha. Saudi Med J 34:806–813
53. Als OS, Riis B, Christiansen C (1987) Serum concentration of vitamin D metabolites in rheumatoid arthritis. Clin Rheumatol 6:238–243
54. Nielen MM, van Schaardenburg D, Lems WF, van de Stadt RJ, de Koning MH, Reesink HW, Habibuw MR, van der Horst-Bruinsma IE, Twisk JW, Dijkmans BA (2006) Vitamin D deficiency does not increase the risk of rheumatoid arthritis: comment on the article by Merlino et al. Arthritis Rheum 54:3719–3720
55. Cutolo M, Otsa K, Laas K, Yprus M, Lehtme R, Secchi ME, Sulli A, Paolino S, Seriolo B (2006) Circannual vitamin d serum levels and disease activity in rheumatoid arthritis: northern versus southern Europe. Clin Exp Rheumatol 24:702–704
56. Grazio S, Naglić ĐB, Anić B, Grubišić F, Bobek D, Bakula M, Kavanagh HS, Kuna AT, Cvijetić S (2015) Vitamin D serum level, disease activity and functional ability in different rheumatic patients. Am J Med Sci 349:46–49
57. Kerr GS, Sabahi I, Richards JS, Caplan L, Cannon GW, Reimold A, Thiele GM, Johnson D, Mikuls TR (2011) Prevalence of vitamin D insufficiency/deficiency in rheumatoid arthritis and associations with disease severity and activity. J Rheumatol 38:53–59
58. Zheng ZH, Gao CC, Wu ZZ, Liu SY, Li TF, Gao GM, Liu ZS (2016) High prevalence of hypovitaminosis D of patients with autoimmune rheumatic diseases in China. Am J Clin Exp Immunol 5:48–54
59. Sainaghi PP, Bellan M, Antonini G, Bellomo G, Pirisi M (2011) Unsuppressed parathyroid hormone in patients with autoimmune/inflammatory rheumatic diseases: implications for vitamin D supplementation. Rheumatology (Oxford) 50:2290–2296
60. Bouillon R, Van Schoor NM, Gielen E, Boonen S, Mathieu C, Vanderschueren D, Lips P (2013) Optimal vitamin D status: a critical analysis on the basis of evidence-based medicine. J Clin Endocrinol Metab 98:1283–1304
61. Sainaghi PP, Bellan M, Nerviani A, Sola D, Molinari R, Cerutti C, Pirisi M (2013) Superiority of a high loading dose of cholecalciferol to correct hypovitaminosis d in patients with inflammatory/autoimmune rheumatic diseases. J Rheumatol 40:166–172
62. Sanders KM, Stuart AL, Williamson EJ, Simpson JA, Kotowicz MA, Young D, Nicholson GC (2010) Annual high-dose oral vitamin D and falls and fractures in older women: a randomized controlled trial. JAMA 303:1815–1822
63. Bischoff-Ferrari HA, Dawson-Hughes B, Orav EJ, Staehelin HB, Meyer OW, Theiler R, Dick W, Willett WC, Egli A (2016) Monthly high-dose vitamin D treatment for the prevention of functional decline: a randomized clinical trial. JAMA Intern Med 176:175–183
64. Hooyman JR, Melton LJ III, Nelson AM, O'Fallon WM, Riggs BL (1984) Fractures after rheumatoid arthritis: a population-based study. Arthritis Rheum 27:1353–1361
65. Peel NF, Moore DJ, Barrington NA, Bax DE, Eastell R (1995) Risk of vertebral fracture and relationship to bone mineral density in steroid treated rheumatoid arthritis. Ann Rheum Dis 54:801–806
66. Kim SY, Schneeweiss S, Liu J, Daniel GW, Chang CL et al (2010) Risk of osteoporotic fracture in a large population-based cohort of patients with rheumatoid arthritis. Arthritis Res Ther 12:R154
67. Haugeberg G, Uhlig T, Falch JA, Halse JI, Kvien TK (2000) Bone mineral density and frequency of osteoporosis in female patients with rheumatoid arthritis: results from 394 patients in the Oslo County rheumatoid arthritis register. Arthritis Rheum 43:522–530
68. Vis M, Haavardsholm EA, Bøyesen P, Haugeberg G, Uhlig T, Hoff M et al (2011) High incidence of vertebral and non-vertebral fractures in the OSTRA cohort study: a 5-year follow-up study in postmeno-

pausal women with rheumatoid arthritis. Osteoporos Int 22:2413–2419

69. EPOS study group (2002) Incidence of vertebral fracture in Europe: results from the European Prospective Osteoporosis Study (EPOS). J Bone Miner Res 17:716–724
70. Nevitt MC, Cummings SR, Stone KL, Palermo L, Black DM, Bauer DC et al (2005) Risk factors for a first-incident radiographic vertebral fracture in women at least 65 years of age: the study of osteoporotic fractures. J Bone Miner Res 20:131–140
71. Geusens PP, Lems WF (2011) Osteoimmunology and osteoporosis. Arthritis Res Ther 13:242
72. Takayanagi H (2009) Osteoimmunology and the effects of the immune system on bone. Nat Rev Rheumatol 5:667–676
73. Rachner TD, Khosla S, Hofbauer LC (2011) Osteoporosis: now and the future. Lancet 377:1276–1287
74. Lories RJ, Luyten FP (2009) Osteoimmunology: Wnt antagonists: for better or worse? Nat Rev Rheumatol 5:420–421
75. Terpos E, Fragiadaki K, Konsta M, Bratengeier C, Papatheodorou A, Sfikakis PP (2011) Early effects of Il-6 receptor inhibition on bone homeostasis. Clin Exp Rheumatol 29:921–925
76. Vis M, Havaardsholm EA, Haugeberg G, Uhlig T, Voskuyl AE, van de Stadt RJ et al (2006) Evaluation of bone mineral density, bone metabolism, osteoprotegerin and receptor activator of the NFkappaB ligand serum levels during treatment with infliximab in patients with rheumatoid arthritis. Ann Rheum Dis 65:1495–1499
77. Canalis E, Mazziotti G, Giustina A, Bilezikian JP (2007) Glucocorticoid-induced osteoporosis: pathophysiology and therapy. Osteoporos Int 18:1319–1328
78. Ohnaka K, Tanabe M, Kawate H, Nawata H, Takayanagi R (2005) Glucocorticoid suppresses the canonical Wnt signal in cultured human osteoblasts. Biochem Biophys Res Commun 329:177–181
79. van Staa TP, Leufkens HG, Cooper C (2002) The epidemiology of corticosteroid-induced osteoporosis: a meta-analysis. Osteoporos Int 13:777
80. Bellan M, Pirisi M, Sainaghi PP (2014) Osteoporosis in rheumatoid arthritis: role of the vitamin D/parathyroid hormone system. Rev Bras Reumatol 55:256–263
81. Ranganathan P (2009) Genetics of bone loss in rheumatoid arthritis – role of vitamin D receptor polymorphisms. Rheumatology 48:342–343
82. Rass P, Pákozdi A, Lakatos P, Zilahi E, Sipka S, Szegedi G et al (2006) Vitamin D receptor gene polymorphism in rheumatoid arthritis and associated osteoporosis. Rheumatol Int 26:964–971
83. Hosomi J, Hosoi J, Abe E, Suda T, Kuroki T (1983) Regulation of terminal differentiation of cultured mouse epidermal cells by 1 alpha, 25-dihydroxyvitamin D3. Endocrinology 113:1950–1957
84. DeLuca GC, Kimball SM, Kolasinski J, Ramagopalan SV, Ebers GC (2013) Review: the role of vitamin D in nervous system health and disease. Neuropathol Appl Neurobiol 39:458–484
85. Li YC, Kong J, Wei M, Chen ZF, Liu SQ, Cao LP (2002) 1,25-Dihydroxyvitamin D(3) is a negative endocrine regulator of the renin-angiotensin system. J Clin Invest 110:229–238
86. Inomata S, Kadowaki S, Yamatani T, Fukase M, Fujita T (1986) Effect of 1 alpha (OH)-vitamin D3 on insulin secretion in diabetes mellitus. Bone Miner 1:187–192
87. Palomer X, González-Clemente JM, Blanco-Vaca F, Mauricio D (2008) Role of vitamin D in the pathogenesis of type 2 diabetes mellitus. Diabetes Obes Metab 10:185–197
88. Norman AW, Frankel JB, Heldt AM, Grodsky GM (1980) Vitamin D deficiency inhibits pancreatic secretion of insulin. Science 209:823–882
89. Rosen CJ, Adams JS, Bikle DD, Black DM, Demay MB, Manson JE, Murad MH, Kovacs CS (2012) The nonskeletal effects of vitamin D: an Endocrine Society scientific statement. Endocr Rev 33:456–492
90. Bhalla AK, Amento EP, Clemens TL, Holick MF, Krane SM (1983) Specific high-affinity receptors for 1,25-dihydroxyvitamin D3 in human peripheral blood mononuclear cells: presence in monocytes and induction in T lymphocytes following activation. J Clin Endocrinol Metab 57:1308–1310
91. Yu XP, Hustmyer FG, Garvey WT, Manolagas SC (1991) Demonstration of a 1,25-dihydroxyvitamin D3-responsive protein in human lymphocytes: immunologic crossreactivity and inverse regulation with the vitamin D receptor. Proc Natl Acad Sci U S A 88:8347–8351
92. Walters MR (1992) Newly identified actions of the vitamin D endocrine system. Endocr Rev 13:719–764
93. Bikle DD (2008) Vitamin D and the immune system: role in protection against bacterial infection. Curr Opin Nephrol Hypertens 179:348–352
94. Xu H, Soruri A, Gieseler RKH, Peters JH (1993) 1,25-Dihydroxyvitamin D3 exerts opposing effects to IL-4 on MHC class II antigen expression, accessory activity, and phagocytosis of human monocytes. Scand J Immunol 38:535–540
95. Zhang Y, Leung DY, Richers BN, Liu Y, Remigio LK, Riches DW, Goleva E (2012) Vitamin D inhibits monocyte/macrophage proinflammatory cytokine production by targeting MAPK phosphatase-1. J Immunol 188:2127–2135
96. Kreutz M, Andressen R (1990) Induction of human monocyte to macrophage maturation in vitro by 1,25-dihydroxyvitamin D3. Blood 76:2457–2461
97. Piemonti L, Monti P, Sironi M, Fraticelli P, Leone BE, Dal Cin E, Allavena P, Di Carlo V (2000) Vitamin D3 affects differentiation, maturation, and

function of human monocyte-derived dendritic cells. J Immunol 164:4443–4451
98. Griffin MD, Lutz WH, Phan VA, Bachman LA, McKean DJ, Kumar R (2000) Potent inhibition of dendritic cell differentiation and maturation by vitamin D analogs. Biochem Biophys Res Commun 270:701–708
99. Gauzzi MC, Purificato C, Donato K, Jin Y, Wang L, Daniel KC, Maghazachi AA, Belardelli F, Adorini L, Gessani S (2005) Suppressive effect of 1alpha,25-dihydroxyvitamin D3 on type I IFN-mediated monocyte differentiation into dendritic cells: impairment of functional activities and chemotaxis. J Immunol 174:270–276
100. Monkawa T, Yoshida T, Hayashi M, Saruta T (2000) Identification of 25-hydroxyvitamin D3 1alpha-hydroxylase gene expression in macrophages. Kidney Int 58:559–568
101. Smith SJ, Hayes ME, Selby PL, Mawer EB (1999) Autocrine control of vitamin D metabolism in synovial cells from arthritic patients. Ann Rheum Dis 58:372–378
102. Fritsche J, Mondal K, Ehrnsperger A, Andreesen R, Kreutz M (2003) Regulation of 25-hydroxyvitamin D3-1 alpha-hydroxylase and production of 1 alpha, 25-dihydroxyvitamin D3 by human dendritic cells. Blood 102:3314–3316
103. Nelson CD, Reinhardt TA, Beitz DC, Lippolis JD (2010) In vivo activation of the intracrine vitamin D pathway in innate immune cells and mammary tissue during a bacterial infection. PLoS One 5:e15469
104. White JH (2012) Regulation of intracrine production of 1,25-dihydroxyvitamin D and its role in innate immune defense against infection. Arch Biochem Biophys 523:58–63
105. Overbergh L, Stoffels K, Waer M, Verstuyf A, Bouillon R, Mathieu C (2006) Immune regulation of 25-hydroxyvitamin D-1alpha-hydroxylase in human monocytic THP1 cells: mechanisms of interferon-gamma-mediated induction. J Clin Endocrinol Metab 91:3566–3574
106. Stoffels K, Overbergh L, Giulietti A, Verlinden L, Bouillon R, Mathieu C (2006) Immune regulation of 25-hydroxyvitamin-D3-1alpha-hydroxylase in human monocytes. J Bone Miner Res 21:37–47
107. Adams JS, Gacad MA (1985) Characterization of 1 alpha-hydroxylation of vitamin D3 sterols by cultured alveolar macrophages from patients with sarcoidosis. J Exp Med 161:755–765
108. Chen SGP, Chen XX, Gu YY, Chen S, Lipsky PE (2007) Modulatory effects of 1,25-dihydroxyvitamin D3 on human B cell differentiation. J Immunol 179:1634–1647
109. Cippitelli M, Fionda C, Di Bona D, Di Rosa F, Lupo A, Piccoli M, Frati L, Santoni A (2002) Negative regulation of CD95 ligand gene expression by vitamin D3 in T lymphocytes. J Immunol 168:1154–1166
110. Rigby WF, Denome S, Fanger MW (1987) Regulation of lymphokine production and human T lymphocyte activation by 1,25-dihydroxyvitamin D3. Specific inhibition at the level of messenger RNA. J Clin Invest 79:1659–1664
111. Cantorna MT, Woodward WD, Hayes CE, DeLuca HF (1998) 1,25-dihydroxyvitamin D3 is a positive regulator for the two anti-encephalitogenic cytokines TGF-beta 1 and IL-4. J Immunol 160:5314–5319
112. Sloka S, Silva C, Wang J, Yong VW (2011) Predominance of Th2 polarization by vitamin D through a STAT6-dependent mechanism. J Neuroinflammation 8:56
113. Joshi S, Pantalena LC, Liu XK, Gaffen SL, Liu H, Rohowsky-Kochan C, Ichiyama K, Yoshimura A, Steinman L, Christakos S, Youssef S (2011) 1,25-dihydroxyvitamin D(3) ameliorates Th17 autoimmunity via transcriptional modulation of interleukin-17A. Mol Cell Biol 31:3653–3669
114. Zhang H, Shih DQ, Zhang X (2013) Mechanisms underlying effects of 1,25-Dihydroxyvitamin D3 on the Th17 cells. Eur J Microbiol Immunol 3:237–240
115. Jeffery LE, Wood AM, Qureshi OS, Hou TZ, Gardner D, Briggs Z, Kaur S, Raza K, Sansom DM (2012) Availability of 25-hydroxyvitamin D(3) to APCs controls the balance between regulatory and inflammatory T cell responses. J Immunol 189:5155–5164
116. Hyppönen E, Läärä E, Reunanen A, Järvelin MR, Virtanen SM (2001) Intake of vitamin D and risk of type 1 diabetes: a birth-cohort study. Lancet 358:1500–1503
117. Ascherio A, Munger KL, White R, Köchert K, Simon KC, Polman CH, Freedman MS, Hartung HP, Miller DH, Montalbán X, Edan G, Barkhof F, Pleimes D, Radü EW, Sandbrink R, Kappos L, Pohl C (2014) Vitamin D as an early predictor of multiple sclerosis activity and progression. JAMA Neurol 71:306–314
118. Pappa HM, Grand RJ, Gordon CM (2006) Report on the vitmin D status of adult and pediatric patients with inflammatory bowel disease and its significance for bone health and disease. Inflamm Bowel Dis 12:1162–1174
119. Müller K, Kriegbaum NJ, Baslund B, Sørensen OH, Thymann M, Bentzen K (1995) Vitamin D3 metabolism in patients with rheumatic diseases: low serum levels of 25-hydroxyvitamin D3 in patients with systemic lupus erythematosus. Clin Rheumatol 14:397–400
120. Zold E, Szodoray P, Gaal J, Kappelmayer J, Csathy L, Gyimesi E, Zeher M, Szegedi G, Bodolay E (2008) Vitamin D deficiency in undifferentiated connective tissue disease. Arthritis Res Ther 10:R123
121. Merlino LA, Curtis J, Mikuls TR, Cerhan JR, Criswell LA, Saag KG, Study I W's H (2004) Vitamin D intake is inversely associated with rheumatoid arthritis: results from the Iowa women's health study. Arthritis Rheum 50:72–77
122. Costenbader KH, Feskanich D, Holmes M, Karlson EW, Benito-Garcia E (2008) Vitamin D intake and

risks of systemic lupus erythematosus and rheumatoid arthritis in women. Ann Rheum Dis 67:530–535
123. Hiraki LT, Arkema EV, Cui J, Malspeis S, Costenbader KH, Karlson EW (2014) Circulating 25-hydroxyvitamin D level and risk of developing rheumatoid arthritis. Rheumatology (Oxford) 53:2243–2248
124. Tizaoui K, Hamzaoui K (2015) Association between VDR polymorphisms and rheumatoid arthritis disease: Systematic review and updated meta-analysis of case-control studies. Immunobiology 220:807–816
125. Fakharan M, Haghighi A, Arabi M, Loghman M (2014) Investigating the levels of serum vitamin d in patients with rheumatoid arthritis referred to rasoul-akram hospital during 2011–2012. Iran J Med Sci 39:476–479
126. Abourazzak FE, Talbi S, Aradoini N, Berrada K, Keita S, Hazry T (2015) 25-Hydroxy vitamin D and its relationship with clinical and laboratory parameters in patients with rheumatoid arthritis. Clin Rheumatol 34:353–357
127. Sabbagh Z, Markland J, Vatanparast H (2013) Vitamin D status is associated with disease activity among rheumatology outpatients. Forum Nutr 5:2268–2275
128. van Hamburg JP, Asmawidjaja PS, Davelaar N, Mus AM, Cornelissen F, van Leeuwen JP, Hazes JM, Dolhain RJ, Bakx PA, Colin EM, Lubberts E (2012) TNF blockade requires 1,25(OH)2D3 to control human Th17-mediated synovial inflammation. Ann Rheum Dis 71:606–612
129. Laragione T, Shah A, Gulko PS (2012) The vitamin D receptor regulates rheumatoid arthritis synovial fibroblast invasion and morphology. Mol Med 18:194–200
130. Luo J, Wen H, Guo H, Cai Q, Li S, Li X (2013) 1,25-dihydroxyvitamin D3 inhibits the RANKL pathway and impacts on the production of pathway-associated cytokines in early rheumatoid arthritis. Biomed Res Int 2013:101805
131. Andjelkovic Z, Vojinovic J, Pejnovic N, Popovic M, Dujic A, Mitrovic D, Pavlica L, Stefanovic D (1999) Disease modifying and immunomodulatory effects of high dose 1 alpha (OH) D3 in rheumatoid arthritis patients. Clin Exp Rheumatol 17:453–456
132. Gopinath K, Danda D (2011) Supplementation of 1,25 dihydroxy vitamin D3 in patients with treatment naive early rheumatoid arthritis: a randomised controlled trial. Int J Rheum Dis 14:332–339
133. Hein G, Oelzner P (2000) Vitamin D metabolites in rheumatoid arthritis: findings – hypotheses – consequences. Z Rheumatol 1:28–32
134. Hansen KE, Bartels CM, Gangnon RE, Jones AN, Gogineni J (2014) An evaluation of high-dose vitamin D for rheumatoid arthritis. J Clin Rheumatol 20:112–114
135. Fairney A, Straffen AM, May C, Seifert MH (1987) Vitamin D metabolites in synovial fluid. Ann Rheum Dis 46:370–374

14 Asthma and Allergy "Epidemic" and the Role of Vitamin D Deficiency

Konstantinos Douros, Barbara Boutopoulou, Sotirios Fouzas, and Ioanna Loukou

Abstract

The increase in asthma and allergies prevalence that has been recorded in many countries during the last decades, and the reemergence of vitamin D (VD) deficiency in many populations worldwide, renders fairly plausible the assumption of an underlying association between these two conditions and justifies the research effort invented in this issue. Indeed, there is growing body of evidence from epidemiological, laboratory, and clinical studies, suggesting that such an association does exist. The hypothesis of low levels of VD leading to compromised fetal programming and impairment of various immune functions involved in asthma and allergic disorders, stands as the most credible explanation of this presumed association. However, the evidence is not yet definite and there are some conflicting results among studies. As a consequence, no safe conclusions can be drawn yet, and more research is required in order to fully clarify the involvement of VD deficiency in the pathogenesis of asthma and allergies, and decide if VD has a role to play in the prevention and therapy of these disorders.

Keywords

Allergies • Asthma • Epidemic • Fetus programming • Vitamin D • VD supplementation

K. Douros (✉) • B. Boutopoulou
3rd Department of Paediatrics, "Attikon" Hospital, University of Athens School of Medicine, Athens, Greece
e-mail: costasdouros@gmail.com

S. Fouzas
Department of Paediatrics, School of Medicine, University of Patras, Patras, Greece

I. Loukou
Cystic Fibrosis Unit, "Aghia Sophia" Children's Hospital, Athens, Greece

S. Ahmad (ed.), *Ultraviolet Light in Human Health, Diseases and Environment*, Advances in Experimental Medicine and Biology 996, https://doi.org/10.1007/978-3-319-56017-5_14

14.1 Introduction

Vitamin D (VD) has gained much attention in recent years and is considered a topic of great scientific interest. Research on VD was in the past decades focused mainly on the osseous-related endocrine system and skeletal health [52]. However, it has now become common knowledge that essentially all tissues bear vitamin D receptors (VDRs) and, in addition, a wealth of evidence supports the vitamin's immunomodulatory role regarding both the innate and adaptive components of the immune system [34]. This, in turn, has led researchers to seek associations between VD deficiency states and a number of immune mediated illnesses, including autoimmune, atopic and infectious conditions. Indeed, there are fairly compelling evidence supporting potential roles of VD deficiency in the pathogenesis of a wide spectrum of mainly chronic and phenomenally unrelated illnesses such as, cardiovascular, autoimmune, respiratory and allergic disorders, metabolic syndrome, dementia, and cancer [38, 95].

The importance of VD can be further emphasized by the fact that in modern societies VD deficiency tends to become a common health problem, probably because of lifestyle changes, such as the more sedentary way of life with more time spend indoors, and the increased use of sun safety practices [52].

The purpose of this chapter is to examine the current evidence for the potential role of VD deficiency on asthma and other allergic disorders focusing mainly on food allergy (FA) and atopic dermatitis (AD).

14.2 Vitamin D Sources and Metabolism

In humans, the majority of VD is synthesized in the skin under the influence of ultraviolet solar radiation which converts 7-dehydrocholesterol into cholecalciferol (vitamin D3). Sunscreens and clothing prevent the conversion of 7-dehydrocholesterol to vitamin D3 [34]. Apart from D3 there is also another bioactive, plant-derived form of VD, named vitamin D2 (ergocalciferol), which has been produced commercially since the early 1920s and is now widely used for food fortification and as a dietary supplement. Structurally, the two forms are similar, with vitamin D2 having an extra methyl group on carbon 24 and an additional double bond between carbons 22 and 23. However, these structural differences have no major consequences on the metabolic activation of VD, and so the two forms are still officially regarded as being equivalent [102]. Dietary VD is only a minor contributor to overall VD used by human organisms, and the exact amount depends on each individual's particular diet. In general however, food sources of VD are limited and include oily fish, cod liver oil, and egg yolks. Fortification of infant formula with VD is mandatory whereas, some dairy products are also fortified with VD. The dietary intake is more important when sunshine exposure is inadequate. People in northern countries are dependent on dietary sources to ensure adequate VD levels, especially during the winter months.

VD is not biologically active and must be converted to its active form. To achieve this, VD from either the skin or the diet, enters the circulation where it is transported by the vitamin D binding protein (DBP) to the liver where it is metabolized to 25 hydroxyvitamin D (25(OH)D; calcidiol) by the enzyme vitamin D 25-hydroxylase. 25(OH)D is the major form of circulating VD; nevertheless, it is practically biologically inactive. 25(OH)D is further transported through DBP to the kidneys where it goes through another hydroxylation in the kidney's mitochondria by the enzyme 25-hydroxyvitamin D-1α-hydroxylase (1-OHase) resulting in $1,25(OH)_2D$ (calcitriol) which is the actual active form of VD. Once formed, $1,25(OH)_2D$ can interact with the specific nuclear VDR to exert its actions.

Kidneys were long thought to be the only organs capable to produce calcitriol. However, we now know that expression of 1-OHase is widespread in extrarenal sites and the conversion of calcidiol to calcitriol can take place in many tissues other than kidneys, such as skeletal and heart muscle, T and B lymphocytes, macrophages, liver, brain, placenta, breast, colon, prostate, and bronchi [12]. It has to be mentioned that

VD is not, strictly speaking, a vitamin but rather a hormone, since vitamins are substances that cannot be synthesized in sufficient quantities by an organism, and must be obtained from its diet.

14.3 Vitamin D Deficiency

14.3.1 Definition of Vitamin D Deficiency

There is a continuing debate on how VD deficiency is defined. Serum concentration of 25(OH) D is considered the best determinant of VD status. However, determining 25(OH)D serum concentrations is not straightforward and considerable interassay discrepancies have been noted in studies comparing different commercially available 25(OH)D tests [25, 114]. In addition to the previous technical problem, there is also scarcity of data regarding the correlation between 25(OH)D serum concentrations with health outcomes, something that makes the definition of VD deficiency on the basis solely of 25(OH)D levels quite challenging. On the other hand, calcitriol is not useful as a marker of VD status or for the correlation with health outcomes because of its short serum half-life [17]. Furthermore, it has been demonstrated that polymorphisms in DBP gene may affect the percentage of bioavailable VD making the issue of determining the normal levels of 25(OH)D even more complicated [96]. Given the many uncertainties, a scientific advisory committee in United Kingdom concluded that the current data are insufficient to clarify relations among VD intake, biochemical status, and chronic disease outcomes [109].

However, despite the many unanswered questions, most authors use 25(OH)D serum levels of 50 nmol/l (25 nmol/l = 10 ng/ml), as the cut-off value of VD deficiency, although, some recommend higher levels, e.g. higher than 75 or even 100 nmol/l [52, 102]. The 2011 report from the Institute of Medicine Committee on dietary requirements for calcium and VD levels concluded that according to available evidence, VD concentration of 20 ng/ml (50 nmol/liter) is adequate for the needs of at least 97.5% of the population [102]. The European Society for Paediatric Gastroenterology, Hepatology and Nutrition (ESPGHAN) has recently re-established the recommendation of a 25(OH)D concentration > 50 nmol/l to indicate sufficiency whereas, severe VD deficiency during childhood, is defined as 25(OH)D concentrations <25 nmol/l [17]. In other pediatric guidelines [81], VD status has been given a more detailed definition and has been schematically divided in 4 categories, as follows: severe deficiency: <12.5 nmol/L (5 ng/mL); deficiency: 12.5–37.5 nmol/L (5–15 ng/mL); insufficiency: 37.5–50 nmol/L (15–20 ng/mL); sufficiency: > 50 nmol/L (20 ng/ml).

14.3.2 Determinants of Hypovitaminosis D

The main determinants of VD status is time spend indoors and sunshine exposure, skin pigmentation, clothing, use of sunscreen, nutrition and use of VD supplements [17]. It has been demonstrated that serum levels of 25(OH)D are not always a characteristic of people living in sun deprived areas. VD levels tend to be higher in Northern than in Southern Europe [123]. This phenomenally "odd" conclusion has been confirmed by subsequent multicenter studies that used a single laboratory for their measurements and the comparison of 25(OH)D levels was straightforward ([67, 74] [122]). The high serum levels of 25(OH) D in Norway and Sweden are probably due to a high intake of fatty fish and cod liver oil, whereas the lower serum 25(OH)D in Spain, Italy and Greece may be the result of more pigmented skin and sun avoidance practices [123].

Obesity has been associated with low serum levels of VD [81], although the underlying mechanism has not still verified. It may occur because of trapping of VD in fat cells [98] or, on the other hand, VD deficiency may not be a consequence but rather a risk factor and obesity may predispose to VD deficiency through upregulation of adiponectin synthesis [115]. Secondary causes of VD deficiency are medications (e.g. rifampicin, anticonvulsants), and diseases causing malabsorption such as celiac disease and cystic fibrosis [78].

14.3.3 Epidemiology of Vitamin D Deficiency

Regardless of the thresholds used, VD deficiency unambiguously comprises a common health problem worldwide. Epidemiological studies suggest that approximately 50% of adults and children worldwide are VD deficient, with serum concentrations of 25(OH)D below 50 nmol/L [13, 73]. Hilger et al., in a recent comprehensive systematic review of the literature, found that 88.1% of the papers reported mean or median values of 25(OH)D levels below 75 nmol/l, 37.3% below 50 nmol/l and 6.7% below 25 nmol/l [51]. Studies included in this review contained data on almost 170,000 participants from 44 countries, representing practically all geographic areas of the earth. The authors found significant differences between the various geographic areas, with values being significantly higher in North America than in Europe or Middle East/Africa region. Age-related differences were not found in Europe and North America. However, in the Asia/Pacific region, children and adolescents had significantly lower 25(OH)D values compared to adults and elderly. In contrast, children and adolescents from Middle East and Africa had significantly higher values than adults and elderly. No significant sex-related differences were observed in any of the regions, although reports in women tended to show lower 25(OH)D values, especially in the Asia/Pacific and Middle East/Africa regions [122]. The later observation could be, at least partially, explained by limited sun exposure of women due to cultural practices.

14.4 Asthma and Allergies: An Epidemic of Modernity?

14.4.1 Epidemiological Evidence

The prevalence of asthma has been increasing since the decade of 1960s in most areas of the world [132]. The highest prevalence of the disease has been observed in developed countries [9]. In 2006, Eder et al. summarized data from registries and epidemiological surveillance studies, and found that increases in the prevalence of asthma had been reported from most of the countries during the second half of the twentieth century. However, this apparent global increase in asthma prevalence was evident only up to the 1990s, whereas after then the temporal pattern was not clear. In some populations, the prevalence of asthma was still rising, while on the contrary in others appeared to be stable or decreasing slightly. In particular, in some areas of the Western world, the prevalence seemed to have reached a plateau [40].

In the same year, another group published a study [5] on global trends in the prevalence of asthma, rhinoconjunctivitis, and eczema in childhood. Researchers repeatedly collected cross-sectional data for two age-groups of school children – 6–7 years and 13–14 years – from many countries as part of phase 3 International Study of Asthma and Allergies in Childhood (ISAAC) and compared data with that of a baseline survey that had been undertaken approximately 7 years before. A large number of centers from around the world took part in this study. In the majority of centers there was a change in prevalence of 1 or more standard errors for at least one disorder, with increases being twice as common as decreases, and increases being more common in the 6–7 year age-group than in the 13–14 year age-group. An exception was observed for asthma symptoms in the older age-group, where decreases were more common at high prevalence. For both age-groups, more centers showed increases in all three disorders more often than showing decreases, but most centers had mixed changes. The changes were greatest for eczema in the younger age group, and for allergic rhinoconjunctivitis in both age-groups.

The ISAAC study also found an increase in AD in both age groups. There was a significant increase in the prevalence of AD, ranging from 0.07 to 1.09% in 48 countries with decrease in 8 countries and little change in 12 countries in the 6–7-year age group and increase in 47 countries, decrease in 32 countries, and little change in 26 countries in the 13–14-year age group. In general, the increase occurred primarily in previous areas of low prevalence, whereas the areas with previous high levels showed leveling off [116].

The prevalence of FA is difficult to be determined with precision because standard criteria for its definition are still lacking. Two recent systematic reviews estimated that FA roughly affects 0.1–6.0% of the European population [90] and more than 1–2% but less than 10% worldwide [26]. In US, from 1997 to 2007, the prevalence of reported FA increased by 18% in children less than 18 years old [18]. Despite the absence of an accurate estimation of its prevalence, available data suggest that there has been a significant increase in the last two decades and the problem has reached almost epidemic characteristics in developed western countries [10, 91, 113, 125]. FA is considered now the main cause of anaphylaxis [72].

14.4.2 Hypotheses for Explaining the "Epidemic" of Asthma and Allergies

Although much research effort has been investigated in the explanation of increasing prevalence of asthma and allergies, a comprehensive and explicit model to interpret this phenomenon is still lacking. The majority of authors try to approach and explain this issue based on the hygiene hypothesis [117] which suggests that children living in modern families of westernized societies have minimal exposure to infectious agents in the first years of life, which in turn leads to inappropriate development of their immune system and its regulatory pathways. As a consequence, the transition from a prevalent Th2 to a balanced Th1/Th2 response that ensues during infancy is not properly and fully implemented, and children develop a propensity for asthma and allergies [101]. However, despite favorable evidence from experimental studies, epidemiological data have not been consistent and sufficient to confirm this hypothesis, and now it is evident that this model cannot explain on its own the many dimensions of the extensive asthma and allergy epidemic [99].

In 2007, Litonjua and Weiss proposed a new approach to explain the issue, based on another consequence of westernized way of living. They argued that, as populations become more prosperous they tend to spend more time indoors and expose themselves less to sunlight [76]. This results in VD deficiency that can lead in a disposition for the development of asthma and allergies. VD deficiency's harmful effect begins as early as prenatal life through its impact on fetal lung and immune system development [64], and is continued postnatally. This hypothesis offers a reasonable alternative explanation for some of the epidemiologic attributes related with the asthma epidemic, namely the increased risks associated with obesity, and recent immigrants to westernized countries. Some of the cons and pros of this theory will be reviewed in the next pages.

14.5 Is Vitamin D a Key Player Allergy and Asthma Epidemic?

14.5.1 The Regulatory Role of Vitamin D in the Immune System

The concept of VD possessing a key role in the regulation of the immune response has emerged from the discovery of a variety of immune cells, such as antigen-presenting cells, natural killer cells, and B and T lymphocytes, that express the intracellular VDR and are responsive to $1,25(OH)_2D$ [36]. Moreover, it is supported by the ability of various types of immune cells, including peripheral blood mononuclear cells, activated T lymphocytes, dendritic cells, and quiescent CD4 cells, to convert 25(OH)D to $1,25(OH)_2D$ [1, 79, 110]. A number of studies have demonstrated the modulatory effects of VD in the function of cells involved in numerous aspects of immunity, affecting both innate and adaptive immunity.

The immune and inflammatory processes that underlie allergic and asthmatic disorders are complex and still not fully understood; however, there is robust evidence that impaired immune regulation plays a pivotal role in the pathogenesis of these disorders [77]. Based on these grounds, it has been argued that VD is directly involved in the pathogenesis of asthma and allergies, and indeed, there is a substantial body of high-quality research in favor of this connection. Some of the evidence supporting a potential role of VD deficiency in the development of asthma and allergies through defective immunoregulation are reviewed below.

Calcitriol decreases the maturation of dendritic cells (DC) and their ability to activate T lymphocytes. Activation of VDR signaling pathways inhibits DC maturation as evidenced by decreased levels of DC markers, such as MHC-class II, co-stimulatory molecules (CD40, CD80, and CD86), and other maturation-induced surface antigens (e.g. CD83) ([8]). Similarly to DC, the antigen-presenting and T cell stimulatory ability of monocytes and macrophages is reduced upon exposure to 1,25$(OH)_2$D [4, 8]. Immunomodulatory pathways involve both direct and indirect effects on the proliferation, differentiation, and function of T cells. VD administration decreases TH1 cytokine secretion and inhibits T-cell proliferation [60, 86]. Matheu et al. found that VD administration in mice could induce IL-4, IL-5, and IL-13 ([80]). A study on naive human cord CD4 and CD8 T cells showed that VD exerts an inhibitory effect on IFN-γ production induced by IL-12, and suppresses IL-4, and IL-13 expression induced by IL-4 [36]. VD stimulates $CD4^+CD25^+Foxp3^+$ Treg cells to increase the secretion of IL-10, which is considered a chief effector molecule in VD–mediated promotion of the immunosuppressive states of T cells [106]. Activation of VDR modulates the humoral immune response by inhibiting IgE expression in B cells and enhancing IL-10 expression [48, 50]. Apart from IL-10, it has been recently demonstrated that 1,25$(OH)_2$D has a profound inhibitory effect on the development of human Th9, a $CD4^+$ T cell subset that is highly associated with asthma, in an IL-10 independent manner [119]. In a study that examined the effects of VD on T cell differentiation it was found that the stimulation of $CD4^+$ $CD25^-$ T cells in the presence of 1,25$(OH)_2$D inhibits the production of pro-inflammatory cytokines including IFN-γ, IL-17, and IL-21, and stimulates expression of high levels of CTLA-4 and FoxP3. CTLA-4 plays an important role in the suppressive activity of Treg, whereas $FoxP3^+$ Treg cells have a central role in the prevention of catastrophic autoimmunity. Through the previous pathway, T cells treated with 1,25$(OH)_2$D were able to suppress proliferation of normally responsive T cells, indicating that they possessed characteristics of adaptive Tregs [59]. In patients with severe asthma, there was a strong correlation of low serum concentrations of 25(OH)D with a paucity of $FoxP3^+$ Treg cells in the peripheral blood. In the same study it was also found that the frequency of circulating $FoxP3^+$ Treg cells were significantly lower in steroid resistant than in steroid sensitive asthmatic patients with comparable disease severity [27]. Effector $CD8^+$ T cells convert from IFN-γ^+ (Th1) to IL-13^+ (Th2) cells as a result of their activation by IL-4-producing $CD4^+$ T cells, a process having been related with the development of asthma [66]. It was also recently shown, that the addition of 25$(OH)_2$D during $CD8^+$ T-cell differentiation prevents IL-4-induced functional conversion of $CD8^+$ T cells from IFNγ to IL-13 producing cells, via a mechanistic link to CYP11A1 activity [108].

VD also regulates the effectiveness of innate immunity in antimicrobial defense mechanisms [83]). This is mainly accomplished through the induction in skin, lung, and monocytes of cathelicidin [47, 107], an antimicrobial peptide that has a pivotal role in the host defence against various microbial infections [44, 88]. Apart from that, VD can augment skin defence mechanisms by increasing locally the production of some other antimicrobial peptides, such as b-defensin 2 and b-defensin 3, epidermal lipid synthesis enzymes, fatty acid synthase, serine C-palmitoyltransferase, and 3-hydroxy-3-methylglutaryl–coenzyme A, as well as the structural proteins of corneocytes, filaggrin and involucrin [53]. All of the above are essential elements for the development of structural and functional integrity of skin.

In general, however, despite the abundance of available data on the immunomodulatory effects of VD and the quite a few proposed mechanisms [62], the synthesis of evidence into a coherent model that can clearly explain how these effects can intervene in the propensity of an individual to develop asthma and allergies has not yet become feasible.

14.5.2 Vitamin D and Asthma

There is an ongoing debate in medical literature as to whether VD deficiency has a causal association with the development of asthma and allergies. Most of the epidemiological and pathophysiological data support the existence of such a link, although there have been contradictory results too [23, 24, 42, 57, 133]. A lot of studies from different geographic areas have shown high prevalence of VD deficiency in asthmatic children. VD insufficiency was very common in North American children with mild-to-moderate persistent asthma and furthermore, the low levels of VD were associated with higher odds of severe exacerbation [20]. In another population-based study it was shown that low serum concentrations of 25(OH)D, were strongly correlated with reduced spirometric indices (FEV_1, and FVC) [14]. A study conducted in Costarican children, demonstrated that VD levels were negatively associated with airway responsiveness, eosinophil counts, IgE levels, and likelihood of hospitalization for asthma exacerbations [19]. In two studies from Italy, low levels of VD in children were associated with reduced lung function and increased reactivity to exercise; only 11% of children with asthma had adequate VD levels [31, 32]. A study from Turkey in preschool aged children found that the frequency of VD deficiency and insufficiency was higher in children with asthma, compared to the controls [120]. An Indian study showed that, compared with controls, children with asthma had lower 25(OH)D concentrations and proportion of Treg cells, whereas the proportions of B cells with IgE receptors (CD23 and CD21) were higher [28]. In a randomized, double-blind, placebo-controlled trial, the daily supplementation with 800 IU of VD, in addition to standard therapy, for 2 months in schoolchildren from Japan with asthma resulted in significant improvement in levels of asthma control [118].

Contrary to the previous results, however, some other studies indicated that supplementation with VD has no effect in asthma. So, in a double-blind, randomized, placebo-controlled trial in Irish children with uncontrolled asthma, the authors assessed the effect of 15 weeks of a relatively high dose of VD supplementation (2000 IU/day) on subjective asthma symptoms, lung function, and biomarkers of inflammation/allergy. They found that apart from decreased school days missed, VD supplementation failed to demonstrate any advantageous changes in asthma parameters [63]. Likewise, a recent meta-nalysis found only relatively weak evidence to support VD supplementation for the reduction of asthma exacerbations, and no real evidence to support any benefits of VD supplementation for other asthma-related outcomes in children [100].

A number of hypotheses have been postulated to explain the possible pathogenetic link between asthma and VD deficiency. One of these hypotheses is based on the protective role of VD against infections with the reasoning being as follows: Viral respiratory tract infections (VRTI) are the most common trigger of asthmatic exacerbations, especially in children, and typically lead to more severe symptoms compared to non-asthmatics [56]. VD has virucidal effects [22, 39, 121] and, in addition, may also exert antiviral effects through alterations in growth and differentiation of airway epithelial cells [21]. Considering the above, it is plausible that VD deficiency may weaken the defense mechanisms against respiratory infections and thus indirectly increase asthma exacerbations and morbidity [16].

Evidence exists for an association between low maternal VD levels during pregnancy and increased risk of wheezing or asthma. In three observational studies from USA, Japan, and Scotland, it was shown that maternal VD supplementation protects offspring from the development of wheeze [23, 24, 37, 82]. In another large prospective cohort study, low levels of 25(OH)D at birth – but not in mid-gestation – were associated with a higher airway resistance measured at the age of 6 years [45]. These results may indicate that VD deficiency could affect the programming of the developing lung resulting in increased disease susceptibility [35, 69]. In support of this assumption there are findings from genetic studies demonstrating regulatory effects of $1,25(OH)_2D$ in airway smooth muscle genes related with morphogenesis, cell growth, and survival, as well as genes encoding structural proteins; all these genes are known to be involved in airway remodelling [11, 15]. These results were strengthened by consistent findings from a study which showed that VD-regulated genes were markedly over-represented in developing human lung transcriptomes; a significant number of these genes were known to be associated with asthma susceptibility [64]. The aforementioned studies support a genomic mechanistic pathway linking low levels of maternal VD with childhood asthma susceptibility, possibly through fetal programming. However, it has to be mentioned, that two recent randomized controlled studies that aimed to investigate the effect of VD supplementation during pregnancy (beyond the usually recommended doses) on asthma-like symptoms in the first 3 years of offspring life, resulted in inconclusive findings [30, 75]. Although, the authors of both trials acknowledge that their studies may have been underpowered to detect a potentially important clinical effect, the results underline the complexity of the issue and the fact that is far from being considered adequately answered [126].

VD, apart from being involved in the pathogenesis of asthma, it may also affect the response to treatment by enhancing the therapeutic response to corticosteroids. Searing et al. showed that in asthmatic children decreased serum VD levels were associated with increased corticosteroid use; the observed correlation may be due to either more severe disease in children with low levels of VD, or to the signaling effects of VD on corticosteroids pathways [111]. Goleva et al. showed that both steroid requirements and in vitro steroid responsiveness were negatively associated with VD status in children; the same trend was also observed in adult patients though it did not reach statistical significance. The later may suggest that the effects of VD were stronger in childhood asthmatics, probably because of the less advanced airway remodeling [46]. The molecular mechanism of steroid resistance is not yet clear but a well-known characteristic of steroid-resistant (SR) asthmatic patients is the inability of their CD4 T cells to increase IL-10 synthesis after in vitro stimulation with corticosteroids [49]. Xystrakis et al. demonstrated that the addition of $1,25(OH)_2D$ and dexamethasone to cultures of CD4 T cells from SR asthmatics, enhanced IL-10 secretion to levels comparable with those produced from cells of steroid-sensitive patients treated only with dexamethasone. They further showed that supplementation with VD of SR asthmatic patients, reversed steroid resistance through induction of IL-10-secreting Tregs [134]. Zhang and colleagues went further and showed that VD could also enhance the anti-inflammatory action of corticosteroids in an IL-10 independent manner through the induction of mitogen-activated protein kinase phosphatase-1 (MKP-1) in blood monocytes. However, they also observed that the responses in patients with SR asthma remained significantly lower compared with patients with steroid-sensitive asthma [135, 136]. In a recent study, Lan et al. showed that pre-treatment of patients with VD enhanced the dexamethasone-induced nuclear translocation of glucocorticoid receptors in airway epithelial cell cultures and monocytes of VD deficient adult patients with severe asthma exacerbations. They also showed that supplemental VD ameliorated the oxidative stress and increased the clinical response to corticosteroid. Their results indicate that VD-deficiency aggravates the oxidative changes observed in severe asthma exacerbation and also imply a possible role for VD therapy in acute asthma [71].

To summarise, although many aspects of the underlying mechanism still remain obscure, there is plethora of data suggesting an aetiological association between VD deficiency and asthma.

14.5.3 Vitamin D and Food Allergy

Epidemiological data have linked FA with sunlight exposure by showing that FA is more prevalent in regions further away from the equator [84, 85]. A recent study from Korea compared VD serum levels and the incidence of food induced anaphylaxis (FIA) between 2 regions of high and low solar radiation; VD levels were lower and concurrently FIA incidence was higher in the region with the low er solar radiation [65]. This kind of studies can lead to assumptions but cannot prove that FA is correlated to VD status and not to any other sunlight-derived, seasonal and/or geographic factor. However, they have provided a stimulating framework for further research on the subject [104]. Indeed, the research hypothesis of a connection between VD and FA has been largely corroborated by cross-sectional and cohort studies assessing VD dietary intake during pregnancy, or measuring maternal, cord blood, neonatal or childhood VD status. Sharief et al. used a large nationally representative sample from United States of more than 3000 children and adolescent, and found that VD deficiency was associated with allergic sensitization to selected foods. This association was not observed in the adult group of the study [112]. Similarly, in a cross-sectional study from Korea that included 226 infants with atopic dermatitis or FA it was found that VD deficiency increased the risk of food allergen sensitization [7]. In another large prospective cohort study conducted in Australia, infants with 25(OH)D < 50 nmol/l at 12 months of age were more likely to suffer from challenge-proven FA – especially to egg and peanut – and were more likely to have multiple food allergies compared with those with adequate VD levels [3]. Interestingly, this association was evident only among infants of Australian-born parents, suggesting a gene–environmental interaction [94].

Nwaru BI et al. showed that maternal intake of VD during pregnancy was associated with decreased risk of food sensitization in the offspring at the age of 5 years [89]. Chiu CY et al. demonstrated an inverse association between cord blood 25(OH)D levels from a birth cohort of Taiwanese children and milk sensitization at the age of 2 years [33]. However, there are also studies that failed to find a valid association between cord blood 25(OH)D levels and FA [29, 61]. Though not proving an association does not necessarily rule out its existence, their results have introduced some scepticism. What is more, a German study reported that maternal and cord blood 25(OH)D was positively associated with children's risk for FA within the first 2 years of life. As a general comment, one could say that the presence of conflicting results reflects the gaps in our knowledge on the exact role of VD in the development of FA.

A proposed model of the induction of FA focuses on the actions of VD on the gastrointestinal tract which is the mucosal site with the richest antigenic exposure in the body. According to this model, VD deficiency not only compromises immune tolerance, but also increases susceptibility to infections and affects gut microbiota composition. The ensuing gastrointestinal infections compromise barrier permeability and other defence mechanisms against dietary and microbial antigens in the intestinal mucosa. These defects synergistically promote dysfunctional responses to food antigens that may manifest as FA in genetically susceptible subjects [124].

14.5.4 Vitamin D and Atopic Dermatitis

AD is characterised by defective skin barrier, skin inflammation, and increased susceptibility to certain skin infections. Given the aforementioned attributes of VD, namely immunomodulatory effects, cathelicidin induction, and impact on the development of skin barrier, it comes as no surprise that VD deficiency is now considered one of the contributing factors in the pathogenesis of AD and indeed, there are many epidemiological reports that have described this association [70, 82, 129, 130]. Besides, there is also direct evidence relating low blood levels of VD in children with the severity of AD [2, 41, 93]. In a large cohort study that sought to examine the associations between early-life VD levels and the development of eczema and other allergy-related outcomes in a population of black and white children from the same region, it was shown that prenatal 25(OH)D levels were inversely associated with eczema. This association was stronger in white children implying that VD plays a role of varying importance depending on other risk factors present in a subject [127]. In another cohort study of children with a family history of allergic disease, cord blood 25(OH)D concentrations found to be negatively associated with risk of eczema in early childhood [92]. A recent metanalysis demonstrated that maternal VD status during pregnancy was associated with childhood eczema [128]. Some authors have even suggested that VD supplementation should be included in the therapeutic armamentarium of atopic dermatitis [58, 105].

14.6 Can Vitamin D Increase the Risk of Asthma and Allergies?

Despite the amount of literature supporting a direct relation of VD deficiency with the development of asthma and allergies, there are also quite a few studies suggesting that VD promotes the development of these disorders. Some of the existing antilogy has already been presented herein and some more will be reviewed below.

In a number of studies, VD supplementation has been linked to allergy and asthma [6, 43, 55, 68]. In an elegant review of the literature on this particular topic, Wjst has sited the evidence and discussed the reasoning underlying the findings [131]. As he postulates, the explanation of the apparent paradox of both VD deficiency and VD supplementation promoting asthma/allergy should be sought in the epigenetic programming in pregnancy and early infancy, where both low and high levels of VD lead to the same end result. In cases of excessive VD exposure in newborn period, either by extra supplements or as an additive to baby foods, it is assumed that different reactions can be induced. The kind of reactions depends on how low or rich the intrauterine environment was in VD. This assumption is based on existing evidence for other disorders showing that their development can be facilitated by large discrepancies between the environments experienced in utero, in early infancy, and later life [131].

Evidence also exists that high VD levels can increase the risk for asthma and allergies independently of dietary supplementation. Hyppönen et al. showed that IgE concentrations were higher in subjects having either low (<25 nmol/L) or very high (>135 nmol/L) levels of 25(OH)D [54]. Rothers et al. demonstrated that both lower (<10 ng/mL) and higher (≥ 20 ng/mL) levels of cord blood 25(OH)D were associated with higher frequency of sensitization to aero-allergens up to the age of 5 years, compared with a reference group (10–19.9 nmol/L) [103]. A Canadian study found that children with both low (≤49 nm/L) and high (≥75 nm/L) 25(OH)D levels had increased risk of asthma [87]. The above results suggest that possibly a U-shaped association between VD levels and asthma/allergies exists, and that there is a threshold in both low and high 25(OH)D levels associated with increased risk of these disorders.

Finally on his chapter on environmental factors and inflamatory non-communicable diseases, Prescott tried to address the complexity of the issue with an alternative but very interesting approach; he proposed that VD might actually be a proxy for UV exposure, which could someway reduce a subject's propensity for inflammation. If that was the case, then VD supplementation would not be enough to overcome the effects of a UV deficiency [97].

14.7 Conclusion

The worldwide epidemic proportions of asthma and other allergic disorders and hypovitaminosis D, have been the subject of extensive research efforts in the last two decades. In an attempt to explain the coexistence of these two synchronous "epidemics", it was proposed that there is a causal link between VD deficiency and the development of asthma and allergies. In support of this is the vast amount of available data showing the important role of VD in regulation of inflammation and modulation of various immune responses. Indeed, most of the studies have verified the hypothesis of VD deficiency predisposing to the development of asthma and allergies. However, there have been some studies with opposing results, supporting that VD supplements and high VD levels may have a detrimental effect on asthma and allergies. It is undoubtedly very challenging to try to interpret phenomenically discordant results and propose a unifying model. And not surprisingly, despite the bulk of available evidence, there are still huge gaps in our knowledge to allow a translation of basic research on immunomodulatory attributes of VD into clinical practice. Clearly, more research, and especially adequately powered randomised control trials are needed in order to extract safe conclusions on the role of VD in the prevention and treatment of asthma and the other allergic disorders. Up to then, it would be prudent in cases of difficult asthma or recalcitrant eczema, to check VD levels and, if low, to supplement with VD.

Conflict of Interest Authors declare no conflicts of interest.

References

1. Adorini L, Penna G, Giarratana N et al (2004) Dendritic cells as key targets for immunomodulation by vitamin D receptor ligands. J Steroid Biochem Mol Biol 89–90:437–441
2. Akan A, Azkur D, Ginis T et al (2013) Vitamin D level in children is correlated with severity of atopic dermatitis but only in patients with allergic sensitizations. Pediatr Dermatol 30:359–363
3. Allen KJ, Koplin JJ, Ponsonby AL et al (2013) Vitamin D insufficiency is associated with challenge-proven food allergy in infants. J Allergy Clin Immunol 131:1109–1116
4. Almerighi C, Sinistro A, Cavazza A et al (2009) 1Alpha,25-dihydroxyvitamin D3 inhibits CD40L-induced pro-inflammatory and immunomodulatory activity in human monocytes. Cytokine 45:190–197
5. Asher MI, Montefort S, Björkstén B et al (2006) ISAAC phase three study group. Worldwide time trends in the prevalence of symptoms of asthma, allergic rhinoconjunctivitis, and eczema in childhood: ISAAC phases one and three repeat multicountry cross-sectional surveys. Lancet 368:733–743
6. Back O, Blomquist HK, Hernell O et al (2009) Does vitamin D intake during infancy promote the development of atopic allergy? Acta Derm Venereol 89:28–32
7. Baek JH, Shin YH, Chung IH et al (2014) The link between serum vitamin D level, sensitization to food allergens, and the severity of atopic dermatitis in infancy. J Pediatr 165:849–854
8. Baeke F, Takiishi T, Korf H et al (2010) Vitamin D: modulator of the immune system. Curr Opin Pharmacol 10:482–496
9. Baïz N, Annesi-Maesano I (2012) Is the asthma epidemic still ascending? Clin Chest Med 3:419–429
10. Berin MC, Sampson HA (2013) Food allergy: an enigmatic epidemic. Trends Immunol 34:390–397
11. Berraies A, Hamzaoui K, Hamzaoui A (2014) Link between vitamin D and airway remodeling. J Asthma Allergy 1(7):23–30
12. Bikle DD (2010) Extrarenal synthesis of 1,25-Dihydroxyvitamin D and its health implications. In: Holick MF (ed) Vitamin D: physiology, molecular biology, and clinical applications. Humana Press, Totowa, pp 277–296
13. Bischoff-Ferrari HA (2012) "Vitamin D – why does it matter?" – defining vitamin D deficiency and its prevalence. Scand J Clin Lab Investig Suppl 243:3–6
14. Black PN, Scragg R (2005) Relationship between serum 25-hydroxyvitamin D and pulmonary function in the third national health and nutrition examination survey. Chest 128:3792–3798
15. Bosse Y, Maghni K, Hudson TJ (2007) 1alpha,25-dihydroxy-vitamin D3 stimulation of brochial smooth muscle cells induces autocrine, contractility, and remodeling processes. Physiol Genomics 29:161–168
16. Bozzetto S, Carraro S, Giordano G et al (2012) Asthma, allergy and respiratory infections: the vitamin D hypothesis. Allergy 67(1):10–17
17. Braegger C, Campoy C, Colomb V et al (2013) ESPGHAN committee on nutrition. Vitamin D in the healthy European paediatric population. J Pediatr Gastroenterol Nutr 56:692–701
18. Branum AM, Lukacs SL (2008) Food allergy among U.S. children: trends in prevalence and hospitalizations. NCHS Data Brief 10:1–8
19. Brehm JM, Celedón JC, Soto-Quiros ME et al (2009) Serum vitamin D levels and markers of severity of childhood asthma in Costa Rica. Am J Respir Crit Care Med 179:765–771
20. Brehm JM, Schuemann B, Fuhlbrigge AL et al (2010) Childhood asthma management program research group. Serum vitamin D levels and severe asthma exacerbations in the childhood asthma management program study. J Allergy Clin Immunol 126:52–58
21. Brockman-Schneider RA, Pickles RJ, Gern JE (2014) Effects of vitamin D on airway epithelial cell morphology and rhinovirus replication. PLoS One 9:86755

22. Busse WW, Lemanske RF Jr, Gern JE (2010) Role of viral respiratory infections in asthma and asthma exacerbations. Lancet 376:826–834
23. Camargo CA Jr, Rifas-Shiman SL, Litonjua AA et al (2007a) Maternal intake of vitamin D during pregnancy and risk of recurrent wheeze in children at 3 y of age. Am J Clin Nutr 85:788–795
24. Camargo CA, Clark S, Kaplan MA et al (2007b) Regional differences in EpiPen prescriptions in the United States: the potential role of vitamin D. J Allergy Clin Immunol 120:131–136
25. Carter GD (2011) Accuracy of 25-hydroxyvitamin D assays: confronting the issues. Curr Drug Targets 12:19–28
26. Chafen JJ, Newberry SJ, Riedl MA et al (2010) Diagnosing and managing common food allergies: a systematic review. JAMA 303:1848–1856
27. Chambers ES, Nanzer AM, Richards DF et al (2012) Serum 25-dihydroxyvitamin D levels correlate with CD4(+)Foxp3(+) T-cell numbers in moderate/severe asthma. J Allergy Clin Immunol 130:542–544
28. Chary AV, Hemalatha R, Murali MV et al (2016) Association of T-regulatory cells and CD23/CD21 expression with vitamin D in children with asthma. Ann Allergy Asthma Immunol 116:447–454
29. Chawes BL, Bønnelykke K, Jensen PF et al (2014) Cord blood 25(OH)-vitamin D deficiency and childhood asthma, allergy and eczema: the COPSAC2000 birth cohort study. PLoS One 9:99856
30. Chawes BL, Bønnelykke K, Stokholm J et al (2016) Effect of vitamin D3 supplementation during pregnancy on risk of persistent wheeze in the offspring: a randomized clinical trial. JAMA 315:353–361
31. Chinellato I, Piazza M, Sandri M et al (2011a) Vitamin D serum levels and markers of asthma control in Italian children. J Pediatr 158:437–441
32. Chinellato I, Piazza M, Sandri M et al (2011b) Serum vitamin D levels and exercise-induced bronchoconstriction in children with asthma. Eur Respir J 37:1366–1370
33. Chiu CY, Yao TC, Chen SH et al (2014) Low cord blood vitamin D levels are associated with increased milk sensitization in early childhood. Pediatr Allergy Immunol 25:767–772
34. Christakos S, Dhawan P, Verstuyf A et al (2016) Vitamin D: metabolism, molecular mechanism of action, and pleiotropic effects. Physiol Rev 96:365–408
35. De Luca G, Olivieri F, Melotti G et al (2010) Fetal and early postnatal life roots of asthma. J Matern Fetal Neonatal Med 23(3):80–83
36. Deluca HF, Cantorna MT (2001) Vitamin D: its role and uses in immunology. FASEB J 15:2579–2585
37. Devereux G, Litonjua AA, Turner SW et al (2007) Maternal vitamin D intake during pregnancy and early childhood wheezing. Am J Clin Nutr 85:853–859
38. Douros K (2016) Cystic fibrosis and vitamin D: the quest for more pieces of the puzzle. Acta Paediatr 105:854
39. Douville RN, Bastien N, Li Y et al (2007) Adult asthmatics display exaggerated IFNgamma responses to human metapneumovirus and respiratory syncytial virus. Biochem Cell Biol 85:252–258
40. Eder W, Ege MJ, von Mutius E (2006) The asthma epidemic. N Engl J Med 355:2226–2235
41. El Taieb MA, Fayed HM, Aly SS et al (2013) Assessment of serum 25-hydroxyvitamin d levels in children with atopic dermatitis: correlation with SCORAD index. Dermatitis 24:296–301
42. Freishtat RJ, Iqbal SF, Pillai DK et al (2010) High prevalence of vitamin d deficiency among inner-city African American youth with asthma in Washington. DC J Pediatr 156:948–952
43. Gale CR, Robinson SM, Harvey NC et al (2008) Princess Anne Hospital Study Group. Maternal vitamin D status during pregnancy and child outcomes. Eur J Clin Nutr 62:68–77
44. Gallo RL, Hooper LV (2012) Epithelial antimicrobial defense of the skin and intestine. Nat Rev Immunol 12:503–516
45. Gazibara T, den Dekker HT, de Jongste JC et al (2016) Associations of maternal and fetal 25-hydroxyvitamin D levels with childhood lung function and asthma: the Generation R Study. Clin Exp Allergy 46:337–346
46. Goleva E, Searing DA, Jackson LP et al (2012) Steroid requirements and immune associations with vitamin D are stronger in children than adults with asthma. J Allergy Clin Immunol 129:1243–1251
47. Gorman S, Judge MA, Hart PH (2010) Immune-modifying properties of topical vitamin D: focus on dendritic cells and T cells. J Steroid Biochem Mol Biol 121:247–249
48. Hartmann B, Heine G, Babina M et al (2011) Targeting the vitamin D receptor inhibits the B cell-dependent allergic immune response. Allergy 66:540–548
49. Hawrylowicz C, Richards D, Loke TK et al (2002) A defect in corticosteroid-induced IL-10 production in T lymphocytes from corticosteroid-resistant asthmatic patients. J Allergy Clin Immunol 109:369–370
50. Heine G, Niesner U, Chang HD et al (2008) 1,25-dihydroxyvitamin D(3) promotes IL-10 production in human B cells. Eur J Immunol 38:2210–2218
51. Hilger J, Friedel A, Herr R et al (2014) A systematic review of vitamin D status in populations worldwide. Br J Nutr 111:23–45
52. Holick MF (2007) Vitamin D deficiency. N Engl J Med 357:266–281
53. Hong SP, Kim MJ, Jung MY et al (2008) Biopositive effects of low-dose UVB on epidermis: coordinate upregulation of antimicrobial peptides and permeability barrier reinforcement. J Invest Dermatol 128:2880–2887
54. Hyppönen E, Berry DJ, Wjst M (2009) Power C. Serum 25-hydroxyvitamin D and IgE – a significant but nonlinear relationship. Allergy 64:613–620
55. Hyppönen E, Sovio U, Wjst M et al (2004) Infant vitamin d supplementation and allergic conditions in

adulthood: northern Finland birth cohort 1966. Ann N Y Acad Sci 1037:84–95
56. Jackson DJ, Johnston SL (2010) The role of viruses in acute exacerbations of asthma. J Allergy Clin Immunol 125:1178–1187
57. Jartti T, Ruuskanen O, Mansbach MJ et al (2010) Low serum 25-hydroxyvitamin D levels are associated with increased risk of viral coinfections in wheezing children. J Allergy Clin Immunol 126:1074–1076
58. Javanbakht MH, Keshavarz SA, Djalali M et al (2011) Randomized controlled trial using vitamins E and D supplementation in atopic dermatitis. J Dermatol Treat 22:144–150
59. Jeffery LE, Burke F, Mura M et al (2009) 1,25-Dihydroxyvitamin D3 and IL-2 combine to inhibit T cell production of inflammatory cytokines and promote development of regulatory T cells expressing CTLA-4 and FoxP3. J Immunol 183:5458–5467
60. Jirapongsananuruk O, Melamed I, Leung DY (2000) Additive immunosuppressive effects of 1,25-dihydroxyvitamin D3 and corticosteroids on TH1, but not TH2, responses. J Allergy Clin Immunol 106:981–985
61. Jones AP, Palmer D, Zhang G et al (2012) Cord blood 25-hydroxyvitamin D3 and allergic disease during infancy. Pediatrics 130:1128–1135
62. Kerley CP, Elnazir B, Faul J et al (2015) Vitamin D as an adjunctive therapy in asthma. Part 1: a review of potential mechanisms. Pulm Pharmacol Ther 32:60–74
63. Kerley CP, Hutchinson K, Cormican L et al (2016) Vitamin D3 for uncontrolled childhood asthma: A pilot study. Pediatr Allergy Immunol 27:404–412
64. Kho AT, Sharma S, Qiu W et al (2013) Vitamin D related genes in lung development and asthma pathogenesis. BMC Med Genet 5(6):47
65. Kim SH, Ban GY, Park HS et al (2016) Regional differences in vitamin D levels and incidence of food-induced anaphylaxis in South Korea. Ann Allergy Asthma Immunol 116:237–243
66. Koya T, Miyahara N, Takeda K et al (2007) CD8+ T cell-mediated airway hyperresponsiveness and inflammation is dependent on CD4+IL-4+ T cells. J Immunol 179:2787–2796
67. Kuchuk NO, van Schoor NM, Pluijm SM et al (2009) Vitamin D status, parathyroid function, bone turnover, and BMD in postmenopausal women with osteoporosis: global perspective. J Bone Miner Res 24:693–701
68. Kull I, Bergström A, Melén E et al (2006) Early-life supplementation of vitamins A and D, in water-soluble form or in peanut oil, and allergic diseases during childhood. J Allergy Clin Immunol 118:1299–1304
69. Kumar R (2008) Prenatal factors and the development of asthma. Curr Opin Pediatr 20:682–687
70. Kuzume K, Kusu M (2007) Before-birth climatologic data may play a role in the development of allergies in infants. Pediatr Allergy Immunol 18:281–287
71. Lan N, Luo G, Yang X et al (2014) 25-Hydroxyvitamin D3-deficiency enhances oxidative stress and corticosteroid resistance in severe asthma exacerbation. PLoS One 9:111599
72. Liew WK, Williamson E, Tang ML (2009) Anaphylaxis fatalities and admissions in Australia. J Allergy Clin Immunol 123:434–442
73. Lips P (2010) Worldwide status of vitamin D nutrition. J Steroid Biochem Mol Biol 121:297–300
74. Lips P, Duong T, Oleksik A et al (2001) A global study of vitamin D status and parathyroid function in postmenopausal women with osteoporosis: baseline data from the multiple outcomes of raloxifene evaluation clinical trial. J Clin Endocrinol Metab 86:1212–1221
75. Litonjua AA, Carey VJ, Laranjo N et al (2016) Effect of prenatal supplementation with vitamin d on asthma or recurrent wheezing in offspring by age 3 years: The VDAART randomized clinical trial. JAMA 315:362–363
76. Litonjua AA, Weiss ST (2007) Is vitamin D deficiency to blame for the asthma epidemic? J Allergy Clin Immunol 120:1031–1035
77. Lloyd CM, Hawrylowicz CM (2009) Regulatory T cells in asthma. Immunity 31:438–449
78. Loukou I, Boutopoulou B, Fouzas S et al (2015) Vitamin D and cystic fibrosis lung disease. Mini-Rev Med Chem 15:974–983
79. Mahon BD, Wittke A, Weaver V et al (2003) The targets of vitamin D depend on the differentiation and activation status of CD4 positive T cells. J Cell Biochem 89:922–932
80. Matheu V, Back O, Mondoc E et al (2003) Dual effects of vitamin D-induced alteration of TH1/TH2 cytokine expression: enhancing IgE production and decreasing airway eosinophilia in murine allergic airway disease. J Allergy Clin Immunol 112:585–592
81. Misra M, Pacaud D, Petryk A et al (2008) Vitamin D deficiency in children and its management: review of current knowledge and recommendations. Pediatrics 122:398–417
82. Miyake Y, Sasaki S, Tanaka K et al (2010) Dairy food, calcium and vitamin D intake in pregnancy, and wheeze and eczema in infants. Eur Respir J 35:1228–1234
83. Muehleisen B, Gallo RL (2013) Vitamin D in allergic disease: shedding light on a complex problem. J Allergy Clin Immunol 131:324–329
84. Mullins RJ, Clark S, Camargo CA Jr (2009) Regional variation in epinephrine autoinjector prescriptions in Australia: more evidence for the vitamin D-anaphylaxis hypothesis. Ann Allergy Asthma Immunol 103:488–495
85. Mullins RJ, Clark S, Camargo CA Jr (2010) Regional variation in infant hypoallergenic formula prescriptions in Australia. Pediatr Allergy Immunol 21:413–420

86. Muthian G, Raikwar HP, Rajasingh J et al (2006) 1,25 Dihydroxyvitamin-D3 modulates JAK-STAT pathway in IL-12/IFNgamma axis leading to Th1 response in experimental allergic encephalomyelitis. J Neurosci Res 83:1299–1309
87. Niruban SJ, Alagiakrishnan K, Beach J et al (2014) Association of vitamin D with respiratory outcomes in Canadian children. Eur J Clin Nutr 68:1334–1340
88. Nizet V, Ohtake T, Lauth X et al (2001) Innate antimicrobial peptide protects the skin from invasive bacterial infection. Nature 414:454–457
89. Nwaru BI, Ahonen S, Kaila M et al (2010) Maternal diet during pregnancy and allergic sensitization in the offspring by 5 yrs of age: a prospective cohort study. Pediatr Allergy Immunol 21:29–37
90. Nwaru BI, Hickstein L, Panesar SS et al (2014) EAACI food allergy and anaphylaxis guidelines group. Prevalence of common food allergies in Europe: a systematic review and meta-analysis. Allergy 69:992–1007
91. Osborne NJ, Koplin JJ, Martin PE et al (2011) Health Nuts Investigators. Prevalence of challenge-proven IgE-mediated food allergy using population based sampling and predetermined challenge criteria in infants. J Allergy Clin Immunol 127:668–676
92. Palmer DJ, Sullivan TR, Skeaff CM et al (2015) DOMINO allergy follow-up team. Higher cord blood 25-hydroxyvitamin D concentrations reduce the risk of early childhood eczema: in children with a family history of allergic disease. World Allergy Organ J 8:28
93. Peroni DG, Piacentini GL, Cametti E et al (2011) Correlation between serum 25-hydroxyvitamin D levels and severity of atopic dermatitis in children. Br J Dermatol 164:1078–1082
94. Peters RL, Allen KJ, Dharmage SC et al (2015) HealthNuts study. Differential factors associated with challenge-proven food allergy phenotypes in a population cohort of infants: a latent class analysis. Clin Exp Allergy 45:953–963
95. Pludowski P, Holick MF, Pilz S et al (2013) Vitamin D effects on musculoskeletal health, immunity, autoimmunity, cardiovascular disease, cancer, fertility, pregnancy, dementia and mortality-a review of recent evidence. Autoimmun Rev 12:976–989
96. Powe CE, Evans MK, Wenger J et al (2013) Vitamin D-binding protein and vitamin D status of black Americans and white Americans. N Engl J Med 369:1991–2000
97. Prescott SL (2013) Early-life environmental determinants of allergic diseases and the wider pandemic of inflammatory noncommunicable diseases. J Allergy Clin Immunol 131:23–30
98. Radhakishun N, van Vliet M, von Rosenstiel I et al (2015) High prevalence of vitamin D insufficiency/deficiency in Dutch multi-ethnic obese children. Eur J Pediatr 174(2):183–190
99. Ramsey CD, Celedon JC (2005) The hygiene hypothesis and asthma. Curr Opin Pulm Med 11:14–20
100. Riverin BD, Maguire JL, Li P (2015) Vitamin D supplementation for childhood asthma: a systematic review and meta-analysis. PLoS One 10:0136841
101. Romagnani S (2004) The increased prevalence of allergy and the hygiene hypothesis: missing immune deviation, reduced immune suppression, or both? Immunology 112:352–363
102. Ross AC, Manson JE, Abrams SA et al (2011) The 2011 report on dietary reference intakes for calcium and vitamin D from the Institute of Medicine: what clinicians need to know. J Clin Endocrinol Metab 96:53–58
103. Rothers J, Wright AL, Stern DA et al (2011) Cord blood 25 hydroxyvitamin D levels are associated with aeroallergen sensitization in children from Tucson, Arizona. J Allergy Clin Immunol 128(5):1093–1099
104. Rudders SA, Camargo CA Jr (2015) Sunlight, vitamin D and food allergy. Curr Opin Allergy Clin Immunol 15:350–357
105. Samochocki Z, Bogaczewicz J, Jeziorkowska R et al (2013) Vitamin D effects in atopic dermatitis. J Am Acad Dermatol 69:238–244
106. Sandhu MS, Casale TB (2010) The role of vitamin D in asthma. Ann Allergy Asthma Immunol 105:191–199
107. Schauber J, Dorschner RA, Yamasaki K et al (2006) Control of the innate epithelial antimicrobial response is cell-type specific and dependent on relevant microenvironmental stimuli. Immunology 118:509–519
108. Schedel M, Jia Y, Michel S et al (2016) 1,25D3 prevents CD8(+)Tc2 skewing and asthma development through VDR binding changes to the Cyp11a1 promoter. Nat Commun 11(7):10213
109. Scientific Advisory Committee on Nutrition (2007) Update on vitaminD. Position statement. The Stationary Office Limited, London. http://www.sacn.gov.uk/pdfs/sacn_position_vitamin_d_2007_05_07.pdf
110. Searing DA, Leung DY (2010) Vitamin D in atopic dermatitis, asthma and allergic diseases. Immunol Allergy Clin N Am 30:397–409
111. Searing DA, Zhang Y, Murphy J et al (2010) Decreased serum vitamin D levels in children with asthma are associated with increased corticosteroid use. J Allergy Clin Immunol 125:995–1000
112. Sharief S, Jariwala S, Kumar J et al (2011) Vitamin D levels and food and environmental allergies in the United States: results from the national health and nutrition examination survey 2005–2006. J Allergy Clin Immunol 127:1195–1202
113. Sicherer SH, Muñoz-Furlong A, Godbold JH et al (2010) US prevalence of self-reported peanut, tree nut, and sesame allergy: 11-year follow-up. J Allergy Clin Immunol 125:1322–1326
114. Snellman G, Melhus H, Gedeborg R et al (2010) Determining vitamin D status: a comparison

between commercially available assays. PLoS One 2010(5):11555
115. Snijder MB, van Dam RM, Visser M et al (2005) Adiposity in relation to vitamin D status and parathyroid hormone levels: a population-based study in older men and women. J Clin Endocrinol Metab 90:4119–4123
116. Spergel JM (2010) Epidemiology of atopic dermatitis and atopic march in children. Immunol Allergy Clin N Am 30:269–280
117. Strachan DP (2000) Family size, infection and atopy: the first decade of the "hygiene hypothesis". Thorax 55:2–10
118. Tachimoto H, Mezawa H, Segawa T et al (2016) Improved control of childhood asthma with low-dose, short-term vitamin D supplementation: a randomized, double-blind, placebo-controlled trial. Allergy 71:1001–1009
119. Takami M, Fujimaki K, Nishimura MI et al (2015) Cutting edge: AhR is a molecular target of calcitriol in human T cells. J Immunol 195:2520–2523
120. Turkeli A, Ayaz O, Uncu A et al (2016) Effects of vitamin D levels on asthma control and severity in pre-school children. Eur Rev Med Pharmacol Sci 20:26–36
121. Urashima M, Segawa T, Okazaki M et al (2010) Randomized trial of vitamin D supplementation to prevent seasonal influenza A in schoolchildren. Am J Clin Nutr 91:1255–1260
122. van der Wielen RP, Lowik MR, van den BH de Groot LC et al (1995) Serum vitamin D concentrations among elderly people in Europe. Lancet 346:207–210
123. van Schoor NM, Lips P (2011) Worldwide vitamin D status. Best Pract Res Clin Endocrinol Metab 25:671–680
124. Vassallo MF, Camargo CA Jr (2010) Potential mechanisms for the hypothesized link between sunshine, vitamin D, and food allergy in children. J Allergy Clin Immunol 126:217–222
125. Venter C, Hasan Arshad S, Grundy J et al (2010) Time trends in the prevalence of peanut allergy: three cohorts of children from the same geographical location in the UK. Allergy 65:103–108
126. von Mutius E, Martinez FD (2016) Inconclusive results of randomized trials of prenatal vitamin D for asthma prevention in offspring: curbing the enthusiasm. JAMA 315:347–348
127. Wegienka G, Havstad S, Zoratti EM et al (2015) Association between vitamin D levels and allergy-related outcomes vary by race and other factors. J Allergy Clin Immunol 136:1309–1314
128. Wei Z, Zhang J, Yu X (2016) Maternal vitamin D status and childhood asthma, wheeze, and eczema: a systematic review and meta-analysis. Pediatr Allergy Immunol 27(6):612–619
129. Weiland SK, Husing A, Strachan DP et al (2004) ISAAC phase one study group. Climate and the prevalence of symptoms of asthma, allergic rhinitis, and atopic eczema in children. Occup Environ Med 61:609–615
130. Willers SM, Devereux G, Craig LC et al (2007) Maternal food consumption during pregnancy and asthma, respiratory and atopic symptoms in 5-year-old children. Thorax 62:773–779
131. Wjst M (2012) Is vitamin D supplementation responsible for the allergy pandemic? Curr Opin Allergy Clin Immunol 12:257–262
132. World Health Organization (2007) Global surveillance, prevention and control of chronic respiratory diseases: a comprehensive approach. World Health Organization, Geneva
133. Wu AC, Tantisira K, Li L et al (2012) Childhood asthma management program research group. Effect of vitamin D and inhaled corticosteroid treatment on lung function in children. Am J Respir Crit Care Med 186:508–513
134. Xystrakis E, Kusumakar S, Boswell S et al (2006) Reversing the defective induction of IL-10-secreting regulatory T cells in glucocorticoid-resistant asthma patients. J Clin Invest 116:146–155
135. Zhang Y, Leung DY, Goleva E (2013) Vitamin D enhances glucocorticoid action in human monocytes: involvement of granulocyte-macrophage colony-stimulating factor and mediator complex subunit 14. J Biol Chem 288:14544–14553
136. Zhang Y, Leung DY, Goleva E (2014) Anti-inflammatory and corticosteroid-enhancing actions of vitamin D in monocytes of patients with steroid-resistant and those with steroid-sensitive asthma. J Allergy Clin Immunol 133:1744–1752

15 Vitamin D Metabolism and the Implications for Atherosclerosis

Amanda L. Bennett and Carl J. Lavie

Abstract

Vitamin D levels and metabolism may play a role in the pathogenesis and treatment of atherosclerosis and subsequent cardiovascular health. Herein, we discuss both normal and disordered vitamin D metabolism as it pertains to atherosclerosis, and we review major clinical trials regarding vitamin D levels and effects of supplementation. Although there are no official recommendations for vitamin D as it applies to atherosclerosis, it is clear that these two entities are linked. Further study of the complex association between vitamin D and atherosclerosis, as well as the effects of supplementation, are recommended.

Keywords

Atherosclerosis • Vitamin D • Vitamin D deficiency • Inflammation • Disclosures • No relevant disclosures

15.1 Introduction

The atherosclerosis disease state is complex with many variables contributing to the appearance of, progression, and stability of the disease. One such variable is vitamin D metabolism. In this chapter, we will review the pathophysiology of atherosclerosis, the natural metabolism of vitamin D, as well as the known effects of vitamin D and its deficiency on the formation of atherosclerosis.

A.L. Bennett
Department of Internal Medicine, Ochsner Clinic Foundation, New Orleans, LA, USA

C.J. Lavie (✉)
Department of Cardiovascular Disease, John Ochsner Heart and Vascular Institute,
1514 Jefferson Hwy, New Orleans, LA, USA, 70121
e-mail: clavie@ochsner.org

S. Ahmad (ed.), *Ultraviolet Light in Human Health, Diseases and Environment*, Advances in Experimental Medicine and Biology 996, https://doi.org/10.1007/978-3-319-56017-5_15

15.2 Vitamin D Metabolism

Multiple forms of vitamin D are present within the body in varying concentrations (Fig. 15.1). Regardless of where this vitamin is derived, all forms must be converted to 25 hydroxy-vitamin D (25OHD) and then activated to 1,25(OH)$_2$D. [1] When the skin is exposed to UV light, the naturally occurring 7-dehydrocholesterol is converted to Vitamin D$_3$ (cholecalciferol, D3). D3 may also be ingested from animal sources. A vast majority of the vitamin D obtained from an omnivorous diet is in the D2 (ergocalciferol) form. D2, as a result of its chemical structure, is cleared from the systemic circulation more readily than D3 forms, limiting the conversion to 25OHD [2] which is produced almost solely in the liver by 25-hydroxylase. 25OHD is converted and activated in the kidney to form 1,25(OH)$_2$D.

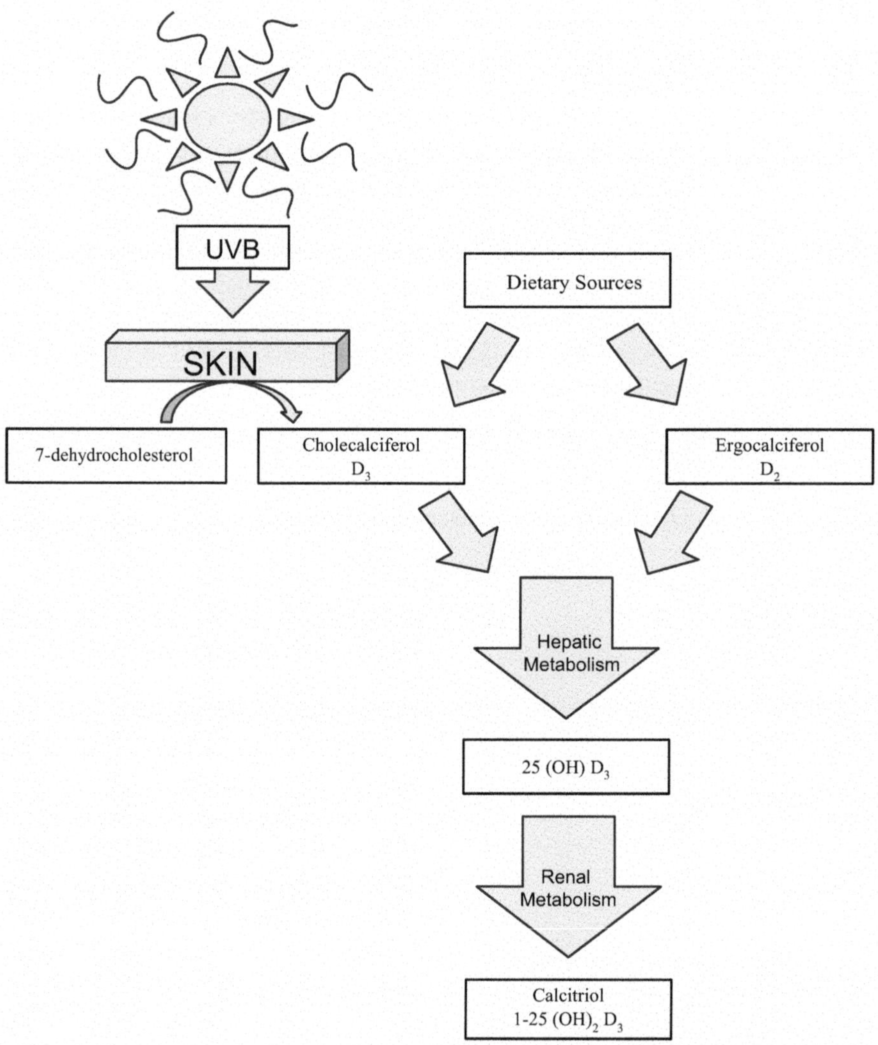

Fig. 15.1 Vitamin D Synthesis
Cholecalciferol is either formed from UVB interaction with 7-dehydrocholesterol in the skin or obtained from dietary sources. In the liver, cholecalciferol and ergocalciferol undergo metabolism and become 25(OH)D3. 25(OH)3D is converted to active 25 (OH)2D3 (calciferol) in the kidney

15.3 Pathophysiology of Atherosclerosis

The initial formation of atherosclerotic plaques occurs through the production of cytokines such as INF-y, TNF-α, TNF-β, and IL-2. Together with T-helper 1 cells, these cytokines bring about macrophage activation, inflammation and vascular activation that drive the inflammatory proatherogenic response. [3] Subsequent migration and proliferation of vascular smooth muscle (VCSM) leads to development of arterial hypertension (HTN). The formation of a vulnerable atherogenic plaque is what poses the greatest threat to patients.

Thin cap fibroatheroma (TCFA) formation is the precursor lesion of unstable plaques and subsequent rupture. TCFAs have a necrotic core with an overlying thin fibrous layer. Infiltration with macrophages and T-lymphocyte leads to thickening and insudation of the lipids accumulated within the atheroma [4–5]. While calcification of the cap potentiates inflammation within the vessel, the lipid cores of these TCFAs become necrotic and unstable leading to intraplaque hemorrhage and destabilization. Several mechanisms potentiate and perpetuate TCFA formation. Destabilization, intimal thickening, calcification, and intra-plaque hemorrhage are the largest contributors to plaque instability. [6].

Table 15.1 Beneficial effects of vitamin D pertaining to atherosclerosis

Beneficial effects of vitamin D
Reduction of endothelial oxidative stress
Inhibition of vascular smooth muscle proliferation and migration
Regulation of vascular smooth muscle tone
Reduction of vasoconstrictor metabolites
Inhibition of pro-inflammatory cytokines
Leads to increased endothelial nitric oxide
Inhibition of platelet and leukocyte aggregation and adhesion
Modulation the immune response

15.4 Effects of Vitamin D in Atherosclerosis

15.4.1 Modulation of Immune Response and Endothelial Response to Oxidative Stress

In the setting of endothelial oxidative stress, T-helper-1 cells are released and trigger inflammatory pathways. As a result of inflammation, macrophages become activated and trigger proatherogenic responses, which may lead to the formation of TCFAs. IL-2 and INF-y expression trigger the T-helper-1 proatherogenic response [6]. D3 decreases the expression of IL-2 and INF-y pro-inflammatory cytokines [7] and causes the T-cell response to shift production from the formation of T-helper-1 to T-helper-2 cells [8]. Additionally, 1,25(OH)2D also promotes a T-helper-2 profile [9]. The T-helper-2 anti-atherogenic immune profile results in secretion of cytokines that are anti-atherogenic, which serves to neutralize the T-helper-1 effect on VCSM [10] (Table 15.1).

Vitamin D also helps in limiting the extent of oxidative stress on VCSM. D3 increases superoxide dismutase (SOD) activity within VCSM through increased expression of IĸB-α [11]. With increased SOD, oxidation of low-density lipoprotein (LDL) is reduced in the vasculature [12]. Reduced LDL oxidation helps decrease the amount of foamy cell formation; consequently, reducing inflammation and reducing TCFA proliferation.

15.4.2 Proliferation and Migration of Vascular Smooth Muscle

Cyclin-dependent kinase 2 (CDK2) is a critical cellular protein involved in endothelin-dependent DNA synthesis. 1,25-OH2D directly inhibits production of endothelin-induced CDK2 production without altering the protein [13]. Serving as a regulatory switch, circulating vitamin D inhibits VCSM proliferation.

15.4.3 Influence on Vascular Smooth Muscle Tone and Vasoconstrictor Metabolites

In animal models, vitamin D derivatives are shown to reduce aortic endothelium-dependent contractions [14]. It has been shown that daily supplementation with 2000 IU of D3 can counterbalance arterial stiffness progression [15]. D3 supplementation also demonstrates improved recovery and viability of VCSM via regulation of heat-shock protein 70 [16].

Circulating vitamin D3 has been shown to correlate with arterial HTN [17]. Shi et al. suggest that one possible mechanism for this vitamin D relationship is the role of its suppression of the calcium ionophore A23187 located in the endothelium [18]. By creating calcium-surge interference, vitamin D helps reduce vasoconstrictor metabolites and consequently reduces endothelium-dependent arterial contraction and reduces VCSM tone [14].

15.4.4 Influence on Pro-inflammatory Cytokines

As previously mentioned, vitamin D is involved in inhibition of pro-inflammatory cytokines. Vitamin D increases 15-hydroxyprostaglandin dehydrogenase expression. By increasing this dehydrogenase, vitamin D promotes inactivation of prostaglandins. Additionally, COX-2 expression is decreased by vitamin D which contributes to prostaglandin receptor down regulation [19].

15.4.5 Effect on Endothelial Nitric Oxide Production

1,25(OH)D directly regulates nitric oxide (NO) synthase transcription within the arterial endothelium [20]. Deficiency of vitamin D leads to decreased bioavailability of NO. Subsequent stiffness may predispose the arterial system to endothelial trauma and initiate pro-atherosclerotic responses.

15.4.6 Aggregation and Adhesion of Platelets and Leukocytes

Platelet-leukocyte interactions are partially regulated through a NO-mediated mechanism in the arterial system [21].Endothelial injury potentiates platelet deposition and further inflammatory response with possible plaque formation. Repeat trauma leads to plaque erosion. Eroded plaques develop irregular surfaces and lose their protective covering of endothelial cells perpetuating further platelet and leukocyte aggregation leading to TFCA formation and destabilization. Vitamin D may serve an important role in decreasing platelet and leukocyte aggregation and adhesion to these eroded plaques may have an impact on preventing further plaque destabilization.

15.5 Effect of Vitamin D on Metabolic Contributors to Atherosclerosis Development and Progression

Atherosclerosis is a complex disease with many contributions in many conditions. The presence of several comorbidities, such as insulin resistance, hypercholesterolemia, hyperparathyroidism, and HTN dramatically increase the risk of atherosclerosis and may lead to worsening of the condition. Vitamin D likewise is a complex bionutrient and cofactor in multiple systemic processes and consequently has many additional roles outside of those already mentioned. In addition to the aforementioned loss of antioxidant properties, alterations in immune responses, and increased vascular reactivity/inflammation, vitamin D helps regulate the Renin-Aldosterone Angiotensin System (RAAS), hypothalamic axis, insulin production and resistance, as well as cholesterol profile all of which may potentiate atherosclerosis (Fig. 15.2.).

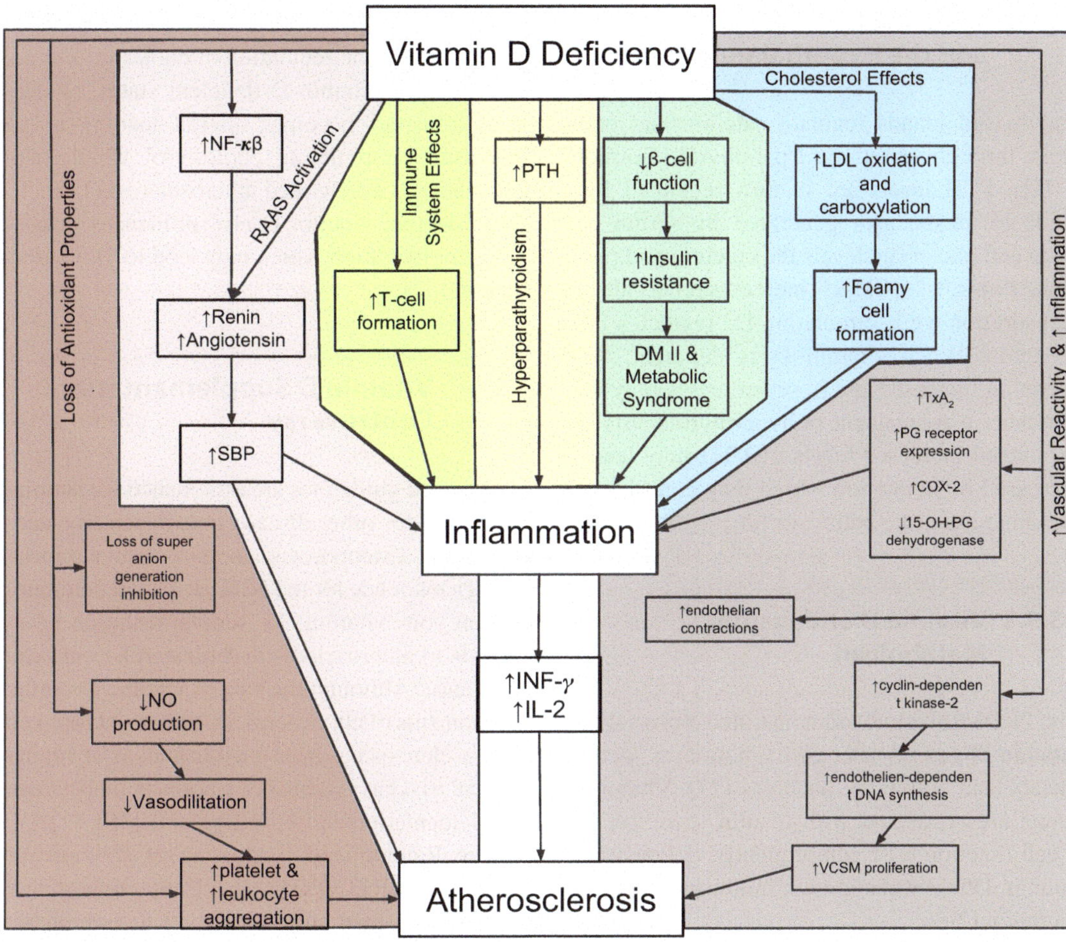

Fig. 15.2. Effects of vitamin D deficiency pertaining to the formation of atherosclerosis
Highlighting the multiple and complex pathways in which deficiency of vitamin D may potentiate atherosclerosis. Vascular Cell Smooth Muscle (*VCSM*), Diabetes Mellitus (*DM*), Low Density Lipoprotein (*LDL*), systolic blood pressure (*SBP*), nitric oxide (*NO*), Renin-Aldosterone Angiotensin System (*RAAS*), Parathyroid hormone (*PTH*)

15.5.1 Vitamin D and RAAS Activation

The RAAS is a major pathway through which the body regulates blood pressure (BP). In a normal system, renin is secreted by the kidney as a response to decreases in renal blood flow. Angiotensinogen is converted to angiotensin I by renin in the plasma. Angiotensin-converting enzyme then converts angiotensin I to angiotensin II. Angiotensin II increases water and sodium reabsorption in the kidney, as well as increases systemic vasoconstriction, leading to increased systemic BP, possibly leading to vascular strain or VCSM remodeling, which can predispose patients to atherosclerosis. Vitamin D may be involved in regulation of RAAS. Studies have shown an inverse relationship between systemic BP and serum 25(OH)D concentration [22]. Similarly, supplementation of vitamin D has been shown to reduce systemic HTN in some patients [23]. 1,25(OH)2D3 suppresses renin gene expression through transcription regulation of cAMP [24].

15.5.2 Vitamin D and the Hypothalamic Axis

Parathyroid glands regulate calcium and phosphate through secretion of parathyroid hormone (PTH). PTH levels are in turn regulated by a feedback mechanism generated by serum calcium and phosphate levels the calcium and phosphate levels. 1,25(OH)2D increases renal calcium reabsorption while circulating D3 regulates PTH secretion. When vitamin D is deficient, PTH secretion increases and in severe cases may lead to rickets, osteomalacia, or hyperthyroidism [25]. Prolonged increased levels of PTH may lead to increased inflammation which may contribute to pro-atherosclerotic pathways [26].

15.5.3 Vitamin D and Insulin Metabolism

The PROMISE study demonstrated a pivotal role vitamin D has on islet cell function as well as metabolism within the pancreas [27]. Vitamin D levels are correlated with insulin secretion and β-cell function [28]. Subsequently, deficiency of vitamin D is correlated with inhibition of insulin secretion [29].

In vitro expression of insulin receptors and subsequent glucose transport are increased by 1,25(OH)2D [30]. Insulin resistance may be accelerated by vitamin D deficiency but is not the sole factor to contribute to this metabolic issue [31]. Ironically, dietary supplementation of vitamin D has not been show in to improve sensitivity or secretion of insulin [32]. In the setting of diabetes mellitus (DM), comorbid deficiency of vitamin D may contribute to the increased incidence of atherosclerosis seen in that population [33].

15.5.4 Vitamin D and Cholesterol Profile

Lipid metabolism is partially regulated by vitamin D; however there is insufficient evidence regarding the relationship between this vitamin and lipid metabolism to recommend supplementation [34]. Hepatocyte receptors for vitamin D are integral in the regulation of cholesterol transport [35]. In vitamin D deficient states, cholesterol transport becomes altered leading to an increase in circulating cholesterol which may contribute to accelerated atherosclerosis [36]. To some degree, vitamin D also influences macrophage polarization which may lead to TCFA formation [37].

15.6 Vitamin D Supplementation: Controversy

While clear guidelines are established for supplementation for other disease states such as osteopenia and osteoporosis, there is not currently enough evidence for the NIH to make recommendations on vitamin D supplementation with regards to atherosclerosis, cholesterol, or insulin resistance. Although the role of vitamin D on the mechanisms of cholesterol, insulin resistance and atherosclerosis has been established, as of yet, no clinical trial has established a correlation between supplementation and improvements [38].

The Randomized Evaluation of Calcium or Vitamin D (RECORD) trial is the largest and most well-known study designed to establish a morbidity/mortality benefit to vitamin D supplementation. RECORD demonstrated that supplementation of D3 had no effect on mortality [39]. Further, recent studies have highlighted serum 25(OH)D deficiency has no benefit of harm in relation to incident HTN and mortality [40]. 25(OH)D has also been shown to have a nonlinear association with carotid intima media thickness in the elderly [41]. The ongoing Vitamin D and Omega-3 Trial (VITAL) seeks to quantify the effects of high-dose vitamin D on cardiovascular disease (CVD) events [42].

Regarding dietary supplementation, a recent systematic review demonstrated that vitamin D supplementation has no effect on myocardial infarction, stroke, hyperlipidemia, insulin resistance or BP [43]. Similarly, a recent meta-analysis of 51 trials pertaining to vitamin D supplementation also showed no significant effect on outcome in these endpoints [43]. While there are well established guidelines available with regard to the treatment of skeletal disorders such as osteoporosis, there are no current recommendations regarding quantity of daily vitamin D supplementation in CVD.

15.7 Conclusion

The effects of vitamin D deficiency are well known. However, the full extent of the impact of this vitamin in health and human disease remains an area of continued study. Throughout this chapter, we have shown several possible mechanisms through which vitamin D and subsequent vitamin deficiency play a role in atherosclerosis. In the current state of literature, no formal supplementations recommendations exist for vitamin D with regards to prevention or treatment of atherosclerosis. Although clear connections are established between various disease states and deficiency of vitamin D, we do not yet fully understand the complex role this nutrient plays in the body.

References

1. Menezes AR, Lamb MC, Lavie CJ, DiNicolantonio JJ (2014) Vitamin D and atherosclerosis. Curr Opin Cardiol 29(6):571–577
2. Houghton LA, Vieth R (2006) The case against ergocalciferol (vitamin D2) as a vitamin supplement. Am J Clin Nutr 84:694–697
3. Frostegard J, Ulfgren AK, Nyberg P et al (1999) Cytokine expression in advanced human atherosclerotic plaques: dominance of pro-inflammatory (Th1) and macrophage-stimulating cytokines. Atherosclerosis 145:33–43
4. Sakakura K, Nakano M, Otsuka F, Ladich E, Kolodgie FD, Virmani R (2013) Pathophysiology of atherosclerosis plaque progression. Heart Lung Circ 22(6):399–411
5. Nakashima Y, Fujii H, Sumiyoshi S, Wight TN, Sueishi K (2007) Early human atherosclerosis: accumulation of lipid and proteoglycans in intimal thickenings followed by macrophage infiltration. Arterioscler Thromb Vasc Biol 27(5):1159–1165
6. Mallat Z, Taleb S, Ait-Oufella H et al (2009) The role of adaptive T cell immunity in atherosclerosis. J Lipid Res 50:S364–S369
7. Bhalla AK, Amento EP, Krane SM (1986) Differential effects of 1,25-dihydroxyvitamin D3 on human lymphocytes and monocyte/macrophages: inhibition of interleukin-2 and augmentation of interleukin-1 production. Cell Immunol 98:311–322
8. Rigby WF, Stacy T, Fanger MW (1984) Inhibition of T lymphocyte mitogenesis by 1,25-dihydroxyvitamin D3 (calcitriol). J Clin Invest 74:1451–1455
9. Hewison M (2012) Vitamin D and the immune system: new perspectives on an old theme. Rheum Dis Clin N Am 38:125–139
10. Young JL, Libby P, Schönbeck U (2002) Cytokines in the pathogenesis of atherosclerosis. Thromb Haemost 88:554–567
11. Kassi E, Adamopoulos C, Basdra EK, Papavassiliou AG (2013) Role of Vitamin D in atherosclerosis. Circulation 128:2517–2531
12. Tukaj S, Trzonkowski P, Tukaj C (2012) Regulatory effects of 1,25-dihydroxyvitamin D3 on vascular smooth muscle cells. Acta Biochim Pol 59:395–400
13. Chen S, Law CS, Gardner DG (2010) Vitamin D-dependent suppression of endothelin-induced vascular smooth muscle cell proliferation through inhibition of CDK2 activity. J Steroid Biochem Mol Biol 118:135–141
14. Wong MS, Delansorne R, Man RY, Vanhoutte PM (2008) Vitamin D derivatives acutely reduce endothelium-dependent contractions in the aorta of the spontaneously hypertensive rat. Am J Physiol Heart Circ Physiol 295:H289–H296
15. Dong Y, Stallmann-Jorgensen IS, Pollock NK, Harris RA, Keeton D, Huang Y, Li K, Bassali R, Guo DH, Thomas J, Pierce GL, White J, Holick MF, Zhu H (2010) A 16-week randomized clinical trial of 2000 international units daily vitamin D3 supplementation in black youth: 25-hydroxyvitamin D, adiposity, and arterial stiffness. J Clin Endocrinol Metab 95:4584–4591
16. Zhu H, Guo D, Li K, Pedersen-White J, Stallmann-Jorgensen IS, Huang Y, Parikh S, Liu K, Dong Y (2012) Increased telomerase activity and vitamin D supplementation in overweight African Americans. Int J Obes 36:805–809
17. Duprez D, de Buyzere M, de Backer T, Clement D (1994) Relationship between vitamin D3 and the peripheral circulation in moderate arterial primary hypertension. Blood Press 3:389–393
18. Shi Y, Feletou M, Ku DD et al (2007) The calcium ionophore A23187 induces endothelium-dependent contractions in femoral arteries from rats with streptozotocin-induced diabetes. Br J Pharmacol 150:624–632
19. Krishnan AV, Feldman D (2010) Molecular pathways mediating the anti-inflammatory effects of calcitriol: implications for prostate cancer chemoprevention and treatment. Endocr Relat Cancer 17:R19–R38
20. Andrukhova O, Slavic S, Zeitz U et al (2014) Vitamin D is a regulator of endothelial nitric oxide synthase and arterial stiffness in mice. Mol Endocrinol 28:53–64
21. Gries A, Herr A, Kirsch S, Günther C, Weber S, Szabo G, Holzmann A, Böttiger BW, Martin E (2003)

Inhaled nitric oxide inhibits platelet-leukocyte interactions in patients with acute respiratory distress syndrome. Crit Care Med 31(6):1697–1704
22. Scragg R, Sowers M, Bell C (2007) Serum 25-hydroxyvitamin D, ethnicity, and blood pressure in the third National Health and nutrition examination survey. Am J Hypertens 20(7):713–719
23. Nasri H, Behradmanesh S, Ahmadi A, Rafieian-Kopaei M (2014) Impact of oral vitamin D (cholecalciferol) replacement therapy on blood pressure in type 2 diabetes patients; a randomized, double-blind, placebo controlled clinical trial. J Nephropathol 3(1):29–33
24. Li YC, Kong J, Wei M, Chen ZF, Liu SQ, Cao LP (2002) 1,25-Dihydroxyvitamin D(3) is a negative endocrine regulator of the renin-angiotensin system. J Clin Invest 110(2):229–238
25. Thacher TD, Clarke BL (2011) Vitamin D insufficiency. Mayo Clin Proc 86(1):50–60. doi:10.4065/mcp.2010.0567
26. Bosworth C, Sachs MC, Duprez D, Hoofnagle AN, Ix JH, Jacobs DR Jr, Peralta CA, Siscovick DS, Kestenbaum B, de Boer IH (2013) Parathyroid hormone and arterial dysfunction in the multi-ethnic study of atherosclerosis. Clin Endocrinol 79(3):429–436. doi:10.1111/cen.12163
27. Kayaniyil S, Retnakaran R, Harris SB et al (2011) Prospective associations of vitamin D with b-cell function and glycemia. The prospective metabolism and islet cell evaluation (PROMISE) cohort study. Diabetes 60:2947–2953
28. Yang Y, Zhang X, Bao M et al (2016) Effect of serum 25-hydroxyvitamin D3 on insulin resistance and β-cell function in newly diagnosed type 2 diabetes patients. J Diabetes Invest 7(2):226–232. doi:10.1111/jdi.12381
29. Norman AW, Frankel BJ, Heldt AM, Grodsky GM (1980) Vitamin D deficiency inhibits pancreatic secretion of insulin. Science 209:823–825
30. Maestro B, Campión J, Dávila N, Calle C (2000) Stimulation by 1,25-dihydroxyvitamin D3 of insulin receptor expression and insulin responsiveness for glucose transport in U-937 human promonocytic cells. Endocr J 47:383–391. pmid:11075718
31. Sung CC, Liao MT, Lu KC, Wu CC (2012) Role of vitamin D in insulin resistance. J Biomed Biotechnol 2012:634195
32. Grimnes G, Figenschau Y, Almås B, Jorde R (2011) Vitamin D, insulin secretion, sensitivity, and lipids. Diabetes Nov 60(11):2748–2757
33. Barrett-Connor EL, Cohn BA, Wingard DL, Edelstein SL (1991) Why is diabetes mellitus a stronger risk factor for fatal ischemic heart disease in women than in men? The rancho Bernardo study. JAMA 265:627–631
34. Bozic M, Guzmán C, Benet M, Sánchez-Campos S, García-Monzón C, Gari E, Gatius S, Valdivielso JM, Jover R (2016) Hepatocyte vitamin D receptor regulates lipid metabolism and mediates experimental diet-induced steatosis. J Hepatol 65(4):748–757
35. Oh J, Riek AE, Darwech I, Funai K, Shao J, Chin K, Sierra OL, Carmeliet G, Ostlund RE Jr, Bernal-Mizrachi C (2015) Deletion of macrophage vitamin D receptor promotes insulin resistance and monocyte cholesterol transport to accelerate atherosclerosis in mice. Cell Rep 10(11):1872–1886
36. Li S, He Y, Lin S, Hao L, Ye Y, Lv L, Sun Z, Fan H, Shi Z, Li J, Feng R, Na L, Wang Y, Li Y, Sun C (2016) Increase of circulating cholesterol in vitamin D deficiency is linked to reduced vitamin D receptor activity via the Insig-2/SREBP-2 pathway. Mol Nutr Food Res 60(4):798–809. doi:10.1002/mnfr.201500425
37. Yin K, You Y, Swier V, Tang L, Radwan MM, Pandya AN, Agrawal DK (2015) Vitamin D protects against atherosclerosis via regulation of cholesterol efflux and macrophage polarization in Hypercholesterolemic swine. Arterioscler Thromb Vasc Biol 35(11):2432–2442
38. Ponda MP, Huang X, Odeh MA et al (2012) Vitamin D may not improve lipid levels: a serial clinical laboratory data study. Circulation 126:270–277
39. Avenell A, MacLennan GS, Jenkinson DJ, McPherson GC, McDonald AM, Pant PR, Grant AM, Campbell MK, Anderson FH, Cooper C, Francis RM, Gillespie WJ, Robinson CM, Torgerson DJ, Wallace WA, RECORD Trial Group (2012) Long-term follow-up for mortality and cancer in a randomized placebo-controlled trial of vitamin D(3) and/or calcium (RECORD trial). J Clin Endocrinol Metab 97:614–622
40. Bischoff-Ferrari HA (2014) Optimal serum 25-hydroxyvitamin D levels for multiple health outcomes. Adv Exp Med Biol 810:500–525
41. van Dijk SC, Sohl E, Oudshoorn C, Enneman AW, Ham AC, Swart KM, van Wijngaarden JP, Brouwer-Brolsma EM, van der Zwaluw NL, Uitterlinden AG, de Groot LC, Dhonukshe-Rutten RA, Lips P, van Schoor NM, Blom HJ, Geleijnse JM, Feskens EJ, Smulders YM, Zillikens MC, de Jongh RT, van den Meiracker AH, Mattace Raso FU, van der Velde N (2015) Non-linear associations between serum 25-OH vitamin D and indices of arterial stiffness and arteriosclerosis in an older population. Age Ageing 44(1):136–142
42. Manson JE, Bassuk SS, Lee IM, Cook NR, Albert MA, Gordon D, Zaharris E, Macfadyen JG, Danielson E, Lin J, Zhang SM, Buring JE (2012) The VITamin D and OmegA-3 TriaL (VITAL): rationale and design of a large randomized controlled trial of vitamin D and marine omega-3 fatty acid supplements for the primary prevention of cancer and cardiovascular disease. Contemp Clin Trials 33:159–171
43. Elamin MB, Abu Elnour NO, Elamin KB, Fatourechi MM, Alkatib AA, Almandoz JP, Liu H, Lane MA, Mullan RJ, Hazem A, Erwin PJ, Hensrud DD, Murad MH, Montori VM (2011) Vitamin D and cardiovascular outcomes: a systematic review and meta-analysis. J Clin Endocrinol Metab 96:1931–1942

Vitamin D and Type 2 Diabetes Mellitus

16

Claire Michael Issa

Abstract

Type 2 diabetes mellitus (T2DM) has become a significant global health care problem and its reported incidence is increasing at an alarming rate. Despite the improvement in therapy and development of new drugs, treatment still remains insufficient especially due to the associated side effects of most available drugs. Efforts are continuing toward disease prevention and search for safer drugs. Conflicting evidence is associating low levels of vitamin D in the body to T2DM and as such studies have been conducted to test the effect of vitamin D levels on incidence of diabetes, diabetic control as well as diabetic complications.

Despite the conflicting evidence, vitamin D replacement seems to have some beneficial effect on the many aspects of diabetes: incidence, control and complications. Further long term and more convincing controlled trials are required in order to draw firmer conclusions on this beneficial role of vitamin D treatment on T2DM.

Keywords

Vitamin D • Type 2 diabetes mellitus • Glycemic control • Diabetic cardiovascular complications • Diabetic retinopathy • Diabetic nephropathy • Diabetic neuropathy

16.1 Introduction

Type 2 diabetes mellitus (T2DM) has become a significant global health care problem and its reported incidence is increasing at an alarming rate. Based on the recent International Diabetes Federation Diabetes Atlas (6th edition) an estimated 382 million global citizens are suffering from diabetes, costing around $1437 USD in

C.M. Issa (✉)
Endocrinology Consultant, Ain Wazein Medical Village, Lebanon- Mount of Lebanon-Shouf Area, Lebanon, PA, Lebanon
e-mail: claireissa_md@yahoo.com

S. Ahmad (ed.), *Ultraviolet Light in Human Health, Diseases and Environment*, Advances in Experimental Medicine and Biology 996, https://doi.org/10.1007/978-3-319-56017-5_16

2013 for each person affected by the condition. Projections based on current trends predict that 592 million people will be living with diabetes by 2035; which means that one in ten people will be affected, with an excessive amount of funding required globally to treat diabetes and manage diabetic complications ($627 billion USD in 2035) [1]. This is due to many factors like the growth of the population, the longer age, sedentariness and, obesity. Despite the improvement in therapy and development of new drugs, treatment still remains insufficient especially due to the associated side effects of most available drugs. For this reason efforts are now increasing towards disease prevention, with main culprits being obesity and sedentariness, changes in lifestyle, particularly weight loss and physical activity are definitely the first to tackle when it comes to offering prevention. Unfortunately weight loss and compliance with exercise are both difficult to achieve and maintain. As such the identification of easily modifiable risk factors becomes urgently needed for the prevention of diabetes.

Recently vitamin D has gained importance as a diabetes risk modifier. Vitamin D deficiency appears to be a global health concern. In a 2008 survey it was estimated that one billion individuals had vitamin D deficiency defined as a 25-hydroxyvitamin D (25 (OH) D) level of <20 ng/ml [2]. With the mounting evidence available depicting the role of vitamin D deficiency in several non-skeletal medical conditions such as multiple sclerosis, some types of cancer and cardiovascular disease, more attention is shifting towards improving the balanced levels of vitamin D to eliminate the deficiency related pathogenesis and eventually better control of T2DM. This was first suggested by the observation of a seasonal variation in the glycemic control where it was perceived to be worse during the winter season [3]. Additional evidence for a role of vitamin D in T2DM comes from a large number of observational and cross-sectional studies that showed an inverse relationship between prevalence of T2DM and a low 25 (OH) D level.

16.2 Vitamin D Physiology

Vitamin D identified by McCollem in 1922 [4] is a secosteroid that is generated in the skin under the influence of sunlight; as such it cannot be considered a true vitamin rather a pre-hormone. Vitamin D (calciferol) is a generic name for a group of fat steroids of which the two major forms are vitamin D2 (ergocalciferol) and vitamin D3 (cholecalciferol). Both forms of vitamin D undergo identical metabolism. During exposure to solar Ultraviolet B (UVB) radiation (290–310 nm), 7-dehydrocholesterol in the skin is converted to pre-vitamin D3, which is converted to vitamin D3 in a heat-dependent non-enzymatic process. A proper sun exposure (mainly of the face and hands) for around 10–15 min/day is probably sufficient to maintain normal levels. Excessive exposure to sunlight degrades pre-vitamin D3 and vitamin D3 into inactive phyto-products (photo-degradation), avoiding vitamin D toxicity in the setting of excess sunlight.

Common human diet is usually poor in vitamin D, exceptions in fatty fish and egg yolks. Vitamin D2 is synthesized by plants and also found in nutrients supplemented with vitamin D (e.g. milk) or dietary supplements, whereas vitamin D3 is mostly from animal source. Human sun exposure is related to the seasonal position of sun, time of day, atmospheric components, clothing, sunscreen (because of the major fear of the carcinogenic effect of sun exposure), and skin pigmentation all compromise vitamin D synthesis [5]. Despite the fact that the sun light is an important source of vitamin D, still a large number of global human populations have been found to suffer from its deficiency. And hence it is important that the sufferers are provided alternative sources of vitamin to maintain healthy living. The bioavailability of vitamin D also depends on its intestinal absorption capacity, liver function and their fat storage.

Adipose tissue easily absorbs vitamin D whether ingested or produced by chemical affinity. Some researchers suggest that the accumulation of vitamin D in adipose tissue is important for its subsequent release during times of reduced production (for example, during winter when the fat storage decreases) [6].

All these factors should be taken into consideration upon evaluation of the studies involving the role of vitamin D. Whether endogenously synthesized or ingested through diet or supplements, vitamin D in the circulation is bound to the vitamin D-binding protein (DBP), which transports it to the liver, where vitamin D is converted by 25-hydroxylase to 25 (OH) D. This form of vitamin D is biologically inactive and must be converted primarily in the kidneys by the 1-alpha hydroxylase to the biologically active form, 1,25-dihydroxyvitamin D 1,25 (OH)2 D. The kidney is the only tissue that can secrete 1,25(OH)2 D into the blood circulation. Another hydroxylase, the 24 hydroxylase is responsible for catabolizing vitamin D and transforming it to its inactive forms 24, 25 (OH)2 D and 1,24,25 trihydroxyvitamin D [7]. The active form of vitamin D exerts its actions by binding to a nuclear receptor {(the vitamin D receptor (VDR)}, first identified and cloned in 1987. Since then, new tissue-specific functions of vitamin D have been discovered. Currently, it is known that the VDR is widely distributed among tissues [8]. This fact reinforces the theory that despite the best known function of vitamin D is maintaining phosphocalcic homoeostasis and normal skeletal function, with most human cells possessing the VDR as well as the one α-hydroxylase and being capable of locally forming 1, 25 (OH)2 vitamin D which remains localized and acts directly on the tissue in an autocrine manner, vitamin D can indeed has many implications in many other pathologies. It was with the discovery of the VDR and vitamin D binding protein in pancreatic tissue and most specifically in β cells that the connection between vitamin D and T2DM has been reinforced [9].

16.3 Classification of Vitamin D Status

The biomarker that is usually used to assess vitamin D status is the blood concentration of 25 (OH) D since it is the major circulating form of vitamin D. There is still no consensus on the thresholds to define deficiency and insufficiency. Most experts and societies suggest a vitamin D insufficiency level as a value between 20 and 30 ng/ml or (50–75 nmol/L) and a vitamin D deficiency level as a level of less than 20 ng/mL or 50 nmol/L [10], except for the institute of medicine (IOM) that didn't agree on the higher levels and defined the deficiency as a value of 25 (OH) D of less than 12 ng/ml or (30 nmol/L) and the insufficiency at a value of between 12–19 ng/ml or (30–50 nmol/L) [11]. These differences can be explained by which populations were targeted by the guidelines and how the evidence was gathered. The IOM guidelines concentrated on the general healthy population and placed more emphasis on interventional studies. They found no convincing evidence linking vitamin D with benefits for non-skeletal outcomes, that's why they adopted lower thresholds that showed evidence mostly in skeletal benefits. In contrast, the Endocrine Society clinical practice guidelines concentrate on people at high risk for vitamin D deficiency and place more emphasis on observational (epidemiologic) studies that showed evidence linking vitamin D deficiency to many chronic diseases. Despite this discrepancy both guidelines agreed that all the recommendations require further reconsideration in the future as additional data from ongoing randomized trials become available.

16.4 Vitamin D Intake Requirements

Since both guidelines differ in their definition of vitamin D deficiency, their recommendations for vitamin D intake are also different. The IOM report on dietary reference intakes for calcium and vitamin D recommends 600 international units (IU) per day of vitamin D for individuals 9–70 years and 800 IU for those older than 70 years as the recommended dietary allowance. Whereas the Endocrine Society clinical practice guidelines conclude that in order to raise the blood level of 25 (OH) D consistently above 30 ng/mL equivalent to an intake of 1500 to 2000 IU/day may be required.

16.5 Possible Mechanisms Linking Vitamin D and T2DM

The role of vitamin D in the pathophysiology of T2DM is a subject of debate in the scientific community. The link between vitamin D deficiency and T2DM was reinforced when both the VDR and the one α-hydroxylase were found to be present in β-cells. There are several mechanisms possibly emphasizing a role of vitamin D deficiency in the pathogenesis of T2DM. The main key players in the pathogenesis of T2DM are: insulin resistance, insulin secretion and inflammation. Thus by affecting any of these key players, vitamin D can affect T2DM pathogenesis.

16.5.1 Effect on Insulin Resistance

Insulin resistance is a well-known culprit in the pathogenesis of T2DM. Any factor that affects insulin resistance will affect glycemic control. There are many possible mechanisms by which vitamin D can affect insulin resistance. The assumed benefit of vitamin D on insulin sensitivity was once thought to be an indirect effect, mostly by increasing and ameliorating muscle mass leading to an improvement in overall body insulin sensitivity. But then other lines of evidence emerged suggesting other possible mechanisms, both direct and indirect. Vitamin D insufficiency has been associated with an increased fat infiltration in skeletal muscle, which appears independent of body mass and is thought to contribute to a decreased insulin action [12]. This effect might be explained by the fact that Vitamin D acts by activating peroxisome proliferator-activated receptor delta (PPAR-δ), which is a transcription factor that regulates the metabolism of fatty acids in skeletal muscle and adipose tissue [13]. The direct effect occurs mostly by stimulating insulin receptors expression, by enhancing the transcriptional activation of the insulin receptor gene. This has been shown in an in-vitro study exposing human promonocytic cells to active vitamin D, leading to an increased expression of mRNA encoding for insulin receptors and a 1.3 fold increase in glucose transport when compared to untreated cells [14].

An indirect effect occurs through increasing extracellular calcium which will lead to a higher intracellular calcium influx that is essential for insulin-mediated intracellular processes, where a narrow range of intracellular calcium is needed for optimal insulin-mediated function [15]. Another indirect effect could be mediated through the renin-angiotensin-aldosterone system (RAAS). Angiotensin II inhibits the action of insulin in vascular and skeletal muscle tissue leading to impaired glucose uptake. Vitamin D suppresses renin formation and local pancreatic RAAS; hence vitamin D could be a negative endocrine regulator of RAAS [16].

16.5.2 Effect on Insulin Secretion and Beta Cell

The effect of vitamin D on the pancreatic β-cell is through the regulation of extracellular calcium concentration and flux through the β-cell. As mentioned above, insulin secretion is a calcium-dependent process therefore alterations in calcium flux could have an effect on insulin secretion [17]. Vitamin D can also act by mediating the activation of β-cell calcium-dependent endopeptidases that convert proinsulin to active insulin [18]. Glucose and sulfonylurea-stimulated insulin secretions were shown to be lower from islets of vitamin D-deficient rats than from islets of vitamin D-sufficient rats or vitamin D deficient rats that were replaced with vitamin D [19].

16.5.3 Secondary Hyperparathyroidism (SHPT) and Stimulation of PTH

Low vitamin D will induce Secondary hyperparathyroidism (SHPT). The raised PTH inhibits insulin synthesis and secretion in β-cells and induces insulin resistance in target cells by regulating intracellular calcium. The SHPT may actually cause a paradoxical increase in intracellular calcium and in turn may impair the calcium signal needed for glucose induced insulin secretion; this is known as the "calcium paradox [20].

16.5.4 Inflammation

Inflammation plays a crucial role in the pathogenesis of T2DM. Vitamin D can promote β-cell survival by modulating the generation and effect of cytokines. Vitamin D regulates calbindin, a cytosolic calcium-binding protein found in β-cells [21]. This latter has been shown to protect β-cell from cytokine-induced cell death [22]. Another way may be through the down-regulation of NF-kB, a major transcription factor for TNF-alpha and other pro-inflammatory molecules [23]. Another pathway that may also mediate the effect of active vitamin D on β-cell function is through counteracting cytokine-induced Fas expression, which in turn will furthermore have anti-apoptotic effects [24]. Moreover vitamin D can have other important effects on inflammatory cytokines by interfering with a number of other cytokine genes or transcription factors involved in cytokine generation [25]. In a recent randomized controlled trial, 118 vitamin D-insufficient diabetic patients were divided into four groups: either receiving high dose vitamin D alone or calcium supplements alone, or both or neither. It has been shown, after adjusting for all confounders that joint calcium-vitamin D supplementation might improve systemic inflammation through decreasing IL-6 and TNF-alpha concentrations in this group of patients [26].

16.5.5 VDR Polymorphisms Genes

More than 25 VDR polymorphism genes have been identified until now. Despite the conflicting results, there seems to be plausible evidence linking VDR polymorphisms to the pathogenesis of T2DM [27]. There are several mechanisms explaining this association: modification of cytokine expression, alteration in calcium metabolism, modulation of insulin secretion and alteration of adipokine function [28].

16.6 Vitamin D and Prevalence of T2DM

Several observational studies have examined the association between vitamin D levels and prevalence of T2DM. The majority has reported an inverse association while others failed to show such association. This discrepancy can be secondary to the many limitations of these studies such as the difficulty in accurately measuring vitamin D exposure since both the dietary intake and the sunlight synthesis are hard to account for, the heterogeneity of the different studies when it comes to seasons, environmental factors, ethnicity, the different assays used to measure vitamin D level, as well as the different levels used to define deficiency, thus leading to different inclusion criteria. Taking these limitations into consideration, there seems to be an association between low levels of vitamin D and diabetes prevalence that should not be ignored, however it is still too early to speculate on whether it's a causality relationship or not.

16.6.1 Observational Studies

Large population-based studies such as The Third National Health and Nutrition Examination survey, a cross-sectional survey, where in 6000 multiethnic participants, vitamin D level as well as insulin and glucose levels were measured, a strong inverse relationship between vitamin D levels and presence of DM 2 after adjusting for other cofounders and risk factors was shown [29]. This is in accordance with the British Birth Cohort study that showed a 74% risk reduction in DM 2 in those patients with a vitamin D level in the highest tertile when compared to the lowest [30]. Again in the Mini-Finland Health Study, 4000 patients were followed up for over 17 years. A statistically significant relative risk of T2DM occurrence of 0.6 has been shown between the highest and the lowest quartile of 25 (OH) D [31].

In a meta-analysis [32] including 21 prospective studies, comparing the highest to the lowest category of 25 (OH) D levels, the summary relative risk for T2DM was 0.62, which was statistically significant and remained so even after adjusting for different confounders, in addition a linear trend analysis showed that with each 10 nmol/L increment in 25 (OH) D levels there was an associated 4% lower risk of T2DM ($P < 0.0001$). In another meta-analysis of 16 studies, Afzal et al. estimated the odds ratio for

T2DM to be 1.5 for the bottom versus the top quartile of 25 (OH) D concentrations [33].

Furthermore, incidence rate of T2DM has been shown to be more than 50% lower in the Grass roots Health cohort with a median serum vitamin D level of 41 ng/ml than in NHANES cohort with a median of 22 ng/ml [34]. A cross-sectional study conducted on Caucasians middle-aged men and women showed that in women, but not in men, low vitamin D levels are independently associated with T2DM (Using a 15 ng/ml of vitamin D as a cutoff, adjusted odds ratios for having newly diagnosed or known T2DM more than doubled only in women with vitamin D levels below the cutoff). These findings suggest possible sex-specific effects of Vitamin D in the pathogenesis of T2DM [35]. Despite the consistency of these results, the observational nature of these studies precludes an assessment of cause and effect because residual confounding cannot be excluded.

16.6.2 Interventional Studies

The effect of vitamin D supplementation on glycaemia or T2DM incident has been reported in several trials with mixed results. Generally speaking, the available studies are mostly limited by lack of randomized, placebo-controlled dosages and a failure to reach sufficient vitamin D concentrations.

An observational study by the Nurses' Health Study [36], that followed-up 83,779 nurses in the USA over 20 years, showed that the incidence of T2DM (2.7%) was lower in patients who were on the higher daily dose of vitamin D (511 IU) when compared to those on the lower dose of 159 IU (5.6%). But this association became non-significant after adjusting for dietary variables, mainly calcium and magnesium intake. In this same study women who reported combined high dose intake of both calcium and vitamin D (1200 mg and 800 IU) had a 33% lower risk of T2DM when compared to women with a low intake (600 mg and 400 IU).

The same association has also been shown in the Women's Health Study [37], the Finnish Mobile Clinic Study [38] and the MRC Ely Study [39]. The potential effect of vitamin D supplementation appears to be more prominent among persons who were already at high risk for diabetes, as shown in the two studies by Pittas et al. [40] and Nazarian et al. [41]. In both studies patients with impaired fasting glucose were included. In the study of Pittas el a calcium and vitamin D intake prevented increases in insulin resistance and glucose levels but only in patients with impaired fasting glucose and not in normal participants. Whereas in the Nazarian et al. study, eight participants with impaired fasting glucose and a 25 (OH) D level < 30 ng/ml were given 20,000 IU/week for 4 weeks. Seven of these patients had improvement in insulin sensitivity.

In a randomized double-blind controlled calcium and vitamin D for diabetes mellitus CaDDM study [42], 92 patients with pre-diabetes were given 2000 IU/day of cholecalciferol and 800 mg /day of calcium carbonate. There was a significant improvement in insulin secretion with no effect on insulin resistance. In another randomized placebo controlled trial [43], where mutli-ethnic vitamin D deficient patients, at risk of diabetes, were given high dose vitamin D in combination with calcium, no beneficial effect was seen on insulin secretion, insulin resistance nor inflammatory markers, but in post-hoc analysis, patients with pre-diabetes showed improvement in insulin secretion when compared to the placebo group.

Contrary to the above-mentioned studies, others showed a negative association between vitamin D replacement and DM 2. A Meta-analysis of 4 prospective cohort studies involving 187,592 participants and 9456 incident cases showed an absence of a significant association between total vitamin D intake and T2DM risk [44]. This has also been shown in the Women's Health Initiative Study [45], as well as the Record Study [46]. But both those latter studies were limited by a low dose supplementation for the first and a low level of target vitamin D reached in the second. Another recent randomized controlled trials [47], conducted on healthy patients given high dose vitamin D for 12 weeks there wasn't any beneficial effect on insulin secretion, resistance or other metabolic markers. There are still ongoing randomized controlled trials to test the hypothesis that vitamin D supplementation lowers the risk of T2DM such as the VITAL study (www.vitalstudy.org) and the D2d study (www.d2dstudy.org).

16.7 Vitamin D and T2DM Control

Many recent randomized controlled trials were published testing the effect of vitamin D replacement on diabetes control showing conflicting results. The SUNNY trial [48], a randomized double blind placebo-controlled trial was conducted on 300 diabetic patients. Those patients were given either a high dose vitamin D (50,000 IU monthly) or placebo and were followed up for 6 months. There was no effect on HbA1c level or on other glycemic control measures.

In accordance, another randomized controlled trial [49] where 50 patients who were diabetic for less than a year, were given high dose vitamin D versus placebo. There was only a transient improvement in glycaemia, but without a measurable change in β-cell function. On the other hand this study was limited by the fact that those patients had a normal level of vitamin D at baseline to start with. In another study, only a positive effect on insulin secretion was seen with no effect on insulin resistance, glycemic control or inflammatory markers [50].

In discordance, another randomized controlled trial [51] tested the effect of high dose vitamin D combined with calcium supplementation on diabetic patients with vitamin D insufficiency and showed positive results on both glycemic control and lipid profile. Other randomized controlled trials also showed a beneficial effect of vitamin D replacement on glycemic control [52–54].

In summary and in the light of those conflicting results it is hard to draw any conclusions about the effect of vitamin D on glycemic control. Those studies cannot be compared sine they differed in the duration of diabetes and its control, the dose of vitamin D given, the level of vitamin D at baseline. In addition some major imitations were the small number of patients included as well as the short duration of follow-up. As such further larger studies with longer follow-ups and more homogenous inclusion criteria are still needed.

16.8 Vitamin D and Diabetic Complications

The best way to prevent diabetic complications is to control glucose levels as well as the other cardiovascular risk factors such as blood pressure and lipid profiles. Unfortunately despite these preventive measures, diabetic complications are still common leading most researchers into seeking other measures of prevention. Vitamin D emerged as a possible promising solution in this matter. Unfortunately, data is still very scarce, but deserves to be mentioned and discussed in the upcoming part.

16.8.1 Diabetic Nephropathy

The effect of vitamin D replacement on diabetic nephropathy seem to be the easiest to study since it is already common practice, as recommended by the guidelines, that all nephropathic patients be on vitamin D replacement.

Sadly T2DM is still the leading cause of end-stage renal disease worldwide. This leads to secondary hyperglycemia which is the main culprit in diabetic nephropathy by promoting glomerular injury through several pathways, including stimulation of profibrotic and proinfalmmatory factors and amplification of the oxidative stress.

The association between low levels of vitamin D and diabetic nephropathy has indeed been observed in many studies.

The NHANES survey found that 25 (OH) D levels were significantly lower in persons with severely decreased glomerular filtration rate (GFR) when compared with healthy individuals. Another recent study showed that patients with diabetic nephropathy had significantly lower levels of vitamin D than patients without diabetic nephropathy [55]. A study by Wolf et al. even suggested that a low 25 (OH) D level in hemodialysis patients was associated with a worse prognosis and a higher mortality rate [56]. In confirmation another study by Fernandez-Juarez [57] et al. showed that vitamin D deficiency is independently associated with a higher risk of >50% increase in baseline serum creatinine, end-stage renal disease or death. There aren't yet sufficient evidence confirming the beneficial role of replacing vitamin D in diabetic nephropathy.

Mice studies [58, 59] have shown great promise when vitamin D replacement added to angiotensin-converting enzyme inhibitors (ACE-I)/ angiotensin receptor blockers (ARB) treatment led to a significant prevention of albuminuria as well as reduction of formation of inflammatory factors and restoration of the glomerular filtration barrier.

There are two main human studies testing the effect of replacing vitamin D in patients with diabetic nephropathy. The first, the VITAL study [60], is a multinational, placebo- controlled, double-blind study conducted on 281 patients with T2DM and albuminuria already receiving ACE-I or ARB. These patients were divided into 3 groups to receive: placebo, 1 μg/day or 2 μg/day of paricalcitol over 6 months. Results have shown a significant decrease in albuminuria in both treatment arms with a higher decrease in the higher dose group when compared to placebo with no major side effects. The second study was an open label prospective study [61], conducted on patients with T2DM and diabetic nephropathy already on ACE/ARB treatment using oral vitamin D 40000 per week for 2 months then monthly if the baseline 25 (OH) D was <16 ng/ml and 40,000 every month if the baseline 25 (OH) D was 16–32 ng/ml. In accordance to the VITAL study, this study has also shown a decrease in albuminuria and TGF-beta 1 after 4 months of treatment, however this study was underpowered and of a short duration.

16.8.2 Cardiovascular Complications

Cardiovascular disease remains the leading cause of death in most patients with diabetes mellitus. There seems to be an association between a low vitamin D level and a higher risk of cardiovascular disease in patients with T2DM. Several possible mechanisms can explain this association: through a negative regulation of the renin-angiotensin system [62], which is well known to be a major player in cardiovascular disease [63], through affecting cardiac contractility, cardiac tissue maturation, collagen content, vascular tone [64] as well as a direct effect on vascular smooth muscle cell calcification and proliferation [65], or through regulation of inflammation which has been shown to increase cardiovascular risk [66].

Indeed many studies have suggested an association between a low 25 (OH) D level and a higher cardiovascular risk. In a cross-sectional analysis from NHANES III and after adjusting for other risk factors, vitamin D deficiency increased the risk of cardiovascular disease by an OR of 1.2 [67]. In another study vitamin D deficiency was associated with higher risk of myocardial dysfunction in diabetic patients with no history of coronary artery disease [68]. In accordance, a study has shown that even after adjustment for possible confounders, diabetic patients with a blood vitamin D concentration < 50 nmol/L had a higher cumulative incidence of macrovascular events than those with levels of >50 nmol/L [69]. A study even suggested adding vitamin D level status to the Framingham Risk Score to improve the assessment of cardiovascular risk factors in diabetic patients [70].

Unfortunately studies on effect of vitamin D replacement on cardiovascular disease in diabetic patients are still very scant. In a randomized controlled trial by Witham et al. [71] comparing the effect of 100,000 IU vs 200,000 IU of vitamin D3 vs placebo on endothelial function (assessed by measuring flow-mediated dilation of the brachial artery to hyperemia) on 61 patients with diabetes over a 4 months period, no effect on endothelial function was found, but there was a significant decrease in blood pressure in both groups vs placebo as well as a decrease in B-type natriuretic peptide levels. But this study was underpowered, of short duration and started with a high cutoff of vitamin D level inclusion (<40 ng/ml). A positive effect on endothelial function has also been shown in another study by Sugden et al. [72]. It was also a randomized double-blind placebo controlled trial where a single high dose of 100,000 IU of vitamin D2 vs placebo was given over the winter period to 34 type 2 DM patients. These patients had lowered 25 (OH) D levels with a cutoff of <20 ng/ml. Even after adjustment for blood pressure changes, treated group showed a significant improvement in endothelial function as well as blood pressure when compared to the placebo group. But this study was of short duration (2 months period) and underpowered. Further studies are still needed before drawing any conclusions on the effect of vitamin D replacement on cardiovascular disease in diabetic patients.

16.8.3 Diabetic Neuropathy

Diabetic neuropathy is a bothersome and very common complication of diabetes that can lead to limb amputation. Unfortunately its management remains frustrating in most cases. There are few—and only contradictory—data concerning the actual correlation between vitamin D deficiency and diabetic neuropathy. In vitro data and the outcomes of animal testing have both confirmed the role played by vitamin D analogues in stimulating and reducing the breakdown of the nerve growth factor that is crucial to the survival of sympathetic and sensory neurons.

The 2001–2004 NHANES [73] showed that low levels of 25 (OH) D were associated with self-reported peripheral neuropathic symptoms even after adjusting for confounders. Shehab et al. [74] also showed a lower 25 (OH) D level in patients with diabetes and typical neuropathic pain. Ahmadieh et al. [75] investigated the relationship between 25 (OH) D levels and microvascular complications in patients with T2DM. Diabetic neuropathy was evaluated using the UK screening score. Mean 25 (OH) D levels were lower in subjects with diabetic neuropathy compared to those without diabetic neuropathy. Furthermore, using a cutoff value of 20 ng/ml, diabetic neuropathy was more prevalent in subjects with vitamin D deficiency than those with levels ≥20 ng/ml (63 vs. 42%, $p = 0.03$). After adjustment for HbA1c, age, smoking, BMI and duration of diabetes in a logistic regression model, diabetes duration and 25 (OH) D levels were significant predictors of diabetic neuropathy.

In a study conducted by Lee and Chen [76] to examine the correlation between vitamin D deficiency and diabetic neuropathy, 51 diabetic patients who had deficient 25 (OH) D serum levels and also diabetic neuropathy were examined. Following 3 months of administering a vitamin D supplement, the score values of neuropathy-induced pain were reduced by 50%. In a prospective, placebo-controlled trial including112 type 2 diabetic patients with peripheral neuropathy and vitamin D deficiency, short-term oral vitamin D 3 supplementation improved vitamin D status and the symptoms of neuropathy [77]. Based on the above suggestive data, administration of vitamin D supplements may prove to be a beneficial adjuvant therapy in mitigating neuropathic pain as well as in blocking the progression of neuronal destruction.

16.8.4 Diabetic Retinopathy

There isn't much evidence on the role of vitamin D in diabetic retinopathy. One study by Aksoy et al. [78] have shown an inverse relationship between the severity of the retinopathy, i.e. neovascularization, and serum 1,25 (OH)2 D3 concentrations, being the lowest in proliferative diabetic retinopathy and the highest in diabetic patients without retinopathy patients. In line, Alcubiere et al. have shown that patients with more advanced stages of retinopathy (grades 2–4) had lower concentrations of 25 (OH) D and were more frequently vitamin D deficient as compared with patients not carrying this eye complication [79]. This is in discordance with a cross-sectional case control study where no association was specifically found between vitamin D deficiency and diabetic retinopathy [80].

16.9 Conclusion

Despite all this conflicting evidence there seems to be a tendency towards an inverse relationship between levels of 25 (OH) D and type 2 diabetes as well as its associated complications. Vitamin D replacement showed a trend towards decreasing incidence of T2DM and ameliorating glycemic control mainly by increasing insulin secretion, decreasing insulin resistance and decreasing inflammation. Despite the fact that the effect of vitamin D replacement on diabetic complications still deserves higher powered long-term studies, vitamin D replacement appears to be beneficial in preventing and improving some diabetic complications mainly nephropathy and cardiovascular diseases. In addition future studies need to address all the previous limitations such as the 25 (OH) D levels used for inclusion, the replacement doses given and their ways of administration, the target levels reached, the level of diabetes control and the duration of diabetes (no conclusions can be drawn on whether their might be a higher benefit in starting vitamin D treatment at the early stages of diabetes before the beta cell function is severely and irreversibly affected).

Awaiting further more homogenous and extensive studies, one thing is sure, vitamin D deficiency is definitely associated with many serious skeletal and non-skeletal complications and as such needs to be tested and replaced accordingly.

References

1. International Diabetes Federation (2013) IDF Diabetes Atlas, 6th edn. International Diabetes Federation, Brussels, pp 32–49
2. Chagas CE, Borges MC, Martini LA, Rogero MM (2012) Focus on vitamin D, inflammation and type 2 diabetes. Forum Nutr 4:52–67
3. Campbell IT, Jarrett RJ, Keen H (1975) Diurnal and seasonal variation in oral glucose tolerance: studies in the Antarctic. Diabetologia 11:139–145
4. DeLuca HF (2004) Overview of general physiologic features and functions of vitamin D. Am J ClinNutr 80(6Suppl):1689S–1696S
5. Tsiaras WG, Weinstoc MA (2011) Factors influencing vitamin D status. Acta Derm Venereol 91:115–124
6. Blum M, Dolnikowski G, Seyoum E, Harris SS, Booth SL, Peterson J, Saltzman E, Dawson-Hughes B (2008) Vitamin D(3) in fat tissue. Endocrine 33:90–94
7. Issa C, Azar S, Zantout M (2014) Vitamin D replacement and type 2 diabetes mellitus. Curr Diabetes Rev 10(5):7–16
8. Ross AC, Abrams SA, Aloia JF, Brannon PM, Clinton SK, Durazo Arvizu RA, Gallagher JC, Gallo RL, Jones G, Kovacs CS et al (2011) Institute of medicine (US) committee to review dietary reference intakes for vitamin D and calcium. In: Ross AC, Taylor CL, Yaktine AL, Valle HBD (eds) Dietary reference intakes for calcium and vitamin D. National Academies Press (US), Washington, DC
9. Cantorna MT, Zhu Y, Froicu M et al (2004) Vitamin D status, 1,25 dihydroxyvitamin D3, and the immune system. Am J ClinNutr 80(6):1717S–1720S
10. Holick MF, Binkley NC, Bischoff-Ferrari HA, Gordon CM, Hanley DA, Heaney RP, Murad MH, Weaver CM (2011) Endocrine society. J Clin Endocrinol Metab 96(7):1911–1930
11. Looker AC, Jhonson CL, Lacher DA, Pfeiffer CM, Schleicher RL, Sempos CT (2001 Mar) Vitamin D status: United states 2001–2006. NCHS Data Brief 59:1–8
12. Gilsanz V, Kremer A, Mo AO et al (2010) Vitamin D status and its relation to muscle mass and muscle fat in young women. J Clin Endocrinol Metab 95:1595–1601. [PubMed: 20164290]
13. Dunlop TW, Vaisanen S, Frank C et al (2005) The human peroxisome proliferator-activated receptor delta gene is a primary target of 1alpha,25-dihydroxyvitamin D3 and its nuclear receptor. J Mol Biol 349:248–260. [PubMed: 15890193]
14. Maestro B, Campion J, Davila N et al (2000) Stimulation by 1,25-dihydroxyvitamin D3 of insulin receptor expression and insulin responsiveness for glucose transport in U-937 human promonocytic cells. Endocr J 47:383–391
15. Draznin B, Sussman K, Kao M et al (1987) The existence of an optimal range of cytosolic free calcium for insulin-stimulated glucose transport in rat adipocytes. J Biol Chem 262:14385–14388. [PubMed: 3312189]
16. Wei Y, Sowers JR, Clark SE, Li W, Ferrario CM, Stump CS (2008) Angiotensin II- induced skeletal muscle insulin resistance mediated by NF-kappaB activation via NADPH oxidase. Am J Physiol Endocrinol Metab 294:E345–E351
17. SergeevIN,RhotenWB(1995)1,25-Dihydroxyvitamin D3 evokes oscillations of intracellular calcium in a pancreatic beta-cell line. Endocrinology 136:2852–2861. [PubMed: 7789310]
18. Billaudel BJ, Faure AG, Sutter BC (1990) Effect of 1,25dihydroxyvitamin D3 on isolated islets from vitamin D3- deprived rats. Am J Phys 258:E643–E648
19. Norman AW, Frankel JB, Heldt AM, Grodsky GM (1980) Vitamin D deficiency inhibits pancreatic secretion of insulin. Science 209(4458):823–825
20. Fujita T, Palmieri GM (2000) Calcium paradox disease: calcium deficiency prompting secondary hyper-

parathyroidism and cellular calcium overload. J Bone Miner Metab 18:109–125
21. Sooy K, Schermerhorn T, Noda M et al (1999) Calbindin-D (28k) controls [Ca(2+)](i) and insulin release. Evidence obtained from calbindin-D (28k) knockout mice and beta cell lines. J Biol Chem 274:34343–34349
22. Kadowaki S, Norman AW (1984) Pancreatic vitamin D-dependent calcium binding protein: biochemical properties and response to vitamin D. Arch Biochem Biophys 233(1):228–236
23. Cohen-Lahav M, Douvdevani A, Chaimovitz C et al (2007) The anti-inflammatory activity of 1,25- dihydroxyvitamin D3 in macrophages. J Steroid Biochem Mol Biol 103:558–562. [PubMed: 17267205]
24. Riachy R, Vandewalle B, Moerman E et al (2006) 1,25-Dihydroxyvitamin D3 protects human pancreatic islets against cytokine-induced apoptosis via down-regulation of the Fas receptor. Apoptosis 11:151–159. [PubMed: 16502254]
25. Gysemans CA, Cardozo AK, Callewaert H et al (2005) 1,25- Dihydroxyvitamin D3 modulates expression of chemokines and cytokines in pancreatic islets: implications for prevention of diabetes in non-obese diabetic mice. Endocrinology 146(4):1956–1964
26. Tabesh M, Azadbakht L, Faghihimani E, Tabesh M, Esmaillzadeh A (2014) Calcium-vitamin D Cosupplementation influences circulating inflammatory biomarkers and Adipocytokines in vitamin D-insufficient diabetics: a randomized controlled clinical trial. JCEM 99(12):E2485–E2493
27. Oh JY, Barrett CE (2002) Association between vitamin D receptor polymorphism and type 2 diabetes or metabolic syndrome in community- dwelling older adults: the Rancho Bernardo study. Metabolism 51(3):356–359
28. Palomer X, Gonzalez-Clemente JM, Blanco-Vaca F, Mauricio D (2008) Role of vitamin D in the pathogenesis of type 2 diabetes mellitus. Diabetes Obes Metab 10(3):185–197
29. Scragg R, Sowers MF, Bell C (2004) Serum 25-hydroxyvitamin D, diabetes and ethnicity in the third National Health and nutrition examination survey. Diabetes Care 27(12):2813–2818
30. Hypponen E, Boucher BJ, Berry DJ et al (2008) 25-hydroxyvitamin D, IGF-1, and metabolic syndrome at 45 years of age: a cross sectional study in the 1958 British birth cohort. Diabetes 57(2):298–305
31. Mattila C, Laaksonen MA, Knekt P et al (2007) Serum 25- hydroxyvitamin D concentration and subsequent risk of type 2 diabetes. Diabetes Care 30(10):2569–2570
32. Song Y, Wang L, Pittas AG, Liana C (2013) Del Gobo. Blood 25-hydroxy vitamin D levels and incident type 2 diabetes. A meta-analysis of prospective studies. Diabetes Care 36:1422–1428
33. Afzal S, Bojesen SE, Nordestgaard BG (2013) Low 25-hydroxyvitamin D and risk of type 2 diabetes: a prospective cohort study and metaanalysis. Clin Chem 59:381–391. [PubMed: 23232064]
34. McDonnell SL, Baggerly LL, French CB, Heaney RP, Gorham ED, Holick MF, Scragg R, Garland CF (2016) Incidence rate of type 2 diabetes is >50% lower in grassroots health cohort with median serum 25–hydroxyvitamin D of 41 ng/ml than in NHANES cohort with median of 22 ng/ml. J Steroid Biochem Mol Biol 155:239–244
35. Stadlmayr A, Aigner E, Huber-Schönauer U, Niederseer D, Zwerina J, Husar-Memmer E, Hohla F, Schett G, Patsch W, Datz C (2015) Relations of vitamin D status, gender and type 2 diabetes in middle-aged Caucasians. Acta Diabetol 52:39–46
36. Pittas AG, Willett WC, Dawson-Hughes B et al (2006) Vitamin D and calcium intake in relation to type 2 diabetes in women. Diabetes Care 29(3):650–656
37. Liu S, Song Y, Ford ES, Manson JE, Buring JE, Ridker PM (2005) Dietary calcium, vitamin D, and the prevalence of metabolic syndrome in middle-aged and older US women. Diabetes Care 28:2926–2932
38. Knekt P, Laaksonen M, Mattila C et al (2008) Serum vitamin D and subsequent occurrence of type 2 diabetes. Epidemiology 19(5):666–671
39. Fourohi NG, Luan J, Cooper A et al (2008) Baseline serum 25-hydroxy vitamin D is predictive of future glycemic status and insulin resistance: the Medical Research Council Ely prospective study 1990-2000. Diabetes 57(10):2619–2625
40. Pittas AG, Harris SS, Stark PC, Wson-Hughes B (2007) The effects of calcium and vitamin D supplementation on blood glucose and markers of inflammation in nondiabetic adults. Diabetes Care 30:980–986
41. Nazarian S, Peter JVST, Boston RC, Jones SA, Mariash CN (2011) Vitamin D3 supplementation improves insulin sensitivity in subjects with impaired fasting glucose. Transl Res 158(5):276–281
42. Mitri J, Dawson-Hughs B, Hu FB, Pittas AG (2011) Effects of viamin D and calcium supplementation on pancreatic beta cell function, insulin sensitivity and glycemia in adults at high risk of diabetes: the calcium and vitamin D for diabetes mellitus (CaDDM) randomized controlled trial. Am J Clin Nutr 94:486–494
43. Gagnon C, Daly RM, Carpentier A, Lu ZX, Shore-Lorenti C, Sikaris K, Jean S, Ebeling PR (2014) Effects of combined calcium and vitamin D supplementation on insulin secretion, insulin sensitivity and b-cell function in multi-ethnic vitamin D-deficient adults at risk for type 2 diabetes: a pilot randomized, placebo-controlled trial. PLoS One 9(10):e109607
44. Zhao L-m, Tian X-q, Ge J-p, Yan-cheng X (2013) Vitamin D intake and type 2 diabetes risk: a meta-analysis of prospective cohort studies. Afr Health Sci 13(4):1130–1138
45. de Boer IH, Tinker LF, Connelly S et al (2008) Calcium plus vitamin D supplementation and the risk of incident diabetes in the Womens health initiative. Diabetes Care 31(4):701–707

46. Avenell A, Cook JA, MacLennan GS, McPherson GC (2009) Vitamin D supplementation and type 2 diabetes: a substudy of a randomized placebo controlled trial in older people (RECORD trial ISRCTN 51647438). Age Ageing 38:606–609
47. Mitchell DM, Leder BZ, Cagliero E, Mendoza N, Henao MP, Hayden DL, Finkelstein JS, Burnett-Bowie S-AM (2015) Insulin secretion and sensitivity in healthy adults with low vitamin D are not affected by high-dose ergocalciferol administration: a randomized controlled trial. Am J Clin Nutr 102:385–392
48. Krul-Poel YHM, Westra S, ten Boekel E, ter Wee MM, van Schoor NM, van Wijland H, Stam F, Lips PTAM, Simsek S (2015 Aug) Effect of vitamin D supplementation on glycemic control in patients with type 2 diabetes (SUNNY trial): a randomized placebo-controlled trial. Diabetes Care 38(8):1420–1426
49. Elkassaby S, Harrison LC, Mazzitelli N, Wentworth JM, Colman PG, Spelman T, Fourlanos S (2014) A randomised controlled trial of high dose vitamin D in recent-onset type 2 diabetes. Diabetes Res Clin Pract 106:576–582
50. Tabesh M, Azadbakht L, Faghihimani E, Tabesh M, Esmaillzadeh A (2014) Effects of calcium–vitamin D co-supplementation on metabolic profiles in vitamin D insufficient people with type 2 diabetes: a randomised controlled clinical trial. Diabetologia 57:2038–2047
51. Kampmann U, Mosekilde L, Juhl C, Moller N, Christensen B, Rejnmark L, Wamberg L, Orskov L (2014) Effects of 12 weeks high dose vitamin D3 treatment on insulin sensitivity, beta cell function, and metabolic markers in patients with type 2 diabetes and vitamin D insufficiency a double-blind, randomized, placebo-controlled trial. Metab Clin Exp 63:1115–1124
52. Strobel F, Reusch J, Penna-Martinez M, Ramos-Lopez E (2014 Jan) Effect of a randomized controlled vitamin D trial on insulin resistance and glucose metabolism in patients with type 2 diabetes mellitus. Horm Metab Res 46(1):54–58
53. Nikooyeh B, Neyestani TR, Farvid M, Alavi-Majd H, Houshiarrad A, Kayali A et al (2011) Daily consumption of vitamin D- or vitamin D + calcium- fortified yogurt drink improved glycemic control in patients with type 2 DM: a randomized clinical trial. Am J Clin Nutr 93:764–771
54. Eftekhari MH, Akbarzadeh M, Dabbaghmanesh MH, Hasanzadeh J (2011) Impact of treatment with oral calcitriol on glucose indices in type 2 diabetes mellitus patients. Am J Clin Nutr 20(4):521–526
55. Peng Y, Li L-j (2015) Serum 25-hydroxyvitamin D level and diabetic nephropathy in patients with type 2 diabetes mellitus. Int Urol Nephrol 47:983–989
56. Wolf M, Shah A, Gutierrez O et al (2007) Vitamin D levels and early mortality among incident hemodialysis patients. Kidney Int 72(8):1004–1013
57. Fernandez-Juarez G, Luno J, BArio V, De Venueza SG (2013 Nov) 25 (OH) vitamin D levels and renal disease progression in patients with type 2 diabetic nephropathy and blockade of the rennin-angiotensin system. Clin J Am Soc Nephrol 8(11):1870–1876
58. Zhang Y, Deb DK, Kong J et al (2009) Long-term therapeutic effect of vitamin D analog doxercalciferol on diabetic nephropathy: strong synergism with AT1 receptor antagonist. Am J Physiol Renal Physiol 297(3):F791–F801
59. Deb DK, Sun T, Wong KE, Zhang Z, Ning G, Zhang Y et al (2010) Combined vitamin D analog and AT1 receptor antagonist synergistically block the development of kidney disease in a model of type 2 diabetes. Int Soc Nephrol 77:1000–1009
60. de Zeeuw D, Agarwal R, Amdahl M, Audhya P, Coyn D, Grimella T (2010) Selective vitamin D receptor activation with paricalcitol for reduction of albuminuria in patients with type 2 diebstes (VITAL study): a randomized controlled trial. Lancet 376:1543–1551
61. Kim MJ, Frankel AH, Donaldson M, Darch SJ, Pusey CD, Hill PD (2011) Oral cholecalciferol decreases albuminuria and urinary TGF-beta 1 in patients with type 2 diabetic nephropathy on established renin-angiotensin-aldostrerone system inhibition. Int Soc Nephrol 80:851–860
62. Li YC, Kong J, Wei M et al (2002) 1,25-Dihydroxyvitamin D(3) is a negative endocrine regulator of the renin-angiotensin system. J Clin Invest 110:229–238
63. Milliez P, Girerd X, Plouin PF et al (2005) Evidence for an increased rate of cardiovascular events in patients with primary aldosteronism. J Am Coll Cardiol 45:1243–1248
64. Achinger SG, Ayus JC (2005) The role of vitamin D in left ventricular hypertrophy and cardiac function. Kidney Int Suppl 95:S37–S42
65. Cardus A, Parisi E, Gallego C et al (2006) 1,25-Dihydroxyvitamin D3 stimulates vascular smooth muscle cell proliferation through a VEGF-mediated pathway. Kidney Int 69:1377–1384
66. Koenig W, Sund M, Frohlich M et al (1999) C-reactive protein, a sensitive marker of inflammation, predicts future risk of coronary heart disease in initially healthy middle-aged men: results from the MONICA (monitoring trends and determinants in cardiovascular disease) Augsburg cohort study, 1984 to 1992. Circulation 99:237–242
67. Kendrick J, Targher G, Smits G, Chonchol M (2009) 25-Hydroxyvitamin D deficiency is independently associated with cardiovascular disease in the third National Health and nutrition examination survey. Atherosclerosis 205:255–260
68. Yan C, Chun-Ting Z, Zhe Z, Arthur W (2014) Tse hung fat, Yiu Kai hang. Association of myocardial dysfunction with vitamin D deficiency in patients with type 2 diabetes mellitus. J Diabetes Complicat 28:286–290
69. Herrmann M, Sullivan DR, Veillard A-S, McCorquodale T, Straub IR, Scott R, Laakso M, Topliss D, Jenkins AJ, Blankenberg S, Burton A,

Keech AC (2015) Serum 25-Hydroxyvitamin D: a predictor of macrovascular and microvascular complications in patients with type 2 diabetes. Diabetes Care 38:521–528. doi:10.2337/dc14-0180

70. Heidari B, Nargesi AA, Hafezi-Nejad N, Sheikhbahaei S, Pajouhi A, Nakhjavani M, Esteghamati A (2015) Assessment of serum 25-hydroxy vitamin D improves coronary heart disease risk stratification in patients with type 2 diabetes. Am Heart J 170:573–579.e5
71. Witham MD, Dove FJ, Dryburgh M, Sigden JA, Morris AD, Struthers AD (2010) The effecr of different doses of vitamin D3 on markers of vascular health in patients with type 2 diabetes: a randomized controlled trial. Diabetologia 53:2112–2119
72. Sugden JA, Davies JI, Witham MD, Mrrist AD, Struthers AD (2008) Vitamin D improves endothelial function in patients with type 2 diabetes mellitus and low vitamin D levels. Diabet Med 25:320–325
73. Soderstrom LH, Johnson SP (2012) Diaz1VA and Mainous AG 3rd. Association between vitamin D and diabetic neuropathy in a nationally representative sample: results from 2001–2004 NHANES. Diabet Med 29:50–55
74. Shehab D, Al-Jarallah K, Mojiminiyi OA, Al Mohamedy H, Abdella NA (2012) Does vitamin D deficiency play a role in peripheral neuropathy in type 2 diabetes? Diabet Med 29:43–49
75. Ahmadieh H, Azar ST, Lakkis N, Arabi A (2013) Hypovitaminosis D in patients with type 2 diabetes mellitus: a relation to disease control and complications. ISRN Endocrinol 22(2013):641098
76. Lee P, Chen R (2008) Vitamin D as an analgesic for patients with type 2 diabetes and neuropathic pain. Arch Intern Med 168:771–772
77. Shehab D, Al-Jarallah K, Abdella N, Olusegun A, Mojiminiyi HAM (2015) Prospective evaluation of the effect of short-term oral vitamin D supplementation on peripheral neuropathy in type 2 diabetes mellitus. Med Princ Pract 24:250–256. doi:10.1159/000375304
78. Aksoy H, Akçay F, Kurtul N, Baykal O, Avci B (2000) Serum 1,25Dihydroxy vitamin D (1,25(OH)2D3), 25 Hydroxy vitamin D (25(OH)D) and Parathormone levels in diabetic retinopathy. Clin Biochem 33(1):47–51
79. Alcubierre N, Valls J, Rubinat E, Cao G, Esquerda A, Traveset A, Granado-Casas M, Jurjo C, Mauricio D (2015) Vitamin D deficiency is associated with the presence and severity of diabetic retinopathy in type 2 diabetes mellitus. J Diabetes Res, Article ID 374178, p 7. http://dx.doi.org/10.1155/2015/374178
80. Bhanuprakash Reddy G, Sivaprasad M, Shalini T, Satyanarayana A, Seshacharyulu M, Balakrishna N, Viswanath K, Sahay M (2015) Plasma vitamin D status in patients with type 2 diabetes with and without retinopathy. Nutrition 31:959–963

Impact of UV Radiation on Genome Stability and Human Health

17

Sujit Roy

Abstract

Gradual depletion of the atmospheric ozone layer during the past few years has increased the incidence of solar UV radiation specifically the UV-C on earth's surface is one of the major environmental concerns because of the harmful effects of this radiation in all forms of life. The solar UV radiation including the harmful wavelength range of UV-B (280–320 nm) represents a significant climatic stress for both animals and plants, causing damage to the fundamental biomolecules such as DNA, proteins and lipids, thus activating genotoxic stress and induces genome instability. When DNA absorbs UV-B light, energy from the photon causes covalent linkages to form between adjacent pyrimidine bases, creating photoproducts, primarily cyclobutane pyrimidine dimers (CPDs) and pyrimidine-6,4-pyrimidinone photoproduct (6,4PPs). Pyrimidine dimers create distortions in the DNA strands and therefore can inhibit DNA replication as well transcription. Lack of efficient repair of UV-induced DNA damage may induce the formation of DNA double stand breaks (DSBs), one of the serious forms of damage in DNA double helix, as well as oxidative damage. Unrepaired DSBs in the actively dividing somatic cells severely affect cell growth and development, finally results in loss of cell viability and development of various diseases, such as cancer in man.

This chapter mainly highlights the incidence of solar UV-radiation on earth's surface along with the formation of major types of UV-induced DNA damage and the associated repair mechanisms as well as methods of detecting DNA damage and finally our present understanding on the impact on solar UV radiation on human health.

S. Roy (✉)
Department of Botany, UGC Centre of Advanced Studies, The University of Burdwan,
Golapbag Campus, Burdwan, West Bengal 713104, India
e-mail: sujitroy2006@gmail.com

S. Ahmad (ed.), *Ultraviolet Light in Human Health, Diseases and Environment*, Advances in Experimental Medicine and Biology 996, https://doi.org/10.1007/978-3-319-56017-5_17

Keywords
Climate change • DNA damage • Environmental stress • Genome instability • Human health • UV radiation

17.1 Introduction

Continuous depletion of the stratospheric ozone layer due to the release of atmospheric pollutants, including chlorofluorocarbons, chlorocarbons, and organobromides has become one of the major global concerns since this has been associated with an increased incidence of solar UV-radiation, specifically C, on earth's surface. UV-radiation acts as one of the most potential genotoxic agents which directly affect the stability and integrity of the genetic material specifically DNA and thus have adverse effects on normal life processes of all living organisms [1, 2–4]. Depletion of ozone layer with the concomitant increase in solar UV radiation on earth's surface has been predicted to persist throughout most of this century, provided the control measures of ozone depletion are not exerted [5].

UV radiation has been broadly classified in three different forms based on their wavelength; UV-C radiation (100–280 nm) is mostly absorbed by the atmospheric ozone layer, which also absorbs ~90% of the UV-B radiation (280–320 nm). Therefore, UV-A (320–400 nm) constitutes for about 95% of the total UV radiation reaching on the earth's surface, while UV-B radiation accounts for the rest of 5% solar UV radiation on earth [6].

The major impact of UV radiation involves the induction of DNA damages in all forms of life [6]. Since absorption maxima of nucleic acids lies within the range of 260 nm, UV-C acts as the most potential form of UV radiation for induction of DNA damage [7]. However, the DNA damage inducing activity of solar UV radiation (UVR) is mainly contributed by UV-A and UV-B since most of the UV-C fails to reach earth surface due to absorption by stratospheric ozone layer [8]. Although earlier studies have indicated absorption of about 90–95% of solar UV-B radiation in the atmospheric ozone layer, the increasing level of depletion of the ozone layer during the recent past years, (as indicated with the discovery of ozone 'hole' in the 1980s), has become one of the major environmental issues in the context of enhanced possibility of interaction of UV-B radiation with the biological system. The magnitude and variability of solar UV-B radiation generally relies on the path length through the various layers of the atmosphere and on the real-time concentration of the ozone layer.

Plethora of studies involving UV-C have also played important roles in our understanding the molecular mechanism of DNA damage and repair systems in human and other organisms including plants. UV-C represents the shortwave of solar UV radiation including the wavelength of 254 nm, constituting one of the important wavelengths of the UV spectra and the most damaging for terrestrial life. Interestingly, this band of UV light is completely absorbed by the ozone layers present in the stratospheric region. But the recent finding of the ozone hole in the polar regions [9, 10] raised concerns over the serious threat imposed for the biological system.

Several lines of evidences have demonstrated the deleterious effect of UV-C irradiation on biomolecules, particularly, nucleic acids and proteins through the production of reactive oxygen species (ROS) and subsequent oxidative damage. The impacts of UV-C irradiation on the conformational and functional aspects for various proteins have already been demonstrated. Furthermore, *in-vitro* experiments have shown that the penetration efficiency UV-C is significantly reduced by chromophores in the upper epidermal layer of human skin [7]. In addition, detection of UV-C mediated DNA damage in the dividing basal layer of human epidermal cells has been shown to be relatively difficult [6], explaining considerably less steep dose response curve for UV-C induced erythema (redness or rash on the skin surface) than UV-B in human skin. However, UV-C exposure from the artificial sources may often cause severe photokeratitis – a burn injury of cornea in the eye.

Like UV-B, UV-C irradiation mainly causes damage to the genetic material, primarily DNA through the generation of different photoproducts like cyclopyrimidine dimers (CPDs) and 6-4-photoproducts (6-4 PPs). However, as compared with UV-B, UV-C induces the production of these dimers at a relatively higher efficiency. CPDs may eventually be transformed into TT tandem mutations via 'dimer bypass' due to UVR generated reactive oxygen species [2, 6]. The formation of TT mutations is more frequent under UV-C stress and therefore routinely utilized for *in vitro* mutagenesis studies.

17.2 The Incidence of Solar UV Radiation and Its Influence on Biological System

The amount of solar UV radiation reaching on the earth's surface generally depends on the season and the location. Some of the variations are generally usual, like the position of sun in the sky during the day time and the year-wise changes in the distance between earth and sun. However, additional variation may occur due to local or wide spread alterations of atmospheric constituents, influencing the transmission of the radiation from the upper part of the atmosphere to the lower surface layers where ozone, clouds, aerosols and additional gases like nitrogen dioxide and sulfur dioxide, especially in the regions with relatively higher levels of pollution, constitute the essential compositions [6].

The part of the solar energy, popularly known as 'albedo', has also been shown to affect the level of UV radiation reaching earth's surface [11]. As a consequence of the higher position of sun in the sky, the level of UV radiation is comparatively higher in the tropical regions of the world, also consistent with the fact of higher intensity of UVR during the mid-day in the summer season. This is because of the influence of solar zenith angle (the relative height of the sun above the horizon) on the incidence of UV radiation, as lower elevation corresponds to longer pathway, causing higher absorption by the atmosphere and eventually lower incidence at the surface level. The solar zenith angel changes with the time of the day and season. Therefore, more UVR reaches on earth's surface during the midday time when sun appears high in the sky, lowering the zenith angel and the solar radiation finds relatively less ozone layer and atmosphere. Furthermore, because of thinner ozone layer, higher altitudes receive relatively higher level of solar UVR. However, recent advances have implicated the importance of cloud cover in regulating the entry of solar UVR for any given altitudes and latitudes. Interestingly, in case of polluted metropolitans, although the aerosols and gases provide some degree of protection against UVR, but may influence scattering of light and therefore, increasing the possibility of exposure to UVR in shaded places [12].

The deleterious effects of chlorofluorocarbons, organobromides and chlorocarbons have been studied extensively in the context of stratospheric ozone layer destruction. Although regulations have been imposed through various global summits, the increased levels of use of these compounds for the past several years and their selective accumulation in the atmosphere have already created alarming amount of damage to the ozone layer [13]. Several biological consequences have been detected alongside the increased exposure to solar UVR, adversely affecting the normal growth and developmental pattern of plants, pathogens, other grazing animals, soils microbes and certainly the harmful effects on human health. The effect of solar UVR on human health has been the subject of extensive investigation worldwide over the last decades in the context of global climate change and increasing incidence of skin cancer.

17.3 The Genotoxic Effects of Solar UV Radiation – Induction of DNA Damage and Genome Instability

DNA damage induced by the solar UVR is one of the crucial cellular signals which probably significantly influences the normal life processes in all living organisms, sometimes without any immediate detectable manifestations. Various harmful endogenous factors, including the ROS

[14], are frequently generated as byproduct of the metabolic processes and due to exposure to exogenous genotoxic agents like UV and ionizing radiations. Eventually the ROS, via oxidative modifications of nucleic acids and DNA strand breaks, interfere with the fundamental cellular processes like DNA replication and transcription, therefore inducing genotoxic effects and genome instability [15]. Damages to the double helical structure of DNA may lead to change in the base-pairing potential of the nucleotides during DNA synthesis and base modification via deamination, which subsequently induces depurination, and depyrimidination [16].

The alkylating agents also cause base modification and thus may generate mutations due to mis-incorporation during replication [17]. In addition, as discussed earlier, direct interaction of ionizing radiations (IR) and UV-B induced ROS with the DNA molecules may result in oxidative damage and DNA intra- and inter-strand cross links [18]. Exposure to UVR and certain genotoxic chemicals may result in single as well as double DNA strand breaks. DNA double strand breaks (DSBs) are considered one of most harmful forms of DNA lesions, resulting in loss of chromosome fragment and thus genetic information, leading to genome instability. High intracellular concentrations of ROS, frequently produced via UV-B mediated genotoxic stress response in eukaryotes, including human cells, cause oxidative damage to membrane lipids, structural and enzymatic proteins and DNA molecules, and found to be associated with various human diseases [19].

Although UV-B radiation accounts for about less than 1% of total solar radiation energy, it represents one of the extremely active parts of solar radiation with the potential capacity for modification in DNA at the structural and chemical levels. Some UV-absorbing pigments, mainly the phenylpropanoid compounds like flavonoids and anthocyanins are produced by various organisms, particularly by the green plants act as the first line of defense against the harmful effects of UVR. However, this mechanism may not be completely effective in blocking the UV light from reaching the surface tissues [20, 21]. Along with the scavengers of oxidative stress generated by ROS, like vitamin C, B, and E and glutathione, certain antioxidant enzymes, including catalase (CAT), peroxidase (POD) superoxide dismutase (SOD) also play key role as part of cellular defense mechanism against UVR [22]. However, for maintaining genome stability, the living cells, particularly the eukaryotes, with large and more complex genomes than the prokaryotes, have developed an extensive and coordinated network of DNA repair mechanisms, such as excision repair, photoreactivation repair (PR), post-replication repair (mismatch repair or MMR, repair of SSBs and DSBs and additional back-up pathways like error-free translesion synthesis (TLS), alongside response at the cellular levels such as activation of cell cycle check-point functions and induction of programmed cell death (PCD) for highly efficient detection, signaling and repair of UV-induced DNA damages [23, 24].

17.4 Solar UV Radiation and Its Impact on Plant Life

Plants, because of their immobile nature, are widely exposed to various environmental stress factors like solar UVR, high salinity, drought, chemical toxicity and endogenous processes, which frequently induce DNA damages via genotoxic stress. DNA damage, if remains unrepaired, perturbs genome stability and thus affecting the normal growth and development in plants and finally crop productivity [17, 25]. UV-B radiation generates various responses in plants at the morphological, physiological, biochemical and cellular levels [26]. However, the mechanisms of action of such orchestrated and complicated network integrating the signals from various levels are relatively less understood. UV induced DNA damages, which mainly include cyclobutane pyrimidine dimers (CPDs) and pyrimidine (6-4) pyrimidone photoproducts (6,4PPs), may usually block DNA and RNA Polymerases [27]. Additionally, UV-B photons may also cause direct damage to proteins [28]. To avert the damaging effects of UV-B radiation, plants have developed two major protective mechanisms; one involves the production of UV absorbing sunscreen compounds, such as flavonoids and anthocyanins and reflection of solar UV radiation by the wax layer present on the leaf surface and other cellular structures [29]. The other UV-protective mechanism recruits direct removal of UV induced DNA lesions by the light dependent pathway called photoreactivation repair or light independent (dark repair) DNA repair pathway, the nucleotide excision repair pathway [27].

With the absorption maxima at 254 nm, DNA is the major target of UV-B radiation. Low doses of UV-B radiation was found to be lethal for mutants lacking specific DNA repair pathways [27, 30].

In flowering plants, the UV absorbing flavonoids and anthocyanins accumulate in the vacuoles of epidermal cells to minimize the absorption of UV radiation of sunlight with the minimal absorption of photosynthetically active radiation [31]. Among the various forms of secondary metabolites produced in plants under abiotic stress through the phenylpropanoid pathway, the flavonoids represent one of the predominant classes of compounds in such biosynthetic pathway. Previous studies have indicated rapid accumulation of flavonoids and anthocyanins in plants in response to UV-radiation [32]. Recent study has also demonstrated effective role of flavonoids in scavenging the r ROS, providing protection against oxidative damage [33].

The light dependent PR pathway is activated in plants under low frequency of UV-induced DNA damage. The PR pathway directly converts the damage DNA into its normal configuration by the activity of enzymes called photolyases [34]. The PR pathway has been reported in various prokaryotic and eukaryotic cells including *E coli*, yeast and some species of plants, but characteristically absent in humans [35]. In contrast, the light independent or dark repair pathway involves removal of UV induced photoproducts by nucleotide excision repair (NER) mechanisms. NER is a general pathway of repair of UV-B induced DNA lesions and more wide spread across the animal and plant kingdom. The homologues of most of the genes involved in NER pathway in mammals, including humans has been identified in *Arabidopsis* genome, indicating the evolutionary conservation of this pathway [36].

17.5 Mechanism of Repair of UV-Induced DNA Damages

DNA repair pathways represent one of the fundamental cellular processes for protecting cells against the damage, and intimately associated with the crucial mechanism to guarantee faithful transfer of genetic information over generations. In general, the DNA repair pathways are highly conserved among the eukaryotes. The *Arabidopsis* (dicot plant with genome 130 Mbp, n = 5) and rice (monocot plant, genome 430 Mbp, n = 12) genome sequence projects have revealed the presence of several repair proteins, homologues to those from human genome, with some key differences [36, 37]. The DNA damages induced by genotoxic stress, such as UVR must be repaired for maintaining the genome stability, growth and productivity. The UV induced non-coding photoproducts, such as pyrimidine dimers, generally inhibits the activity of DNA and RNA polymerases, and thus necessitates their repair for the normal functioning of the living cell. The UV induced photoproducts are repaired via the following major pathway discussed below.

17.6 Direct Repair of UV Photoproducts – Photoreactivation Repair

The high-energy UV-B photons induce modifications in DNA structure, resulting in the generation of lesions, commonly known as UV photoproducts. The cyclobutane CPDs and 6-4PPs are the two most frequently formed photoproducts [38, 39]. CPDs involve the formation of a link between the four-member ring structure with the C5 and C6 on the same strand of DNA. The 6-4PPs are produced due to the linkage of the C6 position of the 5′-pyrimidine to the C4 position of the 3′-pyrimidine in the adjacent pair. The CPDs alone constitute up to 75% of the total UV-induced photoproducts. On the other hand, 6-4PPs may become converted into a Dewar Valence Isomers (DEWs) following absorption of UV-A light, with maximal efficiency at around 320 nm [40, 41]. UV-A has been shown to generate 6-4PPs in human DNA, but comparatively in much lower quantities than CPDs. Recent studies have detected 6-4PPs in the genome of UV-A-irradiated DNA repair deficient human fibroblasts, however not detected in normal fibroblast cells proficient in DNA repair activity [42], suggesting effective repair mechanism of UV-A generated 6-4PPs in human cells. CPDs may arise either due to the absorption of UVA photons or by photosensitization [40]. Photosensitization mechanism involves excitation of endogenous chromophores with the subsequent conversion of long-lived excited triplet states by intersystem crossing, leading to the formation of CPDs [43].

In prokaryotes, yeast and plants, under low frequencies of UV-induced DNA damage, the light-dependent photoreactivation repair pathway is generally activated and recruited. This pathway directly converts the damaged DNA into its normal state through the activity of the enzymes called photolyases which are specialized class of proteins in terms of substrate specificity. All types of photolyases possess two co-factors, one of which mainly the two electron reduced form of FAD (FADH) [44]. Photolyases generally bind specifically to the UV-B induced DNA lesions and remove them directly by absorption of blue light in the wavelengths between 300 and 600 nm, reducing the dimer to monomer pyrimidines with the subsequent release of the enzyme [45].

17.7 Nucleotide Excision Repair Eukaryotes

The light-independent nucleotide excision repair (NER) pathway, also known as dark repair pathway is the more general, flexible and wide spread mechanism of repair of UVR induced DNA damage. NER involves efficient detection of the UV-induced photoproducts and removal of approximately 24–32 oligonucleotides from the damaged DNA strand, followed by repair synthesis and sealing of the nick through ligation [46, 47]. However, defects in NER pathway may not completely remove the UV photoproducts, resulting in the inhibition of fundamental cellular processes, such as DNA replication and transcription, which may eventually lead to accumulation of mutations and cell death.

In general, the NER pathway comprises of two sub-pathways based on the initial detection of the magnitude of UV-induced DNA damage. These sub-pathways include the global genome repair (GGR) and the transcription-coupled repair (TCR). The GGR pathway repairs the UV photoproducts in the DNA on a genome-wide level, while the TCR becomes functional for repairing damage in the transcriptionally active strand of DNA associated with gene expression. In general, the NER pathway involves the participation of approximately 30 proteins which are recruited in a sequential manner for the removal of damaged region of the DNA stand containing the lesion [19, 48]. The GGR and TCR differ in the initial damage recognition steps. While GGR becomes activated following the detection of damage via the activity of Xeroderma pigmentosum group C (XPC)/hHR23B complex [49], TCR pathway is initiated when the movement of RNA polymerase II in the coding stand of DNA is inhibited at the site of DNA damage [50]. After the initial step of damage recognition, in both GGR and TCR, XPA and TFIIH (basal transcription factor IIH) bind the DNA stand at the site of damage, eventually resulting in the unwinding of damaged region, allowing for the subsequent binding of other repair proteins. In the next step, the repair endonucleases ERCC1/XPF (excision repair cross-complementing/XPF) and XPG are recruited at the damaged site, followed by nick of the damaged strand at the 5′ and 3′ sides, with the intact undamaged strand now serving as the template for the repair synthesis. At this step, replication protein A (RPA), an essential component of eukaryotic DNA repair machinery and a heterotrimeric single-stranded DNA-binding protein, participates in both incision reaction and repair synthesis. RPA generally recognizes and binds to the undamaged template strand, protecting the cleavage of the template and also facilitates in stabilizing the open structure. The final step in NER pathway involves repair synthesis of the excised part in the damaged strand by the activity of replicative DNA polymerases, such as DNA polymerase δ and ε, along with other accessory protein factors like RPA, RFC and PCNA, respectively [51, 52].

17.8 UV Radiation, Oxidative Stress and Reactive Oxygen Species

Oxidative damage is one of the major stresses in all forms of life, including the plants and humans. Oxidative stress activates the formation of ROS and other highly reactive free radicals [53]. ROS produced either in response to abiotic and biotic stress or as byproduct of endogenous metabolic processes, are very short-lived reactive free radicals and immediately interact most of the cellular components including DNA causing oxidation of bases, some of which are highly mutagenic. In mammals, including humans, 8-oxo-dG and 1,2-dihydro-2-oxodeoxyadenine (2-OH-dA) are the most commonly generated forms of ROS induced oxidized bases [54]. Presence of 8-oxo-dG lesion in replicating strand may promote misincorporation of A against 8-oxo-dG. On the other hand, existence of 2-OH-dA in the replicating strand strongly inhibits the progression of the replication fork by the replicative DNA polymerases during replication, generating single-stranded DNA (ssDNA) intermediate due to unwinding of DNA strands under replication stress. The ssDNA intermediates are recognized as the signal for replication stress by the activity of ATM (ataxia telangiectasia mutated) and ATR (ATM and Rad3-related) protein kinases. ATR kinases, in particular stimulates cell-cycle checkpoint function under replication stress through phosphorylation of the downstream target Chk1 (check point kinase) which subsequently inhibits activation of origin and blocks the S to G2 phase transition, delaying cell cycle progression, allowing additional time for repairing the damage [55].

Guanine, because of its lowest ionization potential among DNA bases, appears as the major target of oxidative base modification. On the other hand, oxidative modifications of other three bases are generally infrequent. However, the sugar phosphate backbone generally remains unaffected in this case. Charges are readily transferred to guanine for oxidative modification of this purine component of DNA and the effectiveness of charge transfer generally depends on the GC content [56]. As indicated earlier, 8-oxodG is the major product of oxidative DNA damage following exposure to UVR and has also been considered as the molecular marker of oxidative damage in mammalian genomes, including humans, and has been found to be associated with cancer and aging process [57–59]. Plethora of studies have demonstrated the involvement of human DNA Pol λ, an important member of family X-DNA polymerase, in the efficient and error free repair of oxidative DNA damages, including 2-OH-A and 8-oxo-dG lesions via trans-lesion synthesis (TLS) [60–62]. On the other hand, error-prone bypass of 8-oxo-dG by human Pol κ, a member of Y-family DNA polymerase, has been shown to enhance the harmful effects of oxidative damage, thus reduces genome stability.

17.9 Repair of UV-B Mediated Oxidative DNA Damage – The Base Excision Repair Pathway

In human cells, the oxidized bases in the DNA are mainly repaired via base excision repair (BER) pathway. BER involves the elimination of the damaged base by the activity of a specific class of enzyme, known as DNA glycosylase, producing an abasic site, which is subsequently processed by the step-wise action of the APE1 endonuclease (Apurinic/apyrimidinic endonuclease), Pol β and XRCC1-DNA ligase 1 complex to finally seal the nick [63]. In human cells, a specialized repair pathway, known as trans- lesion synthesis (TLS) has been shown to be activated in response to prolonged replication block via checkpoint function, which recruits specialized DNA Pols, mainly members of family X and Y polymerases, capable to bypass the lesion to resume replication. On the other hand, DSBs generated by stalled replication fork or delayed BER, are repaired via homologous recombination (HR) (during the S phase of cell cycle only) and non-homologous end joining (NHEJ) mechanisms.

The effectiveness and accuracy of BER pathway are mainly regulated by various forms of DNA glycosylase, which specifically removes the corresponding types of oxidatively damaged bases via the cleavage of the N-glyosidic linkage between the damaged base and deoxyribose sugar, resulting in the formation of an abasic site (apurinic or apyrimidinic) or single strand breaks [64].

DNA strands with single nucleotide damage are processed and repaired via the short patch BER (SP-BER), while multiple nucleotide damages (two or more nucleotide) are repaired by long patch BER (LP-BER) pathway. After the initial excision of damaged base by a specific glycosylase, the SP-BER pathway recruits abasic site specific endonuclease, which creates incision in the backbone of the DNA at the 5′ end of AP site. The resulting gap is then filled by DNA polymerase β to replace the damaged nucleotide, followed by sealing of the nick by the activity of DNA ligase III and the scaffold protein XRCC1, restoring the intact DNA [65]. In contrast, LP-BER involves nick translation reaction in association with strand displacement in the 5′-3′ direction, producing a flap-like structure. The flap structure is then removed by the activity flap endonuclease FEN-1 together with proliferating cell nuclear antigen (PCNA). DNA polymerase δ/ε fills the gap by insertion of 2–10 nucleotides, followed by nick ligation via DNA ligase I [66].

17.10 Repairing DNA Strand Breaks – The Homologous Recombination (HR) and Non-homologous End Joining (NHEJ) Mechanisms

Several studies have clearly demonstrated the frequent occurrence of DNA SSBs, DSBs in UV-irradiated cells, particularly in replicating DNA. UV-B-induced ROS and photoproducts, such as CPDs and 6-4PPs induce DNA strand breaks due to inefficient repair of primary lesions and prolong replication stress at the site of stalled replication forks containing CPDs, which frequently resulting in the formation of DSBs [67]. Therefore, efficient detection, and rapid repair of DSBs in the genome is crucial for maintaining genome stability and faithful transmission of genetic information to the next generations. The DSBs are repaired by two fundamental mechanisms: the homologous recombination (HR) and the non-homologous-end joining (NHEJ) pathway. The HR pathway is mediated by the proteins of RAD52 epistasis groups (RAD51, RAD52, RAD54, RAD55, RAD57 and the MRN complex, comprising of MRE11, RAD50 and NBS1 [68]. HR pathway utilizes an intact copy of the homologous DNA duplex for the formation of a heteroduplex for repairing the damaged strand using the non-damaged region as a template [69]. DSB repair via HR is common in bacterial and yeast cells, however, in eukaryotes, including humans and other mammals, HR mediated DSB repair is crucial during the early stages of gamete formation in meiotic cells [68].

In human with large and complex genomes, majority of DSBs in somatic cells are repaired via the NHEJ pathway, in which the broken ends of double stranded DNA are directly joined irrespective of sequence homology. Thus, NHEJ repair is error-prone but represents the predominant DSB repair pathway during G1 to early S-phase of cell cycle. However this pathway has also been found to be functional throughout the cell cycle [69]. In NHEJ repair, the KU70/80 complex binds to the DNA ends at the site of DSBs in the double stranded DNA. Broken ends are then processed by the MRN complex for making the ends suitable substrate for joining by the activity of DNA ligase IV and XRCC4. The gap filling synthesis requires involvement of DNA polymerase λ (Pol λ), the sole member of family X DNA polymerase in plants.

17.11 Solar UV Radiation Induced DNA Damage and Biological Impact Assessment

Two main methods, such as the physical detecting and the biological sensing are generally utilized for measurement of solar UV radiation and its biological impact. The physical detectors are of two major types, including the broadband detectors and the spectroradiometers. The broadband detectors are efficient in detecting UV-A and UV-B components (400–315 nm and 315–280 nm) of solar UVR and have been installed at various latitudes in South America [6]. The spectroradiometers are sensitive for measuring individual wavelengths of solar UV light, facilitating efficient detection of shorter UV wavelengths which, in general, remain undetected by other physical detectors like broadband system. However, the physical detector systems are expensive, with the complicated mode of operation and proved inefficient in estimating the biological effects of solar UVR. Considering these limitations, the biosensors have become increasingly popular in recent years for the assessment of fundamental biological impacts of solar UV radiation and during the past years, various biological models were developed as biosensors of sunlight and its UV components [7].

Since UV radiation is known to initiate and induce various harmful effects in all forms of life, particularly its damaging effects on human health, the biosensors have become more useful in directly or indirectly measuring the genome damaging capacity of solar UV radiation. For monitoring the genotoxic effects of sunlight under the conditions of maximum incidence of solar UVR, several experiments have been performed in recent years at Southern hemisphere (5°5′S, 23°3′S and 29°4′S) during the summer using purified DNA preparations. Similar trials were also conducted in parts of Chile during the spring time considering the occurrence of 'ozone hole' at this point of the year. Comparative analysis of combination of data obtained from DNA dosimetry, incidence of UV-A and UV-B radiations, as detected using physical detector systems have revealed sharp increase in daily UV-B doses along with the decrease in latitude. On the other hand, only marginal increase in daily UV-A doses could be detected due to decrease in latitude. Further analysis of DNA damage profile on daily basis have indicated increased frequency of 6-4 PPs with enhanced incidence of UV-B light, reaching the peak phase during the midday and the trough phase during early morning and late afternoon time.

These observations, along with the results from biochemical and immunological approaches have indicated differential effects of sunlight, in terms of induction of DNA damage, depending on the latitude of the locations. Earlier reports have revealed increased frequency of DNA base oxidation with the concomitant increase in latitude and a significant decline in the extent of accumulation of 6-4PPs [6]. In contrast, decrease in latitude favored the formation of 6-4PPs but not oxidized bases. However, UV-induced generation of CPDs did not shown any clear dependence on the latitude of the locations and in fact appears as the most common form of UV-induced DNA lesions. These emerging observations have indicated the importance of 6-4PPs for consideration of a key biomarker for monitoring the biological impact assessment of solar UV radiation in the context of human health and genome stability.

17.12 UV-Induced DNA Damage, Error-Prone Repair and Increased Risk of Cancer in Humans

Several lines of evidences have already established the critical function of DNA repair mechanisms for maintaining the stability of human genome which is under continuous assault from the environmental factors and endogenous processes. Proper functioning of each specific DNA damage repair mechanism eliminates the chances of accumulation of mutations and ensures genome stability. However, defects in DNA damage repair pathway often found to be linked with various health problems in human system. Xeroderma pigmentosum, commonly known as XP, is one of the typical examples of autosomal recessive mutation characterized by an extreme sensitivity to ultraviolet (UV) rays from sunlight, commonly affecting the eyes and exposed areas of skin early in childhood. Patients with XP have defective nucleotide excision repair pathway and therefore deficient in repairing UV-B induced

DNA damage. Patients are hypersensitive to UV light with a remarkably increased risk of developing skin cancer. Overexposure to UV radiation from the sun is considered one of the main causes of skin cancer.

The risk of development of the serious form of skin cancer, melanoma, has been shown to be associated with exposure to sun on long-term and regular basis. The most important defense mechanisms in human skin cells for its protection against UVR are synthesis of melanin in skin and activations DNA repair mechanisms. Low skin pigmentation production capacity, as found in white Caucasians and patients such as of XP, fail to provide such protective shield. Extensive research using animal models have revealed more active role of UV-B in the induction of skin cancer than UV-A. In case of SCC and BCC, UV-induced DNA lesions produce specific mutations, commonly known as 'UV-signature mutation' in predisposed genes. In SCC development, UV-signature mutations in some tumor suppressor genes like p53 have been frequently reported in SCC development. During the last decades, animal models, including genetically engineered mice, and human skin xenografts, have been used to investigate the key role of the DNA repair mechanisms in UV-induced skin cancer. Besides the involvement of NER in UV-induced tumorigenesis of melanoma and nonmelanoma skin cancers, several line of evidences have recently suggested important role of DNA mismatch repair (MMR) system for repairing UV-induced DNA damage in mammals [70].

17.13 Conclusion

In addition to its major damaging effects, solar UV-B radiation at low fluence rate also plays important role in regulating the expression of genes, particularly associated with development, stress, defense and UV-protective responses both in plants and animals, including humans. Although the photoreceptor, which appears to initiate the UV-B mediated effects, has scarcely characterized, various components of the UV-B signaling pathway have been identified in human system.

In plants, the UV resistance locus 8 (UVR8) appears to function as an essential protein for nearly all physiological UV-B mediated responses, suggesting its close association with the putative receptor in the UV-B signaling cascade.. Therefore, it is imperative to critically investigate the mechanisms of UV-B signaling cascades in order to understand how these processes regulate expression of the responsive genes; photoreactivation is one of the major repair pathways of UV-B-induced photoproducts. However, human cell lacks photoreactivation repair and UV-induced DNA damages and oxidized bases are predominantly repaired via NER and BER pathways. On the other hand, DNA DSBs, which represents another major form of DNA damage produced under UV-B stress, are mainly repaired by non-homologous end joining pathway. Defects in any of the components of these crucial inter-linked repair pathways severely affect genome stability and cell viability. Therefore, in depth knowledge on the underlying mechanisms of activation of damage response following detection of UV-induced damage in the genome has become one of the important areas of biomedical research in the context of human health. The challenge for current and future research is to understand the detail molecular mechanisms in human DNA damage repair system in order to get further insight into the damage response signaling and identification of important target points for future utilization in the context of development of potential drugs and the possibility of genome editing as safeguard measures.

Acknowledgements The work was supported by a Start-Up Research Grant to SR from the University Grants Commission, Govt. of India. The author is thankful to the Centre of Advanced Studies at the Department of Botany, The University of Burdwan, West Bengal, India for necessary infrastructural support. Also I sincerely thank Dr. Rajarshi Ghosh, Department of Chemistry, The University of Burdwan, for the critical reading and corrections of the Manuscript.

References

1. Sinha RP, Kumar HD, Kumar A, Hader DP (1995) Effects of UV-B irradiation on growth, survival, pigmentation and nitrogen metabolism enzymes in cyanobacteria. Acta Protozool 34:187–192
2. Norval M, Cullen AP, De Gruijl FR et al (2007) The effects on human health from stratospheric ozone depletion and its interactions with climate change. Photochem Photobiol Sci 6:232–251
3. Solomon KR (2008) Effects of ozone depletion and UV-B radiation on humans and the environment. Atmosphere-Ocean 46:185–202
4. Tabazadeh A, Santee ML, Danilin MY (2000) Quantifying denitrification and its effect on ozone recovery. Science 288:1407–1411
5. Schuch AP, Garcia CC, Makita K, Menck CF (2013) DNA damage as a biological sensor for environmental sunlight. Photochem Photobiol Sci 12:1259–1272
6. Britt AB (2002) Repair of damaged bases. In: The Arabidopsis book.doi:10.1199/tab.0005
7. Yagura T, Makita K, Yamamoto H, Menck CF, Schuch AP (2011) Biological sensors for solar ultraviolet radiation. Sensors (Basel) 11:4277–4294
8. Tyrrell RM, Ley RD, Webb RB (1974) Induction of single-strand breaks (alkali-labile bonds) in bacterial and phage DNA by near UV (365 nm) radiation. Photochem Photobiol 20:395–398
9. Aucamp PJ, Burn LO, Lucas R (2011) Questions and answers about the environmental effects of ozone depletion and its interactions with climate change: 2010 assessment. Photochem Photobiol Sci 10:301–316
10. Norval M, Lucas RM, Cullen AP, de Gruijl FR, Longstreth J et al (2011) The human health effects of ozone depletion and interactions with climate change. Photochem Photobiol Sci 10:199–225
11. Coelho SG, Hearing VJ (2010) UVA tanning is involved in the increased incidence of skin cancers in fair-skinned young women. Pigment Cell Melanoma Res 23:57–63
12. Tevini M, Mandronich S, Blumthaler M, JCvd L, FRd G et al (1993) UV-B radiation and ozone depletion. Effects on humans, animals, plants, microorganisms, and materials, 1st edn. CRC Press, Boca Raton
13. Lubin D, Jensen EH (1995) Effects of clouds and stratospheric ozone depletion on ultraviolet radiation trends. Nature 377:710–713
14. Bailly C, El-Maarouf-Bouteau H, Corbineau F (2008) From intracellular signaling networks to cell death: the dual role of reactive oxygen species in seed physiology. Comptes Rendus Biologies 331:806–814
15. Roy S, Banerjee V, Das KP (2015) Understanding the physical and molecular basis of stability of arabidopsis DNA Pol λ under UV-B and high NaCl stress. PLoS One 10(7):e0133843. doi:10.1371/journal.pone.0133843
16. Cadet J, Mouret S, Ravanat JL, Douki T (2012) Photoinduced damage to cellular DNA: direct and photosensitized reactions. Photochem Photobiol 88:1048–1065
17. Tuteja N, Ahmad P, Panda B, Tuteja R (2009) Genotoxic stress in plants: shedding light on DNA damage, repair and DNA repair helicases. Mutat Res 681:134–149
18. Roy S (2014) Maintenance of genome stability in plants: repairing DNA double strand breaks and chromatin structure stability. Front Plant Sci. doi:10.3389/fpls.2014.00487
19. Perdiz D, Grof P, Mezzina M, Nikaido O, Moustacchi E et al (2000) Distribution and repair of bipyrimidine photoproducts in solar UV-irradiated mammalian cells. Possible role of Dewar photoproducts in solar mutagenesis. J Biol Chem 275:26732–26742
20. Rastogi RP, Richa SP, Singh HDP, Sinha RP (2010) Mycosporine-like amino acids profile and their activity under PAR and UVR in a hot-spring cyanobacterium Scytonema sp.HKAR-3. Aust J Bot 58:286–293
21. Britt AB (1999) Molecular genetics of DNA repair in higher plants. Trends Plant Sci 4:20–25
22. Xie Z, Wang Y, Liu Y, Liu Y (2009) Ultraviolet-B exposure induces photo-oxidative damage and subsequent repair strategies in a desert cyanobacterium Microcoleus vaginatus Gom. Eur J Soil Biol 45:377–382
23. Sinha RP, Hader DP (2002) UV-induced DNA damage and repair: a review. Photochem Photobiol Sci 1:225–236
24. Roy S (2014) DNA polymerase λ – a novel DNA repair enzyme in higher plant genome. Plant Sci Today 1:140–146
25. Roy S, Roy Choudhury S, Sengupta DN, Das KP (2013) Involvement of AtPolλ in repair of high salt and DNA cross linking agent induced double strand breaks in Arabidopsis thaliana. Plant Physiol 162:1195–1210
26. Paul ND, Gwynn-Jones D (2003) Ecological roles of solar UV radiation: towards an integrated approach. Trends Ecol Evol 18:48–55
27. Quaite FE, Takayanagi S, Ruffini J, Sutherland JC, Sutherland BM (1994) DNA damage levels determine cyclobutyl pyrimidine dimer repair mechanisms in alfalfa seedlings. Plant Cell 6:1635–1641
28. Gerhardt KE, Wilson MI, Greenberg BM (1999) Tryptophan photolysis leads to a UVB-induced 66 kDa photoproduct of ribulose-1,5-bisphosphate carboxylase/oxygenase (Rubisco) *in vitro* and *in vivo*. Photochem Photobiol 70:49–56
29. Bornman JF, Reuber S, Cen YP, Weissenbo G (1997) Ultraviolet radiation as a stress factor and the role of protective pigments. In: Lumsden PJ (ed) Plants and UVB: responses to environmental change. Cambridge University Press, Cambridge, UK, pp 157–168
30. Landry LG, Chapple CCS, Last RL (1995) Arabidopsis mutants lacking phenolic sunscreens exhibit enhanced ultraviolet-B injury and oxidative damage. Plant Physiol 109:1159–1166

31. Landry LG, Stapleton AE, Lim J, Hoffman P, Hays JB, Walbot V, Last RL (1997) An Arabidopsis photolyase mutant is hypersensitive to ultraviolet-B radiation. Proc Natl Acad Sci U S A 94:328–332
32. Czemmel S et al (2009) The grapevine R2R3-MYB transcription factor VvMYBF1 regulates flavonol synthesis in developing grape berries. Plant Physiol 151:1513–1530
33. Nakabayashi R (2014) Enhancement of oxidative and drought tolerance in Arabidopsis by overaccumulation of antioxidant flavonoids. Plant J 77:367–379
34. Yasui A, Eker APM (1998) DNA photolyases. In: Nickoloff JA, Hoekstra MF (eds) DNA damage and repair, vol II. Humana Press, Totowa, pp 9–32
35. Todo T (1999) Functional diversity of the DNA photolyase/blue light receptor family. Mutat Res 434:89–97
36. Initiative AG (2000) Analysis of the genome sequence of the flowering plant *Arabidopsis thaliana*. Nature 408:796–815
37. Kimura S, Sakaguchi K (2006) DNA repair in plants. Chem Rev 106:753–766
38. Elledge SJ (1996) Cell cycle checkpoints: preventing an identity crisis. Science 274:1664–1672
39. Esterbauer H, Cheeseman KH (1990) Determination of aldehydic lipid peroxidation products: malonaldehyde and 4-hydroxynonenal. Methods Enzymol 186:407–421
40. Lindahl T, Wood RD (1999) Quality control by DNA repair. Science 286:1897–1905
41. Marnett LJ (1999) Lipid peroxidation-DNA damage by malondialdehyde. Mutat Res 424:83–95
42. Panda S, Isbatan A, Adami GR (2008) Modification of the ATM/ATR directed DNA damage response state with aging and long after hepatocyte senescence induction *in vivo*. Mech Ageing Dev 129:332–340
43. Zhou BB, Elledge SJ (2000) The DNA damage response: putting checkpoints in perspective. Nature 408:433–439
44. Sancar A (2003) Structure and function of DNA photolyase and cryptochrome bluelight photoreceptors. Chem Rev 103:2203–2237
45. Kimura S, Tahira Y, Ishibashi T, Mori Y, Mori T, Hashimoto J, Sakaguchi K (2004) DNA repair in higher plants; photoreactivation is the major DNA repair pathway in non-proliferating cells while excision repair (nucleotide excision repair and base excision repair) is active in proliferating cells. Nucleic Acids Res 32:2760–2767
46. Batty DP, Wood RD (2000) Damage recognition in nucleotide excision repair of DNA. Gene 241:193–204
47. Maillard O, Camenisch U, Blagoev KB, Naegeli H (2008) Versatile protection from mutagenic DNA lesions conferred by bipartite recognition in nucleotide excision repair. Mutat Res 658:271–286
48. Volker M, Mon MJ, Karmakar P, van Hoffen A, Schul W et al (2001) Sequential assembly of the nucleotide excision repair factors in vivo. Mol Cell 8:213–224
49. Thoma BS, Vasquez KM (2003) Critical DNA damage recognition functions of XPChHR23B and XPA-RPA in nucleotide excision repair. Mol Carcinog 38:1–13
50. Mu D, Sancar A (1997) Model for XPC-independent transcription-coupled repair of pyrimidine dimers in humans. J Biol Chem 272:7570–7573
51. Tuteja N, Singh MB, Misra MK, Bhalla PL, Tuteja R (2001) Molecular mechanisms of DNA damage and repair: progress in plants. Crit Rev Biochem Mol Biol 36:337–397
52. Tuteja N, Tuteja R (2001) Unravelling DNA repair in human: molecular mechanisms and consequences of repair defect. Crit Rev Biochem Mol Biol 36:261–290
53. Wise RR, Naylor AW (1987) Chilling-enhanced photooxidation. The peroxidative destruction of lipids during chilling injury to photosynthesis and ultrastructure. Plant Physiol 83:272–277
54. Collins AR (2004) The comet assay for DNA damage and repair: principles, applications and limitations. Mol Biotechnol 26:249–261
55. Barone F, McCulloch SD, Macpherson P, Maga G, Yamada M, Nohmi T, Minoprio A, Mazzei F, Kunkel TA, Karran P, Bignami M (2007) Replication of 2-hydroxyadenine-containing DNA and recognition by human Mut Salpha. DNA Rep 6:355–366
56. Kawai K, Majima T (2011) G-C content independent long-range charge transfer through DNA electronic and magnetic properties of chiral molecules and supramolecular architectures. In: Naaman R, Beratan DN, Waldeck D (eds) Electronic and magnetic properties of chiral molecules and supramolecular architectures. Springer, Heidelberg/Berlin, pp 129–142
57. Degan P, Shigenaga MK, Park EM, Alperin PE, Ames BN (1991) Immunoaffinity isolation of urinary 8-hydroxy-2′-deoxyguanosine and 8-hydroxyguanine and quantitation of 8-hydroxy-2′-deoxyguanosine in DNA by polyclonal antibodies. Carcinogenesis 12:865–871
58. Malins DC, Haimanot R (1991) Major alterations in the nucleotide structure of DNA in cancer of the female breast. Cancer Res 51:5430–5432
59. Fraga CG, Shigenaga MK, Park JW, Degan P, Ames BN (1990) Oxidative damage to DNA during aging: 8-hydroxy-2′-deoxyguanosine in rat organ DNA and urine. Proc Natl Acad Sci U S A 87:4533–4537
60. Maga G, Villani G, Crespan E, Wimmer U, Ferrari E, Bertocci B, Hubscher U (2007) 8-oxo-guanine bypass by human DNA polymerases in the presence of auxiliary proteins. Nature 447:606–608
61. Crespan E, Hubscher U, Maga G (2007) Error-free bypass of 2-hydroxyadenine by human DNA polymerase lambda with proliferating cell nuclear antigen and replication protein a in different sequence contexts. Nucleic Acids Res 35:5173–5181
62. Hubscher U, Maga G (2011) DNA replication and repair bypass machines. Curr Opin Chem Biol 15:627–635

63. Singhal RK, Prasad R, Wilson SH (1995) DNA polymerase beta conducts the gap-filling step in uracil-initiated base excision repair in a bovine testis nuclear extract. J Biol Chem 270:949–957
64. Barnes DE, Lindahl T (2004) Repair and genetic consequences of endogenous DNA base damage in mammalian cells. Annu Rev Genet 38:445–476
65. Matsumoto Y, Kim K (1995) Excision of deoxyribose phosphate residues by DNA polymerase beta during DNA repair. Science 269:699–702
66. Dianov GL, Prasad R, Wilson SH, Bohr VR (1999) Role of DNA polymerase beta in the excision step of long patch mammalian base excision repair. J Biol Chem 274:13741–13743
67. West CE, Waterworth WM, Sunderland PA, Bray CM (2004) *Arabidopsis* DNA double-strand break repair pathways. Biochem Soc Trans 32:964–966
68. Symington LS (2002) Role of RAD52 epistasis group genes in homologous recombination and double-strand break repair. Microbiol Mol Biol Rev 66:630–670
69. Barzel A, Kupiec M (2008) Finding a match: How do homologous sequences get together for recombination? Nat Rev Genet 9:27–37
70. Peters AC, Young LC, Maeda T, Tron VA, Andrew SE (2003) Mammalian DNA mismatch repair protects cells from UVB-induced DNA damage by facilitating apoptosis and p53 activation. DNA Repair (Amst) 2:427–435

Vitamin D, Cardiovascular Disease and Risk Factors

18

Tea Skaaby, Betina H. Thuesen, and Allan Linneberg

Abstract

Observational studies have suggested a possible protective role of vitamin D on the cardiovascular system. The available evidence does not support either cardiovascular benefits or harms of vitamin D supplementation. This chapter provides an overview and discussion of the current knowledge of vitamin D effects from a cardiovascular health perspective. It focuses on vitamin D in relation to cardiovascular disease, i.e. ischemic heart disease, and stroke; the traditional cardiovascular risk factors hypertension, abnormal blood lipids, obesity; and the emerging risk factors hyperparathyroidism, microalbuminuria, chronic obstructive pulmonary diseases, and non-alcoholic fatty liver disease. Meta-analyses of observational studies have largely found vitamin D levels to be inversely associated with cardiovascular risk and disease. However, Mendelian randomization studies and randomized, controlled trials (RCTs) have not been able to consistently replicate the observational findings. Several RCTs are ongoing, and the results from these are needed to clarify whether vitamin D deficiency is a causal and reversible factor to prevent cardiovascular disease.

T. Skaaby (✉) • B.H. Thuesen
Research Centre for Prevention and Health, Capital Region of Denmark,
Glostrup Hospital, Nordre Ringvej 57, DK-2600 Glostrup, Denmark
e-mail: tea.skaaby.01@regionh.dk

A. Linneberg
Research Centre for Prevention and Health, Capital Region of Denmark, Glostrup Hospital, Nordre Ringvej 57, DK-2600 Glostrup, Denmark

Department of Clinical Experimental Research, Rigshospitalet, Glostrup, Denmark

Department of Clinical Medicine, Faculty of Health and Medical Sciences, University of Copenhagen, Copenhagen, Denmark

S. Ahmad (ed.), *Ultraviolet Light in Human Health, Diseases and Environment*, Advances in Experimental Medicine and Biology 996, https://doi.org/10.1007/978-3-319-56017-5_18

Keywords
Vitamin D • Cardiovascular disease • Hypercholesterolemia • Hypertension • Obesity • Diabetes • Fatty liver disease • Microalbuminuria • Hyperparathyroidism • Chronic obstructive pulmonary disease

18.1 Introduction

Vitamin D is produced in the skin when it is exposed to the sun, and some is derived from the diet and dietary supplements. The skin has a great capacity to produce vitamin D [1]. Sensible sun exposure can provide adequate amounts of vitamin D that can be stored in body fat and can be released during the winter, when the strength and availability of the sun is too low to enable enough vitamin D production in the skin [1]. However, vitamin D insufficiency and deficiency are common all over the world [1, 2].

Vitamin D may be considered to be an indicator of a healthy lifestyle. Determinants of low vitamin D status include winter season, smoking, low education, alcohol overconsumption, young and old age, physical inactivity, poor diet, and a high body mass index (BMI) [1, 3–5]. Vitamin D has an important role in preserving skeletal function and integrity, as it regulates calcium homeostasis and bone mineralization. Vitamin D deficiency causes rickets, osteoporosis, and osteomalacia, but a substantial body of evidence has suggested a broader role. Low levels of vitamin D have been found to be associated with mortality [6–11] and suggested to be associated with a number of diseases [10, 12], e.g. cardiovascular disease [10]. In addition, vitamin D receptors have been found in tissues like cardiomyocytes, vascular smooth muscle, and endothelium. Thus, it is believed that vitamin D deficiency may adversely affect the cardiovascular system and hence has been proposed as a modifiable risk factor for cardiovascular disease [1, 13]. Vitamin D may affect the cardiovascular system in a number of ways [1, 14–25]. A selection of these is shown in Table 18.1.

18.2 Cardiovascular Disease

Cardiovascular disease (CVD) is an important cause of mortality and morbidity. Worldwide, ischemic heart disease (IHD) and stroke account for more deaths than any other diseases. The pathogenesis includes forming of atherosclerotic plaques that eventually rupture with superimposed thrombosis. Formation of atherosclerotic plaque is a complicated process that involves a number of pathological pathways. Vitamin D has many possible effects that could have an impact on the cardiovascular system [1]. It affects a large amount of genes, e.g. genes responsible for the regulation of cellular differentiation, proliferation, apoptosis, and angiogenesis [1]. Vitamin D has been suggested to reduce the expansion of atheromatous lesions. It may inhibit the renin–angiotensin system, decrease coagulation, reduce parathyroid hormone levels, and reduce inflammation thereby reducing atherosclerosis, and increase insulin production [1].

Several cross-sectional and prospective studies have examined the association between vitamin D status and CVD, and although somewhat inconsistent, these observational studies often report an inverse association between vitamin D status and CVD [26–28]. Some studies have found an increased risk of CVD mortality with low vitamin D levels among older people [29–31], and some have found an inverse association between vitamin D status and incident cardiovascular disease [32, 33]. Also, Wang and colleagues found an inverse association between vitamin D status and incident CVD in hypertensive participants. However, Messenger et al. found no association between vitamin D levels and incident CVD in older men [26]. Melamed et al. found no

Table 18.1 Possible effects of vitamin D on cardiovascular risk factors [1, 14–25]

Hypercholesterolaemia	Suppression of PTH (PTH can reduce lipolysis)
	Increased calcium levels causing a reduction of hepatic triglyceride formation and secretion
Hypertension	Inhibition of the renin-angiotensin system
	Reduction of vascular calcification
	Reduction of vascular resistance and vasoconstriction
Obesity	Mobilisation of free fatty acids from adipose tissue
	Increase in energy expenditure due to uncoupling of oxidative phosphorylation in adipose tissue
Diabetes	Reduced insulin production and increased insulin resistance
	Reduced insulin sensitivity due to a reduction of osteocalcin
	Protection against cytokine induced β-cell dysfunction and death
Fatty liver disease	Reduce fat accumulation in hepatocytes
	Prevent steatosis
Albuminuria	Prevention of podocyte loss
	Prevention of glomerulosclerosis
COPD	Reduce fibroblast proliferation
	Tissue remodelling

Abbreviations: *COPD* chronic obstructive pulmonary disease, *PTH* parathyroid hormone

statistically significant association between vitamin D levels and CVD mortality [27], and two other studies found no associations with incident IHD or stroke and circulatory disease mortality, respectively, in the general population [8, 9]. Kilkkinen et al. reported no statistically significant association between vitamin D status and coronary death (except for cerebrovascular death) in a general population free from CVD at baseline [28]. Meta-analyses of observational studies of the association between vitamin D status and CVD have generally shown inverse associations between vitamin D and CVD [10, 34–42].

Randomized controlled trials (RCTs) in this area are inconclusive. Most have been designed to examine how vitamin D supplementation affects bone health and often, vitamin D supplementation has been given together with calcium supplementation. Two meta-analyses showed non-significant reductions of CVD events with vitamin D supplementation [32, 43]. In an umbrella review by Theodoratou et al. the relative risk of cardiovascular disease was 0.95 (95% CI: 0.86, 1.05) in the supplemented group [10]. A meta-analysis on 13,033 participants from 21 RCTs on vitamin D supplementation, the hazard ratios and 95% confidence intervals (CIs) for myocardial infarction, and stroke were 0.96 (95% CI: 0.83–1.10), and 1.07 (95% CI: 0.91–1.29), respectively [44]. Another meta-analysis found that vitamin D supplementation had no effect on the incidence of myocardial infarction and ischemic heart disease, or stroke and cerebrovascular disease [45]. Although a Cochrane review found that vitamin D supplementation compared with placebo or no intervention significantly reduced all-cause mortality, vitamin D had no significant effect on cardiovascular mortality [46].

18.3 Cardiovascular Risk Factors

Hypertension, abnormal blood lipids, obesity, type 2 diabetes, and metabolic syndrome, are well-established modifiable risk factors for CVD. Vitamin D may have an effect on all five. Previous meta-analyses of prospective studies have found low vitamin D levels to be associated with higher risk of incident type 2 diabetes but found inconsistent results for vitamin D and metabolic syndrome [47–49]. Vitamin D has also been suggested to be associated with less established cardiovascular risk factors, e.g., hyperparathyroidism, microalbuminuria, chronic obstructive pulmonary diseases (COPD), and non-alcoholic fatty liver disease.

18.3.1 Hypercholesterolemia

Hypercholesterolemia gives a larger influx of low density lipoprotein (LDL) cholesterol. This is released from the LDL particles and oxidized which attracts and stimulates macrophages – an essential step in the process of inflammation [50]. HDL-cholesterol usually removes cholesterol from tissues and brings it back to the liver but if there is insufficient HDL-cholesterol the inflammation process is increased.

Cross-sectional studies reported a higher vitamin D level to be associated with a favorable lipid profile, and a prospective study showed an inverse association between vitamin D status and triglycerides [51, 52]. A systematic review and meta-analysis of the relationship between vitamin and lipid profile in children and adolescents found that a higher vitamin D level was associated with a healthier lipid profile [53]. Jorde et al. summarized the results from the few randomized controlled trials (RCTs) that had examined the effect of vitamin D supplementation on lipid profile as being inconclusive [54]. However, none of the trials were designed to examine the relation between vitamin D and lipids, and they may have had low power [54]. A large meta-analysis reported no effect of vitamin D supplementation on serum lipid levels [55], and another two systematic reviews concluded that the relationship between vitamin D supplementation and blood lipids needs clarification [10, 56].

18.3.2 Obesity

By increasing amounts of cytokines that decrease insulin sensitivity, adipose tissue can turn the system into a pro-inflammatory state that favors the atherosclerotic processes [50]. Some observational studies have reported inverse associations between vitamin D levels and obesity, but RCTs have failed to replicate this: In a meta-analysis of largely cross-sectional studies, a significant association between vitamin D deficiency and obesity was found, but longitudinal data were insufficient [57]. A bi-directional Mendelian randomization study showed that obesity leads to vitamin D deficiency and not vice versa [58]. Likewise, Jorde et al. found no association between a genetically determined higher vitamin D and body mass index [7]. Therefore, the direction of causality and mechanisms need clarification [59]. Meta-analyses of randomized controlled trials of vitamin D supplementation have shown no effect on obesity [10, 60, 61].

In line with the results so far, it may be that the fat-soluble vitamin D is sequestered in adipose tissue, which results in lower levels in obese. This would mean that obesity causes low vitamin D status and not the other way round. An important question for further research is the level of bioavailability of vitamin D in adipose tissue.

18.3.3 Hypertension

Increased force is put on the artery walls in hypertensives. In time, the pressure and oxidative stress will damage the arteries making them more sensitive to the narrowing and plaques formation associated with atherosclerosis [62]. Vitamin D may reduce blood pressure by affecting the renin–angiotensin–aldosterone system. This system regulates electrolyte and volume homeostasis, thereby contributing to the development of hypertension. Most observational studies of vitamin D status and blood pressure are in favor of an inverse association, although the results are somewhat inconsistent. In meta-analysis of cross-sectional and prospective studies, Burgaz et al. found that vitamin D levels were inversely associated with hypertension [63, 64]. Kunutsor et al. meta-analyzed the prospective studies on this matter and reported a statistically significant inverse association between vitamin D status and development of hypertension [65]. Vimaleswaran et al. found that a genetically determined higher vitamin D level was associated with a statistically significant decreased risk of hypertension [66].

RCTs examining a possible effect of vitamin D supplementation or UV radiation (to improve the vitamin D status) on blood pressure have been inconclusive. In two RCTs, there was a blood pressure lowering effect of vitamin D supplementation in hypertensive participants with below normal vitamin D levels, in black Americans and in Danes during winter, respectively [67, 68]. Two meta-analyses of RCTs found weak evidence of a small effect of vitamin D on blood pressure [38, 63]. Vitamin D supplementation may decrease blood pressure in hypertensives with a low vitamin D status rather than in normotensives with a normal vitamin D level. Further studies are needed for clarification.

18.3.4 Hyperparathyroidism

Parathyroid hormone (PTH) is an important regulator of bone health because it helps to maintain normal serum concentrations of calcium and phosphate. PTH is inversely related to glomerular filtration rate (GFR) and is closely related to vitamin D status [14] although a study found an inverse cross-sectional but no prospective association between vitamin D status and PTH [69]. Hagstrom et al. found increased PTH to explain 20% of the population-attributable risk proportion of cardiovascular mortality, which suggests that diseases with increased PTH may be associated with an increased risk of cardiovascular disease and death [70]. Vitamin D regulates the intestinal uptake of calcium. In turn this controls the expression, production and secretion of PTH [14]. Vitamin D deficiency may be able to cause secondary hyperparathyroidism even without substantial hypocalcaemia.

18.3.5 Microalbuminuria

Microalbuminuria refers to the situation where the glomerular permeability for albumin is abnormally high and consequently the kidneys leak out albumin into the urine. Microalbuminuria is an early sign of chronic kidney disease (CKD) usually linked with type 2 diabetes and associated with an increased risk of loss of kidney function and cardiovascular disease. Microalbuminuria can be diagnosed and estimated from the excretion of albumin in a 24-h urine collection or, more frequently, from the urinary albumin creatinine ratio (UACR) in a spot sample. Microalbuminuria is an important therapeutic target because a decrease in urine albumin excretion is associated with a decreased risk of cardiovascular and renal disease.

Microalbuminuria is an important marker of endothelial dysfunction and vascular damage. It may be affected by vitamin D in several ways. Vitamin D may affect UACR by a cellular effect that prevents podocyte loss and glomerulosclerosis. Vitamin D deficiency has been associated adverse effects on diabetes markers and insulin sensitivity, and it could also affect UACR by an effect on diabetes and insulin resistance both are established risk factors of albuminuria. Finally, insufficient vitamin D could contribute to albuminuria through an activation of the renin–angiotensin–aldosterone system leading to albuminuria through both hemodynamic and non-hemodynamic mechanisms.

Previous studies have suggested a possible effect of vitamin D on microalbuminuria [16, 71]. A cross-sectional study by de Boer et al. found increases in the albuminuria prevalence with decreasing vitamin D quartiles [16]. Evidence from prospective studies is also limited. A prospective study by O'Seaghdha et al. found no association between vitamin D status and incident albuminuria [72]. However, another study found a statistically significant association between vitamin D status and both cross-sectional and prospective levels of urine albumin creatinine ratio (UACR) [69]. An RCT of patients with diabetic nephropathy showed a decrease in albuminuria in the group treated with paracalcitol which is a selective activator of the vitamin D receptor [73]. It remains to be proven whether vitamin D deficiency is a causal and reversible factor in the development of albuminuria.

18.3.6 Chronic Obstructive Pulmonary Disease

Chronic obstructive pulmonary disease (COPD) is worldwide the fifth leading cause of death. It is a risk factor for cardiovascular disease [74, 75]. COPD is characterized by an irreversible air flow loss that is suggested to be due to an inflammatory destruction of the airways. The inflammatory destruction is caused by airway irritants and noxious gases, e.g. the components of tobacco smoke. Common symptoms include sputum production, shortness of breath, a productive cough, and acute exacerbations with acute worsening of symptoms. These are typically caused by bacterial or viral infection.

Through the anticipated role in immunity, vitamin D may affect the number of respiratory infections that sets off the exacerbations and the severity of the exacerbations. Observational studies have found a large prevalence of vitamin D deficiency and insufficiency among COPD patients. Both incidences of pulmonary infections and pulmonary function show correlation with vitamin D level. A study found an inverse cross-sectional but no prospective association between vitamin D level and COPD [76].

18.3.7 Non-alcoholic Fatty Liver Disease

Liver diseases include disorders such as viral, autoimmune, and alcoholic hepatitis, fatty liver disease, cirrhosis, and liver cancer. They are important causes of morbidity and mortality, non-alcoholic fatty liver disease has been found to be associated with a higher risk of CVD [77], and fat accumulation in the liver is considered to be the hepatic component of the metabolic syndrome. Low vitamin D levels are frequent among patients with chronic liver diseases. Several studies suggest that vitamin D may play a role in the development of liver disease [78–80]. Barchetta et al. suggested a causal role of vitamin D in the pathogenesis of non-alcoholic fatty liver disease through a dose-dependent effect on fat accumulation in the liver cells [24]. Decreased vitamin D levels associate with increased severity of steatosis, inflammation, and fibrosis in patients with non-alcoholic fatty liver disease. Further studies are needed to determine whether patients in risk of developing impaired liver function, e.g., patients with overweight, diabetes, hypercholesterolemia, fatty liver disease etc. would benefit from screening for vitamin D deficiency.

18.4 Conclusion

Differentiating cause and effect from simple association is the major challenge of investigating vitamin D levels and disease. Although observational studies have found inverse associations between vitamin D level and disease, these associations have shown difficult to replicate in RCTs. Higher vitamin D levels have largely been associated with reduced risk of cardiovascular disease in observational studies [10, 34–42], but RCTs are inconclusive [10, 43, 44, 81]. Observational studies have reported higher vitamin D levels to be associated with healthier lipid profiles [51–53], but RCTs have not supported this [10, 54–56]. Higher vitamin D levels have been found to be associated with lower blood pressure in several observational studies, and although there may be a small effect in RCTs, the evidence is inconclusive [7, 38, 63–68]. Regarding obesity, observational studies have largely shown an inverse association with vitamin D level but RCTs and Mendelian randomization studies have not [7, 10, 57, 58, 60, 61]. Both some observational and RCTs point toward a beneficial role of vitamin D to reduce microalbuminuria, but the evidence is not conclusive [16, 69, 71–73]. Likewise, further studies are needed to investigate the effects of vitamin D on emerging cardiovascular risk factors, e.g., COPD, nonalcoholic fatty liver disease and hyperparathyroidism [69].

Results from RCTs of the effect of vitamin D supplementation on cardiovascular disease are scant, and the evidence is not conclusive so far [10]. It may turn out that vitamin D supplementation benefits certain groups, e.g., elderly or individuals with liver disease. Vitamin D supplementation might only benefit vitamin D deficient individuals rather than the general population. Further studies, preferably RCTs, are needed to clarify whether vitamin D deficiency is a causal and reversible factor to prevent cardiovascular disease. Several large interventional trials with cardiovascular endpoints are in fact ongoing, and the results are expected in 2017–2020 [82, 83].

References

1. Holick MF (2007) Vitamin D deficiency. N Engl J Med 357(3):266–281
2. Hilger J, Friedel A, Herr R, Rausch T, Roos F, Wahl DA et al (2013) A systematic review of vitamin D status in populations worldwide. Br J Nutr 9:1–23
3. Thuesen B, Husemoen L, Fenger M, Jakobsen J, Schwarz P, Toft U et al (2012) Determinants of vitamin D status in a general population of Danish adults. Bone 50(3):605–610
4. Skaaby T, Husemoen LL, Thuesen BH, Pisinger C, Hannemann A, Jorgensen T et al (2015) Longitudinal associations between lifestyle and vitamin D: a general population study with repeated vitamin D measurements. Endocrine 51(2):342–350
5. Levy MA, McKinnon T, Barker T, Dern A, Helland T, Robertson J et al (2015) Predictors of vitamin D status in subjects that consume a vitamin D supplement. Eur J Clin Nutr 69(1):84–89
6. Bjelakovic G, Gluud LL, Nikolova D, Whitfield K, Wetterslev J, Simonetti RG et al (2014) Vitamin D supplementation for prevention of mortality in adults. Cochrane Database Syst Rev 1:CD007470
7. Jorde R, Schirmer H, Wilsgaard T, Joakimsen RM, Mathiesen EB, Njolstad I et al (2012) Polymorphisms related to the serum 25-hydroxyvitamin D level and risk of myocardial infarction, diabetes, cancer and mortality. The tromso study. PLoS One 7(5):e37295
8. Skaaby T, Husemoen LL, Pisinger C, Jorgensen T, Thuesen BH, Fenger M et al (2012) Vitamin D status and cause-specific mortality: a general population study. PLoS One 7(12):e52423
9. Skaaby T, Husemoen LL, Pisinger C, Jorgensen T, Thuesen BH, Fenger M et al (2013) Vitamin D status and incident cardiovascular disease and all-cause mortality: a general population study. Endocrine 43(3):618–625
10. Theodoratou E, Tzoulaki I, Zgaga L, Ioannidis JP (2014) Vitamin D and multiple health outcomes: umbrella review of systematic reviews and meta-analyses of observational studies and randomised trials. BMJ 348:g2035
11. Trummer O, Pilz S, Hoffmann MM, Winkelmann BR, Boehm BO, Marz W et al (2013) Vitamin D and mortality: a Mendelian randomization study. Clin Chem 59(5):793–797
12. Skaaby T, Husemoen LL, Thuesen BH, Pisinger C, Jorgensen T, Roswall N et al (2014) Prospective population-based study of the association between serum 25-hydroxyvitamin-D levels and the incidence of specific types of cancer. Cancer Epidemiol Biomark Prev 23(7):1220–1229
13. Muldowney S, Kiely M (2010) Vitamin D and cardiometabolic health: a review of the evidence. Nutr Res Rev 1:1–20
14. Steingrimsdottir L, Gunnarsson O, Indridason OS, Franzson L, Sigurdsson G (2005) Relationship between serum parathyroid hormone levels, vitamin D sufficiency, and calcium intake. JAMA 294(18):2336–2341
15. Cho HJ, Kang HC, Choi SA, Ju YC, Lee HS, Park HJ (2005) The possible role of Ca2+ on the activation of microsomal triglyceride transfer protein in rat hepatocytes. Biol Pharm Bull 28(8):1418–1423
16. de Boer IH, Ioannou GN, Kestenbaum B, Brunzell JD, Weiss NS (2007) 25-Hydroxyvitamin D levels and albuminuria in the Third National Health and Nutrition Examination Survey (NHANES III). Am J Kidney Dis 50(1):69–77
17. Li YC, Qiao G, Uskokovic M, Xiang W, Zheng W, Kong J (2004) Vitamin D: a negative endocrine regulator of the renin-angiotensin system and blood pressure. J Steroid Biochem Mol Biol 89-90(1-5):387–392
18. Tomaschitz A, Pilz S, Ritz E, Grammer T, Drechsler C, Boehm BO et al (2010) Independent association between 1,25-dihydroxyvitamin D, 25-hydroxyvitamin D and the renin-angiotensin system: the ludwigshafen risk and cardiovascular health (LURIC) study. Clin Chim Acta 411(17-18):1354–1360
19. Zagura M, Serg M, Kampus P, Zilmer M, Eha J, Unt E et al (2011) Aortic stiffness and vitamin D are independent markers of aortic calcification in patients with peripheral arterial disease and in healthy subjects. Eur J Vasc Endovasc Surg 42(5):689–695
20. Shi H, Norman AW, Okamura WH, Sen A, Zemel MB (2001) 1alpha,25-Dihydroxyvitamin D3 modulates human adipocyte metabolism via nongenomic action. FASEB J 15(14):2751–2753
21. Fassina G, Maragno I, Dorigo P, Contessa AR (1969) Effect of vitamin D2 on hormone-stimulated lipolysis in vitro. Eur J Pharmacol 5(3):286–290

22. Chiu KC, Chu A, Go VL, Saad MF (2004) Hypovitaminosis D is associated with insulin resistance and beta cell dysfunction. Am J Clin Nutr 79(5):820–825
23. Dobak J, Grzybowski J, Liu FT, Landon B, Dobke M (1994) 1,25-Dihydroxyvitamin D3 increases collagen production in dermal fibroblasts. J Dermatol Sci 8(1):18–24
24. Barchetta I, Angelico F, Del BM, Baroni MG, Pozzilli P, Morini S et al (2011) Strong association between non alcoholic fatty liver disease (NAFLD) and low 25(OH) vitamin D levels in an adult population with normal serum liver enzymes. BMC Med 9:85
25. Targher G, Bertolini L, Scala L, Cigolini M, Zenari L, Falezza G et al (2007) Associations between serum 25-hydroxyvitamin D3 concentrations and liver histology in patients with non-alcoholic fatty liver disease. Nutr Metab Cardiovasc Dis 17(7):517–524
26. Messenger W, Nielson CM, Li H, Beer T, Barrett-Connor E, Stone K et al (2011) Serum and dietary vitamin D and cardiovascular disease risk in elderly men: a prospective cohort study. Nutr Metab Cardiovasc Dis 22(10):856–863
27. Melamed ML, Michos ED, Post W, Astor B (2008) 25-hydroxyvitamin D levels and the risk of mortality in the general population. Arch Intern Med 168(15):1629–1637
28. Kilkkinen A, Knekt P, Aro A, Rissanen H, Marniemi J, Heliovaara M et al (2009) Vitamin D status and the risk of cardiovascular disease death. Am J Epidemiol 170(8):1032–1039
29. Ginde AA, Scragg R, Schwartz RS, Camargo CA Jr (2009) Prospective study of serum 25-hydroxyvitamin D level, cardiovascular disease mortality, and all-cause mortality in older U.S. adults. J Am Geriatr Soc 57(9):1595–1603
30. Pilz S, Dobnig H, Nijpels G, Heine RJ, Stehouwer CD, Snijder MB et al (2009) Vitamin D and mortality in older men and women. Clin Endocrinol 71(5):666–672
31. Semba RD, Houston DK, Bandinelli S, Sun K, Cherubini A, Cappola AR et al (2010) Relationship of 25-hydroxyvitamin D with all-cause and cardiovascular disease mortality in older community-dwelling adults. Eur J Clin Nutr 64(2):203–209
32. Wang TJ, Pencina MJ, Booth SL, Jacques PF, Ingelsson E, Lanier K et al (2008) Vitamin D deficiency and risk of cardiovascular disease. Circulation 117(4):503–511
33. Hosseinpanah F, Yarjanli M, Sheikholeslami F, Heibatollahi M, Eskandary PS, Azizi F (2011) Associations between vitamin D and cardiovascular outcomes; Tehran Lipid and glucose study. Atherosclerosis 218(1):238–242
34. Brondum-Jacobsen P, Benn M, Jensen GB, Nordestgaard BG (2012) 25-hydroxyvitamin d levels and risk of ischemic heart disease, myocardial infarction, and early death: population-based study and meta-analyses of 18 and 17 studies. Arterioscler Thromb Vasc Biol 32(11):2794–2802
35. Chowdhury R, Kunutsor S, Vitezova A, Oliver-Williams C, Chowdhury S, Kiefte-de-Jong JC et al (2014) Vitamin D and risk of cause specific death: systematic review and meta-analysis of observational cohort and randomised intervention studies. BMJ 348:g1903
36. Fan H, Yu W, Cao H, Li J, Liu B, Wang J et al (2014) Meta-analysis of circulating 25-hydroxyvitamin D levels and risk of cardiovascular and all-cause mortality in elderly population. Int J Cardiol 176(3):1025–1029
37. Grandi NC, Breitling LP, Brenner H (2010) Vitamin D and cardiovascular disease: systematic review and meta-analysis of prospective studies. Prev Med 51(3-4):228–233
38. Pittas AG, Chung M, Trikalinos T, Mitri J, Brendel M, Patel K et al (2010) Systematic review: vitamin D and cardiometabolic outcomes. Ann Intern Med 152(5):307–314
39. Schottker B, Jorde R, Peasey A, Thorand B, Jansen EH, Groot L et al (2014) Vitamin D and mortality: meta-analysis of individual participant data from a large consortium of cohort studies from Europe and the United States. BMJ 348:g3656
40. Sokol SI, Tsang P, Aggarwal V, Melamed ML, Srinivas VS (2011) Vitamin D status and risk of cardiovascular events: lessons learned via systematic review and meta-analysis. Cardiol Rev 19(4):192–201
41. Tomson J, Emberson J, Hill M, Gordon A, Armitage J, Shipley M et al (2013) Vitamin D and risk of death from vascular and non-vascular causes in the Whitehall study and meta-analyses of 12,000 deaths. Eur Heart J 34(18):1365–1374
42. Wang L, Song Y, Manson JE, Pilz S, Marz W, Michaelsson K et al (2012) Circulating 25-hydroxy-vitamin D and risk of cardiovascular disease: a meta-analysis of prospective studies. Circ Cardiovasc Qual Outcomes 5(6):819–829
43. Avenell A, MacLennan GS, Jenkinson DJ, McPherson GC, McDonald AM, Pant PR et al (2012) Long-term follow-up for mortality and cancer in a randomized placebo-controlled trial of vitamin D(3) and/or calcium (RECORD trial). J Clin Endocrinol Metab 97(2):614–622
44. Ford JA, MacLennan GS, Avenell A, Bolland M, Grey A, Witham M (2014) Cardiovascular disease and vitamin D supplementation: trial analysis, systematic review, and meta-analysis. Am J Clin Nutr 100(3):746–755
45. Bolland MJ, Grey A, Gamble GD, Reid IR (2014) The effect of vitamin D supplementation on skeletal, vascular, or cancer outcomes: a trial sequential meta-analysis. Lancet Diabetes Endocrinol 2(4):307–320
46. Bjelakovic G, Gluud LL, Nikolova D, Whitfield K, Wetterslev J, Simonetti RG et al (2011) Vitamin D supplementation for prevention of mortality in adults. Cochrane Database Syst Rev 7:CD007470

47. Song Y, Wang L, Pittas AG, Del Gobbo LC, Zhang C, Manson JE et al (2013) Blood 25-hydroxy vitamin D levels and incident type 2 diabetes: a meta-analysis of prospective studies. Diabetes Care 36(5):1422–1428
48. Khan H, Kunutsor S, Franco OH, Chowdhury R (2013) Vitamin D, type 2 diabetes and other metabolic outcomes: a systematic review and meta-analysis of prospective studies. Proc Nutr Soc 72(1):89–97
49. Ju SY, Jeong HS (2014) Kim dH. Blood vitamin D status and metabolic syndrome in the general adult population: a dose-response meta-analysis. J Clin Endocrinol Metab 99(3):1053–1063
50. Mallika V, Goswami B, Rajappa M (2007) Atherosclerosis pathophysiology and the role of novel risk factors: a clinicobiochemical perspective. Angiology 58(5):513–522
51. Skaaby T, Husemoen LL, Pisinger C, Jorgensen T, Thuesen BH, Fenger M et al (2012) Vitamin D status and changes in cardiovascular risk factors: a prospective study of a general population. Cardiology 123(1):62–70
52. Jorde R, Figenschau Y, Hutchinson M, Emaus N, Grimnes G (2010) High serum 25-hydroxyvitamin D concentrations are associated with a favorable serum lipid profile. Eur J Clin Nutr 64(12):1457–1464
53. Kelishadi R, Farajzadegan Z, Bahreynian M (2014) Association between vitamin D status and lipid profile in children and adolescents: a systematic review and meta-analysis. Int J Food Sci Nutr 65(4):404–410
54. Jorde R, Grimnes G (2011) Vitamin D and metabolic health with special reference to the effect of vitamin D on serum lipids. Prog Lipid Res 50(4):303–312
55. Elamin MB, Abu Elnour NO, Elamin KB, Fatourechi MM, Alkatib AA, Almandoz JP et al (2011) Vitamin D and cardiovascular outcomes: a systematic review and meta-analysis. J Clin Endocrinol Metab 96(7):1931–1942
56. Challoumas D (2014) Vitamin D supplementation and lipid profile: what does the best available evidence show? Atherosclerosis 235(1):130–139
57. Pereira-Santos M, Costa PR, Assis AM, Santos CA, Santos DB (2015) Obesity and vitamin D deficiency: a systematic review and meta-analysis. Obes Rev 16(4):341–349
58. Vimaleswaran KS, Berry DJ, Lu C, Tikkanen E, Pilz S, Hiraki LT et al (2013) Causal relationship between obesity and vitamin D status: bi-directional Mendelian randomization analysis of multiple cohorts. PLoS Med 10(2):e1001383
59. Parikh SJ, Edelman M, Uwaifo GI, Freedman RJ, Semega-Janneh M, Reynolds J et al (2004) The relationship between obesity and serum 1,25-dihydroxy vitamin D concentrations in healthy adults. J Clin Endocrinol Metab 89(3):1196–1199
60. Pathak K, Soares MJ, Calton EK, Zhao Y, Hallett J (2014) Vitamin D supplementation and body weight status: a systematic review and meta-analysis of randomized controlled trials. Obes Rev 15(6):528–537
61. Chandler PD, Wang L, Zhang X, Sesso HD, Moorthy MV, Obi O et al (2015) Effect of vitamin D supplementation alone or with calcium on adiposity measures: a systematic review and meta-analysis of randomized controlled trials. Nutr Rev 73(9):577–593
62. Alexander RW (1995) Theodore Cooper Memorial Lecture. Hypertension and the pathogenesis of atherosclerosis. Oxidative stress and the mediation of arterial inflammatory response: a new perspective. Hypertension 25(2):155–161
63. Witham MD, Nadir MA, Struthers AD (2009) Effect of vitamin D on blood pressure: a systematic review and meta-analysis. J Hypertens 27(10):1948–1954
64. Burgaz A, Orsini N, Larsson SC, Wolk A (2011) Blood 25-hydroxyvitamin D concentration and hypertension: a meta-analysis. J Hypertens 29(4):636–645
65. Kunutsor SK, Apekey TA, Steur M (2013) Vitamin D and risk of future hypertension: meta-analysis of 283,537 participants. Eur J Epidemiol 28(3):205–221
66. Vimaleswaran KS, Cavadino A, Berry DJ, Jorde R, Dieffenbach AK, Lu C et al (2014) Association of vitamin D status with arterial blood pressure and hypertension risk: a mendelian randomisation study. Lancet Diabetes Endocrinol 2(9):719–729
67. Forman JP, Scott JB, Ng K, Drake BF, Suarez EG, Hayden DL et al (2013) Effect of vitamin D supplementation on blood pressure in blacks. Hypertension 61(4):779–785
68. Larsen T, Mose FH, Bech JN, Hansen AB, Pedersen EB (2012) Effect of cholecalciferol supplementation during winter months in patients with hypertension: a randomized, placebo-controlled trial. Am J Hypertens 25(11):1215–1222
69. Skaaby T, Husemoen LL, Pisinger C, Jorgensen T, Thuesen BH, Rasmussen K et al (2013) Vitamin D status and 5-year changes in urine albumin creatinine ratio and parathyroid hormone in a general population. Endocrine 44(2):473–480
70. Hagstrom E, Hellman P, Larsson TE, Ingelsson E, Berglund L, Sundstrom J et al (2009) Plasma parathyroid hormone and the risk of cardiovascular mortality in the community. Circulation 119(21):2765–2771
71. de ZD RG, Parving HH, Keane WF, Zhang Z, Shahinfar S et al (2004) Albuminuria, a therapeutic target for cardiovascular protection in type 2 diabetic patients with nephropathy. Circulation 110(8):921–927
72. O'Seaghdha CM, Hwang SJ, Holden R, Booth SL, Fox CS (2012) Phylloquinone and vitamin D status: associations with incident chronic kidney disease in the Framingham Offspring cohort. Am J Nephrol 36(1):68–77
73. Joergensen C, Tarnow L, Goetze JP, Rossing P (2014) Vitamin D analogue therapy, cardiovascular risk and kidney function in people with Type 1 diabetes mellitus and diabetic nephropathy: a randomized trial. Diabet Med 32(3):374–381
74. Stone IS, Barnes NC, Petersen SE (2012) Chronic obstructive pulmonary disease: a modifiable risk factor for cardiovascular disease? Heart 98(14):1055–1062

75. Sin DD, Man SF (2005) Chronic obstructive pulmonary disease as a risk factor for cardiovascular morbidity and mortality. Proc Am Thorac Soc 2(1):8–11
76. Skaaby T, Husemoen LL, Thuesen BH, Pisinger C, Jorgensen T, Fenger RV et al (2014) Vitamin d status and chronic obstructive pulmonary disease: a prospective general population study. PLoS One 9(3):e90654
77. Targher G, Byrne CD, Lonardo A, Zoppini G, Barbui C (2016) Nonalcoholic fatty liver disease and risk of incident cardiovascular disease: a meta-analysis of observational studies. J Hepatol 65(3):589–600
78. Putz-Bankuti C, Pilz S, Stojakovic T, Scharnagl H, Pieber TR, Trauner M et al (2012) Association of 25-hydroxyvitamin D levels with liver dysfunction and mortality in chronic liver disease. Liver Int 32(5):845–851
79. Skaaby T, Husemoen LL, Borglykke A, Jorgensen T, Thuesen BH, Pisinger C et al (2013) Vitamin D status, liver enzymes, and incident liver disease and mortality: a general population study. Endocrine 47(1):213–220
80. Skaaby T, Husemoen LL, Linneberg A (2013) Does liver damage explain the inverse association between vitamin D status and mortality? Ann Epidemiol 23(12):812–814
81. Bolland MJ, Grey A, Avenell A, Gamble GD, Reid IR (2011) Calcium supplements with or without vitamin D and risk of cardiovascular events: reanalysis of the Women's Health Initiative limited access dataset and meta-analysis. BMJ 342:d2040
82. Kupferschmidt K (2012) Uncertain verdict as vitamin D goes on trial. Science 337(6101):1476–1478
83. Pilz S, Rutters F, Dekker JM (2012) Disease prevention: vitamin D trials. Science 338(6109):883

Part V

UV Light in Sterilization

19 Biofilms: Microbial Strategies for Surviving UV Exposure

Carla C.C.R. de Carvalho

Abstract

Biofilm communities are an ingenious form of protection of microbial cells which have been evolving for billion of years. In general, ultraviolet (UV) radiation presents poor penetration in the matrix of biofilms and only the first few top layers of microbial cells are exposed to its deleterious effects. For further protection against UV radiation, exposed cells can produce specialized compounds such as mycosporine-like amino acids and carotenoid pigments. In this chapter, the adaptation mechanisms presented by biofilms against UV radiation are presented, as well as the application of UV light to monitor and destroy biofilms in man made surfaces.

Keywords

Biofilm detection • Carotenoid • Disinfection • Mycosporine-like amino acid (MAA) • UV protection

19.1 Biofilms Provide Protection from UV Light

Micro-organisms adhere to wet surfaces, and through an exopolymeric matrix, form a slimy film known as a biofilm [7, 16]. Biofilms are so successful that they can be found ubiquitously on Earth, and fossilized biofilms with 3.2–3.4 billion years have been found in the Barberton greenstone belt in South African [38], and in deep-sea hydrothermal rocks of Pilbara Craton in Australia [28]. This suggests that the ability to form biofilms is an ancient property of prokaryotes, which could have provided protection from the extreme and fluctuating temperature, pH and UV radiation values of the primitive earth [16].

This type of protection is still extremely important in environments such as intertidal pools (Fig. 19.1), where biofilms are usually complex agglomerates of both prokaryotes and microbial eukaryotes such as diatoms, protozoa and microalgae [11]. In these pools, the extracellular matrix act as buffer against rapid changes in salinity,

C.C.C.R. de Carvalho (✉)
Department of Bioengineering, iBB-Institute for Bioengineering and Biosciences, Instituto Superior Técnico, Universidade de Lisboa,
Av. Rovisco Pais, 1049-001 Lisboa, Portugal
e-mail: ccarvalho@tecnico.ulisboa.pt

S. Ahmad (ed.), *Ultraviolet Light in Human Health, Diseases and Environment*, Advances in Experimental Medicine and Biology 996, https://doi.org/10.1007/978-3-319-56017-5_19

Fig. 19.1 Examples of microbial biofilms and mats on an intertidal pool (**a**), on a tree (**b**), on a hydrothermal fountain (**c**) and on a marble fountain (**d**)

temperature, desiccation and UV radiation, as binding agent for essential organic molecules and ions for cells, and as anchor against hydrodynamic forces during submersion [5, 34, 37].

Solar radiation includes visible, infrared and UV radiation. The wavelength of UV radiation corresponds to 100–400 nm, and is broadly classified into three wavelength bands: UV-A (315–400 nm), UV-B (280–315 nm) and UV-C (100–280 nm). The UVB radiation, which represents ca. 5% of solar radiation, can damage the DNA of microorganisms, including bacteria, cyanobacteria and phytoplankton, and induce photochemical degradation of dissolved organic matter [19, 33]. UV-A radiation on the other hand is less damaging on its own. Almost all UV-C, the most damaging radiation, is absorbed by the stratospheric ozone layer and cannot reach Earth. In water ecosystems, the effects of UV radiation are dependent of (i) its penetration in the water column, (ii) the period of exposure, and (iii) the vulnerability and repair mechanisms present in the organisms [10, 33].

Among natural photoprotective compounds produced by microorganisms are mycosporine-like amino acids (MAAs) and carotenoid pigments (Table 19.1). MAAs contain a cyclohexenone or cyclohexenimine chromophore conjugated with one or two amino acids, and are low-molecular weight, water-soluble compounds. They are widely distributed in cyanobacteria, fungi and algae and their most important characteristic is the high UV absorption with molar absorptivity (ε) of ca. 40,000 L mol^{-1} cm^{-1} [22]. The carotenoid mol-

Table 19.1 Examples of mycosporine-like amino acids and carotenoids produced by microorganisms

Mycosporine-like amino acids (MAAs)	Carotenoids
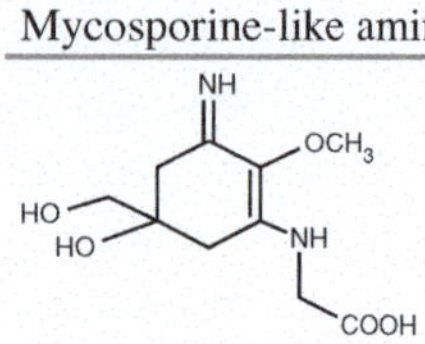	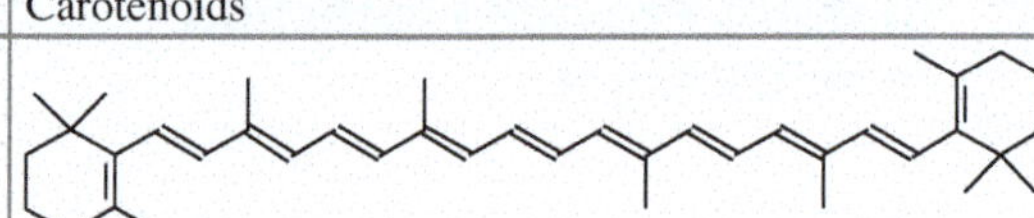
Palythine	β-carotene
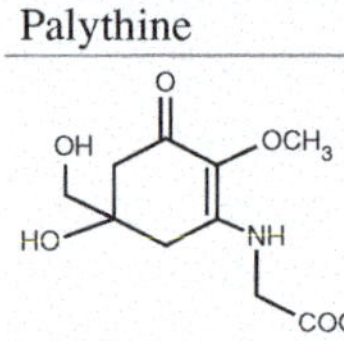	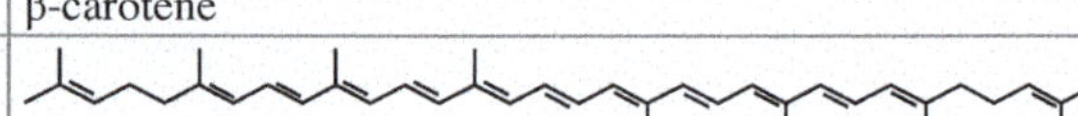
Mycosporine-glycine	Lycopene
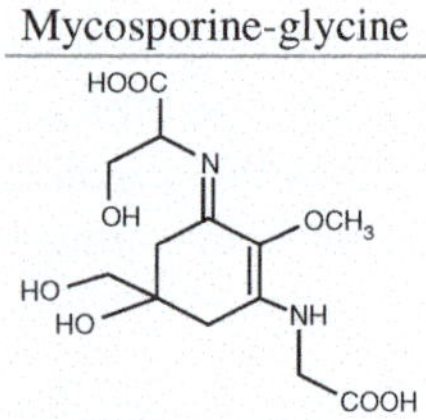	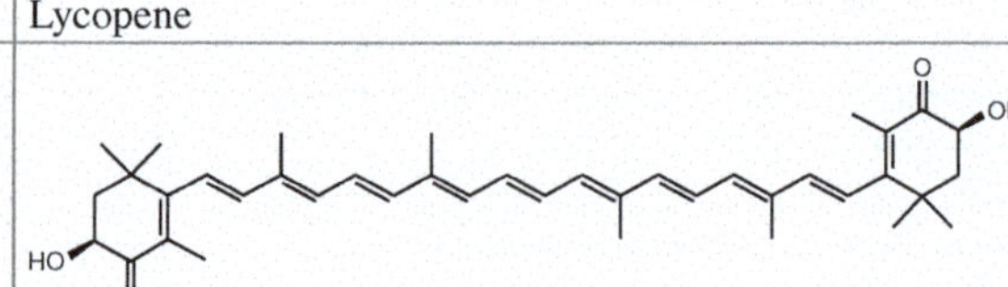
Shinorine	Astaxanthin

ecule is characterized by a long conjugated double-bond system, able to absorb visible light in the 400–500 nm region of the electromagnetic spectrum, which gives the characteristic yellow, orange and red colour to carotenoids. They are widely distributed in plants and algae where they help in the collection of light energy and in its transfer to chlorophyll for photosynthesis while acting as photo-protectors of chlorophyll by dissipating the excessive energy and inhibiting the formation of reactive oxygen species (ROS) [6, 39]. In microorganisms, such as bacteria, carotenoids provide protection against UV radiation, ROS, free radicals, salinity, radioactive compounds, pH and temperature [1, 3, 4, 9].

By using several cut-off filters to select the wavelength of light used to irradiate the cyanobacterium *Nostoc commune* it was possible to demonstrate that MAAs production was induced by UV-B radiation whilst UV-A and photosynthetically active radiation (PAR) had little induction effect [32]. High concentrations of NaCl could not induce MAAs production in this cyanobacterium in the absence of UV-B. On the contrary, high osmotic stress could induce the synthesis of the MAAs in *Chlorogloeopsis* PCC 6912 [26]. While the biosynthesis of shinorine was induced by UV-B, the production of mycosporine-glycine was induced by high salt concentration.

One of most effective sunscreen compounds against UV-A, produced by cyanobacteria, is scytonemin, a yellow to brown, lipid-soluble, dimeric pigment composed by indolic and phenolic subunits, connected through a carbon–carbon bond, and with a molecular mass of 544 g mol^{-1} [14]. This compound is produced in intertidal mats, epilithic biofilms and biological soil crusts as response to UV radiation, although temperature, photo-oxidative stress, periodic desiccation or lack of nitrogen may also influence the levels attained [14, 30].

In a seasonal study at Towra Point in Sydney, Australia, it was found that the cyanobacterial mat growing on the intertidal mangrove sediment was dominated by *Lyngbya* cf. *aestuarii* and *Microcoleus chthonoplastes* [20]. The pigment scytonemin was only produced by *L.* cf. *aestuarii* and was the most important UV-absorbing compound at 140–1300 mg m^{-3}, following the seasonal fluctuating solar intensity contrarily to the areal contents of pterinsc and MAAs.

Using fluorescence imaging to monitor intertidal biofilms, it was possible to observe the rapid migration of diatoms into the sediments to avoid UVB exposure [37]. This behavioral strategy was effective as a short-term UV protection mechanism but long-term exposure led to a reduction in the amount of allocated photosynthetically fixed carbon to colloidal carbohydrate, EPS and glucan. Diatoms are known to migrate vertically in response to light [11], and among the different types of EPS secreted, some are used for motility [17, 36].

Hughes and co-workers studied the transmission of solar UV radiation through artic tundra plants to determine the biological impact of the transmitted UV light on artificial microbial biofilms [18]. The study showed a strong negative correlation between vegetation cover and UV transmission to the soil surface. However, penetration of radiation depended on plant morphology and on the presence of flowers: up to 71.5 and 30.1% of the spores of *Bacillus subtilis* in biofilm that would be killed by ambient UV radiations were inactivated under respectively *Saxifraga oppositifolia* and *Dryas octopetala*, while no UV-induced damage was observed in biofilms beneath *Drepanocladus* sp., *Poa alpina* or *Silene acaulis*.

Photoprotective compounds are mainly produced in aquatic environments but they are also synthesised by e.g. cyanobacterial and algal cells on stone monuments and buildings. Microbial biofilms collected in a district of Bangkok, Thailand, with an average annual maximum temperature of 32 °C, 80% humidity and an UV index of 11–14, contained *Synechocystis* sp., *Scytonema* sp., *Nostoc* sp., *Gloeocapsa* sp. and *Gloeocapsopsis* sp. [29]. The MAAs produced were identified by UV-VIS spectra and ESI-MS as shinorine, porphyra-334, mycosporine-glycine and palythinol whilst two other compounds presenting peaks with maxima at 329 and 320 nm could not be identified. Curiously, biofilms contained up to nearly tenfold of certain MAAs than isolated cyanobacteria, suggesting the presence of photoautotrophs able to produce this compounds and/or induction of MAAs synthesis by environmental stress such as desiccation that is observed in stone monuments. High production levels of scytonemin, mycosporine-like amino acids, and carotenoids was also observed in cyanobacterial biofilms formed on the exterior of three stone monuments in Shantiniketan, India [21].

Carotenoid-producing bacterial cells particularly resistant to UV radiation have been found in radioactive sites [3]. Sunlight-exposed biofilms have also been found to survive high radiation levels [27]. Biofilms from concrete walls or pillars in the Chernobyl area contained mainly Alphaproteobacteria, Bacteroidetes, Acidobacteria and Deinococcales, as well as green algae (Chlorophyta) and ascomycete fungi (Ascomycota) [27]. The diversity of both bacteria and eukaryotes found in the most irradiated samples was comparable to that found in less irradiated samples and in Northern Ireland, although a positive correlation could be found between radiation level and mutation rates. The authors of the study concluded that the organisms in biofilms exposed to UV radiation and desiccation have pre-adaptive mechanisms allowing them to endure high levels of ionizing radiation. In another study, Olsson-Francis *et al.* selected cyanobacteria from a limestone cliff in Beer, Devon, UK, for application in space. To select for extremophilic cyanobacteria from epilithic and endolithic rock-dwelling communities, they were exposed to low Earth orbit conditions (vacuum, 0.133×10^{-6} kPa; temperature, −20 °C to +30 °C; solar radiation, >170 nm) for 10 days. Ground-based exposure experiments to vacuum, desiccation and UV radiation were conducted for comparison. The exposure of the samples to low Earth orbit conditions resulted in the isolation of the single extremophilic cyanobacterium OU_20, which could not be detected after exposure to UV radiation in the ground-based experiments. Curiously, strain OU_13 which survived all ground-level conditions, including UV radiation, could not survive in orbit.

Elasri and Miller developed a bioluminescent assay to study the response of a biofilm to UV radiation [12]. A plasmid-based *recA-luxCDABE* fusion was added to *Pseudomonas aeruginosa* FRD1 which was immobilized in an alginate matrix to simulate a biofilm. The alginate biofilms transmitted only part of the UV radiation: 13% of UV-C, 31% of UV-B and 33% of UV-A. The matrix was thus effective in protecting the cells from UV light exposure, resulting in higher rates of survival of alginate-entrapped cells than of liquid cultures. The production of the extracellular polysaccharide alginate during biofilm formation of *P. aeruginosa* has been found to also confer cell protection from antibiotics and the immune system [23, 31].

19.2 Using UV Radiation to Detect Biofilms

The detection of biofilms in medical and industrial environments is of paramount importance to avoid the spreading of infections and biofouling. Biofilm film-sensors should allow the non-destructive detection of the biofilm to permit further measurements while avoiding the spreading of the cells. Novel UV emitting diodes (UV-LEDs) have been used to study intrinsic protein fluorescence due to the fluorescence excitation of the aromatic amino acids phenylalanine, tyrosine and tryptophan. The main advantages of UV-LEDs are the low power consumption, narrow bandwidth of spectral emission, low cost and nearly instantaneous on/off possibilities with high light intensities [13].

A UV-LED spectroscopy instrument with a wavelength of 280 nm and a narrow spectral bandwidth was made to coincide with the absorption maximum of tryptophan [13]. After the fluorescence of this amino acid was excited by the UV-LED light source, the emitted fluorescence light of the biofilm was collected and guided by fused silica optical fiber to the detector. Appropriate calibration with tryptophan solutions and model biofilms of marine Gram-negative *Pseudoalteromonas carrageenovora* and Gram-positive *Bacillus subtilis*, resulted in a linear signal response, background suppression, wide dynamic range and detection of 4×10^3 cells cm^{-2}. A field experiment conducted in the Baltic Sea for 21 days provided the first continuous observation of biofilm formation dynamics in a natural setting indicating that the sensor could be applied to the shipping industry and deep sea research.

Similarly, it has been shown that the red fluorescence observed with UV-A light in deposits of mature dental plaque on teeth, restorations, or dental appliances due to the presence of a porphyrin compound (mainly protoporphyrin-IX) in bacteria, may be used as a diagnostic tool in dentistry [35]. While the bacteria *Actinomyces odontolyticus*, *Bacteroides intermedius*, *Corynebacterium* spp. and *Candida albicans* emit red light when excited by UV-A radiation, the Gram-positive *Streptococcus mutans*, *Enterococcus faecalis* and various lactobacilli present low or no porphyrin fluorescence in the red spectral region. This indicates that the red fluorescence is mainly the result of the maturity of the dental plaque.

19.3 Using UV Radiation to Destroy Biofilms

Microbial cells in biofilms have developed ingenious ways to survive UV radiation, mainly through the development of thick matrices. UV radiation has a poor penetrating power in biofilms, and thus only affects the first few top layers of cells. Nevertheless, several studies have tested the efficacy of UV radiation to demote microbial biofilms, and UV radiation has been found effective for sterilization when combined with other techniques such as filtration and ozonation [8].

When comparing oxidative and UV-C treatments for inactivating 100-days old bacterial biofilms from groundwater wells, it was found that hydrogen peroxide was the most effective procedure, decreasing by 3.1 orders of magnitude the CFUs [25]. The coupons containing the biofilms were treated with UV dosages of 3000 and 6000 W s cm^{-2} for 30 min or 60 min, respectively, which is one-million times higher than the UV dosage used in aqueous solutions. Although, the CFUs decreased on average by 2.1 orders of magnitude, the effectiveness of UV-C radiation was diminished by high turbidity of the water.

UV radiation was tested as possible disinfectant of medical equipment by using UV-B (296 nm) and UV-C (266 nm) irradiation on *P. aeruginosa* biofilms at different growth stages [2]. A new type of UV LEDs was used. UV-B irradiation was more effective than UV-C and no colony forming units (CFUs) were observed for the UV-B treated biofilms at a dose of 10,000 J m^{-2}. A 3.9 log killing efficacy was observed on mature biofilms at an UV-B irradiation dose of 20,000 J m^{-2}.

Bacterial biofilms of *P. aeruginosa* promoted inside Teflon and silicone catheter tubes for 3 days could be efficiently killed with UV-C LEDs. The control counts were in the range of 5×10^5–1.3×10^9 CFU mL^{-1} and ca. 100% disinfection rates were observed in 10 and 20 cm Teflon tubes exposed for 30 and 300 min, respectively. The latter period was also necessary to achieve the same disinfection level in a 10 cm peritoneal dialysis silicone catheter tube. The ~78 J m^{-2} UV-radiation necessary for a 99.99% killing rate is comparable to that required for the bacterium in planktonic state, indicating that UV-C LEDs may be efficiently used in thin biofilms.

Pulsed ultraviolet light systems allow ultra short duration pulses in the UV-C wavelength. Using up to 21.6 μJ cm^{-2} in biofilm reactors with polyvinyl chloride coupons, Garvey et al. were able to obtain a 7.2 and 5.9 $\log_{10}$ inactivation for *P. aeruginosa* and *Staphylococcus aureus*, respectively, suggesting that this technique could be used for water treatment [15]. Pulsed ultraviolet light has also been efficient in decontaminating *Escherichia coli* O157:H7 and *Listeria monocytogenes* on the surfaces of food packaging materials and fresh produce [24]. After a 10 s treatment at a distance of 4.5 and 8.8 cm, the CFUs of *L. monocytogenes* and *E. coli* O157:H7 in biofilms reduced, respectively, 2.7 log CFU mL^{-1} and 3.9 log CFU mL^{-1} in low-density polyethylene and by 0.6–2.2 log CFU mL^{-1} and 1.1–3.8 log CFU mL^{-1} on lettuce. Microbial inactivation was dependent on e.g. UV radiation dosage, bacterial strain and type of material supporting the biofilm, indicating that further studies are required before general application of the technique.

19.4 Conclusion

UV radiation from the Sun has been affecting nearly all living organisms on Earth and microbes have developed ingenious strategies to survive exposure. These include the formation of biofilms and the production of specialized compounds such as mycosporine-like amino acids and carotenoid pigments. However, technological developments have allowed the utilization of UV radiation to detect and kill harmful microbes. Nevertheless, the efficacy is dependent on microbial organism, UV radiation dosage and local conditions. Further research is needed before a general application of UV radiation for the sterilization of materials may be reached.

Acknowledgements The author acknowledges Fundação para a Ciência e a Tecnologia, I.P. (FCT), Portugal, for financial support under program "FCT Investigator 2013" (IF/01203/2013/CP1163/CT0002).

References

1. Antón J, Rosselló-Mora R, Rodríguez-Valera F et al (2000) Extremely halophilic bacteria in crystallizer ponds from solar salterns. Appl Environ Microbiol 66:3052–3057
2. Argyraki A, Markvart M, Nielsen A et al (2016, April) Comparison of UVB and UVC irradiation disinfection efficacies on Pseudomonas aeruginosa (P. aeruginosa) biofilm. In: Popp J, Tuchin VV, Matthews DL, Pavone FS (eds) Proceedings SPIE 9887, biophotonics: photonic solutions for better health care V, 9887. Brussels, Belgium
3. Asker D, Beppu T, Ueda K (2007) Unique diversity of carotenoid-producing bacteria isolated from Misasa, a radioactive site in Japan. Appl Microbiol Biotechnol 77:383–392
4. Caramujo M-J, de Carvalho CCCR, Silva SJ et al (2012) Dietary carotenoids regulate astaxanthin content of copepods and modulate their susceptibility to UV light and copper toxicity. Mar Drugs 10:998–1018
5. Coelho H, Vieira S, Serôdio J (2009) Effects of desiccation on the photosynthetic activity of intertidal microphytobenthos biofilms as studied by optical methods. J Exp Mar Biol Ecol 381:98–104
6. Cogdell RJ (1978) Carotenoids in photosynthesis. Philos Trans R Soc Lond Ser B Biol Sci 284:569–579
7. Costerton JW, Stewart PS, Greenberg EP (1999) Bacterial biofilms: a common cause of persistent infections. Science 284:1318–1322
8. de Carvalho CCCR (2007) Biofilms: recent developments on an old battle. Recent Pat Biotechnol 1:49–57
9. de Carvalho CCCR, Caramujo MJ (2017) Carotenoids in aquatic ecosystems and aquaculture: a colorful business with implications for human health. Front Mar Sci 4:93
10. de Mora S, Demers S, Vernet M (2000) The effects of UV radiation in the marine environment. Cambridge University Press, Cambridge, UK
11. Decho AW (2000) Microbial biofilms in intertidal systems: an overview. Continent Shelf Res 20:1257–1273
12. Elasri MO, Miller RV (1999) Study of the response of a biofilm bacterial community to UV radiation. Appl Environ Microbiol 65:2025–2031
13. Fischer M, Wahl M, Friedrichs G (2012) Design and field application of a UV-led based optical fiber biofilm sensor. Biosens Bioelectron 33:172–178
14. Gao Q, Garcia-Pichel F (2011) Microbial ultraviolet sunscreens. Nat Rev Micro 9:791–802
15. Garvey M, Rabbitt D, Stocca A et al (2015) Pulsed ultraviolet light inactivation of *Pseudomonas aeruginosa* and *Staphylococcus aureus* biofilms. Water Environ J 29:36–42
16. Hall-Stoodley L, Costerton JW, Stoodley P (2004) Bacterial biofilms: from the natural environment to infectious diseases. Nat Rev Micro 2:95–108
17. Hoagland KD, Rosowski JR, Gretz MR et al (1993) Diatom extracellular polymeric substances: function, fine structure, chemistry, and physiology. J Phycol 29:537–566
18. Hughes KA, Scherer K, Svenøe T et al (2006) Tundra plants protect the soil surface from uv. Soil Biol Biochem 38:1488–1490
19. Working Group IARC (2012) Solar and ultraviolet radiation. In: IARC monographs on the evaluation of carcinogenic risks to humans. International Agency for Research on Cancer, Lyon, pp 35–101
20. Karsten U, Lembcke S, Schumann R (2007) The effects of ultraviolet radiation on photosynthetic performance, growth and sunscreen compounds in aeroterrestrial biofilm algae isolated from building facades. Planta 225:991–1000
21. Keshari N, Adhikary SP (2013) Characterization of cyanobacteria isolated from biofilms on stone monuments at Santiniketan, India. Biofouling 29:525–536
22. La Barre S, Roullier C, Boustie J (2014) Mycosporine-like amino acids (MAAs) in biological photosystems. In: Outstanding marine molecules. Wiley-VCH Verlag GmbH & Co. KGaA, Weinheim, pp 333–360
23. Leid JG, Willson CJ, Shirtliff ME et al (2005) The exopolysaccharide alginate protects *Pseudomonas aeruginosa* biofilm bacteria from IFN-γ-mediated macrophage killing. J Immunol 175:7512–7518

24. Montgomery NL, Banerjee P (2015) Inactivation of *Escherichia coli* o157:H7 and *Listeria monocytogenes* in biofilms by pulsed ultraviolet light. BMC Res Notes 8:235
25. Murray KE, Manitou-Alvarez EI, Inniss EC et al (2015) Assessment of oxidative and UV-C treatments for inactivating bacterial biofilms from groundwater wells. Front Environ Sci Eng 9:39–49
26. Portwich A, Garcia-Pichel F (1999) Ultraviolet and osmotic stresses induce and regulate the synthesis of mycosporines in the cyanobacterium *Chlorogloeopsis* PCC 6912. Arch Microbiol 172:187–192
27. Ragon M, Restoux G, Moreira D et al (2011) Sunlight-exposed biofilm microbial communities are naturally resistant to chernobyl ionizing-radiation levels. PLoS One 6:e21764
28. Rasmussen B (2000) Filamentous microfossils in a 3,235-million-year-old volcanogenic massive sulphide deposit. Nature 405:676–679
29. Rastogi RP, Madamwar D, Incharoensakdi A (2015a) Sun-screening bioactive compounds mycosporine-like amino acids in naturally occurring cyanobacterial biofilms: role in photoprotection. J Appl Microbiol 119:753–762
30. Rastogi RP, Sonani RR, Madamwar D (2015b) Cyanobacterial sunscreen scytonemin: role in photoprotection and biomedical research. Appl Biochem Biotechnol 176:1551–1563
31. Ryder C, Byrd M, Wozniak DJ (2007) Role of polysaccharides in *Pseudomonas aeruginosa* biofilm development. Curr Opin Microbiol 10:644–648
32. Sinha RP, Ambasht NK, Sinha JP et al (2003) Wavelength-dependent induction of a mycosporine-like amino acid in a rice-field cyanobacterium, *Nostoc commune*: role of inhibitors and salt stress. Photochem Photobiol Sci 2:171–176
33. Tedetti M, Sempéré R (2006) Penetration of ultraviolet radiation in the marine environment. A review. Photochem Photobiol 82:389–397
34. Van Colen C, Underwood GJC, Serôdio J et al (2014) Ecology of intertidal microbial biofilms: mechanisms, patterns and future research needs. J Sea Res 92:2–5
35. Walsh LJ, Shakibaie F (2007) Ultraviolet-induced fluorescence: shedding new light on dental biofilms and dental caries. Aust Dent Pract 18:56–60
36. Wang Y, Lu J, Mollet JC et al (1997) Extracellular matrix assembly in diatoms (bacillariophyceae) (ii. 2,6-dichlorobenzonitrile inhibition of motility and stalk production in the marine diatom *Achnanthes longipes*). Plant Physiol 113:1071–1080
37. Waring JEN, Baker NR, Underwood GJC (2007) Responses of estuarine intertidal microphytobenthic algal assemblages to enhanced ultraviolet b radiation. Glob Chang Biol 13:1398–1413
38. Westall F, de Wit MJ, Dann J et al (2001) Early Archean fossil bacteria and biofilms in hydrothermally-influenced sediments from the Barberton greenstone belt, South Africa. Precambrian Res 106:93–116
39. Will OA, Scovel CA (1989) Photoprotective functions of carotenoids. In: Krinsky NI, Mathews-Roth MM, Taylor RF (eds) Carotenoids: chemistry and biology. Springer US, Boston, pp 229–236

UV Induced Mutagenicity in Water: Causes, Detection, Identification and Prevention

20

Roberta (C.H.M.) Hofman-Caris

Abstract

At first it seemed that UV processes for disinfection and advanced oxidation were "harmless", as they didn't involve the addition of "dangerous" chemicals nor seemed to result in the formation of toxic byproducts. However, recently it has become clear that also during UV processes mutagentic/genotoxic byproducts may be formed. It was found that these are nitrogen containing aromatic compounds, which are formed by the reaction of photolysis products of nitrate with (photolysis products of) natural organic matter. Now more has become clear on the formation process of these compounds, it is possible to limit or even prevent their formation during e.g. UV/H_2O_2 processes. Besides, it appears to be possible to remove such byproducts by means of filtration processes. Thus, UV based processes can safely be applied in water treatment.

Keywords

Advanced oxidation • AMES fluctuation assay • Bioassays • Genotoxicity • Mutagenicity • N-BPs • Photolysis • Oxidation • Treshold of toxicological concern • UV/H_2O_2

20.1 Introduction

It has been known for a long time that ultraviolet C (UVC) and UVB irradiation (100–280 nm and 280–315 nm respectively) can damage DNA. Absorption of irradiation causes the formation of thymine dimers, which can inactivate microorganisms. This principle has been applied since the middle of the last century, and it appeared to be a very elegant disinfection method; contrary to the application of chlorine compounds or ozone, it doesn't require the addition of chemicals, and does not seem to generate disinfection byproducts, like chlorinated organic compounds and bromate [1, 2]. Although UVB can cause structural damage to DNA, natural repair processes of the cell seem to correct most of the dam-

R. Hofman-Caris (✉)
KWR Watercycle Research Institute,
P O Box 1072, 3430 BB Nieuwegein,
The Netherlands
e-mail: Roberta.hofman-caris@kwrwater.nl

S. Ahmad (ed.), *Ultraviolet Light in Human Health, Diseases and Environment*, Advances in Experimental Medicine and Biology 996, https://doi.org/10.1007/978-3-319-56017-5_20

age quickly enough to avoid permanent damage. In water treatment in general UVC irradiation, which is more energetic, is applied. Most commonly two types of mercury based UV lamps are used in water treatment: medium pressure (MP) UV lamps, which emit radiation in a range between 200 and 300 nm, and low pressure (LP) UV lamps, which only emit radiation at 253.7 nm.

Since the beginning of this century another application, advanced oxidation processes (AOPs), based on UVC have become more important in water treatment, mostly for the production of drinking water from surface water, river bank filtrate, and also groundwater, but also for wastewater treatment. Sources for drinking water appear to contain increased amounts of organic micropollutants, like pesticides, fungicides, pharmaceuticals, and personal care products. The oxidation processes, e.g. with ozone, can be used to decompose such organic micropollutants, but not every compound is adequately sensitive to oxidation. Since the last 20 years AOPs, characterized by the fact that hydroxyl radicals are formed, are gaining more and more interest. Hydroxyl radicals are very reactive, and can oxidize a broad range of organic compounds.

There are various ways to generate hydroxyl radicals, but often they are based on UV irradiation of reagents. Examples are UV/O_3, UV/H_2O_2, photofenton, and photocatalytic processes (with e.g. TiO_2 as the photocatalyst). In water treatment a lot of research has been done into UV/H_2O_2 processes, which also are applied in full scale installations for drinking water production.

In fact the UV/H_2O_2 process combines two simultaneous processes, which involve the photolysis of chemicals and of micropollutants by UV radiation:

1. Photolysis of micropollutants caused by direct absorption of the UV radiation by the pollutants.
2. Oxidation of micropollutants by hydroxyl radicals, that resulted from the UV photolysis of H_2O_2 (Fig. 20.1).

The absorption spectrum of natural water with H_2O_2 is shown in Fig. 20.1. Water itself is able to absorb UV radiation with a maximum in the vacuum UV (VUV) range (75–185 nm). Although this also results in the formation of radicals, the penetration depth of the radiation is very low, as a result of which practical application of VUV radiation for water treatment purposes still is limited. However, in the range of 200–220 nm still some UV absorption by the water itself can be observed. Furthermore, natural water always contains organic matter ("natural organic matter"; NOM), which, depending on its structure, also may absorb radiation in the lower wavelength range of Fig. 20.1. Other compounds that may be present and cause absorption are carbonate and, at a wavelength ≤240 nm, nitrate.

As H_2O_2 doesn't strongly absorb in this range of wavelengths (200–300 nm), a relatively high UV dose and high H_2O_2 concentration (of about 10 mg/L) are required to obtain an effective AOP. For the generation of·OH both types of lamps are effective, but LP lamps are more energy efficient. However, as MP lamps emit over a broader range of wavelengths, these lamps give a more effective photolysis. For the degradation of some compounds, that are not very sensitive towards oxidation, this may be beneficial.

Although in principle it is possible to mineralize organic micropollutants by means of UV/H_2O_2 processes, this would require an enormous amount of energy and may be expensive. It is believed that partial decomposition of the pollutants will make them better biodegradable, and therefore mostly lower UV doses are applied [3]. However, this may result in the formation of transformation products, the characteristics of which are not yet identified [4].

In recent years it has become clear that sometimes water treated with UV/H_2O_2 causes toxicity [5]. More research into this phenomenon is required, as it is not clear what exactly caused this. For the European drinking water sectors, both removing contaminants from raw water sources and dealing with disinfection byproducts are important challenges [6].

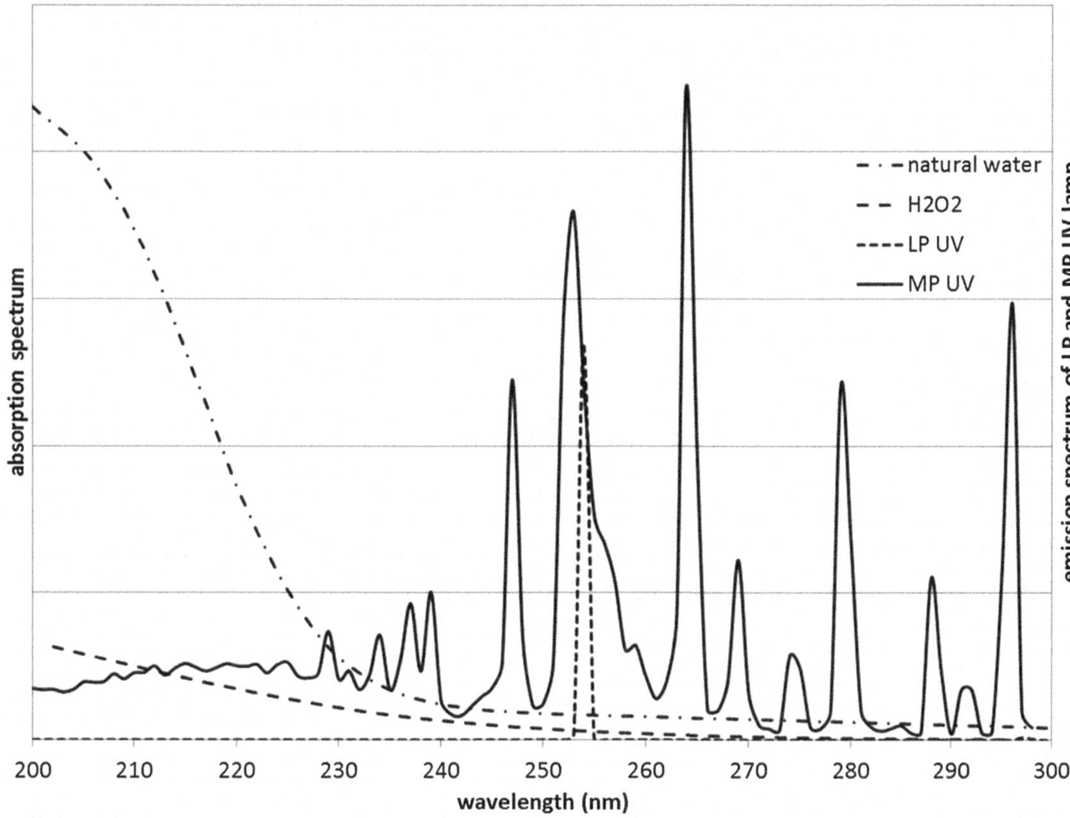

Fig. 20.1 Absorption spectra of H_2O and of natural water with H_2O_2, and emission spectra of LP- and MP-UV lamps

20.2 Toxicity Tests

To determine the toxicity a toxicological risk assessment of putative toxic compounds is required to be carried out taking into account their kinetics too. However, in practice very often it is not known which compound exactly is causing the toxicity. Besides, toxicity may be caused by a mixture of compounds, and depends on concentration and duration of exposure of organisms to the compounds. It is very difficult to obtain relevant information from a sample, as concentrations of the toxic agents may be very low, structure and/or reaction pathways are unknown, and there is a lack of suitable sample preparation and detection methods. Therefore, bioassays have been developed to determine the cumulative effects of chemicals that exhibit the same mode of toxic action (MOA) and thus concentration-additive effects.

20.2.1 In Vivo Assays

In literature a large number of toxicological data on disinfection by-products (BPs) from drinking water production have been published, and also the World Health Organization published a report on this topic (http://apps.who.int/iris/bitstream/10665/42274/1/WHO_EHC_216.pdf) Genotoxicity of BPs has been evaluated in *in vivo* experiments using e.g. the micronuclei test, comet assay, and chromosomal aberration test. Carcinogenicity has been assessed in 2-year dosing studies in rodents. For this purpose mainly various kinds of rats and mice, but also hamsters and gerbils were used. Besides, there are some studies based on the effects in Drosophila. Often the term "mutagenicity" refers to assays that measure a change in DNA sequence (either gene or chromosomal mutation), The term "genotoxicity" refers to mutagenicity as well as DNA damage (like adducts or strand breaks). For "carcinogenicity" the formation of tumors in several different organs was studied. The effects of the DPBs on a broad range of endpoints (like micronucleus formation, sister chromatid exchange, unscheduled DNA synthesis, DNA strand break and sex-linked recessive mutations, DNA methylation, DNA repair, germ-cell mutations, chromosomal aberrations, etc.) have been described [7]. Traditionally *in vivo* bioassays have several drawbacks. Apart from the loss of animal lives involved in tests, they show a high biological variability, and complexity. Besides, *in vivo* tests are expensive and may take time. Therefore, more and more in vitro bioassays are being developed and applied.

20.2.2 In Vitro Toxicity Tests

In vitro bioassays do not show the disadvantages associated with *in vivo* bioassays. They are based on mammalian cells or bacterial species, and can effectively be applied as screening tools. Jia, Escher [8] presented an overview of 36 bioassays covering 18 biological endpoints. They used a battery of assays, including assays for genotoxicity, mutagenicity, estrogenic, glucocorticoid and arylhydrocarbon receptor activities, oxidative stress response, and cytotoxicity. The disadvantage of these bioassays is that they don't give information on the identity and concentration levels of the bioactive toxicants nor take into account kinetics. Furthermore, the type and level of response depend on the specific assay applied. Thus, the results cannot be directly extrapolated to e.g. human health. However, they indicate whether compounds are present, which are capable of inducing toxic effects in living organisms. If a positive response is obtained, more research will be required in order to determine which compounds are causing the effect, and whether or not it may be a threat to human health.

In paragraphs below a short literature overview of some tests that are commonly used for UV/H_2O_2 processes is presented.

20.2.3 Toxicity Tests Used for UV/H_2O_2 Processes

A number of toxicity tests were carried out including tests with *Vibrio fischeri*, *Paracentrotus lividus* sea-urchin, with *Sparus aurata* larvae and Microtox® test on *v fischeri* [9–11]. A detailed description of the Microtox® test protocols can be found in Aguirre-Martínez, Owuor [12]. For *Vibrio fischeri* the toxicity was determined by measuring the bioluminescence intensity of the bacteria after exposure to putative toxic solutions. The authors compared the results obtained with mortality of *Sparus aurata* larvae, treated with the same solutions. The latter test appears to be more sensitive, and thus is recommended for the development of water quality criteria for marine water bodies [11, 13]. With *Paracentrotus lividus* two types of test were performed [10]: a fertilization test, determining the percentage of fertilized gametes, and an embryo-larval development toxicity test, in which the percentage of normal pluteus after exposure is measured. Here also it was observed that the *P. lividus* embryo-larval development test was more sensitive than the *V. fischeri* essay.

Kolkman et al. determined specific endocrine receptor activation by using a panel of CALUX assays for estrogenic, androgenic, glucocorticoid, progestagenic and thyroidogenic agonistic activities [14, 15]. They measured genotoxicity using the Comet Assay and the Ames fluctuation test. In Ames fluctuation assays bacterial strains are used to identify the presence of putative mutagenic and hence carcinogenic or genotoxic compounds. Ames assays use a histidine auxotrophic mutant with certain additional mutations of *Salmonella typhimurium*. Because of their *his*-genetic mutation these bacteria can only grow if histidine is present in the growth medium. However, if there is any reverse mutation, caused by mutagenic compounds, the bacteria can form colonies on histidine deficient agar medium. This enhanced number of colony formation in comparison to the spontaneous number is considered as due to possible mutagenic activity. Often two different strains are applied: TA98, which can show frame shift mutations, and TA100, which is sensitive to base-pair substitution. Spontaneous revertants are used as a control. In order to mimic the effect of metabolism, as the toxicity of a compound may change during metabolism, rat liver S9 extract is usually added. Thus, Ames tests are carried out both with and without addition of S9. In order to quantify the total mutagenic effect the Ames test response was converted into 4-quinoline oxide (4-NQO) equivalents. 4-NQO is a known genotoxic compound that often is used as a positive control in Ames assays.

20.3 UV Induced Toxicity

20.3.1 First Indication of Mutagenic Activity

Guzzella, Feretti [16] studied advanced oxidation processes based on O_3/UV and O_3/H_2O_2/UV, and like Monarca, Feretti [5], observed an increase in mutagenicity impact in treated water. These findings were in accordance with the findings of Heringa, Harmsen [17]. Also Shemer and Linden [18] observed an increase in mutagenicity in drinking water after UV based photolysis and oxidation, although the compounds causing these effects and their parent compounds could effectively be removed by oxidative degradation. However, not all authors observed this mutagenicity effect. Mahmoud, Toolaram [19] studied the phototransformation of thalomide, and, based on Quantitative Structure Activity Relationships (QSARs) expected the formation of mutagenic products. However, surprisingly they didn't observe an increase in the Ames response. It was assumed that the mutagenic products would be present in too low concentrations to observe any effects, or that antagonistic interactions would prevent mutagenicity in this test. Another explanation may be that the bacterial strains used in this test, TA98 and TA100, were not be very sensitive towards the specific compounds formed. Also de Veer, Moriske [20] and Haider, Sommer [21] didn't find a positive Ames test response upon UV irradiation of (ground) water, even though Haider, Sommer [21] increased the UV dose up to 800 mJ/cm^2. These results seem to be contradictory to the results obtained for the UV/H_2O_2 process by other authors [9–11]. These authors applied a multi-barrier treatment, including UV/H_2O_2 with either an LP or an MP UV lamp, to synthetic industrial wastewater. This type of water has a low UV-transmittance (17–43%), and the authors applied H_2O_2: TOC ratios of 1:1–10: 1, resulting in H_2O_2 concentrations of 40–400 mg/L. Compared to concentrations used in drinking water production (ca. 5–10 mg/L), this is very high. The authors found a significant increase in toxicity of the water after the UV/H_2O_2 treatment for both types of UV lamps, which was attributed to the (large) excess of H_2O_2 and possibly the formation of some byproducts during the reaction. However, after filtration over granular activated carbon (GAC), the toxicity appeared to have disappeared.

The removal of mutagenic compounds by means of filtration over activated carbon also had been observed by Heringa, Harmsen [17]. These authors studied both MP and LP UV lamps, and the results suggested that the mutagenicity was caused mainly by photolysis reactions, rather than by oxidation reactions. Because of their broad UV spectrum, MP lamps are more effective for photolysis than UV lamps emitting only one specific wavelength like LP or high pressure (HP) lamps, as were used by Haider, Sommer [21] and de Veer, Moriske [20]. Besides, Parkinson, Barry [22] had shown that also the composition of the water matrix may play a crucial role in the formation of toxic byproducts. They applied UVC and UVC/H_2O_2 processes, using an LP-UV lamp, to remove natural organic matter (NOM) from water. However, degradation of NOM-copper complexes resulted in the release of metal ions, which probably caused the toxicity observed afterwards.

In order to obtain information on possible toxicity generation in disinfection installations and during advanced oxidation UV/H_2O_2 processes research was carried out by Hofman-Caris, Harmsen [23]. These authors identified the main reaction parameters involved in the formation of mutagenic byproducts.

20.3.2 Toxicity Development in Drinking Water Disinfection Installations

For disinfection of drinking water in general relatively low UV-doses of 20–70 mJ/cm^2 are applied. As shown in Table 20.1 samples were taken at full scale drinking water production plants throughout the Netherlands, applying either MP or LP UV-lamps for disinfection purposes.

During treatment the drinking water passes the UV reactor. This is a vessel equipped with UV lamps, placed either in the flow direction or perpendicular to the flow. The dose required is obtained by adjusting the reactor geometry, the number of UV lamps inside the reactor (and their UV output), and the total flow through the reactor [24]. Often a number of such reactors is applied in parallel streets. At the Dutch Drinking Water Utility of PWN in Andijk "Swift 16 L30" UV reactors of Trojan are applied. These reactors are equipped with 16 lamps each, and can apply flows up to 6300 m^3/h each. For advanced oxidation purposes a higher UV-dose, i.e. a longer residence time of the water in the UV reactor is required, and thus the applied flow will be lower. At Andijk three streets, each equipped with four Swift 16 L30 reactors are present, enabling advanced oxidation of about 5000 m^3 of water per hour [25, 26].

Table 20.1 Details of sampling sites at full scale drinking water disinfection processes in the Netherlands [23]

Origin and type of water	Type of UV-lamp	Reduction equivalent UV dose (mJ/cm^2)
River bank filtrate and ground water	MP	25
Ground water	LP	70
Pretreated surface water	LP	42
Pretreated (coagulation, sedimentation, filtrations) surface (Meuse) water	MP	40
		100[a]
		200[a]
Surface water after pretreatment and dune infiltration	LP	40
		100[a]

[a]Dose increased for research purposes

The Ames fluctuation test, applying both TA98 and TA100 bacterial strains, was used to study the water samples from the full scale disinfection installations shown in Table 20.1. It was found that under normal disinfection conditions none of the samples showed an increase in the Ames response suggesting that more likely there are no mutagenic compounds present. However, upon increase of the UV dose in Meuse water to 100 or 200 mJ/cm^2 such an increase could be observed (Fig. 20.2 upper panel). This became even more clear in laboratory experiments, using a "collimated beam set-up" [27] (Fig. 20.2, lower panel).

The fact that in duplicated measurements some differences in the exact number of wells can be seen, shows that a positive response (the maximum number of wells in one plate being 48) is not unusual for bioassays. From these experiments it became clear that after "normal" disinfection conditions no mutagenic compounds could be detected, but that upon increasing the UV dose such compounds may be formed. This means that the (pretreated) Meuse water must contain certain compounds which upon UV irradiation may become mutagenic.

Disinfection experiments with MP UV lamps were also carried out by Martijn, Kruithof [28]. These researchers found a significant Ames test response at an MP UV dose of 40 mJ/cm^2, using IHSS Pony Lake NOM and nitrate. It was suggested that the presence of nitrate is involved in the formation of mutagenic compounds.

20.3.3 UV/H_2O_2 Advanced Oxidation Processes

Hofman-Caris, Harmsen [23] investigated which parameters affect the formation of mutagenic compounds, and derived the following relations:

- Mutagenicity increases with increasing UV-dose, both for LP and for MP UV-lamps
- The formation of mutagenic compounds is significantly higher by MP UV-lamps than by LP UV-lamps.
- Mutagenicity increases with increasing nitrate concentrations.

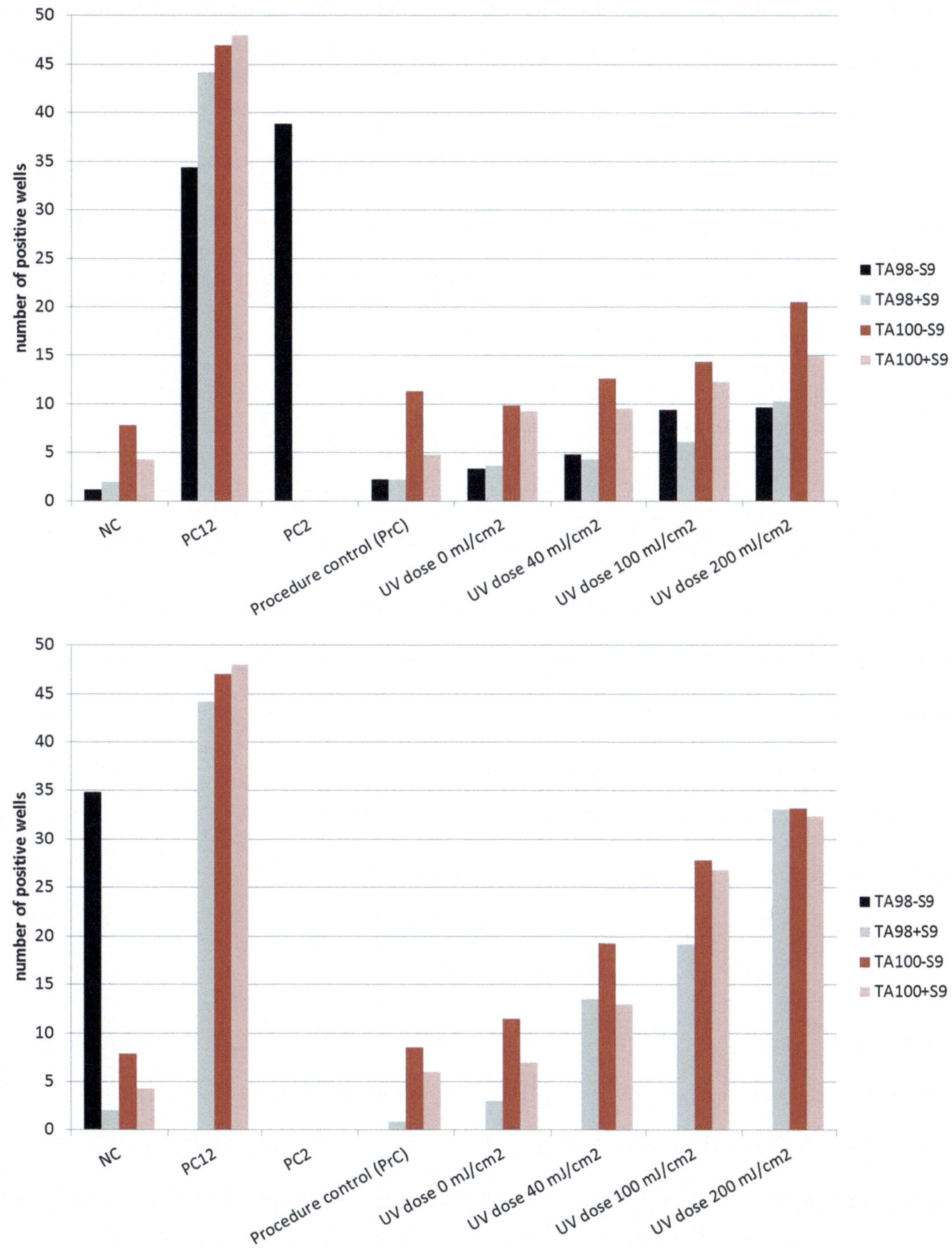

Fig. 20.2 Results of Ames fluctuation assays on pretreated Meuse water at elevated MP UV doses, full scale (*upper panel*) and laboratory (*lower panel*) results. *NC* negative control, *PC* positive control, *PrC* procedure control. Average values of triplicate tests are shown for each sample [23]

- Mutagenicity increases with increasing NOM concentrations (at least in the pretreated Meuse water studied)
- Mutagenicity decreases when H_2O_2 is added.

From the above results it can be concluded that the combination of UV dose, TOC, H_2O_2 and NO_3^- id responsible for 74–87% of the Ames responses obtained, and that these are the main factors involved in the formation of mutagenic byproducts. Based on the experimental data, it was suggested that the photolysis of nitrate plays an important role in the formation of mutagenic compounds. Nitrate has two absorption bands in the UV region, one in the near UV region from 260 to 350 nm (with a maximum at 300 nm), and a more intensive band below 240 nm, with a maximum at 200 nm [29], as shown in Fig. 20.3. The photolysis of nitrate, eventually resulting in the formation of nitrite, is a rather complex process, as shown in Fig. 20.4. On the other hand, UV photolysis of nitrite results in the formation of nitrate.

Buchanan, Roddick [31] suggested that mutagenicity may be caused by the nitrite formed during the UV photolysis process of nitrate. However, Hofman-Caris, Harmsen [23] showed that in the presence of H_2O_2 the nitrite concentration becomes higher than by UV photolysis alone, although the Ames test shows no increase in mutagenecity. This indicates that it is not the nitrite itself, nor the hydroxyl radicals formation which causes mutagenicity. As shown in Fig. 20.4 and by Wols, Hofman-Caris [32] several radicals are formed during photolysis of nitrate; these include probably nitrogen containing aromatics, formed by reaction of intermediate nitro- and nitroso radicals with NOM or with photolysis products of NOM. Thus nitrogen containing disinfection byproducts (N-BPs) can be formed. These findings are in accordance with the findings of other authors [28, 29, 33]. Martijn et al. also concluded that the photolysis of nitrate plays a crucial role in the formation of mutagenic compounds. Several reasons outlined below may explain why in the presence of H_2O_2 less mutagenicity is observed.

1. Part of the UV irradiation is “used” for photolysis of H_2O_2 instead of NO_3^-
2. The hydroxyl radicals formed upon H_2O_2 photolysis will quench the radicals formed during the conversion of NO_3^-
3. The hydroxyl radicals react with aromatic NOM, decomposing the aromatic structure of the molecules, as a result of which less mutagenic compounds (nitrogen containing aromatics) can be formed.

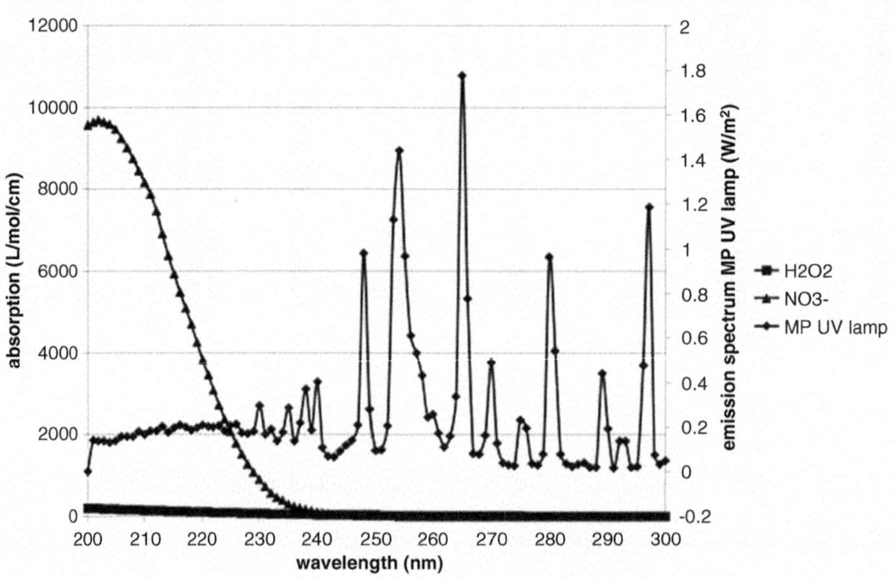

Fig. 20.3 Emission spectrum of an MP UV lamp and molar absorption of nitrate and hydrogen peroxide

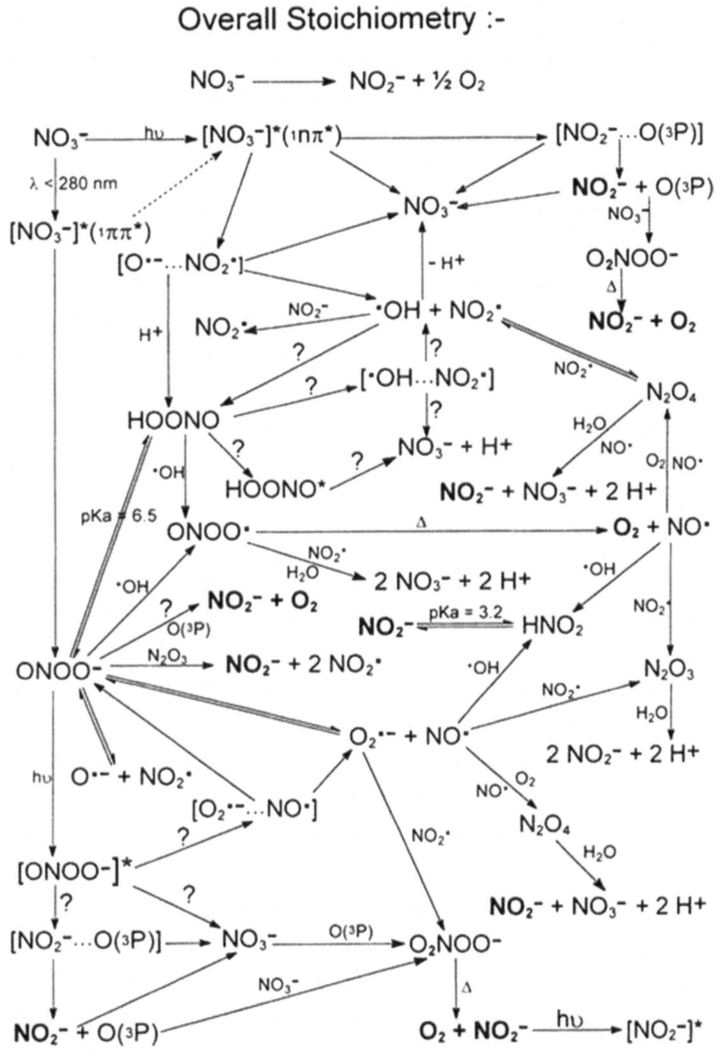

Fig. 20.4 Reaction scheme of the photolysis of nitrate, not including follow-up reactions of nitrite [30]

Martijn et al. [29], 201 showed that the NOM composition plays an important role in the formation of mutagenic compounds. Indeed, they also found that mutagenic compounds seem to be nitro aromatics. This could not be deduced from the inorganic nitrogen mass balance (nitrate versus nitrite). Although a significant Ames test response was obtained, no significant nitrogen deficit could be observed, indicating that only very small amounts of mutagenic compounds are formed. However, when phenol was used as a model compound for aromatic NOM, it could be demonstrated that nitrogen was incorporated in the organic matrix: 2- and 4-nitrophenol and 4-nitrocatechol could be detected. It is known that nitro aromatics may be genotoxic [34, 35], so these findings may account for the increase in toxicity.

As the nitrate content of groundwater is about 4–9 mg/L, and of surface water usually circa 4 mg/L, but possibly, due to agricultural activities, much higher [28], this is something that has to be considered when applying high UV doses for the production of drinking water. It also is possible that organic micropollutants are involved in the formation of mutagenic byproducts; however, Martijn, Kruithof [28] showed that the presence of these micropollutants has no significant effect on the Ames response. This may be explained from the fact that micropollutant's concentrations are in the order of ng or μg/L, whereas NOM is present in mg/L. Therefore, it cannot be excluded that micropollutants are involved but their contribution to mutagenicity will be small in comparison to the contribution of NOM.

20.3.4 Identification of Possible N-BPs

In order to be able to carry out a proper risk assessment, it is essential first to identify the byproducts formed. For this Kolkman, Martijn [14] developed a new technique in which they applied ^{15}N labeled nitrate. After UV irradiation, stable isotope labeled N-BPS identified using high resolution mass spectrometry. In this way they were able to detect 84 N-BPs at concentrations between 1 and 135 ng/L bentazon-d_6 equivalents after MP UV treatment of artificial water containing both NOM (Pony Lake) and nitrate, with a summed concentration of 1.2 μg/L bentazon-d_6 equivalents. Furthermore, they were able to detect the presence of 22 of these N-BPs in real water from a full scale drinking water production plant, applying MP UV/H_2O_2. In a subsequent paper Vughs, Baken [36] were able to identify 14 N-BPs after MP UV treatment, none of which had been listed as (potential) human mutagen or carcinogen. These include nine N-BPs from the former mentioned 22 N-BPs also observed in water from a full scale UV/H_2O_2 process. Applying effect-directed analysis (EDA) the authors were able to identify five N-BPS that are potentially genotoxic and were present in relatively high concentrations in the fractions of treated water, in which mutagenicity had been detected. Two of these had been observed in water from the full scale process as well (Table 20.2).

The identity of a large part of the byproducts has not yet been confirmed, but it seems clear that N-BPS cannot be neglected when UV processes, especially based on MP UV-lamps in nitrate containing water, are applied.

20.3.5 Risk Assessment

For genotoxic compounds in drinking water the concept of a Threshold of Toxicological Concern (TTC) has been proposed [37]. In this concept the human exposure threshold value is established, below which no appreciable health risk occurs. This concept may be applied when the presence of a new contaminant in e.g. food or drinking water is observed, for which no toxicological information is available (e.g. it is not applicable to N-nitroso compounds). For drink-

Table 20.2 Identities of N-BPS [36]

Compound	Formula	potential genotoxicity proven	Present in water from full scale UV/H_2O_2 process
4-nitrophenol	$C_6H_5NO_3$		
4-nitrocatechol	$C_6H_5NO_4$		
4-nitro-1,3-benzenediol	$C_6H_5NO_4$		
2-nitrohydroquinone	$C_6H_5NO_4$		
2-hydroxy-5-nitrobenzoic acid	$C_7H_5NO_5$		
4-hydroxy-3-nitrobenzoic acid	$C_7H_5NO_5$	X	X
2-hydroxy-3-nitrobenzoic acid	$C_7H_5NO_5$		
2,4-dinitrophenol	C6$H_4N_2O_5$		
5-nitrovanillin	$C_8H_7NO_5$		
4-nitrobenzenesulfonic acid	$C_6H_5NO_5S$		
4-nitrophtalic acid	$C_8H_5NO_6$	X	
2-methoxy-4,6-dinitrophenol	$C_7H_6N_2O_6$	X	X
3,5-dinitrosalicylic acid	$C_7H_4N_2O_7$	X	
Dinoterb	$C_{10}H_{12}O_5N_2$	X	

ing water it is suggested that an acceptable TTC is 10 ng/L. This means that only concentrations below this level in general are considered not to pose a significant human health risk [14, 37]. This TTC, however, doesn't apply to carcinogens, which need substance-specific risk evaluation [14].

In order to quantify the total mutagenic effect, the Ames test response was converted into 4-nitroquinoline oxide (4-NQO) equivalents. The response of the Ames assay then is compared with a certain concentration of a known mutagenic compound (4-NQO) that shows a similar Ames response.

Wollin and Dieter [35] derived health based drinking water guideline values for some nitro aromatics. Martijn, Van Rompay [38] used the 4-NQO toxic equivalency factor (TEF) to determine the risk of unknown genotoxic compounds. Based on their results they tried to indicate the risk of MP UV/H_2O_2 treatment via the margin of exposure (MOE) approach. Assuming a body weight of 70 kg and a drinking water consumption of 2 L per day (the only contribution resulting from drinking water consumption), the 4-NQO equivalent concentration should be lower than 80 ng/L associated with a negligible risk. In water from a full scale MP UV/H_2O_2 process, before filtration over GAC, a 4-NQO equivalent concentration of 107 ng/L was observed [38]. However, Vughs, Baken [36] applied a standard allocation of only 20% of the total exposure to drinking water (as part of the exposure may also be caused by e.g. food), resulting in a provisional drinking water guideline of 5.8 mg/L for 2-hydroxy-5-nitrobenzoic acid, 0.01 mg/L for 2,4-dinitrophenol, and 18 mg/L for 5-nitrovanillin. This gives a more realistic estimation of the risks involved. The authors identified five byproducts, which are present at relatively high concentrations in fractions of UV treated water that show mutagenicity. These five byproducts (4-nitrophtalic acid, 4-hydroxy-3-nitrobenzoic acid, 2-methoxy-4,6-dinitrophenol, dinoterb and 3,5-dinitrosalicylic acid) are potentially genotoxic (based on QSARs), but more research will be required to determine the concentrations of these compounds that would be acceptable for drinking water.

20.3.6 How to Prevent Negative Effects from UV/H_2O_2 Processes

As mutagenic compounds may be formed by reaction of photolysis products of nitrate and NOM, this can be prevented by preventing the photolysis of nitrate. In principle it is possible to reduce nitrate concentrations in water by applying ion exchange processes. IEX may also be applied to remove part of the NOM, especially high charge density/high polarity organic fractions such as humic (−like) and fulvic substances [39]. According to these authors, the presence of dissolved organic nitrogen is a key parameter in the formation of N-BPS, and strongly basic ion exchange resins can be applied to remove N-BPs.

Another way is to prevent the photolysis process itself is by applying natural quartz sleeves that cut off UV light at wavelengths below 240 nm [28]. Thus, nitrate will not be able to absorb UV irradiation anymore, and the formation of N-BPs will be hindered or even prevented.

In order to obtain an efficient UV/H_2O_2 process an excess of H_2O_2 has to be added. Then its excess has to be removed, and for this purpose filtration processes can be applied, like filtration over granular activated carbon or dune infiltration. It has been shown that such filtration processes can effectively remove also the mutagenic byproducts formed [16, 17, 40, 41].

20.4 Conclusion

UVC based processes have been shown to be very effective for disinfection purposes, and, e.g. in combination with H_2O_2, for the removal of organic micropollutants. However, recent research has shown that such processes may result in the formation of byproducts which cause a positive response in the Ames test, indicating that they may be mutagenic/genotoxic. It has been proven that these byproducts, N-BPs, are formed during photolysis of nitrate, and consist of aromatic nitrogen compounds. Their formation can be prevented by preventing the photolysis of nitrate and the reaction with NOM, e.g. by using LP instead of MP UV lamps, by removing nitrate or humic acids by means of IEX, or by applying natural quartz sleeves that cut off wavelengths below 240 nm. In case the formation of N-BPs cannot be prevented, they can be effectively removed by a subsequent filtration process. However, this information will have to be taken into account whenever application of a UV process is considered in drinking water treatment.

References

1. Rook JJ (1974) Formation of haloforms during chlorination of natural water. Water Treat Exam 23(2):234–243
2. Kurokawa Y et al (1990) Toxicity and carcinogenicity of potassium bromate – a new renal carcinogen. Environ Health Perspect 87:309–335
3. Oller I, Malato S, Sánchez-Pérez JA (2011) Combination of advanced oxidation processes and biological treatments for wastewater decontamination-a review. Sci Total Environ 409(20):4141–4166
4. Esplugas S et al (2007) Ozonation and advanced oxidation technologies to remove endocrine disrupting chemicals (EDCs) and pharmaceuticals and personal care products (PPCPs) in water effluents. J Hazard Mater 149(3):631–642
5. Monarca S et al (2000) The influence of different disinfectants on mutagenicity and toxicity of urban wastewater. Water Res 34(17):4261–4269
6. Van Der Hoek JP et al (2014) Drinking water treatment technologies in Europe: state of the art – challenges – research needs. J Water Supply Res Technol AQUA 63(2):124–130
7. Richardson SD et al (2007) Occurrence, genotoxicity, and carcinogenicity of regulated and emerging disinfection by-products in drinking water: a review and roadmap for research. Mutat Res Rev Mutat Res 636(1–3):178–242
8. Jia A et al (2015) In vitro bioassays to evaluate complex chemical mixtures in recycled water. Water Res 80:1–11
9. Rueda-Márquez JJ et al (2016) Post-treatment of refinery wastewater effluent using a combination of AOPs (H2O2 photolysis and catalytic wet peroxide oxidation) for possible water reuse. Comparison of low and medium pressure lamp performance. Water Res 91:86–96
10. Rueda-Márquez JJ et al (2015) Combined AOPs for potential wastewater reuse or safe discharge based on multi-barrier treatment (microfiltration-H2O2/UV-catalytic wet peroxide oxidation). Chem Eng J 270:80–90
11. Rueda-Márquez JJ et al (2015) Post-treatment of biologically treated wastewater containing organic contaminants using a sequence of H2O2 based advanced oxidation processes: photolysis and catalytic wet oxidation. Water Res 71:85–96
12. Aguirre-Martínez GV et al (2015) Are standard tests sensitive enough to evaluate effects of human pharmaceuticals in aquatic biota? Facing changes in research approaches when performing risk assessment of drugs. Chemosphere 120:75–85
13. Oliva M et al (2008) Lindane toxicity on early life stages of gilthead seabream (Sparus Aurata) with a note on its histopathological manifestations. Environ Toxicol Pharmacol 25(1):94–102
14. Kolkman A et al (2015) Tracing nitrogenous disinfection byproducts after medium pressure UV water treatment by stable isotope labeling and high resolution mass spectrometry. Environ Sci Technol 49(7):4458–4465
15. Kolkman A et al (2013) Sample preparation for combined chemical analysis and in vitro bioassay application in water quality assessment. Environ Toxicol Pharmacol 36(3):1291–1303
16. Guzzella L, Feretti D, Monarca S (2002) Advanced oxidation and adsorption technologies for organic micropollutant removal from lake water used as drinking-water supply. Water Res 36(17):4307–4318
17. Heringa MB et al (2011) Formation and removal of genotoxic activity during UV/H2O2-GAC treatment of drinking water. Water Res 45(1):366–374
18. Shemer H, Linden KG (2007) Photolysis, oxidation and subsequent toxicity of a mixture of polycyclic aromatic hydrocarbons in natural waters. J Photochem Photobiol A Chem 187(2-3):186–195
19. Mahmoud WMM et al (2013) Identification of phototransformation products of thalidomide and mixture toxicity assessment: an experimental and quantitative structural activity relationships (QSAR) approach. Water Res 49(1):11–22
20. de Veer I, Moriske HJ, Rüden H (1994) Photochemical decomposition of organic compounds in water after UV-irradiation: investigation of positive mutagenic effects. Toxicol Lett 72(1-3):113–119
21. Haider T et al (2002) Genotoxic response of Austrian groundwater samples treated under standardized UV (254nm)-disinfection conditions in a combination of three different bioassays. Water Res 36(1):25–32
22. Parkinson A et al (2001) Preliminary toxicity assessment of water after treatment with UV-irradiation and UVC/H2O2. Water Res 35(15):3656–3664
23. Hofman-Caris RCHM et al (2015) Influence of process conditions and water quality on the formation of mutagenic byproducts in UV/H2O2 processes. Water Res 74:191–202
24. Wols BA et al (2015) Design aspects of UV/H2O2 reactors. Chem Eng Sci 137:712–721
25. Kruithof JC, Kamp PC, Martijn BJ (2007) UV/H2O2 treatment: a practical solution for organic contaminant control and primary disinfection. Ozone Sci Eng 29(4):273–280
26. Martijn BJ, Kruithof JC, Welling M (2006) UV/H2O2 treatment: the ultimate solution for organic con-

taminant control and primary disinfection. In: Water quality technology conference and exposition 2006: taking water quality to new heights, Denver, CO
27. Hofman-Caris CHM et al (2015) Determination of reaction rate constants in a collimated beam setup: the effect of water quality and water depth. Ozone Sci Eng 37(2):134–142
28. Martijn BJ et al (2015) Induced genotoxicity in nitrate-rich water treated with medium-pressure ultraviolet processes. J Am Water Works Assoc 107(6):E301–E312
29. Martijn AJ, Boersma MG, Vervoort JM, Rietjens IMCM, Kruithof JC (2014) Formation of genotoxic compounds by medium pressure ultraviolet treatment of nitrate-rich water. Desalin Water Treat 52(34–36):6275–6281
30. Mack J, Bolton JR (1999) Photochemistry of nitrite and nitrate in aqueous solution: a review. J Photochem Photobiol A Chem 128(1-3):1–13
31. Buchanan W, Roddick F, Porter N (2006) Formation of hazardous by-products resulting from the irradiation of natural organic matter: comparison between UV and VUV irradiation. Chemosphere 63(7):1130–1141
32. Wols BA et al (2013) Degradation of 40 selected pharmaceuticals by UV/H2O2. Water Res 47(15):5876–5888
33. Thorn KA, Cox LG (2012) Ultraviolet irradiation effects incorporation of nitrate and nitrite nitrogen into aquatic natural organic matter. J Environ Qual 41(3):865–881
34. Rosenkranz HS, Mermelstein R (1985) The genotoxicity, metabolism and carcinogenicity of nitrated polycyclic aromatic hydrocarbons. J Environ Sci Health Part C Environ Carcinog Rev 3(2):221–272
35. Wollin KM, Dieter HH (2005) Toxicological guidelines for monocyclic nitro-, amino- and aminonitroaromatics, nitramines, and nitrate esters in drinking water. Arch Environ Contam Toxicol 49(1):18–26
36. Vughs D et al (2016) Application of effect-directed analysis to identify mutagenic nitrogenous disinfection by-products of advanced oxidation drinking water treatment. Environ Sci Pollut Res:1–14
37. Mons MN et al (2013) Use of the threshold of toxicological concern (TTC) approach for deriving target values for drinking water contaminants. Water Res 47(4):1666–1678
38. Martijn BJ et al (2016) Development of a 4-NQO toxic equivalency factor (TEF) approach to enable a preliminary risk assessment of unknown genotoxic compounds detected by the Ames II test in UV/H2O2 water treatment samples. Chemosphere 144:338–345
39. Bazri MM et al (2016) Impact of anionic ion exchange resins on NOM fractions: effect on N-DBPs and C-DBPs precursors. Chemosphere 144:1988–1995
40. Penders EJM et al (2012) Genotoxicity testing of samples generated during UV/H2O 2 treatment of surface water for the production of drinking water using the Ames test in vitro and the comet assay and the SCE test in vivo. J Water Supply Res Technol AQUA 61(7):435–445
41. Martijn AJ, Kruithof JC (2012) UV and UV/H 2O 2 treatment: the silver bullet for by-product and genotoxicity formation in water production. Ozone Sci Eng 34(2):92–100

Role of Ultraviolet Disinfection in the Prevention of Surgical Site Infections

21

Sarah Simmons, Charles Dale, James Holt, Katie Velasquez, and Mark Stibich

Abstract

The role of the environment in surgical site infections is surprisingly understudied. UV disinfection holds promise for reducing the level of contamination in operating rooms and thereby lowering the risk of infection for patients. Issues such as the frequency, amount and locations for UV disinfection to have an impact on the risk of surgical site infection are recently emerging in the literature. As technologies and knowledge improve, UV disinfection will have a role to play in operating rooms in the future.

Keywords

Surgical site infections • Environmental hygiene • Ultraviolet disinfection • Operating rooms

21.1 Introduction

In this chapter, the role of ultraviolet light disinfection in preventing surgical site infections is presented. The root cause of surgical site infections, specifically the role of environmental contamination serving as a fomite in the surgical theater, will be discussed. We will also address the safety considerations for implementing UV disinfection, and review the differences in the currently available technologies. Finally, we will review the emerging evidence correlating enhanced disinfection in the surgical theater with decreases in infection rates and make recommendations for additional research on the topic.

21.1.1 Burden of Surgical Site Infections

Approximately 51.4 million inpatient surgical procedures are performed annually in the

S. Simmons • C. Dale • J. Holt • K. Velasquez
Xenex Disinfection Services, LLC, San Antonio, TX, USA

M. Stibich (✉)
Xenex Disinfection Services, LLC, San Antonio, TX, USA

University of Texas MD Anderson Cancer Center, Houston, TX, USA
e-mail: stibich@gmail.com

S. Ahmad (ed.), *Ultraviolet Light in Human Health, Diseases and Environment*, Advances in Experimental Medicine and Biology 996, https://doi.org/10.1007/978-3-319-56017-5_21

United States alone [1]. Of these patients undergoing inpatient procedures, approximately 1.9% (976,000 patients) develop an infection afterward [2]. Depending on the surgical procedure being performed, this infection risk can be higher or lower [3]. Including procedures performed at outpatient and ambulatory surgery centers would further increase the annual number of infections in the U.S. The mortality rate associated with contracting a surgical site infection (SSI) is 3%, with 75% of associated deaths being directly attributed with the infection at the surgical site [4]. Table 21.1 shows the annual case load for select surgical procedures, along with projected infections and deaths based on reported data. It should be noted that disability, morbidity and other forms of suffering are not presented in this table.

A prevalence study conducted in 2011 found that SSIs are the most common hospital associated infection (HAI), representing 22% of all reported cases [5]. Patients who develop a surgical site infection will spend, on average, 12.1 additional days in the hospital [6].

These additional days in the hospital lead to increased costs for both the patient and the hospital providing the care. The average cost attributed to an SSI is estimated to range from $11,874 to $34,670 [7]. This estimate is an aggregate of costs for all surgical procedures; more invasive procedures such at spinal fusions and vascular surgeries can have substantially higher costs and can exceed $100,000 per case [8].

21.1.2 Causes of Surgical Site Infections

The proximal source of the contamination that results in a SSI is often impossible to identify. During the preparation for and performance of a surgery, there are multiple opportunities for pathogenic organisms to enter the surgical wound and cause infection. The most commonly attributed sources of pathogenic organisms are outlined below:

21.1.2.1 Non-sterile Instruments

The instruments used during surgical procedures are reprocessed between each patient to remove blood, tissue, and microbiologic contamination to assure sterility before use on the next patient. Failures in these processes can lead to the introduction of pathogens into the surgical wound. The first step in the decontamination process is to remove blood and tissue from the instruments. Any residual material left behind can impede the sterilization process and provide a haven for pathogens. After complete removal of residual materials, the instruments are sterilized, typically with a steam sterilizer. Steam sterilizers use steam and pressure to sterilize instruments. If appropriate levels of steam and pressure are not achieved throughout the sterilization cycle, pathogens (especially spores) can remain on the instruments. The final step in preventing instrument contamination is to assure that they are stored in a manner and place that prevents recon-

Table 21.1 Number of surgical procedures annually by procedure type

Procedure	Surgical volume	Projected infections	Projected deaths
Coronary artery bypass graft	395,000	8578	257
Total knee replacement	719,000	7200	216
Total hip replacement	332,000	4669	140
Reduction of fracture	671,000	11,044	331
Hysterectomy	498,000	8847	265
Cesarean section	1,300,000	24,349	730
Excision of large intestine	247,000	14,356	430

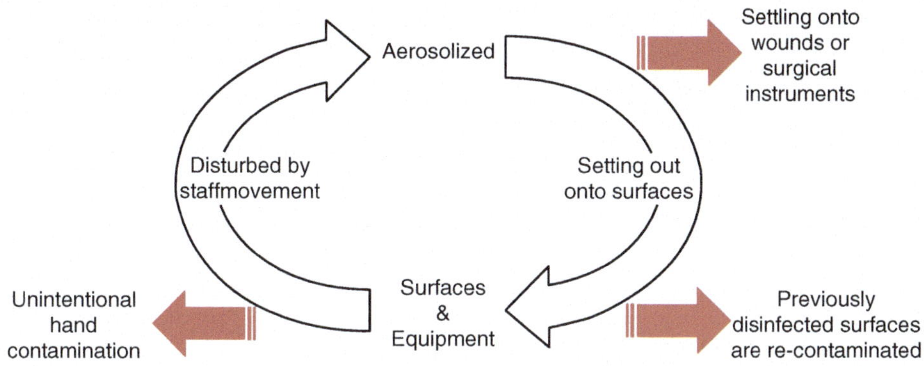

Fig. 21.1 Proposed interaction between surface contamination and airborne contamination for causing surgical site infections

tamination. This is accomplished by assuring that the integrity of the instrument packaging is maintained. The above process can be challenging as more and more surgical instrumentation, such as endoscopes, are becoming more complex devices with imbedded technology, which requires specialized training and processes for sterilization.

21.1.2.2 Patient Factors

The patient themselves can be the source of the organisms that cause infection. Common skin commensals such as *Staphylococcus* spp. can cause infections if skin integrity is compromised at the incision site. Patients with comorbid conditions such as diabetes, obesity, and heart disease are at higher risk of developing an infection. Additionally, the patient's compliance with post-operative wound care measures will impact the risk of developing infection.

21.1.2.3 Environmental Contamination

A contaminated hospital environment can contribute to the transmission of pathogens to patients [9]. Environmental transmission can occur from direct contact with the environment (air or surface) or indirectly from hands that were contaminated by the environment [10]. This interaction of environment and transmission risk could be further complicated in the operating theater, where constant movement of staff members causes air turbulence that disturbs pathogens present on surfaces, causing them to aerosolize. Once the pathogens are in the air, they can resettle onto sterile surgical instruments, previously cleaned surfaces, or even the open surgical wound. See Fig. 21.1.

21.1.2.4 Responsible Pathogens

Magill and colleagues reported on the pathogens associated with 110 surgical site infections identified as part of a multi-state prevalence study. Table 21.2 shows the results of this survey [5]. Additionally, Kramer and colleagues conducted a systematic review which assessed how long pathogenic organisms could persist on inanimate surfaces. This data is presented in Table 21.3 [11].

21.1.3 Effectiveness of Installed UVGI Devices

Ultraviolet germicidal irradiation (UVGI) is a technology that has been used to reduce the microbial contamination in operation rooms (ORs). When installed in ORs, UVGI has proven to be effective in air disinfection. Several studies have shown that the use of a UV device can produce ultraclean (<10 CFU/m^3), or nearly ultraclean, air [12–14]. This is the same level of air quality produced by HEPA air filters [15]. Kowalski suggests that the combination of MERV 13–15 filters and UVGI are equivalent to the effectiveness of HEPA filtration systems with less cost [16]. The same suggestion seems to hold

Table 21.2 Frequency of pathogens attributed to surgical site infections

Pathogen	Number (percent)
Staphylococcus aureus	17 (15.5)
Enterococcus species	16 (14.5)
Klebsiella pneumoniae or *K. oxytoca*	15 (13.6)
Escherichia coli	14 (12.7)
Streptococcus species	8 (7.3)
Pseudomonas aeruginosa	7 (6.4)
Coagulase-negative *Staphylococcus* species	7 (6.4
Enterobacter species	5 (4.5)
Proteus mirabilis	5 (4.5)
Bacteroides species	5 (4.5)
Candida species	3 (2.7)
Acinetobacter baumannii	2 (1.8)
Haemophilus species	2 (1.8)
Peptostreptococcus species	2 (1.8)
Clostridium species other than *C. difficile*	2 (1.8)
Citrobacter species	1 (0.9)
Prevotella species	1 (0.9)
Morganella morganii	1 (0.9)
Other organisms	6 (5.5)
Total	110 (100)

Modified from Magill 2014 [5]

Table 21.3 Persistence of pathogenic organism on inanimate surfaces commonly associated with surgical site infections

Pathogen	Persistence
Staphylococcus aureus	7 days – 7 months
Enterococcus species	5 days – 4 months
Klebsiella pneumoniae or *K. oxytoca*	2 h – >30 months
Escherichia coli	1.5 h – 16 months
Streptococcus species	1 day – 6.5 months
Pseudomonas aeruginosa	6 h – 16 months
Acinetobacter baumannii	3 days – 5 months
Haemophilus species	12 days

Modified from Kramer 2006 [11]

true for laminar air systems as well. In a comparison study done in 1989, the UV lighting system being tested not only worked just as well, if not better, than the laminar air system, but also cost 34 times less [17]. This result has been replicated in multiple studies that have shown that UVGI can reduce airborne bacteria values to a greater degree than laminar air systems [14, 16–18].

In considering installed UVGI as an overall disinfection measure, the system's effect on infection risk must also be taken into account. This subject is closely related to the reduction of airborne microbes. Going as far back as Joseph Lister it has been believed that airborne bacteria represent a significant source of infection, especially in the ORs. [19]. Infection occurs when microbes in the air settle in the operating room, contaminating the wound, the patient, the hospital personnel, and vital medical equipment [16]. It then stands to reason that the fewer bacteria in the room, including in the air, the less risk of infection there is for a patient. One study concluded that the UVGI device was able to disinfect the patients' wounds and possibly operating instruments [14]. The authors stated that this disinfection "negate[d] the argument about the relative effect on air counts. Laminar flow would have to provide considerably cleaner air to produce equally clean wounds." These clean wounds combined with the clean air are the basis of what allows UVGI devices to decrease the risk of infection.

There are many reports of UVGI reducing infection risk in the ORs. An orthopedic study following 5980 joint replacements reported that the odds of infection were 3.1 times greater for patients who had not been operated on under any UV light [17]. The same study reported an infection rate decreased from 1.77% to 0.57%. Others have reported similar findings of reduced infection rate. A study focusing on infection after cardiac operations revealed how using UVC light during operations led to the hospital's overall infection rates being significantly lower than the national averages in the most important risk categories [CDC National Nosocomial Infection Surveillance system, 18]. In 1936 at Duke University Hospital, Hart tried UVGI light after an outbreak of infections in the OR. The infection rate dropped from 11.62% to 0.24%, causing Hart's colleagues to also adopt the practice [20]. UVGI has been recommended as an important tool for operating room personnel to use in order to reduce infection [17, 21]. However, much of the data on the effectiveness of UVGI in ORs are dated and it is not clear the impact that installed UVGI would have for procedures being conducted under current infection control procedures.

21.1.3.1 Installed UVGI Safety

Kraissl et al. took it a step further and not only researched the effectiveness of UVGI on infectious bacteria, but also investigated the safety of UVGI in regards to the patient [22]. The researchers concluded that there is a danger to the exposed visceral tissue of the patient, if a threshold intensity of light is exceeded. However, they also found that there was significant bacterial killing even when using light intensities well below the damaging threshold. This theme of radiation in moderation persists in all safety matters pertaining to UVGI.

UVGI has been proven to be safe provided the proper measures are taken. It is recommended that light intensity be limited in order to protect the patient and hospital personnel in the room [22, 23]. Lidwell stated that the intensity of light should be kept between 25 and 30 $\mu W/cm^2$, but intensities up to 300 $\mu W/cm^2$ did not produce hazardous results [23]. The light must also be placed so that there is no dangerous exposure to the staff, while still allowing for proper irradiation of the room. Some systems are placed above and parallel to the patient, forming a type of barrier that will deactivate bacteria in the air that would fall on the patient. The height of the system keeps the operation room's personnel safe by preventing direct exposure of the UV light. Even with low intensities of light placed in safe positions, operating room staff should follow the safety precautions and wear protective clothing. Items such as disposable caps, drapes, plastic goggles, face masks, and surgical gloves, can greatly reduce the transmission of UVC light to personnel in the room [12]. Studies on staff who took protective measures showed no harmful effects [18, 24]. However, increases in light intensity and noncompliance with safety precautions can lead to injuries such as erythema, photosensitivity, immune system damage, and even cancer [25]. Eye injury in particular is a hazard when the proper face wear is not used [26]. Eye damage includes damage to the cornea and conjunctiva that can lead to temporary blindness, photosensitivity, benign growth, and corneal degeneration [25].

Hospital personnel have continually cited the uncomfortable nature of personal protective equipment as the main reason they do not utilize it [13, 14, 26, 27]. Wearing heavy protective clothing has proven to be too hot for personnel to work in regularly [14]. Other reasons for noncompliance included the lack of necessary supplies, training, and time, as well as increased work difficulty. Due to this noncompliance, there have been cases of basal cell carcinoma, melanoma, and actinic keratosis in operating room personnel [27]. It could be possible for greater compliance to be achieved if safety precautions were less inhibitive for staff. The future of fixed system UVGI may rest on this, as a lack of compliance and an increase in injuries may lead to the abandonment of the system [26].

Beginning in the late 2000s, portable UV systems have been used routinely in ORs for nightly and, in some situations, between cases [28–30]. These mobile devices allow the operator to place them in the room and exit before any human exposure can occur. This removes the need for heavy or difficult protective equipment as well as the cost of installation. These mobile devices may be an effective alternative to the fixed UVGI system.

21.1.4 Portable Room Disinfection UV Technologies

Portable UV technologies available for disinfecting operating rooms must meet the basic requirements of being safe to use, easy to operate, and effective at reducing the number of pathogens on every possible surface. Personal safety is not typically an issue, as (1) germicidal UV light cannot pass through windows or walls [25], and (2) all devices have a mechanism for automatic shutoff if a person enters the room, and are thus considered safe to operate under normal conditions.

Currently, two types of technologies that meet the aforementioned requirements are commonly used in ORs: those using low pressure mercury lamps and others which employ pulsed xenon lamps. Both have been shown to be effective at reducing a large number of pathogens on the surfaces [30–33], and the incidence of infections in the in-patient environment [34–40]. However, only a pulsed xenon device has been demonstrated to reduce SSIs [28, 29].

The main differences between low pressure mercury and pulsed xenon technologies lie in their spectral output, intensity, and operational modes. In the UV range, low pressure mercury lamps produce a narrow spectrum output that is centered at 253.7 nm (Fig. 21.2), while pulsed xenon lamps emit wavelengths covering the entire germicidal range of 200–320 nm (Fig. 21.3). Pulsed xenon produces intense pulses that last for microseconds while low pressure mercury produces lower intensity light but operates in a constant-on mode that allows for effective doses to be delivered over time (Fig. 21.4). The operational differences between these device types may account for the contrasting ways in which they are utilized. For example, pulsed xenon devices have shorter cycle times when used in the OR setting (8–16 min) compared to

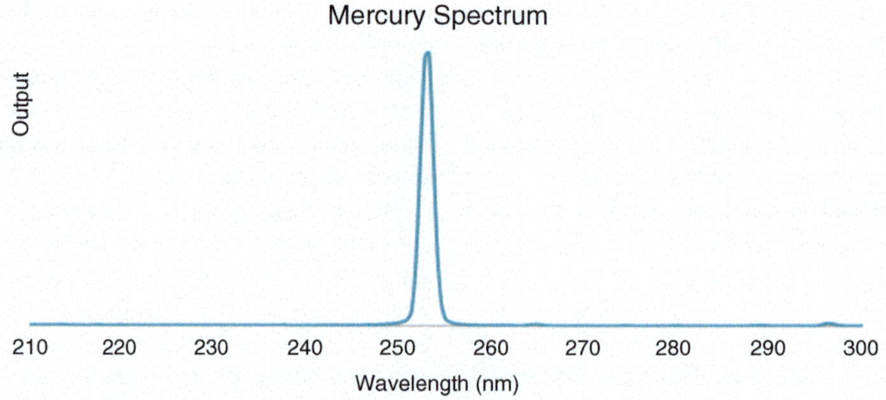

Fig. 21.2 Spectral output of mercury lamp in germicidal UV range

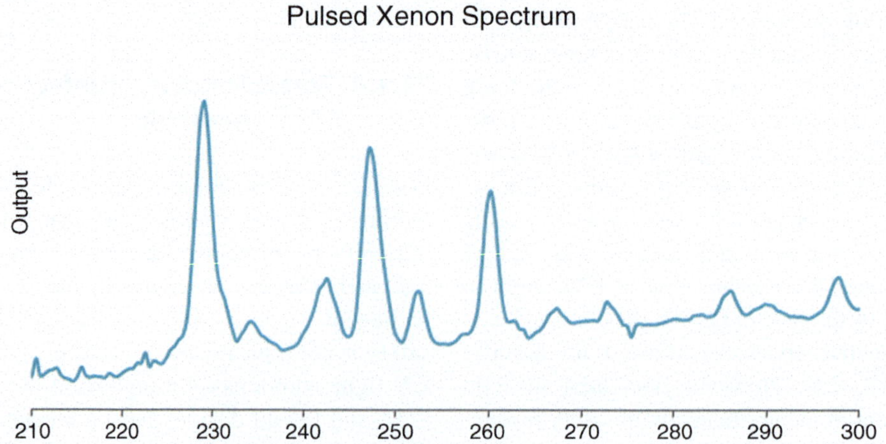

Fig. 21.3 Spectral output of xenon lamp in germicidal UV range

Fig. 21.4 Difference in operational intensity and lamp on-time between mercury and xenon lamps

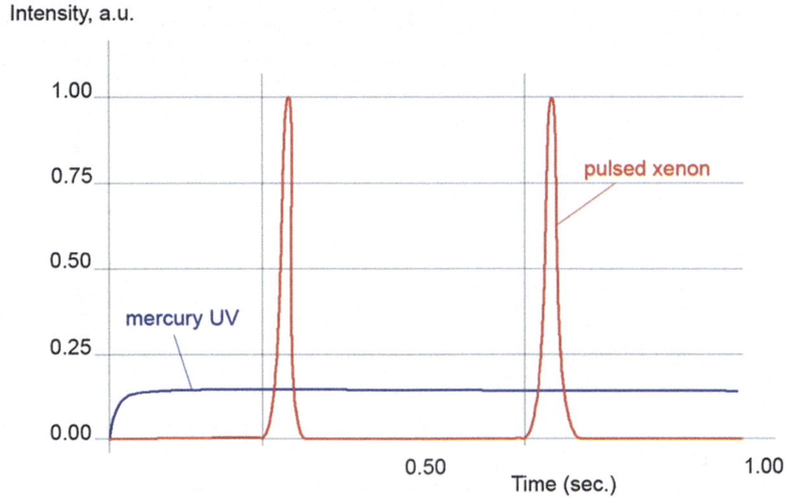

low pressure mercury (OR times are not specified, but a typical patient room cycle time is 45 min).

21.1.5 Materials Damage

In addition to deactivating microbes present on surfaces in the ORs, UV light also interacts with the objects on which these microorganisms reside. When UV is incident upon a surface, one of three things happens: the light is transmitted, absorbed, or reflected [25]. Because UV is not transmitted through most solid objects and there is relatively little reflection, most is absorbed. This absorption can cause photodegradation (the molecular changes due to light) that result in an alteration to the color, texture or mechanical properties of the object. In the OR's setting, this change is primarily evident in the yellowing of white plastics and fading of lighter colored fabrics and most metal objects remain unaffected.

While exposure to any UV device will change the material properties of a susceptible object to some extent, variables such as distance, exposure time, and spectral output make it difficult to predict the effect. More research is needed to fully understand the material's compatibility of portable UV devices commonly used in the ORs.

21.1.6 Use of Portable UV, Cycle Times and Positioning

The aforementioned portable UV disinfection devices are currently deployed in over one hundred OR settings. Two clinical studies demonstrate great success when used following nightly, standard terminal cleaning practices [28, 29]. These devices can be wheeled into a room, plugged into standard electrical outlets, and then set in a fixed position that is proximal to high-touch equipment within. Following completion of the final disinfection cycle, these devices can be moved around the facility to the next area requiring disinfection. Disinfection is made possible by onboard germicidal lamps that contain either mercury vapor or xenon gas. Although different, both technologies have been found to be effective at decreasing environmental bioburden in patient care areas [41, 42].

Regardless of what technology is used, UV disinfection efficacy is highly dependent on the distance between the lamp and the surface being targeted. The propagation of light intensity decreases exponentially with increasing distance from the lamp, so proximity to areas being disinfected will require significantly shorter cycle times. Put simply, doubling the distance between the lamp and the target will quadruple the origi-

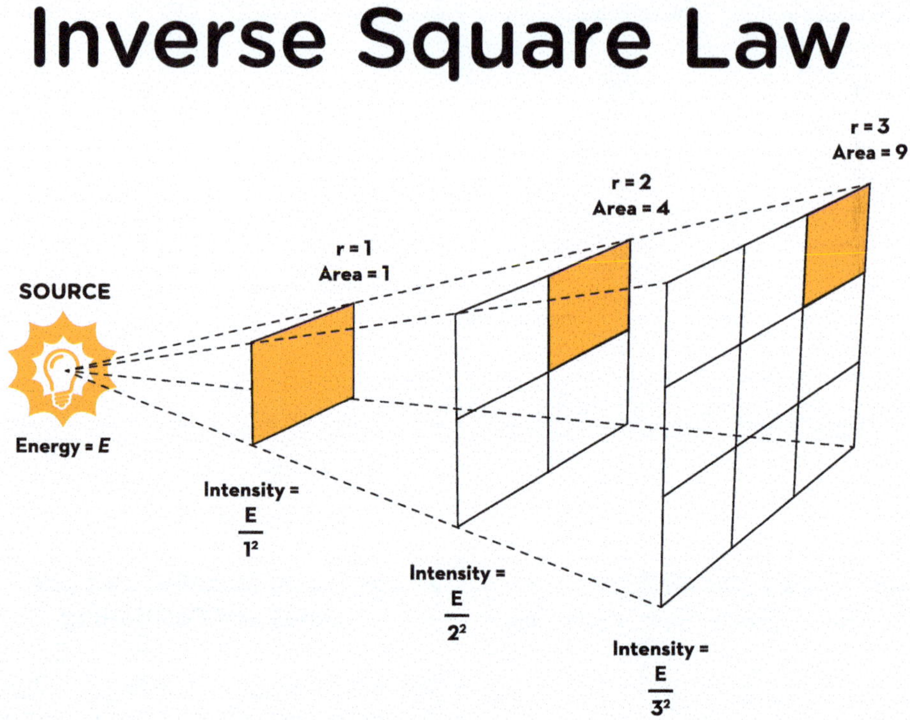

Fig. 21.5 Visual propagation of light following the inverse square law

nal time required for disinfection (See Fig. 21.5). Therefore, if it takes 5 min to disinfect a target 2 meters away, it should take approximately 20 min to produce the same amount of germicidal energy at 4 meters. A publication by Nerandzic and colleagues explores the impact of distance on UV efficacy against both methicillin resistant *S. aureus* (MRSA) and *C. difficile* spores in the laboratory setting [41].

In addition to distance, the reliance on UV reflection to reach targeted areas should also be considered. Common hospital materials are poor at reflecting germicidal UV, with wall paint and linen curtain material reflecting less than 25% of incoming light [25]. Multiple studies have confirmed that reflected light is significantly less effective than direct light at eliminating pathogens when considering the same disinfection time [32, 43]. For these reasons, disinfection will always be best when surfaces are in close proximity and within direct line of site of the lamp.

When considering physical limitations alone, the fastest UV room disinfection would consist of multiple positions and minimal distances from all target areas. However, because user intervention is required for every additional position implemented, this can add burden for the person performing the terminal cleaning. Considering this, two strategies are available for the OR; those using one position, and those using multiple (2–3) positions. Table 21.4 summarizes the pros and cons of each strategy.

One-position devices require minimal user assistance, but require longer periods to disinfect. One manufacturer implementing this strategy uses UV sensors on their devices to detect a set UV germicidal dose [44]. During the disinfection cycle, UV light reflects around the room, and some returns to the sensors. Once the sensors are saturated, the device will consider the room disinfected, and shut itself off. Depending on where the device is placed, and thus the amount of UV sensor activity, cycle times can vary considerably. While not OR-specific, publications report an average median cycle time of 45 min in acute care patient rooms [44, 45]. Despite longer

Table 21.4 Assessing the pros and cons of one versus multiple position UV disinfection

Multiple position devices	One position devices
Pros	Pros
Fast disinfection time	No repositioning required
Known cycle time	Cons
Clinical outcome studies in the OR	Longer disinfection time
Cons	Unknown cycle time
User repositioning required	No known clinical outcomes in the OR

cycles, housekeepers are free to perform other activities such as manual cleaning of other OR suites while UV disinfection is taking place. A handheld tablet tracks the progress of the disinfection taking place.

Multiple position devices are more time efficient, but require some user repositioning. Rather than measuring reflected light, multiple positions allow these systems to rely on direct line of site to disinfect. For this reason, cycle times are known for objects that are within specific distances of the devices. Several publications reporting reductions in SSIs following UV disinfection interventions required only two 5-min cycle times on either side of the OR table to fully disinfect high-touch surfaces within the room [28, 29]. When considering the time to reposition this system, disinfection can be completed in 15 min or less using the multiple position strategy. In addition to the success as an adjunct to terminal cleaning practices, UV disinfection might be a consideration for between case cleaning practices, in particular for quick disinfections when moving from dirty to clean procedures in the same suite.

21.1.7 Evidence for Benefit of Terminal UV Disinfection

Current literature shows that both cleaning and disinfection of the OR environment may be inadequate. An observational study examining OR cleaning found that only 25% (237/946) of fluorescent UV markers were removed from equipment surfaces following terminal cleaning [46]. In another study, only 47% of UV markers (284/600) were removed during the terminal cleaning process [47]. When air and surface microbial cultures were obtained from UV marker sites prior to surgical cases the following morning, 16.6% of surfaces remained contaminated with potentially infectious organisms such as *Pseudomonas spp.*, *Acinetobacter spp.*, *Klebsiella spp.*, and *Enterococcus spp.* [47].

Failure of disinfection practices leaves a potential risk of infection transmission from contaminated surfaces [48, 49]. As described earlier, this transmission risk is exceptionally high in the perioperative setting. Given the high volume of worker traffic, there are many opportunities for transmission between the susceptible patients, hands of healthcare workers, and environmental contamination in air and on surfaces. Multiple publications have reported substantial transfer of bacterial species from the anesthesia work area to intravenous stopcock sets [50, 51]. Furthermore, a recent study confirms that high touch areas of the operating room harbor significant bacterial contamination [52], suggesting greater attention should be paid to disinfecting these areas.

Mobile UV disinfection has demonstrated efficacy beyond what is possible by manual chemical disinfection alone, and can serve as an additional measure to reduce residual contamination. Data on one UV device have been collected from several ORs. For one facility, mean heterotrophic plate count for high-touch surfaces after manual cleaning was 2.73 colony forming units (CFUs) per 25 cm^2 [53]. Following a manual clean plus mobile UV disinfection, mean plate counts decreased 62% to an average of 1.05 CFUs per sample ($p < 0.001$). When comparing contamination levels for select surfaces, researchers determined a 64%, 87% and 94% improvement for the anesthesia cart, OR light and OR table, respectively. In a second study, quick cleaning plus UV disinfection resulted in a 55% and 81% reduction in positive surface cultures and overall bioburden, respectively [54]. UV disinfection also decreased air contamination by 46% during surgical cases when used for between case cleaning, and 100% following terminal cleaning practices [54]. Beyond the OR environment, the efficacy of UV disinfection has been studied in the acute care setting [30, 32, 33, 41–44, 55–60]. Several studies report exceptional decreases in both MRSA and vancomycin resistant *Enterococci* (VRE) contamination following terminal UV disinfection practices [59, 60]. Although the recovery of specific pathogens on OR surfaces is difficult, the laboratory efficacy of UV disinfection against common pathogens has been evaluated, with exceptional efficacy at 1 meter in as short as 5 min for select species [61].

Improvements in SSI rates following terminal UV disinfection interventions have now been published, providing additional evidence that enhanced cleaning with UV is thorough enough to remove exogenous sources of infection from the inadequately cleaned environment. In these studies, a baseline period that involves only manual disinfection for nightly terminal cleans is compared to interventions in which nightly UV disinfection is added in addition to the baseline procedure. Following this disinfection procedure, and as part of a bundled approach including other interventions, one facility reduced the incidence of total-hip and total-knee infections from 7 out of 544 procedures down to 0 out of 585 procedures [29]. In a second study evaluating terminal UV disinfection, SSIs following Class I (clean) procedures were reduced by 46%, contributing to 23 fewer infections over the 21 month intervention period [28]. Infections associated with clean-contaminated Class II procedures did not decrease during the intervention. Class I procedures involve clean incisions, so the wound site has minimal contamination prior to operation. These infections are more likely to be due to environmental transmission routes than infections associated with clean-contaminated Class II procedures, since contamination of the surgical site is already present at the time of the incision in these cases.

Evidence suggests that the risk of surgical site infections caused from the OR environment can be minimized by using terminal UV disinfection. Substantial evidence exists showing the role of environment in SSIs, and the ability of UV disinfection to provide an improvement beyond the capabilities of standard manual cleaning/disinfection. While quasi-experimental studies attribute reductions in SSIs following this application, future research into molecular epidemiology that maps the clonal spread of pathogens from surfaces to patients could provide additional insight into specific transmission dynamics [62].

21.2 Conclusion

UV disinfection holds great promise for improving the safety of the operating room environment. Additional research and improvements in available UV technologies should provide practical, operational solutions for ORs. As reimbursement changes further incentivize reductions in surgical site infection rates, investments in UV technologies should not only make financial sense, but also provide improved outcomes to patients.

References

1. Prevention, C.f.D.C.a. National Center for Health Statistics (2016) July 8, 2016]. Available from: http://www.cdc.gov/nchs/fastats/inpatient-surgery.htm
2. Centers for Disease Control and Prevention, N.H.S.N., Surgical Site Infection (SSI) Event Procedure Associated Module. (2016)
3. Mu Y et al (2011) Improving risk-adjusted measures of surgical site infection for the national healthcare safety network. Infect Control Hosp Epidemiol 32(10):970–986
4. Awad SS (2012) Adherence to surgical care improvement project measures and post-operative surgical site infections. Surg Infect 13(4):234–237
5. Magill SS et al (2014) Multistate point-prevalence survey of health care-associated infections. N Engl J Med 370(13):1198–1208
6. Brandon Savage MDCMO, et al (2011) The cost of healthcare associated infections measured in lives, reputations and dollars, in JAMA 1–8
7. Scott RD, Douglas R (2009) The direct medical costs of healthcare-associated infections in US hospitals and the benefits of prevention. Centers for Disease Control and Prevention, Atlanta
8. Kirkland KB et al (1999) The impact of surgical-site infections in the 1990s: attributable mortality, excess length of hospitalization, and extra costs. Infect Control Hosp Epidemiol 20(11):725–730
9. Chemaly RF et al (2014) The role of the healthcare environment in the spread of multidrug-resistant organisms: update on current best practices for containment. Therapeutic Adv Infect Dis doi:2049936114543287

10. Yezli S, Barbut F, Otter JA (2014) Surface contamination in operating rooms: a risk for transmission of pathogens? Surg Infect 15(6):694–699
11. Kramer A, Schwebke I, Kampf G (2006) How long do nosocomial pathogens persist on inanimate surfaces? A systematic review. BMC Infect Dis 8(December):1–8
12. Berg M, Bergman BR, Hoborn J (1989) Shortwave ultraviolet radiation in operating rooms. J Bone Joint Surg Br 71(3):483–485
13. Berg-Perier M, Cederblad A, Persson U (1992) Ultraviolet radiation and ultra-clean air enclosures in operating rooms. UV-protection, economy, and comfort. J Arthroplast 7(4):457–463
14. Taylor GJ, Bannister GC, Leeming JP (1995) Wound disinfection with ultraviolet radiation. J Hosp Infect 30(2):85–93
15. Dharan S, Pittet D (2002) Environmental controls in operating theatres. J Hosp Infect 51(2):79–84
16. Kowalski W (2008) UVGI for hospital applications. Int Ultraviolet Assoc News 10(4):30–34
17. Ritter MA, Olberding EM, Malinzak RA (2007) Ultraviolet lighting during orthopaedic surgery and the rate of infection. J Bone Joint Surg Am 89(9):1935–1940
18. Brown IW Jr et al (1996) Toward further reducing wound infections in cardiac operations. Ann Thorac Surg 62(6):1783–1789
19. Goldner JL et al (1980) Ultraviolet light for the control of airborne bacteria in the operating room. Ann N Y Acad Sci 353:271–284
20. Reed NG (2010) The history of ultraviolet germicidal irradiation for air disinfection. Public health Rep (Washington DC: 1974) 125(1):15–27
21. Woodhall B, Neill RG, Dratz HM (1949) Ultraviolet radiation as an adjunct in the control of postoperative neurosurgical infection: II clinical experience 1938–1948. Ann Surg 129(6):820–824
22. Kraissl CJ, Cimiotti JG, Meleney FL (1940) Considerations in the use of ultraviolet radiation in operating rooms. Ann Surg 111(2):161–185
23. Lidwell OM (1994) Ultraviolet radiation and the control of airborne contamination in the operating room. J Hosp Infect 28(4):245–248
24. Hart D (1960) Bactericidal ultraviolet radiation in the operating room. Twenty-nine-year study for control of infections. J Am Med Assoc 172:1019–1028
25. Kowalski W (2009) Ultraviolet germicidal irradiation handbook: UVGI for air and surface disinfection. Springer, Berlin/Heidelberg, pp 287–311
26. Sliney D (2013) Balancing the risk of eye irritation from UV-C with infection from bioaerosols. Photochem Photobiol 89(4):770–776
27. Sylvain D, Tapp CL (2009) UV-C exposure and health effects in surgical suite personnel. U.S. Department of Health and Human Services, Centers for Disease Control and Prevention, Boston/Cincinnati
28. Catalanotti A et al (2016) Influence of pulsed-xenon ultraviolet light-based environmental disinfection on surgical site infections. Am J Infect Control 44(6):e99–e101
29. Fornwalt L, Ennis D, Stibich M (2016) Influence of a total joint infection control bundle on surgical site infection rates. Am J Infect Control 44(2):239–241
30. Hosein I et al (2016) Evaluation of a pulsed xenon ultraviolet light device for isolation room disinfection in a United Kingdom hospital. Am J Infect Control 44(9):e157–e161
31. Mahida N, Vaughan N, Boswell T (2013) First UK evaluation of an automated ultraviolet-C room decontamination device (Tru-D). J Hosp Infect 84(4):332–335
32. Rutala WA, Gergen MF, Weber DJ (2010) Room decontamination with UV radiation. Infect Control Hosp Epidemiol 31(10):1025–1029
33. Jinadatha C et al (2015) Can pulsed xenon ultraviolet light systems disinfect aerobic bacteria in the absence of manual disinfection? Am J Infect Control 43(4):415–417
34. Levin J et al (2013) The effect of portable pulsed xenon ultraviolet light after terminal cleaning on hospital-associated Clostridium difficile infection in a community hospital. Am J Infect Control 41(8):746–748
35. Simmons S et al (2013) Impact of a multi-hospital intervention utilising screening, hand hygiene education and pulsed xenon ultraviolet (PX-UV) on the rate of hospital associated meticillin resistant Staphylococcus Aureus infection. J Infect Prev 14(5):172–174
36. Haas JP et al (2014) Implementation and impact of ultraviolet environmental disinfection in an acute care setting. Am J Infect Control 42(6):586–590
37. Miller R et al (2015) Utilization and impact of a pulsed-xenon ultraviolet room disinfection system and multidisciplinary care team on Clostridium difficile in a long-term acute care facility. Am J Infect Control 43(12):1350–1353
38. Nagaraja A et al (2015) Clostridium difficile infections before and during use of ultraviolet disinfection. Am J Infect Control 43(9):940–945
39. Vianna PG et al (2016) Impact of pulsed xenon ultraviolet light on hospital-acquired infection rates in a community hospital. Am J Infect Control 44(3):299–303
40. Napolitano NA, Mahapatra T, Tang W (2015) The effectiveness of UV-C radiation for facility-wide environmental disinfection to reduce health care-acquired infections. Am J Infect Control 43(12):1342–1346
41. Nerandzic MM et al (2015) Evaluation of a pulsed xenon ultraviolet disinfection system for reduction of healthcare-associated pathogens in hospital rooms. Infect Control Hosp Epidemiol 36(2):192–197
42. Nerandzic MM et al (2010) Evaluation of an automated ultraviolet radiation device for decontamination of Clostridium difficile and other healthcare-associated pathogens in hospital rooms. BMC Infect Dis 10:197
43. Boyce JM, Havill NL, Moore BA (2011) Terminal decontamination of patient rooms using an automated mobile UV light unit. Infect Control Hosp Epidemiol 32(8):737–742

44. Anderson DJ et al (2013) Decontamination of targeted pathogens from patient rooms using an automated ultraviolet-C-emitting device. Infect Control Hosp Epidemiol 34(5):466–471
45. Nerandzic M et al (2010) Evaluation of an automated ultraviolet radiation device for decontamination of healthcare-associated pathogens in hospital rooms and on portable medical equipment. SHEA 2010 Decennial, Atlanta
46. Jefferson J et al (2011) A novel technique for identifying opportunities to improve environmental hygiene in the operating room. AORN J 93(3):358–364
47. Munoz-Price LS et al (2012) Decreasing operating room environmental pathogen contamination through improved cleaning practice. Infect Control Hosp Epidemiol 33(9):897–904
48. Dancer SJ (2009) The role of environmental cleaning in the control of hospital-acquired infection. J Hosp Infect 73(4):378–385
49. Otter Ja, Yezli S, French GL (2011) The role played by contaminated surfaces in the transmission of nosocomial pathogens. Infect Control Hosp Epidemiol 32(7):687–699
50. Loftus RW et al (2008) Transmission of pathogenic bacterial organisms in the anesthesia work area. Anesthesiology 109(3):399–407
51. Munoz-Price LS et al (2013) Interactions between anesthesiologists and the environment while providing anesthesia care in the operating room. Am J Infect Control 41(10):922–924
52. Link T et al (2016) Determining high touch areas in the operating room with levels of contamination. Am J Infect Control 44(11):1350–1355
53. Simmons SE et al (2013) Using pulsed xenon ultraviolet to decrease contamination in operating rooms during terminal cleaning. Am J Infect Control 41(6):S34–S35
54. Bruno-Murtha L, Fridman A, Osgood R, Mcallister J (2013) Decreasing operating room contamination of surfaces and air with pulsed xenon ultraviolet disinfection (PX-UVD). APIC 2013 Conference, Cambridge Health Alliance (Harvard Medical School Teaching Affiliate)
55. Sitzlar B et al (2012) Environmental decontamination with ultraviolet radiation to prevent recurrent Clostridium difficile infection in 2 roommates in a long-term care facility. Infect Control Hosp Epidemiol 33(5):534–536
56. Havill NL, Moore BA, Boyce JM (2012) Comparison of the microbiological efficacy of hydrogen peroxide vapor and ultraviolet light processes for room decontamination. Infect Control Hosp Epidemiol 33(5):507–512
57. Wong T et al (2015) Postdischarge decontamination of MRSA, VRE, and Clostridium difficile isolation rooms using 2 commercially available automated ultraviolet-C-emitting devices. Am J Infect Control 44(4):416–420
58. Ghantoji SS et al (2015) Non-inferiority of pulsed xenon ultraviolet light versus bleach versus for reducing environmental Clostridium difficile contamination on high-touch surfaces in Clostridium difficile isolation rooms. J Med Microbiol 64(2):191–194
59. Jinadatha C et al (2014) Evaluation of a pulsed-xenon ultraviolet room disinfection device for impact on contamination levels of methicillin-resistant Staphylococcus Aureus. BMC Infect Dis 14(1):187
60. Stibich M et al (2011) Evaluation of a pulsed-xenon ultraviolet room disinfection device for impact on hospital operations and microbial reduction. Infect Control Hosp Epidemiol 32(3):286–288
61. Stibich M, Stachowiak J (2016) The microbiological impact of pulsed xenon ultraviolet disinfection on resistant bacteria, bacterial spore and fungi and viruses. South Afr J Infect Dis 31(1):12–15
62. Chemaly RF et al (2014) The role of the healthcare environment in the spread of multidrug-resistant organisms: update on current best practices for containment. Ther Adv Infect Dis 2(3–4):79–90

UV Disinfection of Wastewater and Combined Sewer Overflows

22

John Gibson, Jennifer Drake, and Bryan Karney

Abstract

Municipal wastewater contains bacteria, viruses, and other pathogens that adversely affect the environment, human health, and economic activity. One way to mitigate these effects is a final disinfection step using ultraviolet light (UVL). The advantages of UVL disinfection, when compared to the more traditional chlorine, include no chlorinated by-products, no chemical residual, and relatively compact size. The design of most UV reactors is complex. It involves lamp selection, power supply design, optics, and hydraulics. In general, medium pressure lamps are more compact, powerful, and emit over a wider range of light than the more traditional low pressure lamps. Low pressure lamps, however, may be electrically more efficient. In UV disinfection, the fraction of surviving organisms (e.g. *E. coli*) will decrease exponentially with increasing UV dose. However, the level of disinfection that can be achieved is often limited by particle-associated organisms. Efforts to remove or reduce the effects of wastewater particles will often improve UV disinfection effectiveness. Regrowth, photoreactivation, or dark repair after UV exposure are sometimes cited as disadvantages of UV disinfection. Research is continuing in this area, however there is little evidence that human pathogens can photoreactivate in environmental conditions, at doses used in wastewater treatment. The UV disinfection of combined sewer overflows, a form of wet weather pollution, is challenging and remains largely at the research phase. Pre-treatment of combined sewer overflows (CSOs) with a cationic polymer to induce fast settling, and a low dose of alum to increase UV transmittance, has shown promise at the bench scale.

J. Gibson (✉) • J. Drake
Department of Civil Engineering,
University of Toronto, 35 St. George Street,
Toronto, ON M5S 1A4, Canada
e-mail: jh.gibson@mail.utoronto.ca; jenn.drake@utoronto.ca

B. Karney
Division of Environmental Engineering and Energy Systems, Department of Civil Engineering,
University of Toronto, 35 St. George Street,
Toronto, ON M5S 1A4, Canada
e-mail: karney@ecf.utoronto.ca

S. Ahmad (ed.), *Ultraviolet Light in Human Health, Diseases and Environment*, Advances in Experimental Medicine and Biology 996, https://doi.org/10.1007/978-3-319-56017-5_22

Keywords

Disinfection • Photoreactivation • Indicator organisms • Lamp selection • Alum

22.1 Introduction

Ultraviolet light (UVL) is a form of electromagnetic radiation with wavelengths from 10 to 400 nm. Electromagnetic radiation with shorter wavelengths is classified as X-rays. Longer wavelengths include visible light, starting with violet and proceeding to infrared. Because it has a shorter wavelength than visible light, UVL has more energy per photon, and is more likely to initiate chemical reactions.

The first source of UVL observed was from sunlight. UVL from sunlight ranges from 290 to 400 nm. The shorter wavelengths are filtered out by the earth's ozone layer in the atmosphere. UVL is broadly categorized according to wavelength as shown in Table 22.1. Disinfection occurs largely in the UV-C range.

The advent of the mercury vapor lamp in the early 1900s allowed the generation of localized UVL suitable for disinfection. The first use of UVL in water treatment was at the drinking water plant in Marseille, France in 1910 [23]. Using mercury vapor lamps, UVL is still used in drinking water disinfection today, mainly in Europe.

The most significant use of UVL is as a final disinfection step in the treatment of municipal wastewater, with more than a thousand installations. Historically, the goal of wastewater disinfection has been to protect human health and prevent the waterborne disease. The per capita production of wastewater is approximately 400 L/d, so that vast amounts of wastewater must be treated every day in cities around the world.

Table 22.1 Classifications of UV light

Type	Wavelength
UV-A	315–400 nm
UV-B	280–315 nm
UV-C	200–280 nm

22.2 Comparisons with Chlorine Disinfection

Adding chlorine is the predominant method of wastewater disinfection in North America. However, careful dose control is needed to ensure that the chlorine residual itself is not toxic to aquatic organisms. Some jurisdictions require dechlorination, by adding sulfur dioxide for example, before the water is discharged to the environment. This adds to the cost and complexity of disinfection. In addition, wastewater can contain ammonia and other nitrogen compounds, which form chloramines when reacting with chlorine. Chloramines are acutely toxic to aquatic organisms. Chlorination can produce potentially carcinogenic chlorinated disinfection by-products (DBPs), such as tri-halomethanes (THMs) or N-nitrsodimethylamine (NDMA) [29]. Finally, chlorination tends to be less effective against the protozoan pathogens Cryptosporidium and Giardia.

UV disinfection does not form potentially carcinogenic, regulated, chlorinated DBPs. At doses used in UV disinfection of wastewater there is little evidence of by-product formation, with the possible exception of parts per billion levels of aldehydes, such as formaldehyde [23]. Unlike chlorine, infectivity testing has demonstrated that UVL is effective in inactivating cryptosporidium [3].

22.3 UV Disinfection Equipment

The most common equipment used in wastewater treatment is based on mercury vapor lamps. They operate on the principle that mercury vapor and electron collisions result in the excitation of mercury. As the mercury atoms return to the ground state, light in the UV range is emitted. A small amount of mercury is introduced into a lamp usually containing argon carrier gas. Electrodes in the lamp create a plasma, which provides electrons [23]. Since regular glass is opaque to UVL, quartz sleeves are routinely used to contain and protect lamps. Many installations will have a UVL detector in the reactor to monitor effects of lamp aging and fouling.

There are two commercially available lamp types: low pressure (LP) and medium pressure (MP). In the LP lamp, the total carrier gas and mercury vapor pressure is in the range of 10^2–10^3 Pa. In general, LP lamps are monochromatic, emitting light primarily at 254 nm, and tend to produce a low light intensity per unit lamp length (0.3 W/cm). High-intensity LP lamps, based on a mercury-iridium amalgam, are also available. The output of these lamps is approximately double that of conventional LP lamps [29].

In contrast, MP lamps operate at total gas pressures of 10 to 30 MPa, produce polychromatic light, can be modulated, and produce high intensity per unit length (15 W/cm) making for more compact installations. The number of MP installations is rapidly increasing, especially for high flow applications where compactness is important. Lamp features are compared below, based on the analysis of Masschelein [23] and Tchobanoglous [29].

Low Pressure Lamps

- Relatively simple operation: no more complex or onerous power requirements than fluorescent lamps.
- Potential for high electrical efficiency: 85% of the light emitted is at the germicidal wavelength of 254 nm (Some light can be emitted at 185 nm, depending on the lamp sleeve material, though this may not be important to disinfection)
- Can be adapted to fit former chlorine contact chambers, in some cases.
- Monochromatic: single wavelength emission makes design, calibration, and monitoring easier.
- Long life: lamps routinely last 8000 h.
- Low intensity: makes for large installations; cleaning and maintenance of many quartz sleeves can be onerous. Rule of thumb: 40–60 lamps @ 65 W ea. per 150 m^3/h wastewater.
- Higher intensity types are available at the cost of reduced lamp life.
- Monochromatic: no potential benefits of other UVL wavelengths, which may affect proteins, for example.
- Lamps cannot be modulated: though wastewater flows are variable and lamp emission changes over time operate only in ON/OFF mode.

Medium Pressure Lamps

- High intensity makes for compact installation with few lamps to maintain. Suitable for high flows in wastewater treatment.
- Many systems have self-cleaning quartz sleeves.
- Polychromatic: emit across the UV disinfection spectrum from 200 to 380 nm; potential for more effective disinfection.
- Light intensity can be modulated.
- Operate at high temperature: wall temperature approximately 600 °C, containment, cooling, and preventing contact with water is critical.
- Shorter lamp life (≈4000 h) and higher replacement costs.
- Polychromatic: also emit in the visible and infrared range; not useful for disinfection. Visible light may promote algae growth. Approximately 40% of the light emitted is useful for disinfection.

Suppliers of UV equipment generally invest considerable effort into reactor design. Lamp location to maximize UVL exposure is far from straightforward due to the interaction of multiple light sources. In addition, mixing in the reactor is also critical to ensure good exposure of target organisms and prevent dead zones. Minimizing head loss is also a concern. Finally, power supply, lamp containment, and cleaning must be considered. UV reactor design is a highly specialized task.

Other lamp technologies include the pulsed energy broad-band xenon lamp and the narrow-band excimer lamp. The pulsed UV lamp uses a capacitor to store energy which, when released, produces a high temperature (i.e. 10,000 K) plasma. This results in the emission of light in the UV, visible, and infrared range. Excimer lamps produce monochromatic UV radiation at 172, 222 or 308 nm. An excimer is an excited gas dimer [29].

22.4 Indicator Organisms

Certain human pathogens, such as viruses, are difficult to enumerate in wastewater samples. This makes them a poor choice for regulations, which require large numbers of tests. On the other hand, *E. coli* and fecal coliform are abundant in wastewater, and easily cultured in a way that can be quantified. For this reason, they are widely used as indicators of potential wastewater (i.e. sewage) contamination. *E. coli* is not a pathogen per se; it is an indicator that wastewater pathogens could be present. However, a *strain* of *E. coli*, known as O157:H7, is indeed a human pathogen. Culturing methods are imperfect; only a fraction of the bacteria present in the medium can be cultured. Nonetheless, these tests still form the basis of most regulations. As such, *E. coli* counts are an indicator, and not directly proportional to risk. In time, this approach may be replaced by molecular genetics methods.

22.5 Treatment Objectives

The goals for UV disinfection will naturally depend on the type of water being treated, its intended use, and local and federal regulations. In the US, individual states determine disinfection objectives for municipal wastewater. This will depend on factors like the size of the watershed, population, drinking water source protection, and number of swimmers. Many jurisdictions require wastewater disinfection only in the summer months, when swimers might be present. The most common objective is 200 fecal coliforms per 100 mL as determined by the most probable number (MPN) method, but values range from 2.2 to 5000 MPN/100 mL [29].

Where wastewater is being reused for the irrigation of crops, the World Health Organization (WHO) recommends a maximum of 100 total coliforms per 100 mL in 80% of the samples collected. The California Code of Regulations, Title 22 requires a medium value of less than 2.2 total coliforms per 100 mL for wastewater re-used for spray irrigation. A value of 23 total coliforms per 100 mL is permitted once per month.

22.6 Disinfection Mechanisms

In order for a photochemical reaction to occur, the UVL must first be absorbed. In general, aromatic compounds are good absorbers of this light. In particular, the aromatic heterocyclic pyrimidine bases found in DNA are absorbers of UVL. These include cytosine and thymine in DNA, and cytosine and uracil in case of RNA. The resulting reaction produces pyrimidine dimers (i.e. two adjacent bases joined together) in the DNA which can interrupt the replication, and ultimately prevent reproduction. Peak disinfection effectiveness, and absorption of UVL by pyrimidines, both correspond to a wavelength of approximately 254 nm.

Proteins are made up of amino acids. Those amino acids containing aromatic groups, such as tryptophan and phenyl aniline, are potential absorbers of UVL. Since proteins are more abundant in the cell than DNA, photochemical changes in proteins may be important contributors to disinfection or inactivation [23]. There is evidence of improved inactivation of viruses under polychromatic MP light. This has been attributed to greater damage of viral proteins [5].

Other inactivation mechanisms are possible. It has been suggested that nitrates present in wastewater may initiate the formation of the powerfully oxidizing hydroxyl radical under polychromatic light [18].

22.7 Regrowth and Repair

Although UV disinfection has the advantage of forming relatively few disinfection by-products, the potential for regrowth and repair of pathogenic organisms are sometimes cited as a disadvantages. Since UVL from sunlight has been present since the formation of the earth, many organisms have evolved DNA repair mechanisms. Bacteria, such as *E. coli* and fecal coliform, have light-mediated DNA repair mechanism known as photorepair. This repair uses the enzyme photolyase, activated by visible light with wavelengths from 350 to 450 nm, to remove pyrimidine dimers [26]. A second mechanism, known as dark repair, may also occur, but it is believed to be less important [24, 28].

The growth of bacteria resulting from DNA repair is called "photoreactivation". However, the degree of photoreactivation of bacteria decreases as the UV dose is increased [21, 24]. There is evidence that photoreactivation is less than 1% in *E. coli* receiving a typical dose of 40 mJ/cm^2 from either LP or MP lamps [15]. Moreover, photoreactivation of *E. coli* in nature may be unlikely, due to the fact that the disinfecting effects of sunlight may outweigh photoreactivation effects [2]. From a regulatory perspective, compliance sampling is usually based on fecal coliform counts just after UVL exposure, before photoreactivation or regrowth can occur.

Not all organisms are capable of repair, including many pathogens. Pathogenic bacteria such as *Shigella dysenteriae* and *Salmonella typhimurium* show evidence of photoreactivation, although this ability decreases significantly with increasing UV dose [16]. No evidence of light or dark repair in the protozoan pathogen *Cryptosporidium parvum* has been observed [30]. Viruses have no know repair mechanisms [23]. The actual human health risks associated with photoreactivation are not clear.

In addition to DNA repair, undamaged bacteria, shielded by particles for example, may regrow after disinfection, possibly using the contents of UV damaged cells as nutrients [19]. Regrowth in the distribution system can be especially important when the water is reused.

22.8 Disinfection Kinetics

In order to determine the response characteristics of a particular pathogen in the laboratory, it if often desirable to apply a precise UV dose. The standard piece of equipment for this purpose is the collimated beam apparatus (Fig. 22.1). It consists of a UV light, a collimating tube (to ensure the light waves are pointed in the same direction), a mixer for water samples, and a detector to determine the intensity of UV light.

It is customary to call the product of light intensity, usually at 254 nm, and time of exposure the "UV dose". Light intensity is typically reported as mW/cm^2 and time is in seconds, so that common units for UV dose are mJ/cm^2 or J/m^2. Typical doses required in wastewater treatment range from 40 to 100 mJ/cm^2. However, this product is actually the "fluence", since "dose" in photo-chemistry is the light absorbed by a reactant of interest. Nonetheless, the term "UV dose" for the product of light intensity and time is widely used. It is analogous to the product of concentration and time (CT) used in classic chemical disinfection kinetics.

Using this definition of UV dose, it is possible to describe UV disinfection using the Chick-Watson model, used in chlorine disinfection. In this model, the ratio of surviving organisms decreases exponentially with increasing dose. There are two common deviations from this model: shouldering and tailing (Fig. 22.2).

Shouldering occurs at low dose where the number of viable organisms decreases little.

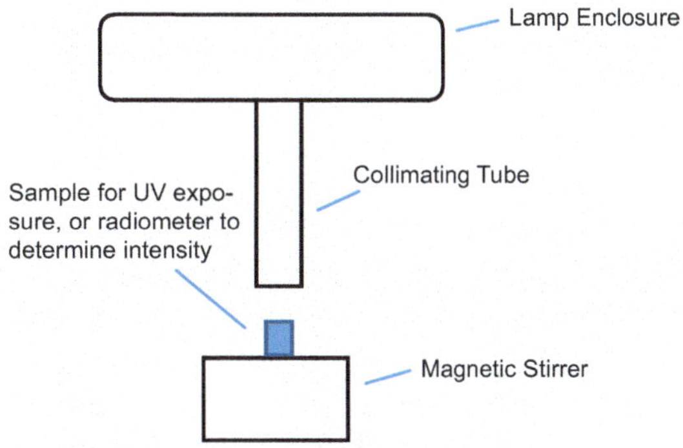

Fig. 22.1 Collimated beam apparatus

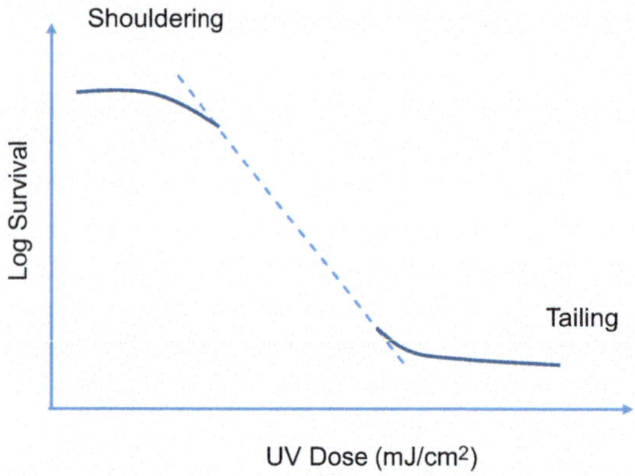

Fig. 22.2 Ideal and non-ideal UV disinfection behavior

Because the doses used in wastewater disinfection tend to be higher, tailing is more often a concern. This phenomenon is attributed to particles in the wastewater, or clumps of organisms. UVL has difficulty penetrating wastewater particles larger than approximately 10 μm in diameter, so that higher doses are required for inactivation [7, 22]. A Chick-Watson disinfection kinetic model that accounts wastewater particles is shown below [6]:

$$N(t) = N_o e^{-kD} + N_p / kD\left(1 - e^{-kD}\right) \quad (22.1)$$

where,

$N(t)$ = number of surviving organisms at time "t" (e.g. fecal coliform)
N_o = initial concentration of organisms
N_p = initial number of particles containing at least one organism
D = UV dose, the product of light intensity and time
k = disinfection rate constant.

Tailing can have a dramatic effect on UV efficacy. According to the above model, when the particle concentration (N_p) is high, the second term dominates and the decay of viable organisms is inversely proportional to the dose. In this case, reducing the number by 50% requires doubling the dose. In the absence of particles, the first term dominates, and ratio of surviving organisms decreases exponentially with increasing dose. As such, steps reduce the concentration of wastewater particles will often improve UV effectiveness. This includes efforts to reduce total suspended solids improving clarifier performance, or membrane treatment. Attempts have been made to disrupt wastewater particles before UV at the lab scale [11, 13]. Upsets in the wastewater treatment process can suddenly increase particle loading, and adversely affect UV effectiveness.

UVL will have no effect if it does not reach the organism of interest. The most important parameter for assessing this ability is the UV transmittance (UVT). This is the fraction of UVL transmitted through 1 cm of water. This measurement is often made by comparing light intensities on either side of a water sample in a quartz cuvette. However, light scattered by small particles will still be available for disinfection, though may not appear at a detector. For this reason, it is desirable to use an integrating sphere spectrophotometer when making UVT measurements. The UVT in secondary wastewater effluents is approximately 60%, but can vary widely. Dyes, humic substances, aromatic organic compounds, and metals, such as iron, in wastewater can all adversely affect UVT [29].

22.9 UV Disinfection of Combined Sewer Overflows

Unlike municipal wastewater treatment, where UV disinfection is widely practiced, the UV disinfection of combined sewer overflows remains largely at the research phase. However, CSOs can have serious adverse health and economic impacts, and are particularly unsuited for chlorine disinfection.

Many cities have sewer systems where rainwater and sanitary sewage are combined. This legacy reaches back hundreds of years, to the early days of wastewater treatment. Early sewer systems discharged directly to creeks, rivers or lakes. Without treatment of the wastewater, disease outbreaks were still common. Over time, pipes were constructed to intercept wastewater and direct it to wastewater treatment facilities.

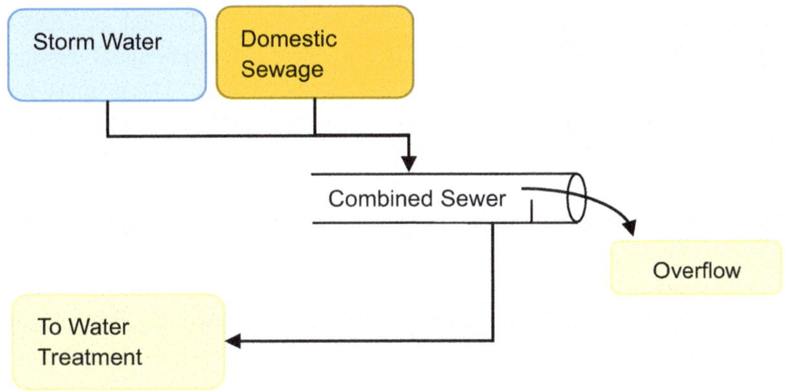

Fig. 22.3 Combined sewer overflow schematic [14]

Cost concerns often prevented interceptor pipes being large enough to collect both storm and sanitary flows during heavy rain [1]. As a result, during heavy rain combined sewers overflow a mixture of storm water and sewage to the environment (Fig. 22.3). Without combined sewer overflows (CSOs) to release excess water, there is a risk of basement flooding or wastewater treatment plant upsets.

From about the middle of the twentieth century many cities have required separate storm and sanitary sewers. However, the legacy of combined sewers remains. For example, 22 billion gallons of untreated wastewater were discharged into the Great Lakes in 2014 [8].

CSOs can increase health risks locally. A study of river sediments impacted by CSOs found high concentrations of fecal coliform, including *Streptococcus* and *Enterococcus* bacteria. Also, fifty percent of the samples contained the protozoan pathogen *Giardia*, and one of sixteen contained the protozoan pathogen *Cryptosporidium* [4]. Links have been made between wet weather pollution and degraded drinking water quality and increased risk of gastrointestinal illness [17, 25].

In addition to health effects, CSOs can have adverse regional socio-economic impacts. Many jurisdictions around the world use *E. coli* as an indicator of fecal contamination in recreational waters. High *E. coli* counts usually result in closure and loss of this resource.

Using chlorine to disinfect CSOs is especially problematic. Chlorine disinfection depends upon oxidizing enough organic matter to produce free chlorine residual. Low quality wastewaters, such as CSOs, increase chlorine demand considerably. Perhaps more importantly, low quality waters increase the chances of forming undesirable chlorinated disinfection by-products. Finally, chlorine disinfection performance is highly pH dependent, something over which there is little control in CSOs.

UV disinfection of CSOs is a challenge. The UVT of CSOs is as low as 30% and the total suspended solids (TSS) can reach 200 mg/L. Recent research has focused on rapid pretreatment of CSOs to reduce TSS. Li et al. [20] had success using high molecular weight cationic polymers to improve the removal of TSS in actual CSOs. Settling column tests were used to quantify the effects of different chemical additions for application in retention basins. Cationic polymers were shown to be effective for high-rate removal of TSS.

Exall and Marsalek [9] used jar tests (a bench-scale test) to study simulated CSOs. Reductions in TSS or dissolved organic carbon (DOC) were used as performance measures. The coagulants alum, ferric chloride, and poly-aluminum chloride (PACl) showed little dependence on the initial TSS, indicating sweep flocculation. All metal coagulants were effective at removing the organic, light-absorbing compounds (i.e. tannins). In contrast, polymer treatment did not reduce the DOC of the treated samples. Using polymers and coagulants together at 229 mg/L and 1 mg/L respectively, was effective in reducing both TSS and DOC.

Gibson et al. [12] used a similar approach with collected CSO samples using TSS reduction and

UVT increase as performance measures. Ferric chloride performed poorly in terms of UVT, reaching approximately 60%. In contrast, high alum doses (e.g. 100 mg/L) consistently resulted in a UVT of 80% or more. Additional mixing to increase flocculation did not increase performance. Cationic polymers acted quickly when compared with metal coagulants, requiring only a few seconds to produce large, fast settling flocs. Consistent with earlier results, polymers had little effect on dissolved, light absorbing organic compounds, reaching a maximum UVT of approximately 60% after settling. There was little evidence of pH depression in these tests, indicating these CSOs were well-buffered.

These results suggest that iron based coagulants are not suitable for use in UVL disinfection of CSOs due to low UVT, as well as propensity for lamp fouling [10, 27]. In contrast, the metal coagulant alum resulted in high UVT, low TSS, and was relatively insensitive to changes in the feed water. Although cationic polymers appear to act more quickly and are more suitable for high-rate processes, disadvantages of these polymers are their inability to remove light absorbing organic compounds. The use of alum and low polymer doses together appears to show promise as pre-treatment for the UV disinfection of CSOs.

22.10 Conclusions

Though most people block it from their minds, wherever there are people there is human waste, and wastewater. This wastewater must be collected by infrastructure that may be centuries old, and treated. Disinfection is an important part of the treatment process designed to prevent the spread of human disease. Since pathogens are so difficult to enumerate, this process is regulated by the ability to remove indicator organisms, such as *E. coli* or fecal coliform. High counts of these indicators in surface waters indicate a risk of pathogens. High *E. coli* counts in surface water can have very serious economic impacts, especially were tourism is important.

Using UVL to disinfect wastewater has a number of advantages when compared to the more traditional chlorine: no chlorinated by-products; no chemical residual; and, relatively compact size. The primary mechanism of UV disinfection is the photo-chemical alteration of the DNA to form pyrimidine dimers. In general, medium pressure (MP) lamps are more compact, powerful, and emit over a wider range than the more traditional low pressure (LP) lamps. There is some evidence that the wider output spectrum of MP can have disinfection benefits, on proteins for example. Low pressure lamps, however, may be electrically more efficient. The design of UV reactors is complex and perhaps best left to the experts.

In UV disinfection, the fraction of surviving organisms (e.g. *E. coli*) will decrease exponentially with increasing UV dose. This is similar to the traditional Chick-Watson kinetics used in chlorine disinfection. However, the level of UV disinfection that can be achieved is often limited by particle-associated organisms. Efforts to remove or reduce the effects of wastewater particles will often improve UV disinfection effectiveness.

Regrowth, photoreactivation, or dark repair after UV exposure are sometimes cited as disadvantages of UV disinfection. Research continues in this area, however, there is little evidence of photoreactivation of human pathogens in environmental conditions at doses used in wastewater treatment. It appears the photoreactivation is unlikely to increase the risk of the spread of many human diseases.

The UV disinfection of combined sewer overflows, a form of wet weather pollution, is challenging, but these waters are especially unsuited for chlorine disinfection. Pre-treatment of combined sewer overflows (CSOs) with a cationic polymer to induce fast settling, and a low dose of alum to increase UV transmittance, has shown promise at the bench scale.

References

1. Adams BJ, Papa F (2000) Urban stormwater management planning with analytical probabilistic models. Wiley, New York
2. Bohrerova Z, Linden KG (2007) Standardizing photoreactivation: comparison of DNA photorepair rate

in Escherichia coli using four different fluorescent lamps. Water Res 41:2832–2838. doi:10.1016/j.watres.2007.03.015

3. Craik SA, Weldon D, Finch GR et al (2001) Inactivation of cryptosporidium parvum oocysts using medium-and low-pressure ultraviolet radiation. Water Res 35:1387–1398
4. Donovan EP, Staskal DF, Unice KM et al (2008) Risk of gastrointestinal disease associated with exposure to pathogens in the sediments of the lower passaic river. Appl Environ Microbiol 74:1004–1018. doi:10.1128/AEM.01203-07
5. Eischeid AC, Linden KG (2011) Molecular indications of protein damage in adenoviruses after UV disinfection. Appl Environ Microbiol 77:1145–1147. doi:10.1128/AEM.00403-10
6. Emerick RW, Loge FJ, Ginn T, Darby JL (2000) Modeling the inactivation of particle-associated coliform bacteria. Water Environ Res 72:432–438
7. Emerick RW, Loge FJ, Thompson D, Darby JL (1999) Factors influencing ultraviolet disinfection performance part II: association of coliform bacteria with wastewater particles. Water Environ Res 71:1178–1187. doi:10.2175/106143097X122004
8. EPA (2016) Report to congress – CSOs into the great lakes basin (EPA 833-R-16-006). U.S. Environmental Protection Agency
9. Exall K, Marsalek J (2013) A coagulant survey for chemically enhanced primary treatment of synthetic CSOs. Water Air Soil Pollut. doi:10.1007/s11270-012-1414-z
10. Gehr R, Wagner M, Veerasubramanian P, Payment P (2003) Disinfection efficiency of peracetic acid, UV and ozone after enhanced primary treatment of municipal wastewater. Water Res 37:4573–4586. doi:10.1016/S0043-1354(03)00394-4
11. Gibson J, Droppo I, Farnood R et al (2012) Hydrodynamic treatment of wastewater effluent flocs for improved disinfection. Water Environ Res 84:387–395. doi:10.2175/106143012X13347678384567
12. Gibson J, Farnood R, Seto P (2016) Chemical pretreatment of combined sewer overflows for improved UV disinfection. Water Sci Technol 73:375–381. doi:10.2166/wst.2015.447
13. Gibson JH, Hon H, Farnood R et al (2009) Effects of ultrasound on suspended particles in municipal wastewater. Water Res 43:2251–2259. doi:10.1016/j.watres.2009.02.024
14. Gooré Bi E, Monette F, Gasperi J, Perrodin Y (2015) Assessment of the ecotoxicological risk of combined sewer overflows for an aquatic system using a coupled "substance and bioassay" approach. Environ Sci Pollut Res 22:4460–4474. doi:10.1007/s11356-014-3650-9
15. Guo M, Hu H, Bolton JR, El-Din MG (2009) Comparison of low- and medium-pressure ultraviolet lamps: Photoreactivation of Escherichia coli and total coliforms in secondary effluents of municipal wastewater treatment plants. Water Res 43:815–821. doi:10.1016/j.watres.2008.11.028
16. Hu X, Geng S, Wang X, Hu C (2012) Inactivation and photorepair of enteric pathogenic microorganisms with ultraviolet irradiation. Environ Eng Sci 29:549–553. doi:10.1089/ees.2010.0379
17. Jagai JS, Li Q, Wang S et al (2015) Extreme precipitation and emergency room visits for gastrointestinal illness in areas with and without combined sewer systems: an analysis of Massachusetts data, 2003–2007. Environ Health Perspect. doi:10.1289/ehp.1408971
18. Keen OS, Love NG, Linden KG (2012) The role of effluent nitrate in trace organic chemical oxidation during UV disinfection. Water Res 46:5224–5234. doi:10.1016/j.watres.2012.06.052
19. Kollu K, Örmeci B (2014) Regrowth potential of bacteria after ultraviolet disinfection in the absence of light and dark repair. J Environ Eng 141:4014069
20. Li JG, Horneck H, Averill D et al (2004) High-rate retention treatment basins for CSO control in Windsor, Ontario. Water Qual Res J Can 39:449–456
21. Lindenauer KG, Darby JL (1994) Ultraviolet disinfection of wastewater: effect of dose on subsequent photoreactivation. Water Res 28:805–817
22. Loge FJ, Emerick RW, Thompson DE et al (1999) Factors influencing ultraviolet disinfection performance part I: light penetration to wastewater particles. Water Environ Res 71:377–381. doi:10.2175/106143097X122176
23. Masschelein WJ (2002) Ultraviolet light in water and wastewater sanitation. Lewis Publishers, Boca Raton
24. Nebot Sanz E, Salcedo Dávila I, Andrade Balao JA, Quiroga Alonso JM (2007) Modelling of reactivation after UV disinfection: effect of UV-C dose on subsequent photoreactivation and dark repair. Water Res 41:3141–3151. doi:10.1016/j.watres.2007.04.008
25. Potera C (2015) After the fall: gastrointestinal illness following downpours. Environ Health Perspect 123:A243–A243. doi:10.1289/ehp.123-A243
26. Sancar A (1994) Structure and function of DNA photolyase. Biochemistry (Mosc) 33:2–9
27. Sehnaoui K, Gehr R (2003) Fouling of UV lamp sleeves: exploring inconsistencies in the role of iron. National Library of Canada= Bibliothèque nationale du Canada
28. Sinha RP, Häder D-P (2002) UV-induced DNA damage and repair: a review. Photochem Photobiol Sci 1:225–236. doi:10.1039/b201230h
29. Tchobanoglous G (2003) Wastewater engineering: treatment and reuse. McGraw-Hill, Boston
30. Zimmer J, Slawson R, Huck P (2003) Inactivation and potential repair of cryptosporidium parvum following low- and medium-pressure ultraviolet irradiation. Water Res 37:3517–3523. doi:10.1016/S0043-1354(03)00238-0

Part VI

UV Light in Phototherapy

Phototherapy in Atopic Dermatitis

23

José María Ortiz-Salvador
and Amparo Pérez-Ferriols

Abstract

Atopic dermatitis (AD) is one of the most common chronic inflammatory skin diseases. Currently management of AD includes avoidance of triggering factors, skin care aiming to compensate the skin barrier defects, anti-inflammatory therapy (mostly topical corticosteroids and topical calcineurin inhibitors). When these first-line approaches are unsuccessful, systemic treatment or phototherapy ought to be carried out as next line of defence. Current phototherapy modalities for AD include broadband UVB (290–320 nm), narrowband UVB (311–313 nm), UVA-1 therapy (340–400 nm), UVA therapy plus 8-methoxypsoralens (PUVA), 308 nm excimer laser (EL) and Full spectrum light (FSL).

Keywords

Atopic Dermatitis • Dermatitis • Eczema • Phototherapy • Narrowband UV-B therapy • PUVA therapy

Currently, narrowband UVB phototherapy is the most employed treatment owing to its availability, security, ease of administration and efficacy. Dose schedules are same used for psoriasis treatment. Phototherapy has been cataloged as "Strength of Recommendation B" and "Level of Evidence II" in the treatment of AD. This second-line treatment may be applied when behavioral measures and topical therapy have failed. Short-term side effects of phototherapy are usually mild. With long-term treatment, photoaging and induction of cutaneous malignancies as potential side effects can be observed. Phototherapy can also be used exceptionally in children with refractory or severe AD. However, risk of long-term photocarcinogenesis is especially significant in this group of patients. In other words, phototherapy represents a secure and effective treatment of AD. It should be used as a second-line treatment when the patient is unresponsive to topical treatment with corticosteroids and calcineurin inhibitors. It can be used as a single treatment or in combination with systemic drugs.

J.M. Ortiz-Salvador (✉) • A. Pérez-Ferriols
Dermatology Department, University General Hospital of Valencia, Av. Tres Cruces n°2, 46014 Valencia, Spain
e-mail: josema.ortiz.salvador@gmail.com

S. Ahmad (ed.), *Ultraviolet Light in Human Health, Diseases and Environment*, Advances in Experimental Medicine and Biology 996, https://doi.org/10.1007/978-3-319-56017-5_23

23.1 Introduction

Atopic dermatitis (AD) is one of the most common chronic inflammatory skin diseases [1]. Characteristic features of AD are pruritus and a chronic or relapsing course, usually beginning during infancy. Acute inflammation of the extensor aspect of extremities and cheeks are common features in infants, while in children and adults there is a shift toward chronic inflammation with hyperkeratosis and lichenification in flexural areas [1]. During the past 3 decades, the prevalence of atopic dermatitis has almost tripled in developed countries.

Abnormally dry skin and a lowered threshold for itching are important features of AD and scratching creates most of the characteristic patterns of the disease. Therefore, agents that promote dryness or increase the desire to scratch worsen AD. Control of these aggravating factors is essential to manage AD successfully.

Bacterium, *Staphylococcus aureus* being the predominant skin microorganism, frequently colonize AD skin lesions. This species also colonize significantly in non-affected part of the skin of atopic patients.

Treatment of AD often based in treating acute flares of the disease with short-term regimens. Nevertheless, recently there has been a change in approach to AD, with proactive treatments and long-term maintenance therapy [2]. Currently management of AD includes attempting to eliminate inflammation and infection, skin care aiming to compensate the skin barrier defects barrier by using emollients, using antipruritic agents to reduce the self-inflicted damage to the involved skin, and controlling exacerbating factors by avoidance of triggering factors. Anti-inflammatory therapy with topical corticosteroids and topical calcineurin inhibitors are the mainstay for mild AD. In moderate or severe AD, adjunctive or complementary modalities may be needed [1, 3]. Most commonly used approach for moderate-severe AD includes phototherapy, systemic corticosteroids, azathioprine, cyclosporine and methotrexate.

Phototherapy denotes the use of ultraviolet (UV) light for the treatment of certain other kind of skin disorders besides AD [4]. Current phototherapy modalities for AD include broadband UVB (290–320 nm), narrowband UVB (311–313 nm), UVA-1 therapy (340–400 nm), UVA therapy plus 8-methoxypsoralens (PUVA), 308 nm excimer laser (EL) and Full spectrum light (FSL) (Table 23.1) [5, 6].

Table 23.1 Phototherapy modalities in atopic dermatitis

UVB-based
Broadband UVB (290–320 nm)
Narrowband UVB (311–313 nm).
Excimer laser (308 nm).
UVA-based
UVA (315–400 nm)
UVA-1 (340–400 nm)
UVA + Psoralens (PUVA)
UVA + UVB combination (280–400 nm)
Full spectrum light (320–5000 nm)

23.1.1 Mechanism of Action of Phototherapy in Atopic Dermatitis

It was empirically known that sun exposure was beneficial for patients with AD and this yield to the first use of broadband UVB in the end of the 1970s [6]. Recent experimental studies have demonstrated that the immunomodulatory effects of phototherapy occur via modified cytokine expression with decreased IL-5, IL-13 and IL-31, induction of T-cell apoptosis and reduction of dendritic cells [7, 8].

Treatment with UVB has also been shown to reduce *Staphylococcus aureus* colonization in the skin of AD patients [9].

Narrowband UV-B (NB UV-B) can damage DNA and induce apoptosis of epidermal T lymphocytes by activating death receptors. It also inhibits the release of cytokines and the Th1 response leading to a Th2 switch [10]. UV-A radiation on the other hand penetrates deeper in the dermis and into the superficial vascular plexus increasing collagen synthesis, inhibiting calcineurin and suppressing tumor necrosis factor-α, IL-12 and interferon-γ. It also induces apoptosis of T-cells and mast cells [7, 8].

The mechanism of action of PUVA phototherapy in AD is not yet fully understood; current concepts support an alteration of lymphocyte function in the dermal infiltrate. It has also been proposed that PUVA reduces pruritus in AD by reducing epidermal hyperinnervation [11].

23.1.2 Phototherapy Modalities

23.1.2.1 Classic Modalities: UV-A, UV-B, UV-AB and PUVA in AD

Broadband UV-B (BB UV-B) was the first modality of phototherapy employed in AD and started to be applied in the late 1970s. Nevertheless, because of its high erythemogenic potential and low efficacy it was soon replaced by UV-A therapy, which demonstrated to be safer as well [12, 13]. Later studies have shown that a combination of UV-A plus UV-B (UV-AB therapy) is superior in almost all aspects of therapy by UV-A or UV-B alone [14, 15]. Currently UV-AB therapy is considered to be the most effective treatment against AD among the classic modalities of phototherapy [5, 6].

UV-AB radiation can be administered with a single device emitting both wavelengths or as two separate simultaneous or subsequent emissions [14–16]. The latter allows to control the doses of UV-A and UV-B separately with better control of treatment [14–16].

23.1.2.2 PUVA in Atopic Dermatitis

The term PUVA refers to the use of UVA in combination with one of the most appropriate psoralen compounds. As 8-methoxypsoralen (8-MOP) has been found to be extremely potent photosensitizing agent which leads to interstrand DNA cross-links, irreversibly damaging DNA unless repaired has been mostly in use [4].

Psoralens can be administered either orally or topically in bath or cream [11]. Bath-PUVA consists of UV-A exposure after 20–30 min of bathing in warm water containing 8-MOP. In cream-PUVA a 0,0006% 8-MOP ointment is applied to specific areas of skin 30–60 min before irradiation [4, 11].

Patients with moderate or severe forms of AD can benefit from PUVA therapy (either topical or systemic) [17–20]. The treatment schemes are virtually the same as for psoriasis [6]. However, as compared to psoriasis, atopic dermatitis is more difficult to treat with a higher number of treatments required [5, 6]. Albeit, there is insignificant support for the use of PUVA in AD [6]. A randomized trial comparing PUVA bath therapy to NB UV-B did not find any significant difference [18]. Other study compared UV-A1 to oral 5-methoxypsoralen (5-MOP) PUVA therapy showed longer remission times and higher AD score improvement in compare to PUVA therapy [19]. In addition as PUVA therapy has been shown to be mutagenic it is reamended that PUVA therapy may be administered only for a short term [16].

23.1.2.3 UV-A1 in Atopic Dermatitis

Although UV-A had been found to be quite effective for AD, its long exposure times remains unacceptable. This problem was overcome with the development of UV-A1 lamps [21]. UV-A1 uses the lower frequencies of UV-A light spectrum (between 340 and 400 nm) avoiding UV-A2 radiation (320–340 nm) and its adverse effects [22].

UV-A1 can be administered either employing a high dose (80–130 J/cm2), medium dose (40–80 J/cm_2) or low dose (<40 J/cm^2) [23, 24]. An issue with UV-A1 at high dose is the excessive heating of the equipment making their use intolerable in many situations [16].

UV-A1 is effective in AD treatment and more effective than UV-AB in several studies. It is at least as effective as topical treatment with fluocortolone [24]. AS no significant difference has been in efficacy or in recurrence time between UVA-A1 at high dose or at medium dose [6, 25], therefore, medium dose of UVA-A1 should be preferred over high dose to reduce adverse side effects and improve tolerability [5, 6]. Low dose UVA-A1 has been shown to be not as effective and hence is barely employed [24]. Usual treatment schedules with UV-A1 at medium dose for AD are 3–5 sessions per week for 3–8 weeks with a maximum dose of 80 J/cm^2. Treatment times can range from 10 min to 1 h per session [5, 25].

Due to its stronger effect, compared with UV-B, it is comparatively more appropriate for patients with acute AD [5]. Still controversy remains about whether UVA-A1 at high dose is more effective than other light sources when treating acute flares of AD [6, 26].

UV-A1 lamps are expensive too and require greater space and dedicated ventilation machinery, making them unaffordable for some centers [16]. Other problem with UV-A1 is the high temperature generated by the lamp [16, 27].

Cold-light UV-A1 uses a filter to eliminate wavelengths above 530 nm and dissipate the excessive heat load generated by UV-A1 generator [27]. It has been found to be more effective than UV-AB than conventional UV-A1 at clearing lesions and reducing duration of AD flares [27].

23.1.2.4 NB-UVB in Atopic Dermatitis

Since around 1990, the NB UV-B has successfully been used to treat AD [28]. The NB UV-B emits highly selective wavelengths of UV-B light between 311 and 313 nm excluding shortwave length UVB radiation [4]. It also has lesser erythemogenic output (sunburning potential) than BB UV-B [4]. Nowadays NB-UVB therapy is considered by most physicians the first-line treatment phototherapy modality owing to its availability, security, ease of administration and efficacy [29].

NB-UVB therapy has been shown to improve the AD scores and reduce the need for potent topical corticosteroids in several randomized trials [5, 6]. These beneficial effects have been demonstrated to persist up to 6 months after the termination of the NB-UVB scheme [30].

Unlike UVA, NB UV-B radiation does not reach the dermis and hence its effect is confined to the epidermis. Because of its limited penetrating potential, NB UV-B has been, albeit controversial, proposed to be more effective in chronic AD [5].

UV-B dosing depends on patient's pigmentation and tolerance to UV radiation. The most used methods for calculating UV-B dose delivery is by determining the "Minimal Erythema Dose" (MED) this being the minimal UV-B radiation able to induce minimal erythema in the patient [31]. Other method widely used is calculating UV-B based in patient's skin phototype [29]. A most innovative technique consists in calculating UV-B dose, on the basis of skin pigmentation, measured by skin reflectance (reflectance-guided UV-B) [32]. In one trial comparing traditional UV-B dose calculation against reflectance-guided UV-B the cumulative UV- B dosage was lower in the reflectance-guided regimen with efficacy being the same as with the classic UV-B dosing protocol [32].

The usual treatment schedule with NB UV-B for AD is 3 sessions per week for 6 weeks [33]. Initially nearly erythemogenic doses of NB UV-B were used but now it has been demonstrated that doses of 50% the MED yields similar results with better tolerability and less carcinogenic risk [6, 29]. NB UV-B dosing according to skin type is shown in Table 23.2.

NB-UVB superiority versus UV-A1 at medium dose is equivocal [5]. Several studies have found that NB-UVB produces better improvement in AD severity scores than UV-A1 [34, 35], while other studies have not found statistically significant differences between UV-A1 and NB UV-B therapies [36]. Furthermore, NB UV-B therapy has been used successfully in conjunction with UV-A1 therapy [37].

23.1.3 Other Phototherapy Modalities

23.1.3.1 Excimer Laser

This lamp is consisting of a coherent single-wavelength light source of 308 nm. Excimer Laser exposure for 10 weeks has been shown to

Table 23.2 Dosing guide for narrowband-UVB

Initial dose according skin type		Dose increase after each treatment (mJ/cm2)	Maximum achievable dose (mJ/cm2)
Skin type	Initial dose (mJ/cm2)		
I	130	15	2000
II	220	25	2000
III	260	40	3000
IV	330	45	3000
V	350	60	5000
VI	400	65	5000

Administered 3–5 times a week

yielded good results in the prurigo form of AD compared versus clobetasol propionate [38].

23.1.3.2 Full Spectrum Light

The emission of this lamp extends 320–5000 nm and used in conjunction with an emollient demonstrating greater improvement in atopic dermatitis severity scores at 4 weeks as compared to the emollient alone [39].

23.1.3.3 Synchronous Balneotherapy

This therapy is a combination of NB UV-B with bathing in 10% Dead Sea salt solution [30]. Synchronous balneophototherapy has yielded a greater reduction in atopic dermatitis severity scores than isolated NB UV-B with the beneficial effects of Synchronous balneotherapy remaining up to 1–6 months after treatment [30].

23.1.4 Integrating Phototherapy in the Management of AD

Phototherapy has been cataloged as "Strength of Recommendation B" and "Level of Evidence II" in the treatment of AD [5, 6]. Nevertheless, it should be emphasized that phototherapy is a second-line treatment and should be reserved for cases where behavioral measures and topical therapy have failed [1, 40].

Phototherapy has some limitations. Equipment is expensive and requires qualified personnel. Patients must be compliant enough to undergo frequent treatment. Some body areas are difficult or may even be impossible to be treated with phototherapy (i.e. hairy areas, folds etc.) [16].

A randomized trial has compared 1% pimecrolimus cream to NB UV-B in patients between the ages of 5–17 years. Both interventions were beneficial, and concomitant application of both treatments was not superior to NB-UVB alone or pimecrolimus alone [41].

In one study comparing cyclosporine A to UV-AB results were in favor of cyclosporine [42]. The mean number of days in remission was 186 after cyclosporine A compared with 114 after UV-AB. Both the patients and the researchers rated cyclosporine A treatment more effective than UV-AB phototherapy [42].

When comparing UV-AB to UV-AB plus topical fluticasone or topical hydrocortisone butyrate, significant improvement was seen in both groups [43]. In patients who received a corticosteroid, fewer phototherapy sessions were required and the total mean UV-B dose was lower without influencing the duration of remissions or the frequency of adverse effects [43].

23.1.5 Side-Effects of Phototherapy in Atopic Dermatitis

Short-term side effects of phototherapy (Table 23.3) are usually mild, being the most frequents skin burning (usually associated with errors in dosage or unwise treatment schedules) skin pruritus and tenderness. Other short-term side effects are skin light-induced eruption or inducing flares of lupus or herpes simplex infections. With long-term treatment, photo-aging and induction of cutaneous malignancies as potential side-effects can be observed [29]. The side-effect

Table 23.3 Side effects of phototherapy

Common	
Short-term	**Long-term**
Burning	Actinic damage
Stinging	Skin aging
Pruritus	Dyspigmentation
Heat-induced flares	
Skin erythema and tenderness	
Claustrophobia	
Uncommon	
Short-term	**Long-term**
Polymorphous light eruption	Non-melanoma skin cancer
Lupus flare	Melanoma
Herpes simplex reactivation	Ocular toxicity
Photosensitive eruptions	Lentigines
Folic acid depletion	
B6 vitamin deficiency	
Photo-onycholysis	
Hepatotoxicity	

profile of phototherapy is favorable when compared to other systemic immunosuppressive agents used in the treatment of AD, phototherapy being a well-tolerated treatment with relatively fewer and mild adverse events [29, 44].

Systemic PUVA treatment is associated with short-term general toxicity, including nausea, vomiting and hepatotoxicity, as well as long-term photosensitivity, cataract and possibly skin cancer. Topical PUVA can lessen or avoid these problems [11, 20].

When phototherapy is used in AD, flares and recurrences are common events after finishing a treatment schedule and multiple treatment cycles may be needed, with an increased risk of photoaging and photo-carcinogenesis [30]. This is why maintenance therapy with long term exposure should always be avoided and especially in younger patients [44].

Most trials have confirmed the effectiveness and security of phototherapy in children with AD being usually well tolerated [2, 44]. However, risk of long-term photocarcinogenesis is specially significant in this group of patients [44]. For this and for practical reasons (e.g. lack of cooperation) it is advised to avoid phototherapy in children [5, 30]. Nevertheless, phototherapy may be used exceptionally in children with refractory or severe AD [41, 44]. In this cases PUVA is usually avoided and NB UV-B is the preferred therapy [2, 44].

Due to the lack of randomized trials of phototherapy in pregnant women with AD there is no evidence to support the use of phototherapy during pregnancy [16]. Also due to the time and effort required to travel several times a week to receive phototherapy may be troublesome for some patients with attendance problems at school or work. In this cases, home phototherapy devices have been proposed to be useful [16].

23.1.6 Future Trends

There are few studies comparing phototherapy with systemic immunosuppressive therapies in AD [6, 42]. Furthermore, these studies do not include the modalities for which the strongest evidence is available (UV-A1 and NB UV-B). There are no studies comparing phototherapy versus oral corticosteroids [6].

As AD being a chronic and disabling disease, life-quality impact measures should be emphasized when studying AD treatment options and when comparing different treatment schemes [5].

AD severity assessment criteria, irradiation techniques, and assessment scales ought to be standardized. The Harmonizing Outcome Measures for Eczema initiative was born with the aim of providing quality evidence in the treatment of AD [6].

23.2 Conclusions

Phototherapy represents a secure and effective treatment of AD. It ought to be used as a second-line treatment when the patient is unresponsive to topical treatment with corticosteroids and calcineurin inhibitors. It can be used as a single treatment or in combination with systemic drugs [1].

References

1. Sidbury R, Davis DM, Cohen DE, Cordoro KM, Berger TG, Bergman JN et al (2014) Guidelines of care for the management of atopic dermatitis: section 3. Management and treatment with phototherapy and systemic agents. J Am Acad Dermatol 71(2):327–349
2. Song E, Reja D, Silverberg N, Rothe MJ (2015) Phototherapy: kids are not just little people. Clin Dermatol 33(6):672–680
3. Chan IHY, Murrell DF (2016) Itch Management: physical approaches (UV phototherapy, acupuncture). Curr Probl Dermatol 50:54–63
4. Carrascosa JM, Gardeazábal J, Pérez-Ferriols A, Alomar A, Manrique P, Jones-Caballero M et al (2005) Consensus document on phototherapy: PUVA therapy and narrow-band UVB therapy. Actas Dermosifiliogr 96(10):635–658
5. Meduri NB, Vandergriff T, Rasmussen H, Jacobe H (2007) Phototherapy in the management of atopic dermatitis: a systematic review. Photodermatol Photoimmunol Photomed 23(4):106–112
6. Pérez-Ferriols A, Aranegui B, Pujol-Montcusí JA, Martín-Gorgojo A, Campos-Domínguez M, Feltes RA et al (2015) Phototherapy in atopic dermatitis: a systematic review of the literature. Actas Dermosifiliogr 106(5):387–401

7. Bogaczewicz J, Malinowska K, Sysa-Jedrzejowska A, Wozniacka A (2016) Medium dose ultraviolet A1 phototherapy and mRNA expression of interleukin 8, interferon γ, and chemokine receptor 4 in acute skin lesions in atopic dermatitis. Postępy Dermatol Alergol 33(3):170–175
8. Morita A, Werfel T, Stege H, Ahrens C, Karmann K, Grewe M et al (1997) Evidence that singlet oxygen-induced human T helper cell apoptosis is the basic mechanism of ultraviolet-A radiation phototherapy. J Exp Med 186(10):1763–1768
9. Faergemann J, Larkö O (1987) The effect of UV-light on human skin microorganisms. Acta Derm Venereol 67(1):69–72
10. Tintle S, Shemer A, Suárez-Fariñas M, Fujita H, Gilleaudeau P, Sullivan-Whalen M et al (2011) Reversal of atopic dermatitis with narrow-band UVB phototherapy and biomarkers for therapeutic response. J Allergy Clin Immunol 128(3.) 583-593-4
11. Morison WL, Parrish JA, Fitzpatrick TB (1978) Oral psoralen photochemotherapy of atopic eczema. Br J Dermatol 98(1):25–30
12. Pérez Ferriols A, Aguilera J, Aguilera P, de Argila D, Barnadas MA, de Cabo X et al (2014) Determination of minimal erythema dose and anomalous reactions to UVA radiation by skin phototype. Actas Dermosifiliogr 105(8):780–788
13. Jekler J, Larkö O (1991) UVA solarium versus UVB phototherapy of atopic dermatitis: a paired-comparison study. Br J Dermatol 125(6):569–572
14. Jekler J, Larkö O (1990) Combined UVA-UVB versus UVB phototherapy for atopic dermatitis: a paired-comparison study. J Am Acad Dermatol 22(1):49–53
15. Midelfart K, Stenvold S-E, Volden G (1985) Combined UVB and UVA phototherapy of atopic eczema. Dermatology 171(2):95–98
16. Patrizi A, Raone B, Ravaioli GM (2015) Management of atopic dermatitis: safety and efficacy of phototherapy. Clin Cosmet Investig Dermatol 8:511–520
17. Grau-Salvat C, Vilata-Corell JJ, Azón-Massoliver A, Pérez-Ferriols A (2007) Use of psoralen plus UV-A therapy in the autonomous community of Valencia, Spain. Actas Dermosifiliogr 98(9):611–616
18. Der-Petrossian M, Seeber A, Hönigsmann H, Tanew A (2000) Half-side comparison study on the efficacy of 8-methoxypsoralen bath-PUVA versus narrow-band ultraviolet B phototherapy in patients with severe chronic atopic dermatitis. Br J Dermatol 142(1):39–43
19. Tzaneva S, Kittler H, Holzer G, Reljic D, Weber M, Hönigsmann H et al (2010) 5-Methoxypsoralen plus ultraviolet (UV) A is superior to medium-dose UVA1 in the treatment of severe atopic dermatitis: a randomized crossover trial. Br J Dermatol 162(3):655–660
20. Garritsen FM, MWD B, Limpens J, Spuls PI (2014) Photo(chemo)therapy in the management of atopic dermatitis: an updated systematic review with implications for practice and research. Br J Dermatol 170(3):501–513
21. Attili SK, Dawe RS, Ibbotson SH (2016) Ultraviolet A1 phototherapy: one center's experience. Indian J Dermatol Venereol Leprol 23
22. Polderman MCA, Wintzen M, le Cessie S, Pavel S (2005) UVA-1 cold light therapy in the treatment of atopic dermatitis: 61 patients treated in the Leiden University Medical Center. Photodermatol Photoimmunol Photomed 21(2):93–96
23. Krutmann J, Czech W, Diepgen T, Niedner R, Kapp A, Schöpf E (1992) High-dose UVA1 therapy in the treatment of patients with atopic dermatitis. J Am Acad Dermatol 26(2):225–230
24. Krutmann J, Diepgen TL, Luger TA, Grabbe S, Meffert H, Sönnichsen N et al (1998) High-dose UVA1 therapy for atopic dermatitis: results of a multicenter trial. J Am Acad Dermatol 38(4):589–593
25. Tzaneva S, Seeber A, Schwaiger M, Hönigsmann H, Tanew A (2001) High-dose versus medium-dose UVA1 phototherapy for patients with severe generalized atopic dermatitis. J Am Acad Dermatol 45(4):503–507
26. Dittmar HC, Pflieger D, Schöpf E, Simon JC (2001) UVA1 phototherapy. Pilot study of dose finding in acute exacerbated atopic dermatitis. Hautarzt Z Für Dermatol Venerol Verwandte Geb 52(5):423–427
27. von Kobyletzki G, Pieck C, Hoffmann K, Freitag M, Altmeyer P (1999) Medium-dose UVA1 cold-light phototherapy in the treatment of severe atopic dermatitis. J Am Acad Dermatol 41(6):931–937
28. George SA, Bilsland DJ, Johnson BE, Ferguson J (1993) Narrow-band (TL-01) UVB air-conditioned phototherapy for chronic severe adult atopic dermatitis. Br J Dermatol 128(1):49–56
29. Rodenbeck DL, Silverberg JI, Silverberg NB (2016) Phototherapy for atopic dermatitis. Clin Dermatol 34(5):607–613
30. Heinlin J, Schiffner-Rohe J, Schiffner R, Einsele-Krämer B, Landthaler M, Klein A et al (2011) A first prospective randomized controlled trial on the efficacy and safety of synchronous balneophototherapy vs. narrow-band UVB monotherapy for atopic dermatitis. J Eur Acad Dermatol Venereol 25(7):765–773
31. Pérez-Ferriols A (2013) The minimal erythema dose (MED) project: in search of consensus on phototesting. Actas Dermosifiliogr 104(7):541–542
32. Selvaag E, Caspersen L, Bech-Thomsen N, Wulf HC (2005) Optimized UVB treatment of atopic dermatitis using skin reflectance measurements. A controlled, left–right comparison trial. Acta Derm Venereol 85(2):144–146
33. Reynolds NJ, Franklin V, Gray JC, Diffey BL, Farr PM (2001) Narrow-band ultraviolet B and broad-band ultraviolet A phototherapy in adult atopic eczema: a randomised controlled trial. Lancet 357(9273):2012–2016
34. Gambichler T, Othlinghaus N, Tomi NS, Holland-Letz T, Boms S, Skrygan M et al (2009) Medium-dose

ultraviolet (UV) A1 vs. narrowband UVB phototherapy in atopic eczema: a randomized crossover study. Br J Dermatol 160(3):652–658
35. Legat FJ, Hofer A, Brabek E, Quehenberger F, Kerl H, Wolf P (2003) NArrowband uv-b vs medium-dose uv-a1 phototherapy in chronic atopic dermatitis. Arch Dermatol 139(2):223–223
36. Majoie IML, Oldhoff JM, van Weelden H, Laaper-Ertmann M, Bousema MT, Sigurdsson V et al (2009) Narrowband ultraviolet B and medium-dose ultraviolet A1 are equally effective in the treatment of moderate to severe atopic dermatitis. J Am Acad Dermatol 60(1):77–84
37. Fernández-Guarino M, Aboin-Gonzalez S, Barchino L, Velazquez D, Arsuaga C, Lázaro P (2016) Treatment of moderate and severe adult chronic atopic dermatitis with narrow-band UVB and the combination of narrow-band UVB/UVA phototherapy. Dermatol Ther 29(1):19–23
38. Brenninkmeijer EEA, Spuls PI, Lindeboom R, Van Der Wal AC, Bos JD, Wolkerstorfer A (2010) Excimer laser vs. clobetasol propionate 0·05% ointment in prurigo form of atopic dermatitis: a randomized controlled trial, a pilot. Br J Dermatol 163(4):823–831
39. Byun HJ, Lee HI, Kim B, Kim MN, Hong H, Choi Y et al (2011) Full-spectrum light phototherapy for atopic dermatitis. Int J Dermatol 50(1):94–101
40. Dogra S, Mahajan R (2015) Indian association of dermatologists, venereologists and leprologists. Phototherapy for atopic dermatitis. Indian J Dermatol Venereol Leprol 81(1):10–15
41. Tzung T-Y, Lin C-B, Chen Y-H, Yang C-Y (2006) Pimecrolimus and narrowband UVB as monotherapy or combination therapy in children and adolescents with atopic dermatitis. Acta Derm Venereol 86(1):34–38
42. Granlund H, Erkko P, Remitz A, Langeland T, Helsing P, Nuutinen M et al (2001) Comparison of cyclosporin and UVAB phototherapy for intermittent one-year treatment of atopic dermatitis. Acta Derm Venereol 81(1):22–27
43. Valkova S, Velkova A (2004) UVA/UVB phototherapy for atopic dermatitis revisited. J Dermatol Treat 15(4):239–244
44. Jury CS, McHenry P, Burden AD, Lever R, Bilsland D (2006) Narrowband ultraviolet B (UVB) phototherapy in children. Clin Exp Dermatol 31(2):196–199

Phototherapy of Psoriasis, a Chronic Inflammatory Skin Disease

24

Emőke Rácz and Errol P. Prens

Abstract

Phototherapy is an effective treatment modality for several skin diseases which has been in use from the era of the Egyptians. Insight into its mode of action has gradually accumulated over the past decades. A crucial biological effect of ultraviolet radiation is the induction of apoptosis in T lymphocytes and in keratinocytes in the epidermis. Via this mechanism inflammation-induced pathological changes characteristic of psoriasis are counteracted.

Phototherapy remains the only therapeutic option for certain patient groups where modification of the systemic immune reactions is contraindicated, such as by HIV, internal malignancy or pregnancy. UVB treatment is highly cost-effective, which is important in this age of increasing health care costs.

Keywords

Narrow band UVB • Broad band UVB • PUVA • Phototherapy • Psoriasis

24.1 Introduction

Phototherapy is used in dermatology for an array of skin diseases that show improvement upon exposure to natural sunlight or man-made lamps. Natural light in combination with herbal extracts has been in use for the treatment of skin disease from the era of the ancient Egyptians. The most common disease for which phototherapy has been in use is psoriasis. Consequently, most experience gained and research has been carried out is on phototherapy for psoriasis. Other skin diseases where phototherapy has been in use are atopic dermatitis, vitiligo, mycosis fungoides, morphoea, pruritus, lichen planus and cutaneous mastocytosis. Certain photodermatoses, such as

E. Rácz
Department of Dermatology, University of Groningen, University Medical Center Groningen, Groningen, The Netherlands

E.P. Prens (✉)
Department of Dermatology, Erasmus University Medical Center, Rotterdam, The Netherlands
e-mail: e.prens@erasmusmc.nl

S. Ahmad (ed.), *Ultraviolet Light in Human Health, Diseases and Environment*, Advances in Experimental Medicine and Biology 996, https://doi.org/10.1007/978-3-319-56017-5_24

polymorphous light eruption or solar urticaria, can be controlled or even prevented with the help of phototherapy [1].

Artificial light sources have been used for the treatment of psoriasis since 1920s. The most frequently applied regimen for psoriasis was the combination of topical coal tar and subsequent UV radiation, introduced by Goeckerman in 1925 [2]. The broad-band UV-B (wavelengths 280–320 nm) has been in use since the 1970s and the Narrow-band UV-B phototherapy, using Philips TL-01 fluorescent lamps (Eindhoven, The Netherlands), emitting light of 311–313 nm, was introduced in 1988 for the treatment of psoriasis [2]. In 1970s PUVA (psoralen + UVA of 320–400 nm) therapy was introduced. 8-methoxypsoralen, one of a group of psoralen compounds, is plant-derived photosensitizer, first applied topically or taken orally and subsequent UVA irradiation causes a therapeutically beneficial phototoxic reaction in the skin. PUVA therapy has anti-inflammatory and antiproliferative effects, and is highly efficacious in the treatment of psoriasis, inducing psoriasis area and severity index (PASI) improvement rates from 74% to 100%. PUVA is thereby one of the most effective treatment options for psoriasis; however, it is less well tolerated than UV-B phototherapy, and there is more evidence of its carcinogenic potential [3].

24.2 Mechanism of Action

24.2.1 Primary Molecular Targets

The epidermis is the primary target of UV-B radiation. UV-B radiation is absorbed to the greatest extent by chromophores (light absorbing molecules) in the upper layers of the skin, mostly in the horny layer of the epidermis [4]. Light absorption by chromophores induces structural changes, thereby changing their functionality. Molecules that undergo light-induced structural modifications are called photoproducts. The most important chromophores in the skin are DNA, urocanic acid (UCA), aromatic amino acids, retinol esters, and melanin. When DNA absorbs UV-B radiation, different types of photoproducts are formed, the most frequent being cyclobutyl pyrimidine dimers (CPD) and (6–4)-photoproducts [5]. These UV-B signature molecules have been demonstrated in keratinocytes and Langerhans cells after exposure to UV-B radiation. CPD were shown to be involved in UV-B-induced apoptosis, inflammation, immunosuppression, and photocarcinogenesis [6]. Presence of CPD's throughout the psoriatic epidermis and the papillary dermis was demonstrated even after a single irradiation with 70% MED [7].

UCA is generated in the skin from histidine, and accumulates in the stratum corneum of the epidermis. The major source of UCA in the epidermis is filaggrin, a histidine-rich basic protein. On UV radiation, the naturally occurring trans-UCA isoform converts to cis-UCA. UCA was first identified as a chromophore responsible for UV-B-induced suppression of contact hypersensitivity [8]. Cis-UCA is detectable in the skin and in the urine of persons exposed to UV-B radiation [9]. The involvement of UCA and DNA damage with UV-B has been evaluated in psoriasis in addition to expression of cytoprotective enzymes. The epidermal cis-UCA concentration was found to be increased by heliotherapy of psoriasis, from a mean initial value of 0.2 nmol/cm^2 to a mean final value of 2.9 nmol/cm^2. Clinical response of psoriasis to heliotherapy, however, appeared to be independent of UCA isomer levels [10].

Active cellular metabolism in the presence of oxygen results in the formation of reactive oxygen species (ROS), such as superoxide anion ($O_2^{\cdot -}$), hydrogen peroxide (H_2O_2), hydroxyl radicals ($^{\cdot}OH$) and singlet oxygen ($^1O_2$). UV-B radiation on its own or UVA+ a photosensitive (such as psoralen) agent can induce production of ROS and cells response with upregulation of ROS scavenging enzyme synthesis and increased activities. As a result, these radiations drive the cells into a complex stress response, leading to the immune responded inflammation [11].

UV-B radiation also leads to clustering and internalization of cell membrane receptors for epidermal growth factor, tumor necrosis factor, and interleukin (IL)-1, resulting in ligand-independent activation of members of the MAPK family [12]. Furthermore, CD95 or FAS, another cell surface receptor, is also activated by UV-B radiation on a ligand-independent manner, playing a role in UV-B-induced apoptosis [13].

24.2.2 Functional Changes and Apoptosis of T-Cells and Keratinocytes

Narrow band UVB (NB-UVB) phototherapy reverses several pathologic alterations in psoriasis lesions, especially keratinocyte proliferation. Interestingly, it has been demonstrated that keratinocyte apoptosis can be sufficient for the clearance of psoriatic plaques.

The number of T lymphocytes in the epidermis and dermis has also been shown to decrease, likely caused by UV-B-induced apoptosis. Epidermal and dermal T cell numbers were significantly reduced by NB-UV-B than by Broad band. Although the decrease in epidermal T cells correlated well with clinical improvement, this was not the case for dermal T cell numbers [14, 15].

T lymphocytes, in vitro, are 10-fold more sensitive to the cytotoxic effects of UV-B than keratinocytes, which explains their depletion from the epidermis on UV-B phototherapy. In addition, whereas hyperplastic keratinocytes in untreated psoriatic plaques do not express CD95L/FASL on their plasma membrane, after NB-UV-B treatment there is strong and diffuse keratinocyte FASL/CD95L expression that coincides in a temporal fashion with depletion of intra-epidermal T cells, indicating a role for FASL in epidermal T-cell apoptosis. Keratinocyte and lymphocyte apoptosis also play a critical role in the mode of action of PUVA in psoriasis [16].

T cells that remain in the skin lesions, 4 weeks after NB-UV-B treatment, produce less interferon-y and IL-12 and more IL-4. A single dose of BB-UV-B radiation resulted in decreased interferon-y production and increased IL-4 production in psoriatic skin and interestingly, neutrophils were found to be the source of the increased IL-4 production [17, 18]. UVB increases the number of forkhead box P3 (Foxp3)-positive regulatory T cells (Treg) in psoriatic skin lesions, thereby enhancing Treg stability and reducing proinflammatory T cell-derived cytokines in the lesions [19].

Clearance of psoriasis by NB-UVB is associated with suppression of type I and type II interferon signaling and downregulation of the Th1 pathway in the lesional epidermis. NB-UVB inhibits the phosphorylation of signal transducer and activator of transcription 3 (STAT3), resulting in reduced expression of its transcriptional targets (e.g., the antimicrobial peptide human β-defensin 2) [7]. Recently, ubiquitination and downregulation of the type I interferon receptor chain IFNAR1, by UV light, was shown to mediate UV-response in the imiquimod-induced psoriasis model in mice [20]. As opposed to wildtype mice, psoriasiform inflammation in IFNAR1 deficient mice did not show improvement by UV light. In analogy, IFNAR1 was also ubiquitinated and downregulated by UV light in human psoriatic skin.

Both UVB and PUVA phototherapy also affect circulating T lymphocytes in patients with psoriasis. They reduce the number of circulating Th1 cells, and restore the impaired regulatory T-cell function in psoriasis [21]. When peripheral blood mononuclear cells of patients with psoriasis, isolated before NB-UVB therapy and weekly thereafter, were stimulated with superantigen in vitro, reduced production of IL-1lbeta, IL-2, IL-5, and IL-6 and increased production of IL-10 was detected [22]. Gene expression levels of IL-6 and TNF-α in circulating PBMC were decreased after bath-PUVA therapy, whereas NB-UVB also suppressed IL-17A expression [23].

After phototherapy, vitamin D levels were found to be dramatically increased by UV-B (but not by PUVA) in patients with psoriasis and in control subjects. Serum vitamin D levels in patients with psoriasis showed less increase with NB-UVB than with BB-UVB phototherapy [24]. PASI improvement does not correlate with the increase in vitamin D levels upon NB-UVB therapy [25]. Interestingly, a single nucleotide polymorphism in the vitamin D receptor gene was found to be a predictive factor for responsiveness to NB-UVB treatment [26].

24.3 Phototherapy in Practice

24.3.1 Efficacy and Clinical Theatre for the Treatment of Psoriasis

Phototherapy is a standard second line treatment option for psoriasis, generally used when topical treatment modalities fail or are contra-indicated or non-practical, such as in extensive guttate psoriasis. Phototherapy may lead to the clearance of psoriasis in 5 to 8 weeks and has one of the highest treatment satisfaction rates compared with other treatment modalities [27]. PASI 75 scores are 70% for UVB and 80% for PUVA, which are comparable or better than outcomes of treatments with several other biological treatments [28]. Phototherapy may be the only treatment option for patients with severe psoriasis who cannot tolerate systemic treatments or where these are contra-indicated due to comorbidities (such as HIV or recent malignancy), drug interactions or toxicity [1, 29].

24.3.2 Treatment Regimens

Phototherapy is mostly applied in a clinical setting of the UV chamber where patients stand for few seconds to a few minutes, two to five times a week. The starting dose of phototherapy is ideally based on the minimal erythema dose (MED) and in the case of UV-B treatment, or the minimal phototoxic dose in case of PUVA therapy (see Table 24.1). MED is defined as the lowest radiation dose that produces just perceptible erythema on exposed area of skin after 24 h. Common MEDs reported for NB-UVB and BB-UVB are shown in Table 24.2. Thus, at least five to ten times higher doses of NB-UVB, compared with BB-UVB, are needed for the induction of erythema. NB-UVB doses required for the induction of hyperplasia, edema, sunburn cell formation, and Langerhans cell depletion are 5 to 10 times higher than equally effective BB-UVB doses.

A more convenient approach is to base the starting dose on the skin type of the patient, although MED-based therapy is thought to be the safest regimen for the patient. Maintenance of a slight asymptomatic erythema throughout the treatment can result in optimal clinical efficacy. Treatments are continued until total remission is reached or until no further improvement can be seen with continued phototherapy. The median number of treatments needed for clearance with UV-B is between 25 and 30 and for PUVA between 17 and 19 [30] [31]. The median duration of remission was reported to be 288 days for NB-UVB and 231 days for PUVA therapy [32]; however, a systematic review found that more patients are still in remission 6 months after completing PUVA therapy than after NB-UVB therapy [33]. The duration of remission seems to correlate with the PASI score at the end of the treatment [34].

For the treatment of chronic localized psoriatic plaques, localized phototherapy is available in the form of hand-held non-laser UV-B (light-emitting diode) lamps, and the 308-nm excimer laser. The excimer laser emits monochromatic light equivalent to that of NB-UVB with similar biologic and clinical effects [35]. Localized phototherapy was

Table 24.1 Treatment regimens

			PUVA	
		UVB	Oral	Bath
Initial dose determination		Reading after 24 h	Reading after 72–96 h	Reading after 96–120 h
Initial dose		70% MED	75% MPD	30% MPD
Treatment frequency		2–5 times weekly	2–4 times weekly	
Dose adjustment during treatment	No erythema	Increase by 30–40%	Increase by 30% max 2 times weekly	
	Minimal erythema	Increase by 20%	No increase	
	Persistent asymptomatic erythema	No increase		
	Painful erythema	Break in therapy		
Resume therapy after symptoms fade		Reduce last dose by 50%, further increase by 10%		

Adapted from Pathirana D et al. European S3-guidelines on the systemic treatment of psoriasis vulgaris [50]

Table 24.2 Minimal erythema dose with narrow-band and broad-band UV-B

Study	Fitzpatrick skin type	MED BB-UVB (mJ/cm²)	MED NB-UVB (mJ/cm²)
Van Weelden et al.	II	76	410
Johnson et al.	II	100	500
Karvonen et al.	II	230	970
Storbeck et al.	II	114	1034
Srinivas et al.	IV	21	300
Tejasvi et al.	III–V	–	1000
Youn et al.	III–V	–	750–1075
Morita et al.	IV	–	700

From Racz E, Prens EP. Phototehrapy and Photochemotherapy of psoriasis [51]

shown to be less efficacious than total body irradiation, but is a practical solution for adjunctive home treatment of localized psoriasis, such as scalp, hand, nail or foot psoriasis [36].

Photochemotherapy can also be applied locally by using psoralen-containing gels or solutions (topical PUVA); this form of treatment is most often used for the treatment of psoriasis of the palms and soles.

24.3.3 Combination Treatment

Combination of phototherapy with topical crude coal tar (Goeckerman therapy) is a very effective treatment for psoriasis, that can be applied even after failure of biological treatment [37].

According to a meta-analysis, addition of topical vitamin D-derivatives to standard UVB-treatment does not lead to better efficacy of the phototherapy [38]. However, topical vitamin A-derivatives do improve clearance efficacy when added to either UVB or to PUVA therapies [38]. The addition of oral retinoids is also a highly effective treatment and leads to clearance in a shorter time than any of the two treatments alone [39].

As of combinations with systemic treatments, UVB and alefacept, UVB and adalimumab, and UVB and methotrexate combinations are more effective than UVB monotherapy [40, 41]. Caution is required with combinations of immunosuppressants and UVB and especially PUVA, because of the increased risk of skin cancer in the long-term. The combination of cyclosporine and phototherapy is therefore contraindicated.

24.3.4 Contraindications and Adverse Effects

Absolute contraindications are mutation(s) with an increased sensitivity for light (such as xeroderma pigmentosum or porphyria) or with an increased risk of skin cancer (such as Gorlin Goltz syndrome or epidermodysplasia verruciformis). Patients with lupus erythematosus in their medical history should also not enroll for this type of phototherapy. PUVA therapy is also contraindicated during pregnancy and lactation.

Due to the history of skin cancer and effects of certain reasons causing skin cancer (skin type I, dysplastic melanocytic nevi, high cumulative UVA dose previously administered) also require caution in the usage of phototherapy. Ask patients suffering from epilepsy and claustrophobia and non-compliant patients will also not benefit from phototherapy [42, 43].

The necessary use of photosensitizing or phototoxic medication or the medical history of photodermatoses or photosensitive disorders comprises a relative contraindication. Interestingly, some photodermatoses might allow the successful completion of phototherapy, such as recently reported by Nakamura et al. in a patient with iatrogenic polymorphous light eruption [44].

Five to 24 percent of all psoriatic patients report worsening of their psoriasis upon exposure to UV light [45]. For this group of patients the term photosensitive psoriasis is used. Even in this group of patients, PUVA might be an effective treatment of psoriasis.

Adverse effects on the short term are redness, itch and blistering, such as seen in a sunburn reaction. Long-term adverse effect might be the development of skin cancer, which has been shown after more than 300 sessions of PUVA therapy, induction of multiple lentigines, photoaging, and cataract formation [45].

24.3.5 Cost-Effectiveness

Outpatient office-based phototherapy is more cost-effective than systemic treatment, although phototherapy can be inconvenient for patients because of travel time and costs, and the costs of absence from work [46]. Home NB-UVB therapy can solve these problems for many patients. A Dutch study demonstrated equal efficacy and tolerability of hospital- and home-based NB-UVB phototherapy; in this study home phototherapy was not more expensive than hospital-based phototherapy, and was preferred by patients [47, 48].

Arzpayma et al. created a computerized modeling tool for the planning of the provision of home and hospital-based phototherapy [49]. Including e.g. the driving time, travel costs to calculate accessibility they could establish whether a certain area could be sufficiently provided by a hospital-based phototherapy unit and if so how many units were necessary to make phototherapy available for all patients with moderate to severe psoriasis, or whether home phototherapy units alone were a better option, e.g. for an area without a dermatology service.

24.4 Conclusion

Phototherapy is a highly effective treatment modality for the chronic inflammatory skin disease psoriasis, as well certain other ailments. Ultraviolet irradiation leads to apoptosis of T lymphocytes and of keratinocytes in the epidermis, and leads to immune counteract the pathological changes that characterize psoriasis.

Although treatment options for moderate to severe psoriasis are increasing due to novel treatments, their use are often limited by high costs and side-effects. Phototherapy remains first line treatment for certain patient groups where modification of the systemic immune reactions is contraindicated, such as by HIV or in the case of internal malignancy. Several studies point out the cost-effectiveness of UVB treatment, which remains important to consider in this age of constant increasing health care costs.

Continued awareness and education need to ensure the maintenance, improvement and optimization of the current phototherapy practices and the availability of this treatment option for patients with psoriasis worldwide.

References

1. Lim HW et al (2015) Phototherapy in dermatology: a call for action. J Am Acad Dermatol 72(6):1078–1080
2. Bolognia J, Jorizzo J, Rapini R (2003) Dermatology, vol 2. Mosby, Edinburgh/London/New York/Oxford/Philadelphia/St Louis/Sidney/Toronto
3. Stern R (2007) Psoralen and ultraviolet a light therapy for psoriasis. N Engl J Med 357(7):682–690
4. McGregor J, Hawk J (2003) Acute effects of ultraviolet radiation on the skin. In: Freedberg I et al (eds) Fitzpatrick's dermatology in general medicine. The McGraw-Hill Companies, Inc., New York
5. Maccubbin AE et al (1995) DNA damage in UVB-irradiated keratinocytes. Carcinogenesis 16(7):1659–1660
6. Jans J et al (2006) Differential role of basal keratinocytes in UV-induced immunosuppression and skin cancer. Mol Cell Biol 26(22):8515–8526
7. Rácz E et al (2011) Effective treatment of psoriasis with narrow-band UVB phototherapy is linked to suppression of the IFN and Th17 pathways. J Invest Dermatol 131:1547
8. De Fabo EC, Noonan FP (1983) Mechanism of immune suppression by ultraviolet irradiation in vivo. I. Evidence for the existence of a unique photoreceptor in skin and its role in photoimmunology. J Exp Med 158(1):84–98
9. Kammeyer A et al (1997) Prolonged increase of cis-urocanic acid levels in human skin and urine after single total-body ultraviolet exposures. Photochem Photobiol 65(3):593–598
10. Snellman E et al (1992) Effect of psoriasis heliotherapy on epidermal urocanic acid isomer concentrations. Acta Derm Venereol 72(3):231–233
11. Schade N, Esser C, Krutmann J (2005) Ultraviolet B radiation-induced immunosuppression: molecular mechanisms and cellular alterations. Photochem Photobiol Sci 4(9):699–708
12. Rosette C, Karin M (1996) Ultraviolet light and osmotic stress: activation of the JNK cascade through multiple growth factor and cytokine receptors. Science 274(5290):1194–1197
13. Aragane Y et al (1998) Ultraviolet light induces apoptosis via direct activation of CD95 (Fas/APO-1) independently of its ligand CD95L. J Cell Biol 140(1):171–182
14. Krueger JG et al (1995) Successful ultraviolet B treatment of psoriasis is accompanied by a reversal of

keratinocyte pathology and by selective depletion of intraepidermal T cells. J Exp Med 182(6):2057–2068
15. Weatherhead SC et al (2011) Keratinocyte apoptosis in epidermal remodeling and clearance of psoriasis induced by UV radiation. J Invest Dermatol 131(9):1916–1926
16. El-Domyati M et al (2013) Evaluation of apoptosis regulatory proteins in response to PUVA therapy for psoriasis. Photodermatol Photoimmunol Photomed 29(1):18–26
17. Walters IB et al (2003) Narrowband (312-nm) UV-B suppresses interferon gamma and interleukin (IL) 12 and increases IL-4 transcripts: differential regulation of cytokines at the single-cell level. Arch Dermatol 139(2):155–161
18. Piskin G et al (2003) IL-4 expression by neutrophils in psoriasis lesional skin upon high-dose UVB exposure. Dermatology 207(1):51–53
19. Zhang D et al (2016) Ultraviolet irradiation promotes FOXP3 transcription via p53 in psoriasis. Exp Dermatol 25(7):513–518
20. Gui J et al (2016) Therapeutic elimination of the type 1 interferon receptor for treating psoriatic skin inflammation. J Invest Dermatol 136:1990
21. Furuhashi T et al (2013) Photo(chemo)therapy reduces circulating Th17 cells and restores circulating regulatory T cells in psoriasis. PLoS One 8(1):e54895
22. Sigmundsdottir H et al (2005) Narrowband-UVB irradiation decreases the production of pro-inflammatory cytokines by stimulated T cells. Arch Dermatol Res 297:39
23. Batycka-Baran A et al (2016) The effect of phototherapy on systemic inflammatory process in patients with plaque psoriasis. J Photochem Photobiol B 161:396–401
24. Osmancevic A et al (2009) Vitamin D production in psoriasis patients increases less with narrowband than with broadband ultraviolet B phototherapy. Photodermatol Photoimmunol Photomed 25(3):119–123. 2009. 25(3): p. 119-23
25. Gupta A et al (2016) Efficacy of narrowband ultraviolet B phototherapy and levels of serum vitamin D3 in psoriasis: a prospective study. Indian Dermatol Online J 7(2):87–92
26. Ryan C et al (2010) Clinical and genetic predictors of response to narrowband ultraviolet B for the treatment of chronic plaque psoriasis. Br J Dermatol 163(5):1056–1063
27. Callis Duffin K et al (2014) Patient satisfaction with treatments for moderate-to-severe plaque psoriasis in clinical practice. Br J Dermatol 170(3):672–680
28. Miller DW, Feldman SR (2006) Cost-effectiveness of moderate-to-severe psoriasis treatment. Expert Opin Pharmacother 7(2):157–167
29. Nast A et al (2015) European S3-Guidelines on the systemic treatment of psoriasis vulgaris – Update 2015 – Short version – EDF in cooperation with EADV and IPC. J Eur Acad Dermatol Venereol 29(12):2277–2294
30. Chen X et al (2013) Narrow-band ultraviolet B phototherapy versus broad-band ultraviolet B or psoralen-ultraviolet A photochemotherapy for psoriasis. Cochrane Database Syst Rev 10:CD009481
31. Lapolla W et al (2011) A review of phototherapy protocols for psoriasis treatment. J Am Acad Dermatol 64(5):936–949
32. Markham T, Rogers S, Collins P (2003) Narrowband UV-B (TL-01) phototherapy vs oral 8-methoxypsoralen psoralen-UV-A for the treatment of chronic plaque psoriasis. Arch Dermatol 139(3):325–328
33. Archier E et al (2012) Efficacy of psoralen UV-A therapy vs. narrowband UV-B therapy in chronic plaque psoriasis: a systematic literature review. J Eur Acad Dermatol Venereol 26(Suppl 3):11–21
34. Coimbra S et al (2013) Principal determinants of the length of remission of psoriasis vulgaris after topical, NB-UVB, and PUVA therapy: a follow-up study. Am J Clin Dermatol 14(1):49–53
35. Gerber W et al (2003) Ultraviolet B 308-nm excimer laser treatment of psoriasis: a new phototherapeutic approach. Br J Dermatol 149(6):1250–1258
36. Almutawa F et al (2015) Efficacy of localized phototherapy and photodynamic therapy for psoriasis: a systematic review and meta-analysis. Photodermatol Photoimmunol Photomed 31(1):5–14
37. Fitzmaurice S, Bhutani T, Koo J (2013) Goeckerman regimen for management of psoriasis refractory to biologic therapy: the University of California San Francisco experience. J Am Acad Dermatol 69(4):648–649
38. Bailey EE et al (2012) Combination treatments for psoriasis: a systematic review and meta-analysis. Arch Dermatol 148(4):511–522
39. Ruzicka T et al (1990) Efficiency of acitretin in combination with UV-B in the treatment of severe psoriasis. Arch Dermatol 126(4):482–486
40. Asawanonda P, Nateetongrungsak Y (2006) Methotrexate plus narrowband UVB phototherapy versus narrowband UVB phototherapy alone in the treatment of plaque-type psoriasis: a randomized, placebo-controlled study. J Am Acad Dermatol 54(6):1013–1018
41. Wolf P et al (2011) 311 nm ultraviolet B-accelerated response of psoriatic lesions in adalimumab-treated patients. Photodermatol Photoimmunol Photomed 27(4):186–189
42. Nast A et al (2012) German S3-guidelines on the treatment of psoriasis vulgaris (short version). Arch Dermatol Res 304(2):87–113
43. Menter A et al (2010) Guidelines of care for the management of psoriasis and psoriatic arthritis: section 5. Guidelines of care for the treatment of psoriasis with phototherapy and photochemotherapy. J Am Acad Dermatol 62(1):114–135
44. Nakamura M, Bhutani T, Koo JY (2016) Narrowband UVB-induced iatrogenic polymorphous light erup-

tion: a case and suggestions to overcome this rare complication. Dermatol Online J 22(6)
45. Wolf P et al (2016) Desired response to phototherapy versus photo-aggravation in psoriasis: what makes the difference? Exp Dermatol 25(12):937–944
46. D'Souza LS, Payette MJ (2015) Estimated cost efficacy of systemic treatments that are approved by the US Food and Drug Administration for the treatment of moderate to severe psoriasis. J Am Acad Dermatol 72(4):589–598
47. Koek MB et al (2009) Home versus outpatient ultraviolet B phototherapy for mild to severe psoriasis: pragmatic multicentre randomised controlled non-inferiority trial (PLUTO study). BMJ 338:b1542
48. Koek MB et al (2010) Cost effectiveness of home ultraviolet B phototherapy for psoriasis: economic evaluation of a randomised controlled trial (PLUTO study). BMJ 340:c1490
49. Arzpayma P et al (2016) Creation and assessment of a computerized modeling tool for optimizing planning of home and hospital-based phototherapy. Br J Dermatol 176:1390
50. Pathirana D et al (2009) European S3-guidelines on the systemic treatment of psoriasis vulgaris. J Eur Acad Dermatol Venereol 23(Suppl 2):1–70
51. Racz E, Prens EP (2015) Phototherapy and photochemotherapy for psoriasis. Dermatol Clin 33(1):79–89

Ultraviolet Irradiation of Blood: "The Cure That Time Forgot"?

25

Michael R. Hamblin

Abstract

Ultraviolet blood irradiation (UBI) was extensively used in the 1940s and 1950s to treat many diseases including septicemia, pneumonia, tuberculosis, arthritis, asthma and even poliomyelitis. The early studies were carried out by several physicians in USA and published in the American Journal of Surgery. However with the development of antibiotics, UBI use declined and it has now been called "the cure that time forgot". Later studies were mostly performed by Russian workers and in other Eastern countries and the modern view in Western countries is that UBI remains highly controversial.

This chapter discusses the potential of UBI as an alternative approach to current methods used to treat infections, as an immune-modulating therapy and as a method for normalizing blood parameters. No resistance of micro-organisms to UV irradiation has been reported, and multi-antibiotic resistant strains are as susceptible as their wild-type counterparts. Low and mild doses of UV kill microorganisms by damaging the DNA, while any DNA damage in host cells can be rapidly repaired by DNA repair enzymes. However the use of UBI to treat septicemia cannot be solely due to UV-mediated killing of bacteria in the blood-stream, as only 5–7% of blood volume needs to be treated with UV to produce the optimum benefit. UBI may enhance the phagocytic capacity of various phagocytic cells (neutrophils and dendritic cells), inhibit lymphocytes, and oxidize blood lipids. The oxidative nature of UBI may have mechanisms in common with ozone therapy and other oxygen therapies. There may be some similarities to

M.R. Hamblin (✉)
Wellman Center for Photomedicine, Massachusetts General Hospital, Boston, MA 02114, USA

Department of Dermatology, Harvard Medical School, Boston, MA 02115, USA

Harvard-MIT Division of Health Sciences and Technology, Cambridge, MA 02139, USA
e-mail: Hamblin@helix.mgh.harvard.edu

S. Ahmad (ed.), *Ultraviolet Light in Human Health, Diseases and Environment*, Advances in Experimental Medicine and Biology 996, https://doi.org/10.1007/978-3-319-56017-5_25

extracorporeal photopheresis (ECP) using psoralens and UVA irradiation. However there are differences between UBI and ECP in that UBI tends to stimulate the immune system, while ECP tends to be immunosuppressive. With the recent emergence of bacteria that are resistant to all known antibiotics, UBI should be more investigated as an alternative approach to infections, and as an immune-modulating therapy.

Keywords
Ultraviolet C • Knott hemo-irradiator • UBI • DNA repair • Blood cells • Antigen-presenting cells • Infections • Cytokines

25.1 Historical Introduction

Ultraviolet (UV) radiation is part of the electromagnetic spectrum with a wavelength range (100–400 nm) shorter than that of visible light (400–700 nm), but longer than x-rays (<100 nm). UV radiation is divided into four distinct spectral areas including vacuum UV (100–200 nm), UVC (200–280 nm), UVB (280–315 nm) and UVA (315–400 nm). Only part of UVB and UVA can reach on earth, because wavelengths shorter than 280 nm are filtered out by the atmosphere especially by the "ozone layer".

In 1801 Johann Wilhelm Ritter, a Polish physicist working at the University of Jena in Germany discovered a form of light beyond the violet end of the spectrum that he called "Chemical Rays" and which later became "Ultraviolet" light [1]. In 1845, Bonnet [2] first reported that sunlight could be used to treat tuberculosis arthritis (a bacterial infection of the joints).

In the second half of the nineteenth century, the therapeutic application of sunlight known as heliotherapy gradually became popular. In 1855, Rikli from Switzerland opened a thermal station in Veldes in Slovenia for the provision of heliotherapy [3]. In 1877, Downes and Blunt discovered by chance that sunlight could kill bacteria [4]. They noted that sugar water placed on a window-sill turned cloudy in the shade but remained clear while in the sun. Upon microscopic examination of the two solutions, they realized that bacteria were growing in the shaded solution but not in the one exposed to sunlight.

In 1904, the Danish physician Niels Finsen was awarded the Nobel Prize in Physiology or Medicine for his work on UV treatment of various skin conditions. He had a success rate of 98% in thousands of cases, mostly the form of cutaneous tuberculosis known as lupus vulgaris [5]. Walter H Ude reported a series of 100 cases of erysipelas (a cutaneous infection caused by *Streptococcus pyogenes)* in the 1920s, with high cure rates using irradiation of the skin with UV light [6].

Emmett K Knott (Fig. 25.1) in Seattle, WA reasoned that the beneficial effects of UV irradiation to the skin obtained by Ude, might (at least

Fig. 25.1 Emmett K Knott

partly) be explained by the irradiation of blood circulating in the superficial capillaries of the skin. With his collaborator Edblom, an irradiation chamber was constructed to allow direct exposure of the blood to UV. The irradiation chamber was circular and contained a labyrinthine set of channels that connected the inlet and outlet ports. All these channels were covered with a quartz window that formed the top of the chamber. The irradiation chamber was so designed as to provide maximum turbulence of the blood flowing through (see Fig. 25.2). This was done in order to: (a) prevent the formation of a thin film of blood on the chamber window that would absorb and filter out much of the UV light; (b) insure that all the blood passing through the chamber was equally exposed to UV [7].

Knott and co-workers then carried out a series of experiments using UV irradiation of blood extracted from dogs that had been intravenously infected with *Staphylococcus aureus* bacteria and hemolytic *Streptococcus* species, and then the treated blood was reinfused into the dogs. They found that it was unnecessary to deliver a sufficient exposure of UV light to the blood to directory kill all the bacteria in the circulation. It was also found unnecessary to expose the total blood volume in the dogs. The optimum amount of blood to be irradiated was determined to be only 5–7% of the estimated blood volume or approximately 3.5 mL per kg of body weight. Exceeding these limits led to loss of the benefits of the therapy. All the dogs that were treated with the optimized dose of UV to the blood, recovered from an overwhelming infection (while many dogs in the control group died). None of the dogs that were treated and survived, showed any long-term ill effects after 4 months of observation [7].

The first treatment on a human took place in 1928 when a patient was determined to be in a moribund state after a septic abortion complicated by hemolytic streptococcus septicemia. UBI therapy was commenced as a last resort, and the patient responded well to the treatment and made a full recovery [7]. She proceeded to give birth to two children.

Hancock and Knott [8] had similar success in another patient suffering from advanced hemolytic streptococcal septicemia. These workers noted that in the majority of cases, a marked cyanosis (blue tinge to the skin caused by a lack of oxygenated blood flow) was present at the time of initiation of UBI. It was noted that during (or immediately following) the treatment a rapid relief of the cyanosis occurred, with improvement in respiration accompanied by a noticeable flushing of the skin, with a distinct loss of pallor.

Fig. 25.2 The Knott Hemo-Irradiator

These observations led to application of UBI in patients suffering from pneumonia. In a series of 75 cases in which the diagnoses of pneumonia were confirmed by X-rays, all patients responded well to UBI showing a rapid decrease in temperature, disappearance of cyanosis (often within 3–5 min), cessation of delirium if present, a marked reduction in pulse rate and a rapid resolution of pulmonary consolidation. A shortening of the time of hospitalizations and accelerated convalescence was regularly observed.

The knowledge gained in these successful studies led to the redesign of the irradiation chamber to allow a more thoroughly uniform exposure of the circulating blood, and led to the development of the "Knott Technic of Ultraviolet Blood Irradiation." A number of irradiation units were manufactured and placed in the hands of physicians interested in the procedure, so that more extensive clinical data could be accumulated [7]. The Knott technique involved removing approximately 3.5 mL/kg venous blood, citrating it as an anticoagulant, and passing it through the radiation chamber. The exposure time per given unit of blood was approximately 10 s, at a peak wavelength of 253.7 nm (ultraviolet C) provided by a mercury quartz burner, and the blood was immediately re-perfused [7].

George P Miley at the Hahnemann Hospital, Philadelphia, PA published a series of articles on the use of the procedure in the treatment of thrombophlebitis, staphylococcal septicemia, peritonitis, botulism, poliomyelitis, non-healing wounds, and asthma [9–22].

Henry A Barrett at the Willard Parker Hospital in New York City in 1940 reported on 110 cases including a number of different infections. Twenty-nine different conditions were described as being responsive, including the following: infectious arthritis, septic abortion, osteoarthritis, tuberculosis glands, chronic blepharitis, mastoiditis, uveitis, furunculosis, chronic paranasal sinusitis, acne vulgaris, and secondary anemia [23, 24].

EV Rebbeck at the Shadyside Hospital in Pittsburgh, PA, reported the use of UBI in *Escherichia coli* septicemia, post-abortion sepsis, puerperal sepsis, peritonitis, and typhoid fever [25–29] and Robert C Olney at the Providence Hospital, Lincoln, NE, treated biliary disease, pelvic cellulitis and viral hepatitis [30–32].

In this chapter, we will discuss the mechanisms and the potential of UBI as an alternative approach to infections and as a new method to modulate the immune system. Our goal is to remind people to continue to do more research and explore more clinical uses. The topics include the efficacy of UBI for infections (both bacterial and viral), to cure autoimmune disease, disease, and the similarities and differences between UBI, and intravenous ozone therapy, and extracorporeal psoralen-mediated photochemotherapy (photophoresis).

25.2 Mechanisms of Action of UBI

One of the major obstacles that UBI has consistently faced throughout the almost 90 years since the first patient was treated has been the lack of understanding of the mechanisms of action. Over the years its acceptance by the broad medical community has been hindered by this uncertainty. Confusion has been caused by the widely held idea that since UV is used for sterilization of water and surgical instruments; therefore its use against infection must also rely on UV-mediated direct destruction of pathogens. Another highly confusing aspect is the wide assortment of diseases, which have been claimed to be successfully treated by UBI. It is often thought that something that appears to be "too good to be true" usually is.

UBI affects various functions of red blood cells and various different leukocytes as has been proven in various in vitro studies. A common model is stimulator cells in mixed leukocyte cultures; another is helper cells in mitogen-stimulated cultures. UV also reversed cytokine production and blocked cytokine release. UV can also disturb cell membrane mobilization (Fig. 25.3).

25.2.1 Effects on Red Blood Cells

Anaerobic conditions strongly inhibited the process by which long wave UV light induces the loss of K^+ ions from red blood cells. Kabat proved that UV-irradiation could affect the osmotic properties of red blood cells, the submicroscopic structure and the metabolism of adenine nucleotides. Irradiation times (60, 120, 180, 240 and 300 minutes) were used; during the irradiation, ATP decreased while the amounts of ADP, AXP, adenine compounds all increased. UV also increased hypotonic Na + and K+ ion exchange and the hematocrit value increased [33].

When Rh-positive blood was irradiated with UV light there was a significant increase in immunosorption activity. Vasil'eva et al. [34] studied varying UV irradiation conditions on both red blood cells and leucocyte-thrombocyte suspensions. Immunosorption activity increased immediately after irradiation in whole blood and red blood cells; however the immunosorption capacity in leucocyte–thrombocyte suspensions was lost after 2 days.

A two-phase polymer system containing polydextran was used to show that the cell surface of circulating erythrocytes was reduced after UV

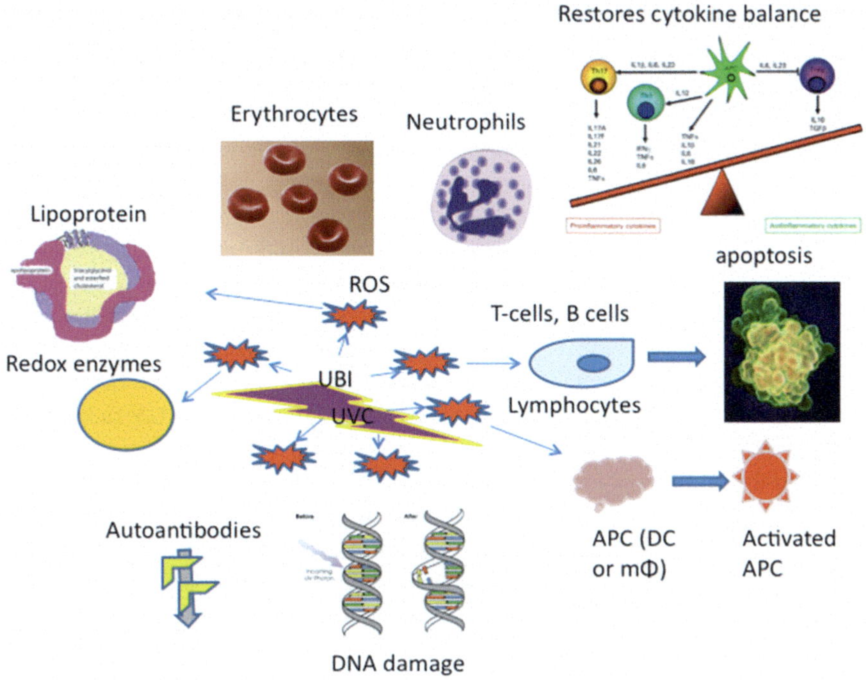

Fig. 25.3 Proposed mechanisms of UBI

irradiation. This contributed to the prolongation of survival of transfused erythrocytes and was suggested to explain the more effective therapeutic activity of autotransfused blood [35]. Snopov et al. suggested that some structural alterations in the erythrocytes, particularly in the glycocalyx were related to the improved effect of autotransfused blood after UV-irradiation [36]. Ichiki et al. showed that the cellular volume and the membrane potential of erythrocytes could be changed by UV irradiation. However an excessive dose of UV could decrease the production of H_2O_2 [37].

25.2.2 Effects on Neutrophils

Lower doses of UV (<0.1 J/cm^2) increased the production of peroxides (H_2O_2) by polymorphonuclear leukocytes (which is the largest amongst all the different blood cells). The ability of UBI to increase the production of reactive oxygen species (ROS) by neutrophils could be inhibited by addition of arachidonic acid or lysophosphatidylcholine (LPC), as well as the anti-oxidant, α-tocopherol [38]. In chronic inflammatory diseases, the concentration of large IC--IgG, IgM, and small IC–IgM showed an inverse linear correlation with increased UBI dose delivered to autotransfused blood [39].

Artiukhov suggested that the generation of nitric oxide (NO) by photomodified neutrophils was due to the activation of the iNOS enzyme. De novo NO synthesis was increased by UV-irradiation, which also affected TNF-alpha production. Irradiation with lower dose (75.5 J/m2) allowed the maintenance of the physiological homeostasis. While higher dose (755 and 2265 J/m2) delivered to neutrophils led to potential damage, by increasing the concentration of NO metabolites. When UV-irradiated cells were incubated with the transcriptional inhibitor of protein synthesis, cycloheximide the activation of iNOS and NO synthesis was prevented. High doses of UV-irradiation (755 J/m2) on neutrophils, showed a positive correlation between NO and TNF-alpha concentrations [40].

Zor'kina carried out a 30-day rabbit experiment, and suggested that the chronic stress produced with a combination of hypodynamia and UBI, affected neutrophils and eliminated coagulation. UBI contributed to improvement in the body's abilities to resist long-term hypodynamia and ameliorated chronic stress. UBI enhanced the adaptive process through activated neutrophils, prevented disseminated intravascular coagulation, and changed the atherogenic metabolic profile [41].

25.2.3 Effects on Lymphocytes (T-Cells and B-Cells)

UBI generally decreases lymphocyte viability. UVC irradiation is the most effective among the three UV spectral regions. UVB and UVC irradiation can abolish the proliferative and stimulatory ability as well as the accessory/antigen-presenting ability of lymphocytes in vitro. The cell-surface properties, calcium mobilization, cytokine production and release, and other sub cellular processes could all be changed by UV irradiation [42]. Areltt et al. used the "Comet "assay to detect DNA-strand breakage (single cell gel electrophoresis) as an indicator of excision repair to prove that circulating human T–lymphocytes were exquisitely hypersensitive to the DNA-damaging and lethal effects of UV-B radiation, raising the possibility that UV-B may make a contribution to immunosuppression via a direct effect on extracapilliary T-lymphocytes [43].

Teunissen et al. suggested that UVB radiation neither selectively affects either Th1 or Th2 nor CD4 or CD8 T-cell subsets. Compared with different dose of UVB irradiation, although the phototoxic effect was not immediately apparent, a low dose of UVB (LD50: 0.5–1 mJ/cm^2) irradiation was sufficient to kill most T cells after 48–72 h [44]. There was a dose dependent reduction in all measured cytokines (IL-2, IL-4, IL-5, IFN-ϒ, TNF-a) in the same way 72 h after irradiation. This fall in production was indicated by a remarkable correlation between loss of viability and reduction of cytokine production that may be caused directly by cell death. However, CD4+ or CD8+ T cell subsets, expression of CD4 and CD8 as well as the CD4/CD8 ratio compared with the non-irradiated control, was not altered by UVB, suggesting that none of the T-cell subsets was selectively affected.

Schieven et al. observed that UV-induced tyrosine phosphorylation in B cells after surface immunoglobulin cross-linking. This observation was very similar to the production of Ca^{2+} signals in T cells. It means that UV irradiation of lymphocytes could induce both tyrosine phosphorylation and Ca^{2+} signals. Ca^{2+} channels in lymphocyte membranes are sensitive to UV irradiation; UV radiation causes DNA damage through the activation of cellular signal-transduction processes. UV radiation (depending on dose and wavelength) not only induces tyrosine phosphorylation in lymphocytes but also Ca^{2+} signals in Jurkat T cells. Furthermore, the pattern of surface immunoglobulin cross-linking was similar to the UV-induced tyrosine phosphorylation in B cells. The UBI effect on lymphocyte function may play an important role in tyrosine phosphorylation and Ca^{2+} signals, which can escape from normal receptor control. They showed that both CD4+ and CD8+ T cells (normal human lymphocytes) gave strong reactions during UV-irradiation [45].

In a similar study, Spielberg et al. found that UV-induced lymphocyte inhibition showed a similar course in disruption of Ca^{2+} homeostasis by comparing UV with gamma irradiation, which have different effects on lymphocyte membranes [46]. Furthermore, the presence of Ca^{2+} channels in lymphocyte membranes that are sensitive to UV irradiation was shown by indo-1 staining and cytofluorometry. Intracellular calcium $[Ca^{2+}]$i kinetics was measured in UVC or UVB-exposed human peripheral blood leukocytes (PBL) and Jurkat cells were in parallel with functional assays. The UV-induced i$[Ca^{2+}]$ rise was predominantly due to an influx of extracellular calcium, and it was more pronounced in T-cells than in non-T cells. It was observed that $[Ca^{2+}]$i increased within 2–3 h of irradiation; these increases were UV-dose dependent and reached maxima of 240% and 180% above the baseline level (130 nM) for UVB and UVC. UV induced a bigger $[Ca^{2+}]$i rise in T-cells than in non-T cells, due to the influx of extracellular calcium. UV-induced calcium shifts, and UV irradiation on the plasma membrane decreased the sensitivity to respond to phytohemagglutinin (PHA) in mixed leukocyte cultures.

A series of studies confirmed that UVR irradiated lymphocytes were not able to induce allogeneic cells in the mixed lymphocyte culture (MLC) as first reported by Lindahl-kiessling. [47–49]. Clusters formed by specialized accessory cells after mitogenic or allogenic stimulation, with dendritic cells (DC) are necessary for lymphocyte activation to occur. Aprile found that UV irradiation of DC before culture completely abrogated accessory activity was capable to block both cluster formation and no lymphocyte proliferation occurred [50].

Kovacs et al. [51] found that induction of DNA repair mechanisms was dependent on the dose of UVC light between 2 and 16 J/cm^2. It was evaluated in irradiated and non-irradiated lymphocytes in 51 healthy blood donors. UVC irradiation (253.7 nm) at doses of 2, 4, 8 and 16 J/m^2, by measuring [^{3}H] thymidine incorporation in the presence of 2 mM hydroxyurea added 30 min before irradiation to inhibit DNA-replication synthesis. No significant age-related difference was found in donors between 17 and 74 years.

UV-induced differentiation in human lymphocytes, and accelerated the intensity of DNA repair in these cells [52]. Exposure to UV irradiation was more effective than methyl methane sulfonate (MMS) in increasing unscheduled DNA synthesis, especially when MMS was added prior to the UV-irradiation, at 2 h or 26 h before UVC, because MMS affects DNA repair by alkylating the DNA polymerase [53]. Photo-modification of HLA-D/DR antigens could be a trigger mechanism for activation of immunocompetent cells by UV-irradiation. Lymphocytes were isolated from non-irradiated blood, irradiated blood and a mixture of the two in different ratios (1:10,1:40,1:160) [54].

UBI before transfusion can inhabit immune recognition and prevent bone marrow graft rejection in vivo. After 9.2 Gy of total body irradiation (TBI) and $2.8 \pm 2.1 \times 10^8$/Kg donor marrow cells were infused, whole blood was exposed for 30 minutes to UV light at a dose of 1.35 J/cm^2, and then injected into the recipient dogs. The control group, which was transfused with sham-exposed blood, rejected the bone marrow grafts, while no rejection was found in the group, which received UV-exposed blood before the transplanted marrow. UV irradiation on blood inhibited lymphocyte activation by eliminating a critical DC-dependent signal [55].

Oluwole et al. suggested that transfusion of UV-irradiated blood into recipients prior to heart transplantation could be carried out, in order to inhibit immune response, and reduce lymphocyte-mediated rejection [56]. Three sets of different rat strains (ACI, Lewis, W/F) were used for heart transplantation in his research. In the series where ACI rats received a Lewis heart, 1 mL transfusion of donor-type blood with or without UV-irradiation was transfused at 1, 2, and 3 weeks prior to the transplantation. A mixed lymphocyte reaction showed that ACI lymphocytes were weaker responders to Lewis lymphocytes, and the same as the other two series of different type heart transplantations. UV irradiation of donor rhesus-positive blood can be used to increase the therapeutic effect of blood exchange transfusion in children with rhesus-conflict hemolytic disease [57].

25.2.4 Effects on Monocytes, Macrophages and Dendritic Cells

All these types of blood cells including monocytes, macrophages and dendritic arise from the myelocytic lineage of hematogenous stem cells, and act as phagocytes and antigen presenting cells. The phagocytic capacity of UV-B irradiated mononuclear cells derived from human peripheral blood could be enhanced by all four types of deoxyribonucleoside supplementation [58].

Stimulation of phagocytic activity (PhA) appears to be one of the earliest mechanisms in immuno-correction by UV-irradiation of blood therapy. In Samoĭlova's research, non-irradiated blood, mixed with 1:10 irradiation blood, were tested for PhA of monocytes and granulocytes. Increase of 1.4–1.7 times of PhA compare with non-supplemented blood, because monocytes and granulocytes could be increase by adding UV-irradiated blood into healthy adults. The enhancement of PhA depends on its initial level and may occur simultaneously with structural changes of the cell surface components [59].

UV-irradiation increased the phagocytic activity of human monocytes and granulocytes, and the "integrated phagocytic index" increased in proportion to the irradiation dose, while a lower initial level would increase more than a higher initial rate after UV-irradiation [60].

Simon et al. [61] concluded that UVB could convert Langerhans cells (LC) or splenic adherent cells (SAC) from an immunogenic phenotype into a tolerogenic phenotype, as far as antigen presenting cells were concerned (LC or SAC). In his research, a single dose of irradiation (200 J/cm^2) was delivered to LC and SAC. The loss of responsiveness was found when UV-LC or UV-SAC were incubated with Th1 cells that had been pre-incubated with keyhole limpet hemocyanin (KLH). Furthermore, such loss of responsiveness was not related to the release of soluble suppressor factors, but was Ag-specific, MHC-restricted, and long-lasting. The hypothesis to explain these results was that delivery of a costimulatory signal(s) had been interfered with by UVB irradiation, because unresponsiveness by UVB-LC or UVB-SAC could not be induced by non-irradiated allogeneic SAC.

25.2.5 Effects on Platelets

H_2O_2 production in platelets is low at very low UV dose, but it increased suddenly as the dose increased above 0.4 J/cm^2. Pamphilon reported that platelet concentrates (PC) could become non-immunogenic after UVR and after being stored for 5 days in DuPont Stericell containers. Lactate levels, β-thromboglobulin and platelet factor were higher after UV, while glucose levels decreased with an irradiation dose of 3000 J/m^2 at a mean wavelength of 310 nm applied in DuPont Stericell bags [62]. Ultraviolet B (UVB) irradiation of platelet concentrate (PC) accelerated downregulation of CD14 and nonspecifically increased the loss of monocytes by inhibiting the upregulation of ICAM-1 and HLA-DR [63]. However, UV irradiation of platelet concentrates produced a reduction of immunological response in a cell suspension [64–66].

25.2.6 Effects on Low Density Lipoprotein (LDL) and Lipids

Roshchupkin et al. found that UV irradiation played a core role in lipid peroxidation in the membranes of blood cells [67]. UV irradiation of blood could stimulate arachidonic acid to be metabolized by cyclooxygenase, and could induce dark lipid autoperoxidation into free radicals and direct photolysis of photooxidants. UV contributed to lipid photoperoxidation producing lipid hydroperoxides.

UV irradiated lipid emulsion greatly enhanced the production of reactive oxygen species (ROS) by monocytes, and highly atherogenic oxidized LDL could be generated in the blood circulation. UV light-oxidized lipofundin (a parenteral lipid emulsion designed for injection) was injected into rabbits, then blood samples were taken from the ear vein with EDTA (before and 6 h after) lipofundin treatment. Although UV-oxidized lipofundin induced less chemiluminescence from monocytes compared with Fe^{3+}-oxidized lipofundin, the effect lasted 2.3 times longer. UV–oxidized lipofundin could more effectively stimulate H_2O_2 production than monocyte-oxidized LDL, even with the same concentration of thiobarbituric acid reactive substances (TBARS) in the preparations. Six hours after injection of oxidized lipofundin, the lipid peroxide content was significantly increased, however the neutral lipids in LDL isolated from rabbit plasma showed no significantly difference to the monocyte-oxidized human LDL [68].

Salmon found that UVB (280–315 nm) irradiation could easily damage LDL and also the tryptophan (Trp) residues in high density lipoprotein (HDL) [69]. The TBARS assay was used to measure the photooxidation of tryptophan residues which accompanied the peroxidation of low and high density lipoprotein unsaturated fatty acids. Vitamin E and carotenoids were also rapidly destroyed by UVB. However UVA radiation could not destroy tryptophan residues and cause lipid peroxidation.

UV radiation (wavelength range 290–385 nm) easily oxidized the lipoproteins contained in the suction blister fluid of healthy volunteers, which is a good model of the interstitial fluid feeding the epidermal cells. Apolipoprotein B of LDL and apolipoprotein A-I and II of HDL were all altered in a similar way under UV irradiation. Irradiation with wavelengths in the range 290–385 nm altered the single Trp (tryptophan) residue of serum albumin which is susceptible to photo-oxidation. UVA irradiation of undiluted suction blister fluid induced A-I aggregation; however purified lipoproteins were not degraded. During UV irradiation of suction blister fluid, antigenic apolipoprotein B is fragmented and polymerized. Reactive oxygen radicals in the suction blister fluid were derived from lipid peroxidation occurring in HDL. UV-light irradiation could play an important role in triggering inflammation and degeneration by inducing lipoprotein photo-oxidation which could have systemic effects [70].

25.2.7 Redox Status

Artyukhov et al. [71] discovered that dose-dependent UV-irradiation could activate the myeloperoxidase (MPO) and the NADPH-oxidase systems in donor blood. Two doses of UV-light were used (75.5 and 151.0 J/m^2) and the higher dose activated more free radicals and H_2O_2 than the lower dose, another two groups were divided by the type of relationship between MPO activity and UV light dose (from 75.5 to 1510 J/m^2), low enzyme activity (group 1) increased under the effect of UV exposure at doses of 75.5 and 151.0 J/m^2, while in group 2 this parameter (MPO activity) decreased. MPO activity showed the same results in dose dependent UV-irradiation, however, increasing the dose to 1510 J/m^2 could not increase the activity of MPO. In the next experiments, lipid peroxidation (LPO) was evaluated after UV exposure of the blood. Two groups of donors were distinguished by the relationship between blood content of LPO products and UV exposure dose. UV irradiation at low doses (75.5–151.0 J/m^2) decreased initially high LPO values and increased initially low LPO levels. In phagocytes, NADPH-oxidase plays one of the most important roles as a photoacceptor for UV light. NADPH oxidase causes increased superoxide ($O_2^{\bullet -}$) production after UV-irradiation of blood by activation of the enzyme complex. UV irradiation also decreases intracellular pH caused by activation of the NADPH-oxidase complex.

UBI can also protect against free radical damage by elevating the activity of various antioxidants after spinal cord injury in rabbits, 186 rabbits were randomly divided into 4 groups, (control, blood transfusion, injured and UV treatment). UV irradiation (wavelength 253.7 nm, 5.68 mW/m^2) was used in the treatment group at 48–72 h after surgery for spinal cord injury. Free radical signals (FR), malondialdehyde (MDA), superoxide dismutase (SOD) and glutathione peroxidase (GSH-PX) were measured. In the treatment group, superoxide dismutase and glutathione peroxidase were much increased and showed significant differences compared with the other groups, while FR and MDA decreased significantly compared to other groups. Because UV irradiation of blood decreased the MDA and FR content in spinal cord tissue; they also suggested that these two factors contributed to higher SOD activity and increased GSH-PX [72].

25.2.8 Conclusions Regarding Mechanisms

UBI has always caused much confusion, both in the general public and also in some medical professionals, because germicidal UV light (UVC) is used to sterilize water, disinfect surfaces, and as an aid to infection control in operating rooms, and food processing and packaging plants. Many people therefore assume that UBI must act by killing pathogens (bacteria, viruses or other microorganisms) circulating in the bloodstream. However there is no evidence that this is actually the case. Therefore the mechanisms of action must lie in some other action of UV on the various components of blood. Although the entire body of evidence on the mechanisms of action of UBI is very complex, as can be seen from the foregoing material, we can attempt to draw some general conclusions. Firstly UBI is clearly an example of the well-known phenomenon called "hormesis" or "biphasic dose response'. This phenomenon has been well reviewed by Edward Calabrese from U Mass Amherst [73, 74]. The basic concept states that any toxic chemical substance or drug, or any physical insult (such as ionizing radiation, hyperthermia, or oxidative stress) can be beneficial, protective or even therapeutic, ***provided the dose is low enough.*** If the dose is increased, the beneficial or protective effects disappear, and if the dose is even further increased, then the detrimental effects of the treatment become very evident. This is clearly shown by Knott's original experiments on dogs that led to the establishment of only 5–7% of total blood volume as the optimal amount of blood to be irradiated.

UBI appears to have three broadly different classes of effects on different blood components. In the case of neutrophils, monocytes, macrophages, and dendritic cells, UBI can activate phagocytosis, increase the secretion of NO and reactive nitrogen species, and convert the DC phenotype from an immunogenic one into a tolerogenic one, thus perhaps lessening the effects of a "cytokine storm" as is often found in sepsis. In the case of lymphocytes, the effects of UBI are to inhibit (or in fact kill) various classes of lymphocytes. This is not perhaps very surprising, considering the well-established cell-death pathways and apoptotic signaling found in lymphocytes. However it is not impossible, that the killing of circulating lymphocytes could reduce systemic inflammation, which would again be beneficial in cases of sepsis. It is also clear that UBI can oxidize blood lipids and lipoproteins, and therefore increase oxidative stress. However it is also possible that a brief burst of oxidative stress, may be beneficial, whereas continued chronic levels of oxidative stress have been generally considered as detrimental. Many antioxidant defenses are up-regulated by brief exposure to oxidative stress, and this has been postulated to be one of the fundamental mechanisms responsible for may aspects of hormesis. The oxidative nature of UBI has encouraged us to draw parallels with ozone therapy and other forms of 'oxygen therapy".

25.3 Ozone Therapy

Since UBI is generally considered to be controversial, then ozone therapy is even more controversial. Ozone therapy consists of the introduction of ozone (O_3) into the body via various methods, usually involving its mixture with various gases and liquids before injection, with potential routes including the vagina, rectum, intramuscular, subcutaneously, or intravenously Ozone can also be introduced via a process called "autohemotherapy", in which blood is drawn from the patient, exposed to ozone and re-injected into the patient [75]

The United States Food and Drug Administration initially stated in 1976, and reiterated its position in 2006, "that when inhaled, ozone is a toxic gas which has no demonstrated safe medical application", though their position statements primarily deal with its potential for causing inflammation and pulmonary edema in the lungs. Moreover there exist additional types of "oxygen therapy" involving hyperbaric oxygen, hydrogen peroxide, and various kinds of "oxygenated water".

25.4 Extracorporeal Photochemotherapy (ECP)

Extracorporeal photochemotherapy (ECP) involves the addition of a photosensitizing drug 8-methoxypsoralen (8-MOP) into blood that is then treated with UVA light (320–360 nm). ECP was originally derived from the use of PUVA (psoralen and UVA) to treat psoriasis and other skin diseases. In the case of dermatology the psoralen was administered either orally (pills) or as a bath therapy. Often the whole body was exposed to light in a "PUVA box" containing UVA emitting fluorescent tube lights. ECP has been widely used as immunotherapy for cutaneous T cell lymphoma (CTCL) since it received US Food and Drug Administration (FDA) approval in 1988. As an apheresis-based immunomodulatory therapy which involves UVA irradiation of autologous peripheral blood mononuclear cells (PBMCs) exposed to the 8-MOP, there are a numbers of features of ECP that distinguish it from other immunologic therapy, which are beneficial in immune-stimulation against cancer and in the transplant setting as an immune-modulator; for induction of antigen presenting cells (APCS), to extracorporeal sequester and modify processed leukocytes, and so on. [76] It has used for treatment of other autoimmune-mediated disorders and organ allograft rejection, and is especially beneficial for cutaneous T-cell lymphoma (CTCL) and graft-versus host disease (GVHD). Both these indications require killing of lymphocytes.

25.4.1 ECP Therapy Treatment

The standard schedule of ECP treatment involves 2 successive days at 4 week intervals Tens of thousands of patients afflicted with CTCL, organ transplant rejection, GVHD, Crohn's disease and type 1 diabetes [77–82], have received benefits from treatment with ECP since the first report of the systemic efficacy of by Edelson [83]. In his studies, he carried out treatment of skin manifestations in patients with cutaneous T-cell lymphoma (CTCL) and achieved a response rate of greater than 70% compared with other forms of treatment. Wollnia tested ECP in fourteen patients (all male) aged 38–72 years with CTCL of the mycosis fungoides type, stage IIa/IIb, and achieved a total response rate of 56% [84].

25.4.2 Mechanism of ECP

It is known that both UVC and UVB can damage DNA strands, as well as UVA activated 8-MOP. However the types of DNA lesions produced are very different for these two different kinds of UV-mediated DNA damage (Fig. 25.4). UVC and UVB both produce defined UV photoproducts which are mainly the cyclobutane pyrimidine dimers (particularly TT dimers [85]) and pyrimidine-pyrimidone (6–4) photoproducts [86]. On the other hand, PUVA or ECPBM as it is known today cross-links the pyrimidine bases of DNA in complementary sister strands (inter-strand cross-links). These two different mechanisms of action are shown in Fig. 25.2. DNA damage by whatever means it is caused is likely to cause apoptosis of the extracorporally targeted lymphocytes [87]. ECP can treat erythrodermic CTCL by killing malignant CD8 T-cells but also by stimulating an immune response against thee malignant cells [88]. Two major effects of ECP have been well-confirmed: one is its immunostimulatory effects against neoplastic cells in CTCL; the other is its immunosuppressive effects

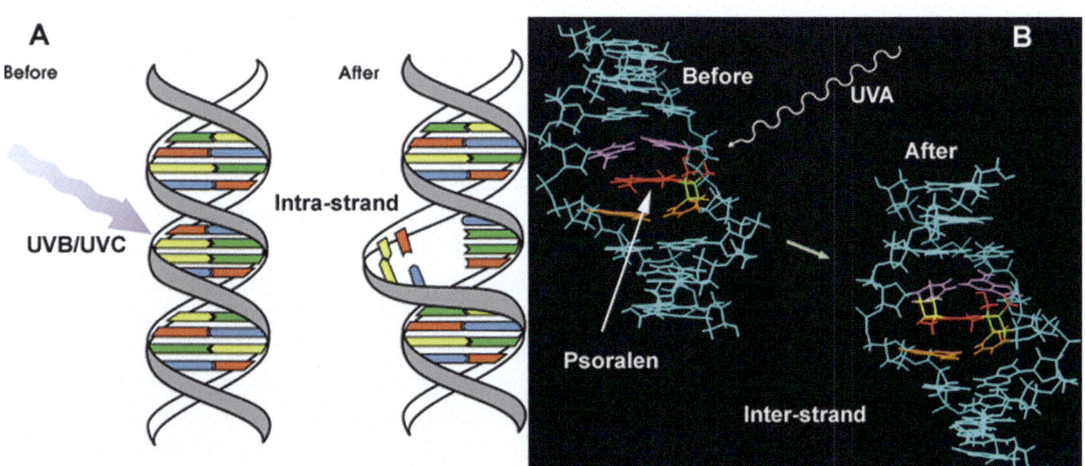

Fig. 25.4 Comparison of DNA damage produced by (**a**) UVB or UVC (intra-strand cross-links), and (**b**) DNA damage produced by psoralens and UVA (ECP, inter-strand cross-links)

against T-cell-mediated disorders such as GVHD and rejection in organ transplantation [89].

25.5 Modern Devices to Carry Out UBI

Although it is often said that UBI is "the cure that time forgot" [90, 91], it has not actually been completely forgotten. There are several companies, organizations and devices existing at the present time, which are being used or proposed (on a rather small scale) to carry out UBI, or as it often called "Photoluminescence Therapy (PT)". Several websites provide information on UBI and PT. Perhaps one of the most comprehensive is (http://www.mnwelldir.org/docs/uv_light/uv_light3.htm) that provides a listing of practitioners located in USA that offer UBI to patients. UBI medical (http://ubimedical.com/about-us.html) also has a lot of information available. The website entitled "Infections cured" (http://infections-cured.com) is also worth checking out. Physicians UBI Awareness Center (http://drsubi.com) even has a video posted online comparing different kinds of UBI machines.

25.6 Conclusion

UV irradiation of blood was hailed as a miracle therapy for treating serious infections in the 1940s and 1950s. In an ironic quirk of fate, this historical time period coincided with the widespread introduction of penicillin antibiotics, which were rapidly found to be an even bigger medical miracle therapy. Moreover another major success of UBI, which was becoming increasingly used to treat polio, was also eclipsed by the introduction of the Salk polio vaccine in 1955 [91]. UBI had originally been an American discovery, but then was transitioned to being more studied in Russia and other eastern countries, which had long concentrated on physical therapies for many diseases, which were more usually treated with drugs in the West.

However in the last decade the problem of multi-antibiotic resistant bacteria has grown relentlessly. Multidrug-resistant (MDR) and pandrug resistant (PDR) bacterial strains and their related infections are emerging threats to public health throughout the world [92]. These are associated with approximately two-fold higher mortality rates and considerably prolonged hospital admissions [93]. The infections caused by antibiotic resistant strains are often exceptionally hard to treat due to the limited range of therapeutic options [94]. Recently in Feb 2015, the Review on Antimicrobial Resistance stated "Drug-resistant infections could kill an extra 10 million people across the world every year by 2050 if they are not tackled. By this date they could also cost the world around $100 trillion in lost output: more than the size of the current world economy, and roughly equivalent to the world losing the output of the UK economy every year, for 35 years" [95].

Sepsis is an uncontrolled response to infection involving massive cytokine release, widespread inflammation, which leads to blood clots and leaky vessels. Multi-organ failure can follow. Every year, severe sepsis strikes more than a million Americans. It is estimated that between 28–50% percent of these people die. Patients with sepsis are usually treated in hospital intensive care units with broad-spectrum antibiotics, oxygen and intravenous fluids to maintain normal blood oxygen levels and blood pressure. Despite decades of research, no drugs that specifically target the aggressive immune response that characterizes sepsis have been developed.

We would like to propose that UBI be reconsidered and re-investigated as a treatment for systemic infections caused by multi-drug resistant Gram-positive and Gram-negative bacteria in patients who are running out of (or who have already run out) of options. Patients at risk of death from sepsis could also be considered as candidates for UBI. Further research is required into the mechanisms of action of UBI. The present confusion about exactly what is happening during and after the treatment is playing a large role in the controversy about whether UBI could ever be a mainstream medical therapy, or must remain side-lined in the "alternative and complementary" category where it has been allowed to be forgotten for the last 50 years.

References

1. Frercksa J, Weberb H, Wiesenfeldt G (2009) Reception and discovery: the nature of Johann Wilhelm Ritter's invisible rays. Stud Hist Philos Sci Part A 40:143–156
2. Bonnet A (1845) Traite des Maladies des Articulations. Bailliere, Paris
3. Barth J, Kohler U (1992) Photodermatologie in Dresden-ein historischer Abriss. Festschrift anlasslich des 75. Geburtstages von Prof. Dr. Dr. Dr. h.c. H.-E. Kleine-Natrop (1917–1985). Dresden
4. Downes A, Blunt TP (1877) Researches on the effect of light upon bacteria and other organisms. Proc R Soc Lond 26:488–500
5. Finsen NR (1901) Phototherapy. Edward Arnold, London
6. Ude WH (1929) Ultraviolet radiation therapy in erysipela. Radiology 13:504
7. Knott EK (1948) Development of ultraviolet blood irradiation. Am J Surg 76(2):165–171
8. Hancock VK, Knott EK (1934) Irradiated blood transfusion in the treatment of infections. Northwest Med 200(33)
9. Miley G, Christensen JA (1947) Ultraviolet blood irradiation further studies in acute infections. Am J Surg LxxIII(4):486–493
10. Miley G Uv irradiation non healing wounds. Am J Surg LXV(3):368–372, 1944
11. Miley GP (1946) Recovery from botulism coma following ultraviolet blood irradiation. Rev Gastroenterol 13:17–19
12. Miley GP, Seidel RE, Christensen JA (1946) Ultraviolet blood irradiation therapy of apparently intractable bronchial asthma. Arch Phys Med Rehabil 27:24–29
13. Miley G (1943) The control of acute thrombophlebitis with ultraviolet blood irradiation therapy. Am J Surg 60:354–360
14. Miley G (1944) Efficacy of ultraviolet blood irraidation therapy in the control of staphylococcemias. Am J Surg 64:313–322
15. Miley G (1944) Ultraviolet blood irraidation therapy in acute poliomyelitis. Arch Phys Ther 25:651–656
16. Miley G (1943) Disapperance of hemolytic staphylococcus aureus septicemia following ultraviolet blood irradiation therapy. Am J Surg 62:241–245
17. Miley G (1942) The knott technic of ultraviolet blood irradiation in acute pyogenic infections. New York state Med 42:38–46
18. Miley G (1944) Present status of ultraviolet blood irradiation (Knott technic). Arch Phys Ther 25:368–372
19. Miley G (1942) Ultravilet blood irradiation. Arch Phys Ther 536(23)
20. Miley G (1942) Ultraviolet blood irradiation therapy (knott technic) in acute pyogenic infections. Am J Surg 493(57)
21. Miley G (1943) The knott technic of ultraviolet blood irradiation as a control of infection in peritonitis. Rev Gastroenterol 1(10)
22. Miley GP, Seidel, R.E, Christensen, J.A (1943) Preliminary report of results observed in eighty cases of intractable bronchial asthma. Arch Phys Ther 533(24)
23. Barrett HA (1940) The irradiation of autotransfused blood by ultraviolet spectral energy. Result of therapy in 110 cases. Med clin N Am 721(24):1040
24. Barrett HA (1943) Five years' experience with hemoirradiation according to the Knott technic. Am J Surg 61(1):42–53
25. Rebbeck EW (1942) Double septicemia following prostatectomy treated by the knott technic of ultraviolet blood irradiation. Am J Surg 57(3):536–538
26. Rebbeck EW (1943) Preoperative hemo-irradiations. Am J Surg 61(2):259–265
27. Rebbeck EW (1941) Ultraviolet irradiation of autotransfused blood in the treatment of puerperal sepsis. Am J Surg 54(3):691–700
28. Rebbeck EW (1942) Ultraviolet irradiation of autotransfused blood in the treatment of postabortional sepsis. Am J Surg 55(3):476–486
29. Rebbeck EW (1943) Ultraviolet irradiation of blood in the treatment of escherichia coli septicemia. Arch Phys Ther 24:158–167
30. Olney RC (1946) Ultraviolet blood irradiation in biliary disease; Knott method. Am J Surg 72:235–237
31. Olney RC (1947) Ultraviolet blood irradiation treatment of pelvic cellulitis; Knott method. Am J Surg 74(4):440–443
32. Olney RC (1955) Treatment of viral hepatitis with the Knott technic of blood irradiation. Am J Surg 90(3):402–409
33. Kabat IA, Sysa J, Zakrzewska I, Leyko W (1976) Effect of UV-irradiation of shifts of energy-rich phosphate compounds: ADP, ATP and AXP in human red blood cells represented by a trigonometrical polynomial. Zentralbl Bakteriol Orig B 162(3–4):393–401
34. Vasil'eva ZF, Samoilova KA, Shtil'bans VI, Obolenskaia KD, Vitiuk NG (1991) Changes of immunosorption properties in the blood and its components at various times after UV-irradiation. Gematol Transfuziol 36(5):26–27
35. Samoilova KA, Snopov SA, Belisheva NK, Kukui LM, Ganelina IE (1987) Functional and structural changes in the surface of human erythrocytes after irradiation by different wave lengths of UV rays. III. The immediate effect of the autotransfusion of UV-irradiated blood. Tsitologiia 29(7):810–817
36. Snopov SA, Aritsishevskaia RA, Samoilova KA, Marchenko AV, Dutkevich IG (1989) Functional and structural changes in the surface of human erythrocytes following irradiation with ultraviolet rays of various wave lengths. V. Modification of the glycocalyx in autotransfusions of UV-irradiated blood. Tsitologiia 31(6):696–705
37. Ichiki H, Sakurada H, Kamo N, Takahashi TA, Sekiguchi S (1994) Generation of active oxygens, cell deformation and membrane potential changes upon

UV-B irradiation in human blood cells. Biol Pharm Bull 17(8):1065–1069
38. Savage JE, Theron AJ, Anderson R (1993) Activation of neutrophil membrane-associated oxidative metabolism by ultraviolet radiation. J Invest Dermatol 101(4):532–536
39. Ivanov EM, Kapshienko IN, Tril NM (1989) Effect of the UV irradiation of autologous blood on the humoral link in the immune response of patients with chronic inflammatory processes. Vopr Kurortol Fizioter Lech Fiz Kult 1:45–47
40. Artiukhov VF, Gusinskaia VV, Mikhileva EA (2005) Level of nitric oxide and tumor necrosis factor-alpha production by human blood neutrophils under UV-irradiation. Radiats Biol Radioecol 45(5):576–580
41. Zor'kina AV, Inchina VI, Kostin Ia V (1996) Effect of UV-irradiation of blood on the course of adaptation to conditions of hypodynamia. Patol Fiziol Eksp Ter 2:22–24
42. Deeg HJ (1988) Ultraviolet irradiation in transplantation biology. Manipulation of immunity and immunogenicity. Transplantation 45(5):845–851
43. Arlett CF, Lowe JE, Harcourt SA et al (1993) Hypersensitivity of human lymphocytes to UV-B and solar irradiation. Cancer Res 53(3):609–614
44. Teunissen MB, Sylva-Steenland RM, Bos JD (1993) Effect of low-dose ultraviolet-B radiation on the function of human T lymphocytes in vitro. Clin Exp Immunol 94(1):208–213
45. Schieven GL, Ledbetter JA (1993) Ultraviolet radiation induces differential calcium signals in human peripheral blood lymphocyte subsets. J Immunother Emphasis Tumor Immunol 14(3):221–225
46. Spielberg H, June CH, Blair OC, Nystrom-Rosander C, Cereb N, Deeg HJ (1991) UV irradiation of lymphocytes triggers an increase in intracellular Ca2+ and prevents lectin-stimulated Ca2+ mobilization: evidence for UV- and nifedipine-sensitive Ca2+ channels. Exp Hematol 19(8):742–748
47. Pamphilon DH, Corbin SA, Saunders J, Tandy NP (1989) Applications of ultraviolet light in the preparation of platelet concentrates. Transfusion 29(5):379–383
48. Lindahl-Kiessling K, Safwenberg J (1971) Inability of UV-irradiated lymphocytes to stimulate allogeneic cells in mixed lymphocyte culture. Int Arch Allergy Appl Immunol 41(5):670–678
49. Slater LM, Murray S, Liu J, Hudelson B (1980) Dissimilar effects of ultraviolet light on HLA-D and HLA-DR antigens. Tissue Antigens 15(5):431–435
50. Aprile J, Deeg HJ (1986) Ultraviolet irradiation of canine dendritic cells prevents mitogen-induced cluster formation and lymphocyte proliferation. Transplantation 42(6):653–660
51. Kovacs E, Weber W, Muller H (1984) Age-related variation in the DNA-repair synthesis after UV-C irradiation in unstimulated lymphocytes of healthy blood donors. Mutat Res 131(5–6):231–237
52. Genter EI, Zhestianikov VD, Mikhel'son VM, Prokof'eva VV (1984) DNA repair in the UV irradiation of human peripheral blood lymphocytes (healthy donors and xeroderma pigmentosum patients) in relation to the dedifferentiation process in phytohemagglutinin exposure. Tsitologiia 26(5):599–604
53. Genter EI, Mikhel'son VM, Zhestianikov VD (1989) The modifying action of methylmethane sulfonate on unscheduled DNA synthesis in the UV irradiation of human peripheral blood lymphocytes. Radiobiologiia 29(4):562–564
54. Volgareva EV, Volgarev AP, Samoilova KA (1990) The effect of UV irradiation and of UV-irradiated autologous blood on the functional state of human peripheral blood lymphocytes. Tsitologiia 32(12):1217–1224
55. Deeg HJ, Aprile J, Graham TC, Appelbaum FR, Storb R (1986) Ultraviolet irradiation of blood prevents transfusion-induced sensitization and marrow graft rejection in dogs. Blood 67(2):537–539
56. Oluwole SF, Iga C, Lau H, Hardy MA (1985) Prolongation of rat heart allografts by donor-specific blood transfusion treated with ultraviolet irradiation. J Heart Transplant 4(4):385–389
57. Vasil'eva ZF, Shtil'bans VI, Samoilova KS, Obolenskaia KD (1989) The activation of the immunosorptive properties of blood during its UV irradiation at therapeutic doses. Biull Eksp Biol Med 108(12):689–691
58. Green MH, Waugh AP, Lowe JE, Harcourt SA, Cole J, Arlett CF (1994) Effect of deoxyribonucleosides on the hypersensitivity of human peripheral blood lymphocytes to UV-B and UV-C irradiation. Mutat Res 315(1):25–32
59. Samoilova KA, Obolenskaia KD, Freidlin IS (1987) Changes in the leukocyte phagocytic activity of donor blood after its UV irradiation. II. Simulation of the effect of the autotransfusion of UV-irradiated blood. Tsitologiia 29(9):1048–1055
60. Obolenskaia KD, Freidlin IS, Samoilova KA (1987) Changes in the leukocyte phagocytic activity of donor blood after its UV irradiation. I. Its relation to the irradiation dose and initial level of phagocytic activity. Tsitologiia 29(8):948–954
61. Simon JC, Tigelaar RE, Bergstresser PR, Edelbaum D, Cruz PD Jr (1991) Ultraviolet B radiation converts Langerhans cells from immunogenic to tolerogenic antigen-presenting cells. Induction of specific clonal anergy in CD4+ T helper 1 cells. J Immunol 146(2):485–491
62. Pamphilon DH, Potter M, Cutts M et al (1990) Platelet concentrates irradiated with ultraviolet light retain satisfactory in vitro storage characteristics and in vivo survival. Br J Haematol 75(2):240–244
63. Fiebig E, Lane TA (1994) Effect of storage and ultraviolet B irradiation on CD14-bearing antigen-presenting cells (monocytes) in platelet concentrates. Transfusion 34(10):846–851
64. Kahn RA, Duffy BF, Rodey GG (1985) Ultraviolet irradiation of platelet concentrate abrogates lymphocyte activation without affecting platelet function in vitro. Transfusion 25(6):547–550
65. Andreu G, Boccaccio C, Klaren J et al (1992) The role of UV radiation in the prevention of human leukocyte antigen alloimmunization. Transfus Med Rev 6(3):212–224

66. Tandy NP, Pamphilon DH (1991) Platelet transfusions irradiated with ultraviolet-B light may have a role in reducing recipient alloimmunization. Blood Coagul Fibrinolysis 2(2):383–388
67. Roshchupkin DI, Murina MA (1998) Free-radical and cyclooxygenase-catalyzed lipid peroxidation in membranes of blood cells under UV irradiation. Membr Cell Biol 12(2):279–286
68. Gorog P (1991) Activation of human blood monocytes by oxidized polyunsaturated fatty acids: a possible mechanism for the generation of lipid peroxides in the circulation. Int J Exp Pathol 72(2):227–237
69. Salmon S, Maziere JC, Santus R, Morliere P, Bouchemal N (1990) UVB-induced photoperoxidation of lipids of human low and high density lipoproteins. A possible role of tryptophan residues. Photochem Photobiol 52(3):541–545
70. Salmon S, Haigle J, Bazin M, Santus R, Maziere JC, Dubertret L (1996) Alteration of lipoproteins of suction blister fluid by UV radiation. J Photochem Photobiol B 33(3):233–238
71. Artyukhov VG, Iskusnykh AY, Basharina OV, Konstantinova TS (2005) Effect of UV irradiation on functional activity of donor blood neutrophils. Bull Exp Biol Med 139(3):313–315
72. Dong Y, Shou T, Zhou Y, Jiang S, Hua X (2000) Ultraviolet blood irradiation and oxygenation affects free radicals and antioxidase after rabbit spinal cord injury. Chin Med J 113(11):991–995
73. Calabrese EJ, Dhawan G, Kapoor R, Iavicoli I, Calabrese V (2015) HORMESIS: a fundamental concept with widespread biological and biomedical applications. Gerontology 62:530
74. Calabrese EJ (2014) Hormesis: from mainstream to therapy. J Cell Commun Signal 8(4):289–291
75. Zaky S, Kamel SE, Hassan MS et al (2011) Preliminary results of ozone therapy as a possible treatment for patients with chronic hepatitis C. J Altern Complement Med 17(3):259–263
76. Edelson RL (2014) Mechanistic insights into extracorporeal photochemotherapy: efficient induction of monocyte-to-dendritic cell maturation. Transfus Apher Sci 50(3):322–329
77. Child FJ, Ratnavel R, Watkins P et al (1999) Extracorporeal photopheresis (ECP) in the treatment of chronic graft-versus-host disease (GVHD). Bone Marrow Transplant 23(9):881–887
78. Atta M, Papanicolaou N, Tsirigotis P (2012) The role of extracorporeal photopheresis in the treatment of cutaneous T-cell lymphomas. Transfus Apher Sci 46(2):195–202
79. De Waure C, Capri S, Veneziano MA et al (2015) Extracorporeal Photopheresis for second-line treatment of chronic graft-versus-host diseases: Results from a health technology assessment in Italy. Value Health 18(4):457–466
80. Patel J, Klapper E, Shafi H, Kobashigawa JA (2015) Extracorporeal photopheresis in heart transplant rejection. Transfus Apher Sci 52(2):167–170
81. Reinisch W, Knobler R, Rutgeerts PJ et al (2013) Extracorporeal photopheresis (ECP) in patients with steroid-dependent Crohn's disease: an open-label, multicenter, prospective trial. Inflamm Bowel Dis 19(2):293–300
82. Ludvigsson J, Samuelsson U, Ernerudh J, Johansson C, Stenhammar L, Berlin G (2001) Photopheresis at onset of type 1 diabetes: a randomised, double blind, placebo controlled trial. Arch Dis Child 85(2):149–154
83. Edelson R, Berger C, Gasparro F et al (1987) Treatment of cutaneous T-cell lymphoma by extracorporeal photochemotherapy. Preliminary results. N Engl J Med 316(6):297–303
84. Wollina U, Looks A, Meyer J et al (2001) Treatment of stage II cutaneous T-cell lymphoma with interferon alfa-2a and extracorporeal photochemotherapy: a prospective controlled trial. J Am Acad Dermatol 44(2):253–260
85. Niggli HJ, Rothlisberger R (1988) Cyclobutane-type pyrimidine photodimer formation and induction of ornithine decarboxylase in human skin fibroblasts after UV irradiation. J Invest Dermatol 91(6):579–584
86. Vendrell-Criado V, Rodriguez-Muniz GM, Lhiaubet-Vallet V, Cuquerella MC, Miranda MA (2016) The (6-4) Dimeric Lesion as a DNA Photosensitizer. Chem Phys Chem 17(13):1979–1982
87. Santella RM, Dharmaraja N, Gasparro FP, Edelson RL (1985) Monoclonal antibodies to DNA modified by 8-methoxypsoralen and ultraviolet A light. Nucleic Acids Res 13(7):2533–2544
88. Heald P, Rook A, Perez M et al (1992) Treatment of erythrodermic cutaneous T-cell lymphoma with extracorporeal photochemotherapy. J Am Acad Dermatol 27(3):427–433
89. Hart JW, Shiue LH, Shpall EJ, Alousi AM (2013) Extracorporeal photopheresis in the treatment of graft-versus-host disease: evidence and opinion. Ther Adv Hematol 4(5):320–334
90. Rowen RJ (1996) Ultraviolet blood irradiation therapy (Photo-Oxidation) the cure that time forgot. Int J Biosocial Med Research 14(2):115–132
91. Wu X, Hu X, Hamblin MR (2016) Ultraviolet blood irradiation: is it time to remember "the cure that time forgot"? J Photochem Photobiol B 157:89–96
92. Kraus CN (2008) Low hanging fruit in infectious disease drug development. Curr Opin Microbiol 11(5):434–438
93. Munoz-Price LS, Poirel L, Bonomo RA et al (2013) Clinical epidemiology of the global expansion of Klebsiella pneumoniae carbapenemases. Lancet Infect Dis 13(9):785–796
94. Yoneyama H, Katsumata R (2006) Antibiotic resistance in bacteria and its future for novel antibiotic development. Biosci Biotechnol Biochem 70(5):1060–1075
95. O'neill J (2015) Review on antimicrobial resistance: tackling a global health crisis. Initial Steps

From UV Protection to Protection in the Whole Spectral Range of the Solar Radiation: New Aspects of Sunscreen Development

26

Leonhard Zastrow, Martina C. Meinke, Stephanie Albrecht, Alexa Patzelt, and Juergen Lademann

Abstract

Sunscreens have been constantly improving in the past few years. Today, they provide an efficient protection not only in the UVB but also in the UVA spectral region of the solar radiation. Recently it could be demonstrated that 50% of all free radicals induced in the skin due to solar radiation are formed in the visible and infrared spectral region. The good protective efficacy of sunscreens in the UV region prompts people to stay much longer in the sun than if they had left their skin unprotected. However, as no protection in the visible and infrared spectral region is provided, high amounts of free radicals are induced here that could easily exceed the critical radical concentration.

This chapter describes how the effect of sunscreens can be extended to cover also the visible and infrared spectral region of the solar radiation by adding pigments and antioxidants with high radical protection factors to the sunscreen formulations.

Keywords

Sun protection • Skin cancer • Reactive oxygen species • Lipid peroxide radicals • Critical radical concentration • Antioxidants

26.1 Introduction

Solar radiation has been an essential source for the development of life on earth. It is the basis for vitamin D synthesis in the organism and indispensable for human wellbeing [1]. However, next to genetic preconditions, solar radiation is also the main reason for premature skin ageing [2]. High doses of sun radiation incident on the human skin can trigger

L. Zastrow (✉) • M.C. Meinke • S. Albrecht
A. Patzelt • J. Lademann
Center of Experimental and Applied Cutaneous Physiology, Department of Dermatology, Venerology and Allergology, Charité – Universitätsmedizin Berlin, 10117 Berlin, Germany
e-mail: leonard.zastrow@gmail.com

S. Ahmad (ed.), *Ultraviolet Light in Human Health, Diseases and Environment*, Advances in Experimental Medicine and Biology 996, https://doi.org/10.1007/978-3-319-56017-5_26

processes leading to skin damage extending from sunburn via immunosuppression to even skin cancer [3, 4]. Among the reasons for the development of these partially severe skin damages is the production of free radicals that has been underestimated, so far. These highly reactive molecules are vital for signalling processes in our body. However, it is well established that at high concentrations of excess free radicals, molecules that are essential for the function of the cells can be destroyed, e.g. DNA, RNA, proteins and lipids [4].

Sunscreens contain filter substances with strong protective efficacy against ultraviolet B (UVB 280–320 nm) and adequate protection against formation of free radicals in ultraviolet A (UVA 320–400 nm). Recently, it could be demonstrated that 50% of the free radicals are produced by solar radiation in the visible (VIS) and infrared (IR) spectral region [5]. Therefore, ultimately for preventing skin cancerogenesis, new concepts need to focus on protecting the human skin in the complete range of the earth's solar spectrum, i.e. the UV, the VIS and IR spectral regions. The basics of these developments are described hereinafter.

26.2 Free Radical Action Spectrum

In 2009, Zastrow et al. [5] (experimentally determined the action spectrum of free radical formation proving for the first time that radical formation is the general biophysical response to radiation in the range between UVB (280 nm) and near infrared (NIR 1600 nm). Also the NIR's radical forming capacity (700–1600 nm) was evidenced in separate EPR (electron paramagnetic resonance) spectrometry experiments, whereby it was found that the radical generation (RG) depends not only on the applied NIR irradiation dose but also on the increase of the skin temperature related thereto. The wavelength induced mixtures [6] of short-lived reactive oxygen species (ROS) and long-lived lipid peroxide radicals (LOS) together behave exactly like the UVA induced free radicals, which are classified as being carcinogenic. The action spectrum is demonstrated in Fig. 26.1. Although the concentration of the free radicals in the VIS/IR spectral region is lower than that in the UV region, its wavelength range is distinctly larger leading to the for-

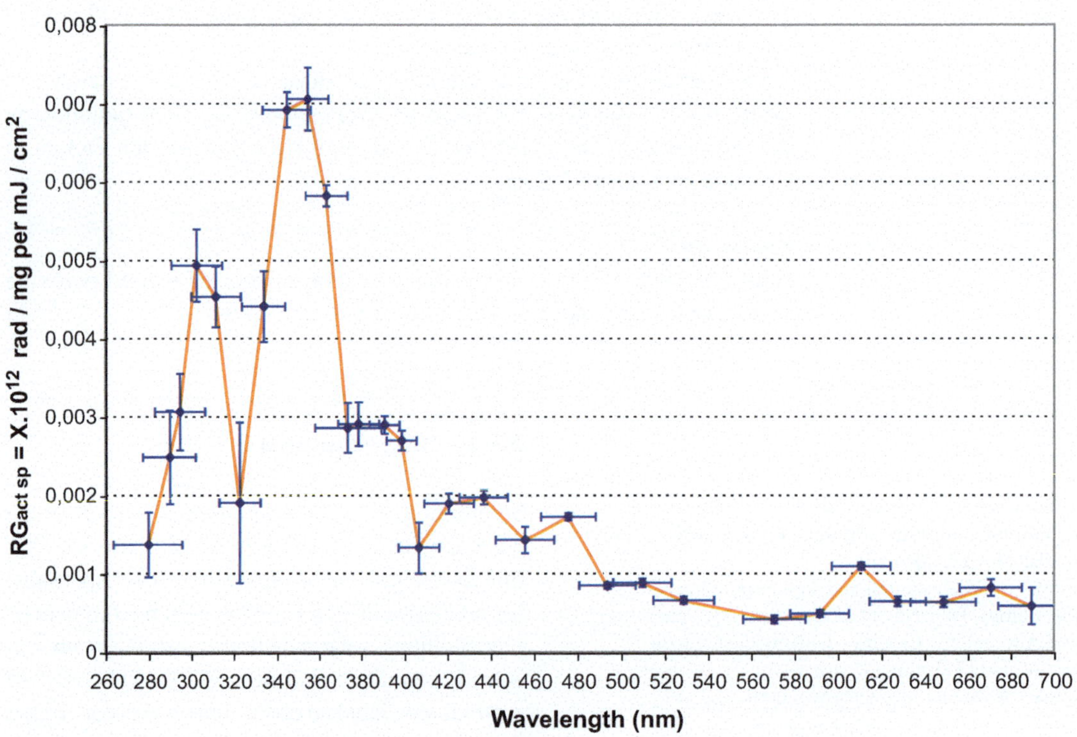

Fig. 26.1 Action spectrum – wavelength dependence for free radical formation in human skin [5]

mation of substantial amounts of free radicals. At least 50% of the free radicals produced by the sun originate from the VIS/IR spectral region.

26.3 Critical Radical Concentration (Free Radical Threshold Value)

Using the free radical action spectrum, we calculated in 2009 (3) the amount of free radicals induced by the dose of natural sunlight which is necessary to produce a recommended amount of vitamin D. For estimated 1000 IU daily, a dose of about a quarter of the minimum erythemal dose (MED) leads to a number of roughly 3.5×10^{12} radicals per mg skin tissue. Based on this standard vitamin dose, this amount of $\sim 3.5 \times 10^{12}$ rad/mg ROS/LOS represents the tolerated number of free radicals in skin. At that concentration the ROS/LOS mixture does not show clinically detectable damage indicating that the excess free radicals are still under the control of the antioxidative defense system. We called this critical radical concentration that exists on an essential biological endpoint Free Radical Threshold Value (FRTV). Recently the existence of the FRTV, separating the "beneficially" from "deleteriously" acting free radicals, could be confirmed by quantitative EPR x-band spectroscopy [6]. As described in [5–7], spin trap PBN and spin labeled PCA were used to quantify the free radicals. The spin traps DMPO and DEPMPO allowed distinguishing ROS from carbon-centred LOS.

Figure 26.2 summarizes the results. With increasing UV + VIS doses a rising number of ROS/LOS is generated. Up to $\sim 3.5 \times 10^{12}$ rad/mg this increase is nearly linear. Once this concentration is reached, the gradient of the curve changes – still being linear though clearly lower. Further measurements with UV and VIS light alone (2) led to RG values between $\sim 2.8 \times 10^{12}$ rad/mg to $\sim 4.0 \times 10^{12}$ rad/mg, all in the range of the calculated FRTV.

Moreover it could be demonstrated that below the critical radical concentration (FRTV) the mixture of ROS/LOS is dominated by short-lived ROS. Above the FRTV RG ~ 3.5×10^{12} rad/mg the long-lived LOS are dominant. They are well-known as measurable signs of cell destructive processes. The two circles in Fig. 26.2 show exemplarily this dynamic switch in the free radical mixture.

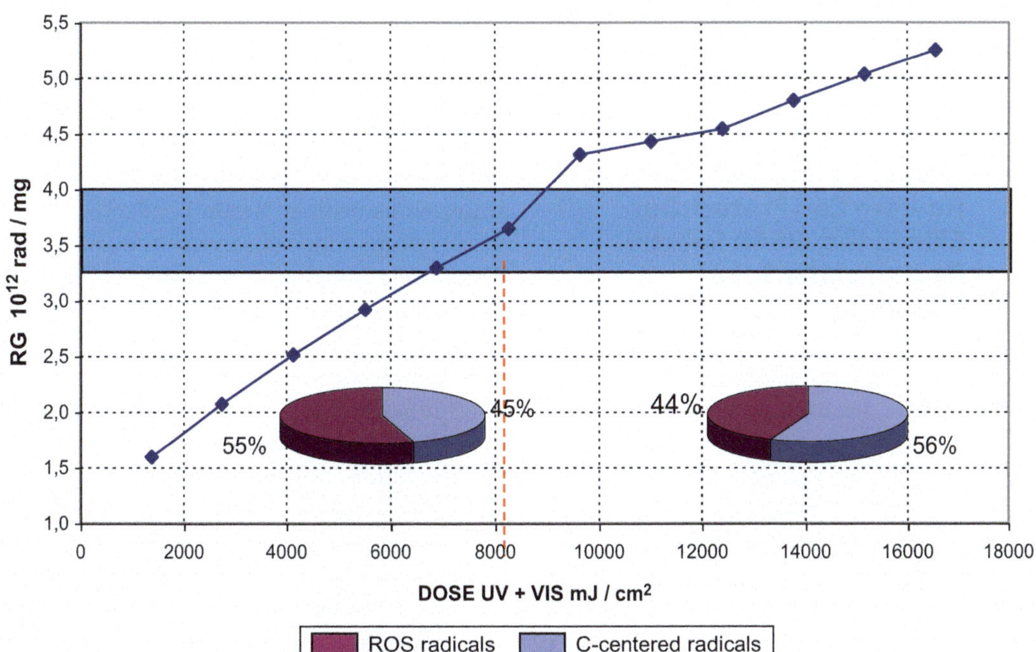

Fig. 26.2 Radical generation (RG) depending on the UV + VIS doses and the radical composition [6]

26.4 Influence of Sunscreen Application on the Free Radical Formation in Human Skin

Standard commercial sunscreens contain only UVB and UVA filters aimed to suppress the development of sunburn and at the end of a long chain of biomolecular processes, skin cancer. Today, sunscreens with very high sun protection factors (SPF) are available. People exposing themselves to intense direct sunlight after having applied sunscreens with an SPF of 30 or 50, imagine that they can stay much longer in the sun without being affected by solar-induced damage as their skin shows no signs of warning over an extended period of time. In addition, the high protective efficiency of these sunscreens in the UV spectral region leads to a strongly reduced free radical ROS/LOS mixture. In the absence of sunburn as an unmistakeable warning, applicants use to over-exposing themselves to solar radiation, sometimes up to 10 times longer. As the standard commercial sunscreens, however, provide no protection in the VIS/IR spectral region, the free radicals being formed in this spectral region can easily overcome the critical radical concentration (FRTV). Thus, it is not surprising that skin cancer incidence is still rising although the use of sunscreens exhibiting high SPF values has become increasingly popular.

26.5 How We Can Protect Our Skin in the VIS/IR Spectral Region of the Solar Radiation

In the UV spectral region, highly efficient UVB and UVA filters are used to protect the human skin from being damaged. This strategy is impracticable in the VIS/IR spectral region as suitable filter substances are lacking. Developing a new concept – sunscreen products should reflect the skin natural light defense systems. Three natural protective mechanisms of the skin have to be taken into consideration. The first mechanism is the formation of light callosity as an expression of an adaption of the optical skin parameters to intense light exposure. This means that after exposure of the skin to solar radiation the stratum corneum becomes thicker thus increasing the absorption and scattering of the sunlight. The second protective mechanism is the reactive increase of melanin production. It is well known that people exposing their skin to intense solar radiation get tanned and are therefore better protected due to increased melanin quantities. The third protective mechanism is supportable by the uptake of antioxidants with a diet rich in fruits and vegetables [8]. These antioxidants strengthen the skin's own anti-oxidative defence system, which neutralizes free radicals before they start damaging the skin.

In principle, these protective mechanisms are – often nearly unintentionally – integrated in modern sunscreens. To our knowledge a commercial product stimulating light callosity does not yet exist. However, sunscreens often contain pigments, so called physical filters, e.g. titanium dioxide or zinc oxide, to increase the SPF values. These pigments act as micro mirrors in the human skin, reflecting the sunlight not only in the UV but also in the VIS/IR spectral regions [9, 10]. In addition, modern sunscreens contain antioxidants to protect the UV filters from damage caused by solar radiation supporting simultaneously the skin's antioxidative defense system. To elucidate the exact light protective efficacy of commercial sunscreens, the Department of Dermatology of the Charité – Universitätsmedizin Berlin conducted a study investigating four sunscreen products, two of which were bought in drugstores and the other two in supermarkets, for their radical protection efficiency in the IR spectral region [11]. The experiments were carried out with a water-filtered infrared lamp at physiological doses. In these experiments, the radical formation after IR irradiation was investigated for untreated and for skin treated with standard COLIPA sunscreens containing neither pigments nor antioxidants. Under the same conditions the four commercial products were investigated for their radical formation during exposure to IR radiation. The results are presented in Fig. 26.3.

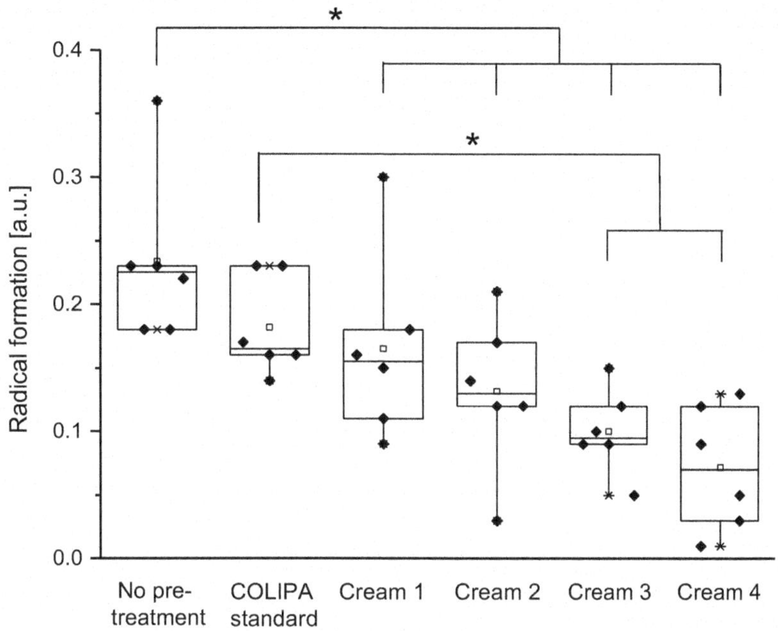

Fig. 26.3 Amount of free radicals produced by IR irradiation in untreated skin, treated with standard COLIPA sunscreen and four commercial sunscreens [11]

Table 26.1 Radical protection factor of the sunscreens [11]

	RPF
Standard	22
Cream 1	47
Cream 2	61
Cream 3	119
Cream 4	40

To explain the obtained results, the radical protection factors (RPF) of the five sunscreens were investigated. The radical protection factor (RPF) describes the number of the by test substance reduced test radicals with RPF = N × 10^{14} DPPH radicals/mg test substance. For practical reasons it is useful to present only the real number N. The results are presented in Table 26.1.

Furthermore, the optical properties, i.e. scattering, reflectance and absorbance of the sunscreens were investigated. These results are summarized in Fig. 26.4.

As can be seen from Fig. 26.3, the best results were obtained for sunscreen 4. Although this sunscreen exhibited not the best RPF, its optical properties proved highly efficient. As the pigments are highly reflecting and scattering, only a few photons of the solar radiation could penetrate the skin. The situation for cream 3, which ranged second best, is different. This cream yielded high RPF values but strongly inferior optical properties compared to cream 4.

These results reveal that protection in the VIS/IR spectral region of the solar radiation can be obtained by application of pigments and antioxidants. While the application of pigments is limited to a concentration of less than 8% of the formulation, antioxidants are applicable even at high concentrations with high RPF values.

26.6 Integrating VIS/IR Protection into Sunscreens

Taking into consideration the above basics and results, there is no doubt that future sunscreens must provide protection not only in the UV but also in the VIS/IR spectral regions. Interestingly, while we are discussing the epidemic increase of skin cancer, in some cultures, sun protection has been performed in an optimized way since ancient times. Asians, e.g. often use sun umbrellas to protect themselves against solar radiation.

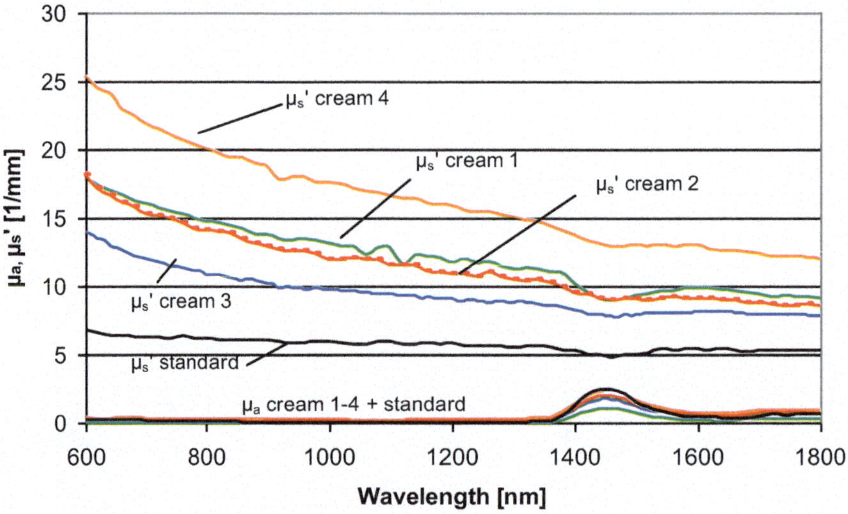

Fig. 26.4 Optical properties of the five investigated sunscreens [11]

These sun umbrellas shield their users against solar radiation in the complete spectral range. In Arab countries people are wearing garments which cover their skin almost completely. These people have been practising light protection instead of UV protection.

Taking into account the results of the above Charité study [11] it may be highlighted that formulators of sunscreens have broad room for designed products with advanced protection in VIS/NIR area. An important role for further progress in this field plays the EPR spectroscopy, the only method able to detect directly free radicals [12]. This technique allows optimisation and standardisation of sunscreens not only in the UV but also the VIS/IR spectral regions. It is encouraging that there is a variety of commercial sunscreens meanwhile being available which claim to provide also protection in the VIS/IR spectral region.

26.7 Standardisation of SPF Values in the VIS/IR Spectral Region

Since the first sunscreens were developed in 1933, the efficiency of sunscreen products has always been improving. In 1956, the SPF values, which are derived from the biological response of the skin in form of erythema (minimal erythemal dose MED), were standardised. Later on, it could be demonstrated that the skin has to be protected also against UVA radiation as this type of radiation create excess ROS/LOS which can induce skin cancer [2]. Even with persistent pigment darkening (PPD) as measurable parameter for the biological response in this specific region, the determination of the efficiency of UVA protection is a topic of ongoing research. Therefore, the European Commission recommended in 2006 that sunscreens must provide a 3:1 absorption in the UVB and UVA fraction of the solar spectrum to be labelled as sunscreen products [13]. This was a foresighted decision, indeed, as the generally known and well accepted SPF should continue to be used in the future. But how can the protection against VIS/IR radiation be included in this concept? Specific legislation will have to be adopted, whereby the protective function in the VIS/IR spectral region will be determined by EPR spectrometry.

As described earlier in this chapter, people applying sunscreens with a high SPF use to stay longer in the sun so that in the VIS/IR spectral region free radicals can be produced at concentrations exceeding the critical radical concentration (FRTV). For a sunscreen product with a declared SPF of 10, it is to be supposed that applicants stay up to 10 times longer in the sun. Now it has to be checked if the free radicals (ROS/LOS) produced in the VIS/IR spectral region for an irradiation dose of 10 MED are below or above the FRTV. If the concentration proves to be below the FRTV, the SPF value could remain unchanged. If the radical concentration is found to exceed the critical concentration value, the FRTV value of 3.5×10^{12} rad/mg of tissue has to be divided by the actual free radical concentration. Subsequently the SPF will be divided by the obtained quotient.

If the radical prevention or protection power of the respective sunscreen is insufficient, this "VIS/IR corrected SPF" will reduce its SPF, thus providing a much more realistic recommendation to the customer.

However, it remains to be emphasized that light protection throughout the entire solar spectrum will not be successful unless customers improve their compliance drastically. It is absolutely necessary that sunscreens are applied to the skin at the correct amount, i.e. 2 mg/cm^2, which is the basis for calculating the SPF. New technical solutions, from application forms to packaging, supporting correct client behaviour are available. For instance, sunscreens are equipped with dispenser systems that inform their users about the applied amounts.

In summary, it could be shown that the common biophysical answer of the skin to solar exposure in the wavelength range between 280–1600 nm is the generation of an identical free ROS/LOS radical mixture. If these ROS/LOS exceed the critical radical concentration (FRTV) of roughly 3.5×10^{12} rad/mg, they are starting to develop carcinogenic action as it is known for UVA-borne free radicals.

26.8 Conclusion

To successfully fight the still rising skin cancer incidence a fast transition from today's legally required UVB/UVA protection to total light protection is necessary. Therefore there is an urgency for new innovative concepts, including intelligent behaviour in the sun, mimicking natural protection mechanisms, e.g. by specific modification of skin reflectance and scattering. The focus of the innovation needs to be put on systems for controlling the prevention of free radical formation above the FRTV.

The protection efficacy, especially in the VIS/IR should, as proposed above, find entrance in an "Integral SPF" giving a realistic exposure recommendation to the applicants.

References

1. Zittermann A, Gummert JF (2010) Sun, vitamin D, and cardiovascular disease. J Photoch Photobio B 101:124–129
2. Haywood R, Wardman P, Sanders R, Linge C (2003) Sunscreens inadequately protect against ultraviolet-A-induced free radicals in skin: implications for skin aging and melanoma? J Invest Dermatol 121:862–868
3. Agbai ON, Buster K, Sanchez M, Hernandez C, Kundu RV, Chiu M et al (2014) Skin cancer and photoprotection in people of color: A review and recommendations for physicians and the public. J Am Acad Dermatol 70:748–762
4. Cadet J, Douki T, Ravanat JL (2015) Oxidatively generated damage to cellular DNA by UVB and UVA radiation. Photochem Photobiol 91:140–155
5. Zastrow L, Groth N, Klein F, Kockott D, Lademann J, Renneberg R et al (2009) The missing link – light-induced (280–1600 nm) free radical formation in human skin. Skin Pharmacol Phys 22:31–44
6. Zastrow L, Doucet O, Ferrero L, Groth N, Klein F, Kockott D et al (2015) Free radical threshold value: a new universal body constant. Skin Pharmacol Physiol 28:264–268
7. Albrecht S, Ahlberg S, Beckers I, Kockott D, Lademann J, Paul V et al (2016) Effects on detection of radical formation in skin due to solar irradiation measured by EPR spectroscopy. Methods 109:44
8. Heinrich U, Gartner C, Wiebusch M, Eichler O, Sies H, Tronnier H et al (2003) Supplementation with beta-carotene or a similar amount of mixed carotenoids protects humans from UV-induced erythema. J Nutr 133:98–101
9. Lademann J, Schanzer S, Jacobi U, Schaefer H, Pflucker F, Driller H et al (2005) Synergy effects between organic and inorganic UV filters in sunscreens. J Biomed Opt 10:14008
10. Syring F, Weigmann HJ, Schanzer S, Meinke MC, Knorr F, Lademann J (2016) Investigation of model sunscreen formulations comparing the sun protection

factor, the universal sun protection factor and the radical formation ratio. Skin Pharmacol Phys 29:18–23
11. Meinke MC, Syring F, Schanzer S, Haag SF, Graf R, Loch M et al (2013) Radical protection by differently composed creams in the UV/VIS and IR spectral ranges. Photochem Photobiol 89:1079–1084
12. Lohan SB, Muller R, Albrecht S, Mink K, Tscherch K, Ismaeel F et al (2016) Free radicals induced by sunlight in different spectral regions – in vivo versus ex vivo study. Exp Dermatol 25:380–385
13. Hojerova J, Medovcikova A, Mikula M (2011) Photoprotective efficacy and photostability of fifteen sunscreen products having the same label SPF subjected to natural sunlight. Int J Pharm 408:27–38

Safety and Efficacy of Phototherapy in the Management of Eczema

27

Annalisa Patrizi, Beatrice Raone, and Giulia Maria Ravaioli

Abstract

Atopic Dermatitis (AD), a common skin disease, can occur in patients of all age, gender and ethnicity. It is an inflammatory affection, characterized by chronic and highly debilitating behavior. First-line interventions against AD include environmental measures and topical emollients, corticosteroids or calcineurin inhibitors. When these measures are not sufficient, phototherapy represents an efficient second-line option of treatment; it can be administered on its own, or in the most severe cases combined with systemic medicaments such as corticosteroids.

Different types of light therapy, including photochemotherapy, have been tested in the past and in recent years for AD: in particular, ultraviolet A1 (UVA1) and narrow band ultraviolet B (NB-UVB) have been reported in the literature as the most effective resources, respectively for acute and chronic AD. However, to date, no guidelines have been realized concerning the use of phototherapy for AD, as no light form has been defined superior to the others. The most reliable protocols and dosimetry are standardized within the American Academy of Dermatology (AAD) psoriasis guidelines.

In adults and children over 12 years (8 years for NB-UVB) phototherapy is recommended with strength B and level of evidence II (excluding home phototherapy, which is recommended with strength C and level of evidence III). It is usually safe and well tolerated; however its short- and

A. Patrizi (✉) • B. Raone • G.M. Ravaioli
Department of Specialized, Diagnostic and Experimental Medicine, Division of Dermatology, University of Bologna, via Massarenti 1, 40138 Bologna, Italy
e-mail: annalisa.patrizi@unibo.it; beatrice.raone@libero.it; giuliamaria.ravaioli@gmail.com

S. Ahmad (ed.), *Ultraviolet Light in Human Health, Diseases and Environment*, Advances in Experimental Medicine and Biology 996, https://doi.org/10.1007/978-3-319-56017-5_27

long-term adverse effects are the same as those observed when light therapy is performed for other pathologic conditions. Erythema and photodamage are in particular quite frequent; moreover it has not been clarified whether UV radiation may induce neoplastic cellular transformation. For all these reasons, the use of phototherapy must be chosen only after a comprehensive and careful evaluation of the patient's features and compliance, as well as of the limitations of the procedure due to costs and availability.

Keywords
Atopic dermatitis • Atopic eczema • Phototherapy • PUVA • UVA1 • NB-UVB • Balneophototherapy

Abbreviations

8-MOP	8-methoxypsoralen
AAD	American Academy of Dermatology
AD	atopic dermatitis
AE	atopic eczema
DLQI	Dermatology Life Quality Index
EASI	Eczema Area and Severity Index
ECP	extracorporeal photopheresis
FLG	filaggrin
HD	high dose
IgE	immunoglobulin E
LD	low dose
LDH	lactate dehydrogenase
MD	medium dose
MED	minimal erythema dose
mRNA	messenger ribonucleic acid
NB-UVB	narrow band ultraviolet B
PUVA	psoralen plus ultraviolet A
QoL	quality of life
SCORAD	Severity Scoring of Atopic Dermatitis
TARC/CCL17	thymus and activation-regulated chemokine/chemockine (C-C motif) ligand 17
TEWL	transepidermal water loss
Th	T helper
UVA	ultraviolet A
UVA1	ultraviolet A1
UVA2	ultraviolet A2
UVAB	ultraviolet AB
UVB	ultraviolet B
VAS	visual analog scale

27.1 Introduction

27.1.1 Atopic Dermatitis: An Epidemiologic and Clinical Overview

Atopic Dermatitis (AD), a chronic pruritic inflammatory cutaneous disease, affects all genders and ethnicities and is very common in children, where it represents the most widespread cutaneous skin disease, but it can occur also during adulthood. Known also as Atopic Eczema (AE), AD is a chronic inflammatory skin disease, resulting in protracted symptoms that usually have a relapsing course, with fluctuating remissions and flares. AD is frequently encountered in clinical practice and has a highly debilitating impact on the quality of life (QoL) of the patients. For all these reasons, together with its constantly increasing incidence, AD has become a major social and economic issue worldwide [1].

AD clinically presents with eczematous pruritic lesions, with specific morphology and typical distribution, commonly symmetric. In particular, acute lesions include erythema, papules, vescicules, and exudation, whereas the most common chronic lesions are lichenification, scales, crusts and prurigo. The distribution of the eczematous lesions varies with age: infants are usually affected on the scalp and face, and then on the limbs and trunk; school-age children commonly show lesions on the flexural surfaces of the upper and lower limbs, whereas adolescents and adults are most frequently affected on the face (especially periocular, perioral, periauricular regions), neck and upper body [2].

Dermatitis is only one of the multiple clinical manifestations that characterize atopic individuals: others are asthma, rhinitis, conjunctivitis and food allergy. Moreover, in the personal or family history of a patient the clinical coexistence of one or more of these conditions as well as an over the limit immunoglobulin E (IgE) antibodies serum level is defined atopic diathesis [3].

27.1.2 Pathogenesis

The pathogenesis of AD is multifactorial and still debated. A primary cause is a reduced barrier function of the epidermis, resulting in enhanced transepidermal water loss (TEWL). This may be caused by different defects of the stratum corneum, for example alterations of intercellular lipids, such as ceramides, or of cellular proteins, such as filaggrin (FLG). As a consequence, the impaired corneum layer is more susceptible to external damage due to allergens or a specific irritants that enhance inflammation, for example pH variations or excessive dryness.

Atopic skin is also characterized by a marked epidermal hyperplasia and by a predominantly T helper (Th) 2/"22" immune response to allergens, that leads to excessive IgE production. In the advanced chronic phases of the disease, the immune activation may switch toward a Th1 profile [4–8].

27.1.3 Diagnosis

Several clinical algorithms for the diagnosis of AD have been proposed in the last three decades, including major and minor criteria. Almost all of them include the presence of eczema with pruritus, chronic or relapsing course and typical features and location [3].

When a diagnosis of AD is made, the personal and family history of the patient should be investigated, moreover an accurate clinical examination should be performed to exclude the presence of associated disorders or complications and to define the severity of the disease. The most indicative laboratory tests include high serum IgE level, over the limit peripheral blood eosinophil count and enhanced lactate dehydrogenase (LDH) and thymus, and activation-regulated chemokine/chemokine (C-C motif) ligand 17 (TARC/CCL17) level [3].

The severity of AD can be defined as mild, moderate or severe. The most used severity scores of AD worldwide are the Severity Scoring of Atopic Dermatitis (SCORAD) and the Eczema Area and Severity Index (EASI), whose maximum scores are respectively 103 and 72 [9, 10]. Some adjunctive helpful scoring systems are the Skindex-16 and the Dermatology Life Quality Index (DLQI), for evaluating the QoL, and the visual analog scale (VAS), for measuring the pruritus [11–13].

27.1.4 Management and Treatment

Age-related remission in children with AD is possible, but not constant. The treatment of AD is mandatory and the therapeutic choice depends on the severity of the disease and on the age and compliance of the patients.

As the clinical course of the disease is usually relapsing, with periods of remission followed by flares, the goals of the therapy are to reduce the symptoms and to reach a stable state with absent or controlled symptoms and signs using the least amount of medicament possible, and finally to maintain this state and a satisfying QoL.

First and foremost, environmental measures are mandatory, in particular the avoidance of external irritants, food, air and contact allergens or other trigger factors. Another baseline fundamental intervention is the use of topical therapies:

emollients, moisturizers, corticosteroids or calcineurin inhibitors.

Moisturizers and emollients in particular play an important role in the treatment of AD; they are well tolerated and effective, as they act respectively to promote the hydration of the stratum corneum and to restore or replace the function of the epidermal barrier [2, 14]. Furthermore their frequent application is fundamental to reduce the dose of steroid needed to maintain remission and to prevent acute flares of AD. The viscosity of the emollient and other topical products varies from oil-in-water emulsions to ointments, creams and rich-in-water preparations. The vehicle should be chosen accurately considering the degree of dryness of the lesions.

For acute flares topical corticosteroids and/or calcineurin inhibitors such as tacrolimus or pimecrolimus are usually required. Topical corticosteroids are, together with emollients, the mainstay of treatment for AD. Steroids should be chosen considering the right rank and vehicle, and establishing the precise volume and frequency of application. Moreover their long-term use should be avoided to limit the risk of side effects. On the other hand, calcineurin inhibitors are a second-line choice for mild to moderate AD [2].

In general, severe cases and acute exacerbations require the use of systemic corticosteroids and other anti-inflammatory drugs [15, 16]. Moreover, for maintaining clinical remissions in severe cases, phototherapy or immunosuppressants such as oral cyclosporine are often required.

Phototherapy has an important beneficial potential for those patients with deterioration of the QoL caused by the impact of AD [2]. When integrated in a comprehensive treatment plan, phototherapy can be a curative second-line option or an additional resource to conventional treatments for AD. It is optimal for chronic and acute cases, for children over 12 years (8 years for NB-UVB) and adults and in particular for cases with extensive disease [2, 17].

The compliance of the patients is essential for realizing an appropriate light therapy, as this technique requires periodic and well-scheduled hospital sessions to reach good clinical results. On the other hand, home topical therapy requires less time and money, so it is usually preferred for cases with limited extension.

Indications, dosages, safety and efficacy of use of phototherapy in AD have not been systematically reviewed yet in the literature, therefore this technique must be performed with caution and always taking into account a comprehensive overview of the patient [2, 18].

27.2 Phototherapy for Atopic Dermatitis: Types and Indication

27.2.1 The Role of Phototherapy in AD

Since the early twenties of the last century an improvement of the symptoms had been observed in several AD patients concurrently with sun exposure and sea climate, until Nexman in 1948 documented the beneficial effect of carbon arc lamps UV radiations on AD patients [19].

In the following 30 years many lamps with different UV emission spectra and specific light wavelength were developed and tested for AD, alone or associated with systemic agents (photochemotherapy). The work of Morison et al. in particular is considered a milestone in the use of phototherapy for AD, as it first described the successful use of oral psoralen and UV radiation [20].

Multiple beneficial effects of light radiations on atopic skin have been reported. First, they exert positive immunosuppressive effects on the cutaneous inflammatory cells by altering their cytokine production, by inhibiting the antigen-presenting function of Langerhans cells and finally by inducing apoptosis of infiltrating T cells [2, 21–23].

In the second instance, UV radiations protect the skin by inducing a thickening of the stratum corneum; this enhances the epidermal barrier function, reduces the entry of allergens and irritants and limits the eczematous processes [2, 24]. Moreover, light and especially NB-UVB radiations exert an antibacterial activity, both by inhibiting superantigen production and by influencing messenger ribonucleic acid (mRNA) levels of antimicrobial peptides. In this way, they prevent infections due to Staphylococcus Aureus and Pityrosporum Orbiculare, which are quite common in AD individuals [25, 26].

Not all the forms of light therapy have the same degree of beneficial action for AD: the most effective are UVA1, narrowband (NB)-UVB and psoralen plus ultraviolet A (PUVA). Secondary options may be broadband (BB)-UVB or ultraviolet AB (UVAB) therapies, whereas less used or standardized resources include full spectrum light, natural sunlight, balneophototherapy and extracorporeal photopheresis (ECP) [2, 21, 23, 27–29]. Finally, new emerging techniques that are today under development, showing good potential, are pulsed-dye laser and 308-nm monochromatic eccimers light [18, 25].

27.2.2 Types of Phototherapy Used for AD

The main types of phototherapy used for AD are summarized in Table 27.1.

Table 27.1 Applications of phototherapy for AD: different types of radiation

	Modality of administration and efficacy	Limitations, adverse effects
UVB (NB-UVB and BB-UVB)	BB-UVB was the first radiation used in Dermatology. Today it has been substituted by NB-UVB, which is the first choice light therapy for adult and pediatric (over eight years of age) chronic AD cases.	Erythema, sunburning, blisters, xerosis and long-term epidermal photodamage. Carcinogenic risk has not been excluded but it seems to be very low.
		Side effects with NB-UVB are milder or less frequent.
UVA and UVA1 (HD-UVA1, MD-UVA1, LD-UVA1)	MD-UVA1 radiation is a first choice for acute AD cases. When not available, it can be substituted by full-spectrum UVA with satisfactory results.	HD-UVA1 is often not well tolerated unless a "cold-light UVA1" system is used, whereas LD-UVA1 is not effective for AD.
		UVA needs longer exposures than UVB radiation and frequently causes dermal adverse effects: lentigines, folliculitis, hypertrichosis, pruritus, herpes simplex virus reactivations, hyperpigmentation, redness, polymorphous skin eruption and cataract.
PUVA (systemic and topical)	It combines the assumption of psoralens with UVA radiation. Some good results have been obtained in adult AD patients, but further studies are required.	Systemic toxicity, folliculitis, photoonycolysis are quite frequent.
		Moreover its long-term use has been associated with a carcinogenic effect.
		Therefore it must be avoided in children younger than twelve years.
Balneophototherapy (UVA and NB-UVB variants)	An emerging technique with encouraging results in AD cases. In the UVA variant psoralen is dissolved in warm water, while in the UVB variant salt water solution bathing is combined with NB-UVB radiation.	Better tolerated than PUVA, with only mild but frequent adverse effects reported, in particular with the use of the UVB variant.
UVAB	Subsequent or simultaneous emission of UVA and UVB radiation. Rarely used today.	Side effects of UVA and UVB radiation.

27.2.2.1 BB-UVB (280–315 nm)

Nexman tested BB-UVB on AD patients for the first time in 1948 [19], while Hannuksela et al. and Jekler and Larkö confirmed its efficacy during the following decades. This type of radiation was traditionally obtained with fluorescent and mercury arc lamps (Psorilux 9050©), which today have been replaced by more specific devices to eliminate ultraviolet A (UVA) wavelengths from the emitted spectrum (Philips TL01©). Nevertheless, NB-UVB has now almost totally substituted the use of BB-UVB [18, 24, 30–32].

27.2.2.2 NB-UVB (311–313 nm)

NB-UVB radiation is much more effective and less erythemogenic than BB-UVB, as it excludes shortwave lengths from the emission. This restricts the penetration of the UV radiation within the epidermis. This superficial effect prevents the radiation from reaching the dermal layer and is optimal for chronic AD cases [18, 26]. Moreover, it induces epidermal lymphocytes apoptosis and Th2 switch in the immune response by inhibiting Th1 activity, which is usually hyperactivated in chronic AD. Although the oncogenic risk of narrowband may be higher than broadband radiation, NB-UVB usually achieves faster clinical remission in shorter exposure, due to its higher efficacy. The result is a reduction in the overall carcinogenic risk [2, 18].

27.2.2.3 UVA (315–400 nm) and UVA1 (340–400 nm)

In the past UVA therapy was difficult to use in AD patients, due to the long exposure times needed to obtain effective dosages. The recent advent of UVA1 lamps has overcome many difficulties related with its use; the elimination of ultraviolet A2 (UVA2) wavelengths from the emission has allowed high doses to be performed, preventing the majority of adverse effects. The multiple biological actions of UVA1 radiation are mainly concentrated in the deep dermis and in the superficial blood vessels plexus; T lymphocytes and immature mast cells apoptosis, cytokines suppression, collagen synthesis and calcineurin inhibition [18, 33–38]. This deep dermal effect makes UVA1 therapy more beneficial for acute AD cases.

UVA1 can be administrated in a high dose (HD, 80–130 J cm^{-2}), a medium dose (MD, 40–80 J cm^{-2}) or a low dose (LD, <40 J cm^{-2}), in quite long treatment sessions that may last from 10 min to 1 h [23, 39].

High doses of UVA1 produce intolerable heat, overcome only with the use of "UVA1 cold light" lamps, which filter the infrared radiation with a cooling ventilation machine [18]. Essential requirements like special cabins, machinery and space, make the lamps selective for UVA1 expensive and difficult to manage.

Therefore conventional UVA lamps are today still the most accessible and least expensive choice when selective UVA1 radiation cannot be administered, since their emission includes 90% of UVA1 radiation [38, 39].

27.2.2.4 UVAB (280–400 nm)

UVAB radiation includes both wavelengths of UVA and ultraviolet B (UVB) in the same emission by a single device (Metec Helarium©) or by combined simultaneous or subsequent emission. Its use for AD started with Jekler and Larkö, however its use is rare today, since more specific emissions have taken over [18, 24, 30].

27.2.2.5 Photochemotherapy (PUVA)

PUVA technique consists in the administration of UVA combined with psoralens, for example 8-methoxypsoralen (8-MOP). These are orally taken or topically applied respectively in systemic PUVA and in cream-PUVA or in bath-PUVA. Cream-PUVA, in particular, consists in applying in occlusion 0.0006% of psoralen in oil-in-water ointment on limited body areas, 30 min up to 1 h before UVA irradiation. It is very useful for those cases with localized disease. Bath-PUVA, also defined balneophototherapy, acts on the whole body and consists in a 30-min bathing in 0.5–1.0 mg/l of psoralen dissolved in warm water, immediately followed by UVA exposure. Balneophototherapy is often performed also with NB-UVB radiation, but usually in this case a salt in water solution is used instead of psoralens (photo-brine therapy), and the two agents are administered synchronously [2, 18, 40, 41].

The beneficial action of PUVA for AD is mediated by photoadducts formation, inhibition of cell proliferation and T lymphocytes death. A decrease of hyperinnervation of the epidermis may also lead to pruritus reduction [18, 42–45].

The administration of PUVA should be limited in time due to its possible cutaneous carcinogenic side effects [18, 42].

27.3 Recommendations and Efficacy

Phototherapy is recommended for the treatment of AD with strength B and its efficacy has been verified with level II of evidence [17, 20, 24, 27, 46–52]. It is a second-line treatment and is very useful for cases not controlled with first-line interventions such as behavioral measures and topical products, for example emollients, moisturizers, calcineurin inhibitors and steroids. It can be administrated as a monotherapy or, more often, in addition to the other first-line measures, except topical immunomodulators.

The combined use has a considerable corticoid-sparing effect and may also reduce the amount of topical immunomodulators needed, although topical calcineurin inhibitors should be carefully prescribed with light therapy [17, 32, 53].

No type of light radiation has been defined as superior to the others: all forms of phototherapy have similar estimated efficacy except for home phototherapy, which has strength of recommendation C and level of evidence III [2, 17, 53, 54].

In numerous reports of the literature, phototherapy achieved improved SCORAD and long-lasting remission in AD patients, both adults and children. Light radiation lamp emission is demonstrably more effective than natural sunlight [27], with NB-UVB as the most commonly used light resource due to its proven efficacy, good tolerability and poor side effects [17, 24, 30–32].

The SCORAD reduction obtained with the use of NB-UVB in AD patients was generally >50%, up to 68%, after a 12 weeks treatment made 3 times weekly. This was verified mainly in adults, but also in some pediatric trials, with significant improvement of the clinical scores and QoL, reduced extension of the disease and satisfactory rate of remission (ranging from 40 to 68%) [48, 52, 55–57]. Moreover, the oncogenic risk associated with NB-UVB has not yet been verified. These evidences make UVB phototherapy a valid therapeutic option in pediatric patients, as an alternative to immunosuppressant treatment, although it may be limited by poor compliance of the child to remain inside the cabin for the duration of the light emission. In clinical practice 8 and 12 years are the minimum age limits considered for performing phototherapy in children, respectively NB-UVB and PUVA. Finally, most recent studies have correlated the SCORAD improvement with histopathologic and molecular modifications, evidencing Th2, "T22" and Th1 pathways suppression, with reversal of the disease activity and normalization of the epidermal barrier function [2].

UVA1 is also significantly effective in improving the clinical symptoms of AD. It is more efficient than UVAB and its action is very rapid: the clinical response has been reported to start within only six applications [58–60].

However, the use of UVA1 is markedly limited by heat development which can be reduced by eliminating >530 nm wavelengths with cold-light UVA1 lamps, and by preferring Medium-dose UVA1 over high-dose UVA1 to achieve better tolerability [31, 48, 60]. The clinical results documented with MD- and HD-UVA1 were in fact similar in most studies; on the contrary low-dose UVA1 was not sufficiently effective for AD treatment in the performed studies [61, 62].

UVA1 treatment for moderate-to-severe AD requires on average courses of 15 exposures of MD-UVA1 [38, 40]. As already mentioned above, in the clinical practice UVA1 is not usually available and so it is often replaced by conventional UVA radiation, with satisfactory results.

The clinical improvements of AD observed in the literature were similar using UVA1 or NB-UVB, however the clinical practice suggests that the former is optimal for acute cases and the latter is more effective and suitable for chronic cases [22, 39, 63]. In fact acute and chronic AD are caused by differently located pathogenic processes: respectively in the dermis and in the epidermis [23, 25, 58, 62].

To date only few studies have tested the efficacy of photochemotherapy; for this reason, systemic or topical PUVA and balneophototherapy are considered as second-choice modalities. However topical PUVA is often considered a good option for the cases of localized disease [64]. The best results in SCORAD reduction, comparable with the effect of the first-choice light emission, were obtained with balneotherapy in salt solution combined with synchronous NB-UVB therapy [20, 55, 65–67].

UVAB was very used in the past and can also be an option today, although it is rarely chosen, due to the more frequent side effects and limited availability. Its beneficial action is milder than the effect of cyclosporine and of corticosteroids, although satisfactory [23, 24, 30, 58–60, 68].

BB-UVB and full-spectrum light (320–5000 nm) are scarcely recommended for the treatment of AD, due to their limited efficacy and to the availability of more specific radiations [28, 31].

Last-choice treatments are old phototherapy types, like Goeckerman therapy and heliothalassotherapy, whose role remains uncertain [23].

Home phototherapy may be an alternative option for patients that do not comply with the standard settings; however it is rarely performed as it is expensive and requires careful medical supervision [54].

It is remarkable that to date the good and side effects of phototherapy for AD have not been sufficiently studied: some additional larger, homogeneous, controlled, randomized trials should be realized to assess its tolerability and efficacy with certainty and precision. Some authors reviewed the heterogeneous, scarcely comparable and often small-sized series available in the literature, with the conclusions that we described above, but no comprehensive and totally reliable meta-analyses have been realized yet [2, 23].

27.4 Dosage and Therapeutic Protocols

Phototherapy has some limitations, above all the compliance: light therapy requires frequent courses of UV emission, usually at least 2–5 sessions per week, up to 2–3 months. This may be limiting and incompatible with the daily activities, for example, of workers and students. UV lamps and machinery are complex and expensive, in particular when equipped with cooling systems, as in the case of cold-light UVA1. In addition their use requires specialized medical staff and dedicated spaces [2, 18]. Moreover, some body areas like the skin folds or the scalp are scarcely reached by radiations due to their protected location or to the presence of hairs. In addition phototherapy should be avoided in the genitals, as the semimucosas in these regions are highly exposed and so more susceptible to short- and long-term light radiation damages.

The correct therapeutic radiation, including type and dosage, must be chosen considering multiple factors: the characteristics of the patient and of the lesions (acute or chronic), the severity and the location of the disease. In addition the availability, costs and benefits of the therapeutic decisions must be evaluated and balanced.

The characteristics of the patient are fundamental, therefore an accurate anamnesis and careful clinical examination are recommended before embarking on phototherapy [17]. In particular, the assumption of photosensitizing medicaments must be excluded.

Parameters and dosing protocols may vary, since the use of phototherapy for the treatment of AD is largely empirical and has not yet been uniquely regulated. In clinical practice dermatologists refer to the specific UVA and UVB dosages recommended within the AAD psoriasis guidelines [46], which are also suitable for AD cases.

Treatment begins with an initial dose that is progressively increased during each phototherapy session. Traditionally, the initial dose is calculated in relation to the patient's skin surface (mJ/cm^2) and I-VI Fitzpatrick skin type. However, this method has been superseded in recent years in favor of the determination of the minimal erythema dose (MED). In fact, while the phototype evaluation is a subjective procedure, prone to a wide margin of error, phototesting the skin of the patient is the most reliable and objective technique to determine its UVA and UVB sensitivity. After the MED has been calculated both for UVA and UVB radiations, the therapeutic dose of light treatment is calculated. For UVB the initial dose must be set to a 30% lower value than the MED. Monitoring the development of erythema in the 24 h following each UVB exposure thus becomes an optimal dosimeter. The dosage is increased proportionally to the tolerance of the patient, if no or mild erythema has developed: respectively by 40% and by 20%. If mild asymptomatic erythema persists, dosage is not increased, whereas if painful and/or severe erythema occurs, phototherapy is stopped. The therapy can be restarted when symptoms subside, initially with half of the last dosage and then with an increase of 10% during the following sessions [69].

For maintenance, phototherapy can be used for AD either intermittently or continuously [17, 47].

27.5 Safety and Adverse Effects

Phototherapy is usually safe and well tolerated. Adverse effects are rare and mild in most cases, although they must always be considered before starting treatment. The main side effects of the different types of phototherapy are summarized in Table 27.1 [27, 47, 48].

In particular, both UVA and UVB light therapy can be administered to pediatric patients, where they meet good compliance and applicability. By involving the family in the decision making process and gradually introducing the children into the care setting, phototherapy is usually accepted and well tolerated also by the youngest of patients, so that it is becoming a mainstay of the therapy of pediatric AD. The age restriction may be a limit, which could be overcome only by high cooperation of the child to remain in the cabin for the duration of the phototherapy session [47, 48, 50–56].

Side effects of phototherapy at any age can be short-term, mainly related to improper doses, or long-term.

Acute adverse effects are quite frequent and usually develop within 24 h after UV exposure. The most frequent are: erythema, sunburning, blisters and xerosis. Other less common reactions are pruritus, hypertrichosis, hyperpigmentation, polymorphous skin eruption and different manifestations of photosensitivity.

In addition, some adverse effects are more related to the use of particular light modalities: for example photoonycholysis and systemic toxicity with PUVA, pruritus, herpes simplex virus reactivations and redness with UVA1, folliculitis with UVA1 and PUVA, cataract with UVA. On the other hand, erythema is more common with UVB use, and this effect has been the main limit of this type of radiation since its first applications in Dermatology [2]. These differences are due to the different levels of penetration of different UV radiations through the skin layers, where they concentrate their therapeutic activity but also cause side effects: UVB rays carry out their action within the epidermis, while UVA rays reach the deep dermis [38].

Long-term adverse effects include photodamage, lentigines and actinic keratosis. The risk of carcinogenesis represents the most fearsome long-term adverse effect of phototherapy, which could induce non-melanoma and melanoma skin cancers [17]. However, except for PUVA, this risk has not yet been proved nor quantified [18, 62, 69, 70].

For example, a carcinogenic action was initially related to UVB in addition to the erythemogenic effect, but was then resized. Many studies documented an absent or only mild increase of non-melanoma skin cancers, and no increased risk for melanoma. The carcinogenic risk of NB-UVB has been estimated as 50% higher than BB-UVB at comparable dose; however the global carcinogenic risk due to NB-UVB is reduced by less exposure needed, together with milder and less frequent acute side effects.

The carcinogenic risk related to the use of UVA1 has been suspected, although not proved [2, 71]. On the other hand, PUVA has a proven carcinogenic effect. An increased incidence of squamous cell carcinoma, basal cell carcinoma and melanoma was documented in cohorts of psoriatic patients with a history of long-term use, after >20 years of follow-up [18, 42, 72–75].

Other inconveniences of systemic PUVA, which can be reduced by using topical PUVA, are symptoms of general toxicity, for example nausea, vomiting, hepatotoxicity. Moreover longer-term effects like photosensitivity and cataract have been reported.[13]

Finally, balneophototherapy usually meets good patient compliance, despite its quite frequent mild adverse effects [67].

When considering the studies that have investigated the possibility of an oncogenic effect associated with the use of phototherapy, resulting in uncertain and controversial data, one possible bias should be underlined: almost all of the observed patients were adult and concomitantly, previously or even later had assumed immunosuppressants for their disease, which could probably have enhanced their risk for skin tumors. The same risk in children has not been verified; therefore NB-UVB and UVA are permitted in younger patients, while only PUVA must be avoided. However, pending further studies, some authors do not recommend phototherapy in children, while most of them suggest reserving it only for refractory or severe cases. In general, phototherapy should be always prescribed conscientiously and with caution in children, and in particular it should be avoided in younger children [2, 25].

27.6 Conclusion

Phototherapy is a first-choice treatment for AD, both in the adult and pediatric population. It is recommended with level of evidence II, strength of recommendation B, as a second-line intervention for patients that have not benefited from environmental measures and topical treatment. It can be used alone or else combined with corticosteroids, sparing the dose needed, or other systemic agents.

UVA1 and NB-UVB are the most effective and preferred types of light therapy, used respectively for acute and chronic AD.

Phototherapy is usually safe and well tolerated. However short- and long-term adverse effects are possible, and carcinogenic risk has not been excluded. Therefore, the use of phototherapy must be careful and should be related to the patient's overall condition, especially in children.

Conflict of Interest All authors give their consent to the publication of the article and declare no conflict of interest.

References

1. Bieber T (2010) Atopic dermatitis. Ann Dermatol 22(2):125–137
2. Patrizi A, Raone B, Ravaioli GM (2015) Management of atopic dermatitis: safety and efficacy of phototherapy. Clin Cosmet Investig Dermatol 8:511–520. doi:10.2147/CCID.S87987
3. Saeki H, Nakahara T, Tanaka A, Kabashima K, Sugaya M, Murota H et al (2016) Clinical practice guidelines for the management of atopic dermatitis 2016. J Dermatol. doi:10.1111/1346-8138.13392
4. Elias PM, Hatano Y, Williams ML (2008) Basis for the barrier abnormality in atopic dermatitis: outside-inside-outside pathogenic mechanisms. J Allergy Clin Immunol 121(6):1337–1343
5. Lee HJ, Lee SH (2014) Epidermal permeability barrier defects and barrier repair therapy in atopic dermatitis. Allergy, Asthma Immunol Res 6(4):276–287
6. Dainichi T, Hanakawa S, Kabashima K (2014) Classification of inflammatory skin diseases: a proposal based on the disorders of the three-layered defense systems, barrier, innate immunity and acquired immunity. J Dermatol Sci 76(2):81–89
7. Guttman-Yassky E, Lowes MA, Fuentes-Duculan J, Zaba LC, Cardinale I, Nograles KE et al (2008) Low expression of the IL-23/Th17 pathway in atopic dermatitis compared to psoriasis. J Immunol 181:7420–7427
8. Levy ML (2007) Atopic dermatitis: understanding the disease and its management. Curr Med Res Opin 23:3091–3103
9. Kunz B, Oranje AP, Labrèze L, Stalder JF, Ring J, Taïeb A (1997) Clinical validation and guidelines for the SCORAD index: consensus report of the European task force on atopic dermatitis. Dermatology 195(1):10–19

10. Hanifin JM, Thurston M, Omoto M, Cherill R, Tofte SJ, Graeber M (2001) The eczema area and severity index (EASI): assessment of reliability in atopic dermatitis. EASI Evaluator Group. Exp Dermatol 10(1):11–18
11. Furue M, Ebata T, Ikoma A, Takeuchi S, Kataoka Y, Takamori K et al (2013) Verbalizing extremes of the visual analogue scale for pruritus: a consensus statement. Acta Derm Venereol 93(2):214–215. doi:10.2340/00015555-1446
12. Chren MM, Lasek RJ, Sahay AP, Sands LP (2001) Measurement properties of Skindex-16: a brief quality-of-life measure for patients with skin diseases. J Cutan Med Surg 5(2):105–110. Epub 2001 Mar 21
13. Finlay AY, Khan GK (1994) Dermatology Life Quality Index (DLQI) – a simple practical measure for routine clinical use. Clin Exp Dermatol 19(3):210–216
14. Boguniewicz M, Leung DY (2010) Recent insights into atopic dermatitis and implications for management of infectious complications. J Allergy Clin Immunol 125:4–13. quiz 4-5
15. Eichenfield LF, Tom WL, Berger TG, Krol A, Paller AS, Schwarzenberger K et al (2014) Guidelines of care for the management of atopic dermatitis: section 2. Management and treatment of atopic dermatitis with topical therapies. J Am Acad Dermatol 71(1):116–132
16. Akdis CA, Akdis M, Bieber T, Bindslev-Jensen C, Boguniewicz M, Eigenmann P et al (2006) Diagnosis and treatment of atopic dermatitis in children and adults: European Academy of Allergology and Clinical Immunology/American Academy of Allergy, Asthma and Immunology/PRACTALL Consensus Report. Allergy 61(8):969–987
17. Sidbury R, Davis DM, Cohen DE, Cooper KD, Silverman RA, Berger TG et al (2014) Guidelines of care for the management of atopic dermatitis. J Am Acad Dermatol 71(2):327–349
18. Grundmann SA, Beissert S (2012) Modern aspects of phototherapy for atopic dermatitis. J Allergy (Cairo) 2012:121797
19. Nexman PH 1948 Clinical Studies' of Besnier's Prurigo, Dissertation. Copenhagen: Rosenkilde & Bagger
20. Morison WL, Parrish J, Fitzpatrick TB (1978) Oral psoralen photo-chemotherapy of atopic eczema. Br J Dermatol 98:25–30
21. Gambichler T (2009) Management of atopic dermatitis using photo(chemo)therapy. Arch Dermatol Res 301:197–203
22. Majoie IM, Oldhoff JM, van Weelden H, Laaper-Ertman M, Bousema MT, Sigurdsson V et al (2009) Narrowband ultraviolet B and medium-doseultraviolet A1 are equally effective in the treatment of moderate to severe atopic dermatitis. J Am Acad Dermatol 60:77–84
23. Garritsen FM, Brouwer MWD, Limpens J, Spuls PI (2014) Photo(chemo)therapy in the management of atopic dermatitis: an updated systematic review with implications for practice and research. Br J Dermatol 170:501–513
24. Jekler J, Larkö O (1990) Combined UVA-UVB versus UVB phototherapy for atopic dermatitis: a paired-comparison study. J Am Acad Dermatol 22(1):49–53
25. Ring J, Alomar A, Bieber T, Deleuran M, Fink-Wagner A, Gelmetti C et al (2012) Guidelines for treatment of atopic eczema (atopic dermatitis) Part II. J Eur Acad Dermatol Venereol 26(9):1176–1193
26. Gambichler T, Skrygan M, Tomi NS, Altmeyer P, Kreuter A (2006) Changes of antimicrobial peptide mRNA expression in atopic eczema following phototherapy. Br J Dermatol 155(6):1275–1278
27. Meduri NB, Vandergriff T, Rasmussen H, Jacobe H (2007) Phototherapy in the management of atopic dermatitis: a systematic review. Photodermatol Photoimmunol Photomed 23(4):106–112
28. Byun HJ, Lee HI, Kim B, Kim MN, Hong H, Choi Y et al (2011) Full-spectrum light phototherapy for atopic dermatitis. Int J Dermatolog 50:94–101
29. Dennis M, Bhutani T, Koo J, Liao W (2013) Goeckerman therapy for the treatment of eczema: a practical guide and review of efficacy. J Dermatolog Treat 24:2–6
30. Jekler J, Larkö O (1991) Phototherapy for atopic dermatitis with ultraviolet A (UVA), low-dose UVB and combined UVA and UVB: two paired-comparison studies. Photodermatol Photoimmunol Photomed 8(4):151–156
31. Jekler J, Larkö O (1991) UVA solarium versus UVB phototherapy of atopic dermatitis: a paired-comparison study. Br J Dermatol 125(6):569–572
32. Hannuksela M, Karvonen J, Husa M, Jokela R, Katajamäki L, Leppisaari M (1985) Ultraviolet light therapy in atopic dermatitis. Acta Derm Venereol Suppl (Stockh) 114:137–139
33. Morita A, Werfel T, Stege H, Ahrens C, Karmann K, Grewe M et al (1997) Evidence that singlet oxygen-induced human T helper cell apoptosis is the basic mechanism of ultraviolet-A radiation phototherapy. J Exp Med 186(10):1763–1768
34. Breuckmann F, von Kobyletzki G, Avermaete A, Pieck C, Kreuter A, Brockmeyer NH et al (2002) Mononuclear cells in atopic dermatitis in vivo: immunomodulation of the cutaneous infiltrate by medium-dose UVA1 phototherapy. Eur J Med Res 7(7):315–322
35. Breuckmann F, von Kobyletzki G, Avermaete A, Kreuter A, Altmeyer P (2002) Efficacy of ultraviolet A1 phototherapy on the expression of bcl-2 in atopic dermatitis and cutaneous T-cell lymphoma in vivo: a comparison study. Photodermatol Photoimmunol Photomed 18(5):217–222
36. Breuckmann F, Pieck C, Kreuter A, Bacharach-Buhles M, Mannherz HG, Altmeyer P et al (2001) Opposing effects of UVA1 phototherapy on the expression of bcl-2 and p53 in atopic dermatitis. Arch Dermatol Res 293(4):178–183

37. Beissert S, Granstein RD (1996) UV-induced cutaneous photobiology. Crit Rev Biochem Mol Biol 31(5–6):381–404
38. Zandi S, Kalia S, Lui H (2012) UVA1 phototherapy: a concise and practical review. Skin Therapy Lett 17(1):1–4
39. (a) Darsow U, Wollenberg A, Simon D, Taïeb A, Werfel T, Oranje A et al (2010) ETFAD/EADV eczema task force 2009 position paper on diagnosis and treatment of atopic dermatitis. J Eur Acad Dermatol Venereol 24(3):317–328; (b) Legat FJ, Hofer A, Brabek E, Quehenberger F, Kerl H, Wolf P (2003) Narrowband UV-B vs medium-dose UV-A1 phototherapy in chronic atopic dermatitis. Arch Dermatol 139(2):223-224
40. Fitzpatrick TB, Pathak MA (1984) Research and development of oral psoralen and longwave radiation photochemotherapy: 2000 B.C.-1982 A.D. Natl Cancer Inst Monogr 66:3–11
41. Stege H, Berneburg M, Ruzicka T, Krutmann J (1997) Cream PUVA photochemotherapy. Hautarzt 48(2):89–93. German
42. Lim JL, Stern RS (2005) High levels of ultraviolet B exposure increase the risk of non-melanoma skin cancer in psoralen and ultraviolet A-treated patients. J Invest Dermatol 124(3):505–513
43. Marks DI, Fox RM (1991) Mechanisms of photochemotherapy-induced apoptotic cell death in lymphoid cells. Biochem Cell Biol 69(10–11):754–760
44. Johnson R, Staiano-Coico L, Austin L, Cardinale I, Nabeya-Tsukifuji R, Krueger JG (1996) PUVA treatment selectively induces a cell cycle block and subsequent apoptosis in human T-lymphocytes. Photochem Photobiol 63(5):566–571
45. Tominaga M, Tengara S, Kamo A, Ogawa H, Takamori K (2009) Psoralen-ultraviolet A therapy alters epidermal Sema 3A and NGF levels and modulates epidermal innervation in atopic dermatitis. J Dermatol Sci 55(1):40–46
46. Menter A, Korman NJ, Elmets CA, Feldman SR, Gelfand JM, Gordon KB et al (2010) Guidelines of care for the management of psoriasis and psoriatic arthritis: section 5. Guidelines of care for the treatment of psoriasis with phototherapy and photochemotherapy. J Am Acad Dermatol 62(1):114–135
47. Tay YK, Morelli JG, Weston WL (1996) Experience with UVB phototherapy in children. Pediatr Dermatol 13(5):406–409
48. Clayton TH, Clark SM, Turner D, Goulden V (2007) The treatment of severe atopic dermatitis in childhood with narrowband ultraviolet B phototherapy. Clin Exp Dermatol 32(1):28–33
49. Rombold S, Lobisch K, Katzer K, Grazziotin TC, Ring J, Eberlein B (2008) Efficacy of UVA1 phototherapy in 230 patients with various skin diseases. Photodermatol Photoimmunol Photomed 24(1):19–23
50. Yoshiike T, Aikawa Y, Sindhvananda J, Ogawa H (1993) A proposed guideline for psoralen photochemotherapy (PUVA) with atopic dermatitis: successful therapeutic effect on severe and intractable cases. J Dermatol Sci 5(1):50–53
51. Atherton DJ, Carabott F, Glover MT, Hawk JL (1988) The role of psoralen photochemotherapy (PUVA) in the treatment of severe atopic eczema in adolescents. Br J Dermatol 118(6):791–795
52. Jury CS, McHenry P, Burden AD, Lever R, Bisland D (2006) Narrowband ultraviolet B (UVB) phototherapy in children. Clin Exp Dermatol 31(2):196–199
53. Koek MB, Sigurdsson V, van Weelden H, Steegmans PH, Bruijnzeel-Koomen CA, Buskens E (2010) Cost effectiveness of home ultraviolet B phototherapy for psoriasis: economic evaluation of a randomised controlled trial (PLUTO study). BJM 340:c1490. doi:10.1136/bmj.c1490
54. Cameron H, Yule S, Dawe RS, Ibbotson SH, Moseley H, Ferguson J (2014) Review of an established UK home phototherapy service 1998–2011: improving access to a cost-effective treatment for chronic skin disease. Public Health 128(4):317–324
55. George SA, Bilsland DJ, Johnson BE, Ferguson J (1993) Narrow-band (TL-01) UVB air-conditioned phototherapy for chronic severe adult atopic dermatitis. Br J Dermatol 128(1):49–56
56. Tan E, Lim D, Rademaker M (2010) Narrowband UVB phototherapy in children: a New Zealand experience. Australas J Dermatol 51(4):268–273
57. Darné S, Leech SN, Taylor AE (2014) Narrowband ultraviolet B phototherapy in children with moderate-to-severe eczema: a comparative cohort study. Br J Dermatol 170(1):150–156
58. Krutmann J, Diepgen TL, Luger TA et al (1998) High-dose UVA1 therapy for atopic dermatitis: results of a multicenter trial. J Am Acad Dermatol 38(4):589–593
59. Krutmann J, Czech W, Diepgen T, Niedner R, Kapp A, Schöpf E (1992) High-dose UVA1 therapy in the treatment of patients with atopic dermatitis. J Am Acad Dermatol 26(2 Pt 1):225–230
60. von Kobyletzki G, Pieck C, Hoffmann K, Freitag M, Altmeyer P (1999) Medium-dose UVA1 cold-light phototherapy in the treatment of severe atopic dermatitis. J Am Acad Dermatol 41(6):931–937
61. Abeck D, Schmidt T, Fesq H, Strom K, Mempel M, Brockow K, Ring J (2000) Long-term efficacy of medium-dose UVA1 phototherapy in atopic dermatitis. J Am Acad Dermatol 42(2 Pt 1):254–257
62. Tzaneva S, Seeber A, Schwaiger M, Hönigsmann H, Tanew A (2001) High-dose versus medium-dose UVA1 phototherapy for patients with severe generalized atopic dermatitis. J Am Acad Dermatol 45(4):503–507
63. Gambichler T, Othlinghaus N, Tomi NS, Holland-Letz T, Boms S, Skrygan M et al (2009) Medium-dose ultraviolet (UV) A1 vs. narrowband UVB phototherapy in atopic eczema: a randomized crossover study. Br J Dermatol 160(3):652–658
64. Institute for Quality and Efficiency in Health Care. Balneophototerapy (2004) IQWiG Reports – Commission No. N04–04

65. Tzaneva S, Kittler H, Holzer G et al (2010) 5-Methoxypsoralen plus ultraviolet (UV) A is superior to medium-dose UVA1 in the treatment of severe atopic dermatitis: a randomized crossover trial. Br J Dermatol 162(3):655–660
66. Der-Petrossian M, Seeber A, Hönigsmann H, Tanew A (2000) Half-side comparison study on the efficacy of 8-methoxypsoralen bath-PUVA versus narrow-band ultraviolet B phototherapy in patients with severe chronic atopic dermatitis. Br J Dermatol 142(1):39–43
67. Heinlin J, Schiffner-Rohe J, Schiffner R et al (2011) A first prospective randomized controlled trial on the efficacy and safety of synchronous balneophototherapy vs. narrow-band UVB monotherapy for atopic dermatitis. J Eur Acad Dermatol Venereol 25(7):765–773
68. Granlund H, Erkko P, Remitz A et al (2001) Comparison of cyclosporin and UVAB phototherapy for intermittent one-year treatment of atopic dermatitis. Acta Derm Venereol 81(1):22–27
69. (a) Bolognia JL, Jorizzo JL, Schaffer JV (2012) Dermatology text book. 3rd edn. Elsevier Publishing; (b) Lee E, Koo J, Berger T (2005) UVB phototherapy and skin cancer risk: a review of the literature. Int J Dermatol 44(5):355–360
70. von Thaler AK, Kamenisch Y, Berneburg M (2010) The role of ultraviolet radiation in melanomagenesis. Exp Dermatol 19(2):81–88
71. Kirke SM, Lowder S, Lloyd JJ, Diffey BL, Matthews JN, Farr PM (2007) A randomized comparison of selective broadband UVB and narrowband UVB in the treatment of psoriasis. J Invest Dermatol 127(7):1641–1646. Epub 2007 Mar 22
72. Nijsten TE, Stern RS (2003) The increased risk of skin cancer is persistent after discontinuation of psoralen+ultraviolet A: a cohort study. J Invest Dermatol 121(2):252–258
73. Stern RS, Liebman EJ, Väkevä L (1998) Oral psoralen and ultraviolet-A light (PUVA) treatment of psoriasis and persistent risk of nonmelanoma skin cancer. PUVA Follow-up Study. J Natl Cancer Inst 90(17):1278–1284
74. Stern RS (2001) PUVA follow up study. The risk of melanoma in association with long-term exposure to PUVA. J Am Acad Dermatol 44(5):755–761
75. Stern RS, Nichols KT, Väkevä L (1997) Malignant melanoma in patients treated for psoriasis with methoxsalen (psoralen) and ultraviolet A radiation (PUVA). The PUVA follow-up study. N Engl J Med 336(15):1041–1045

Part VII

Tanning Saloon

UV Driven Tanning Salons: Danger on Main Street

28

Jean-François Doré and Marie-Christine Chignol

Abstract

Appearing in the early 1980s, at a time when UVA was considered as relatively safe, the tanning industry has substantially developed in occidental countries, especially in Northern European countries. In Europe, the erythemally-weighted irradiance of a modern sunbed should not exceed 0.3 W/m^2, equivalent to an UV index of 12, i.e. to a tropical midday sun, but increased in recent years, the UV spectrum emitted by sunbeds had evolved towards higher UVA irradiance and solariums UV had become even less similar to natural sun.

Keywords

UV • Sunbeds • Prevalence • Melanoma • Skin cancer

Motivation for indoor tanning is mainly a cosmetic one: the desire to be more attractive, healthy, with a "good looking attractive" tan. Prevalence of ever sunbed use is rather low in general populations (e.g. 5.6% of adults in a recent USA survey), but concentrates in specific sub-populations: white-skinned populations from Northern Europe, and in women younger than 30 years. The highest prevalence rates being reported among US University students: up to 69%. Prevalence of sunbed use by adolescents is usually low before the age of 15, but the age of first use may be very young (e.g. <13 years), and the highest rates are observed among US high school students and in Scandinavia among girls and teenagers 15–18 years old (up to 43%). However, in Denmark, recent surveys show that the prevalence of sunbed use in the age group 15–19 years is currently substantially declining.

A long list of epidemiological studies have consistently showed that sunbed exposure is a risk factor for cutaneous melanoma (RR =1.20 (1.08–1.34) according to a recent meta-analysis of 27 studies), more especially when first exposure takes place at a younger age (RR = 1.59 (1.36–1.85) for age at exposure <35 years old, from 13 studies). Recent meta-analyses showed

J.-F. Doré (✉) • M.-C. Chignol
Centre de Recherche en Cancérologie de Lyon, Univ Lyon, Université Claude Bernard Lyon 1, INSERM1052, CNRS 5286, Centre Léon Bérard, 28 rue Laënnec, Cedex 08, 69373 Lyon, France
e-mail: jean-francois.dore@lyon.unicancer.fr

S. Ahmad (ed.), *Ultraviolet Light in Human Health, Diseases and Environment*, Advances in Experimental Medicine and Biology 996, https://doi.org/10.1007/978-3-319-56017-5_28

that sunbed exposure is also a risk factor for squamous cell carcinoma (RR = 2.23 (1.39–3.57), 5 studies) and basal cell carcinoma (RR = 1.09 (1.01–1.18), 6 studies), the risk being higher when first exposure takes place at an early age. The increase in melanoma risk appears as relatively modest in the general population (+16 to 20%, according to the more recent meta-analyses), but concentrates among specific sub-populations (e.g. younger women: +59% for first exposure before the age of 30 years, and even +200% for frequent use in the 10–39 years period). Sunbed use is associated with early onset melanoma: the fraction of risk attributable to sunbed use in patients diagnosed with a melanoma before the age of 30 may be very high: 76% in Australia among those who had ever used a sunbed and were diagnosed between 18–29 years of age. It has been estimated that in 2008, in Europe, 5.4% of 63,942 new cases of melanoma diagnosed each year may be related to sunbed use, women representing most of this burden (6.9% of all melanomas in women). And that about 498 women and 296 men may die each year from a melanoma as a result of being exposed to indoor tanning. Public Health authorities should at least strongly discourage use of sunbeds, particularly by young women, and even consider imposing a total ban as in Brazil and Australia.

28.1 Introduction

She is walking down Main Street. She is a pretty young woman, in her early twenties, with blond hairs and a pale complexion. Today, she has decided to get a tan, to have a more attractive and healthier appearance. She enters a tanning salon, and minutes later ... she has substantially increased her personal risk to develop a cutaneous melanoma!

In our grandmas' days, in the occidental societies, being tanned was regarded as belonging to lower social classes. Peasants, outdoor workers were chronically exposed to sun and as a consequence were tanned. Before World War I, women from the upper classes made any effort to preserve their pale complexion and never went outdoor without their hat and umbrella. Attitudes towards sun exposure began to change shortly before World War II. Some even say that this was due to Coco Chanel and her friends who imported the fashion of tanning into the Parisian society! After the War, the development of mass tourism and holidays in sunny resorts led to a change in the social perception of tanning which now is no longer considered as a symbol of peasant origin, but as symbols of beauty, sensuality, success and good health. Nowadays, in our occidental world, almost everybody wants to get a summer tan, a marker of successful holidays: an attitude that Asians who avoid tanning and prefer pale skins do not understand! However, there appear to be some signs of change, and in certain societies (e.g. California), tanning is no longer considered fashionable.

The desire of a "good looking" tan all the year round fueled the development of indoor artificial tanning. The first tanning devices appeared in the early 1980s, at a time when UVA were considered as relatively safe. And the indoor tanning industry has grown substantially over the last three decades, not only in Europe and North America, especially more in the sun-deprived Northern countries, Iceland, Scandinavia, but also in more sunny countries such as Italy. In the USA, according to the Indoor Tanning Association (a professional organization of indoor tanning manufacturers, distributors, facility owners), indoor tanning is today a strong part of the American small business community, with a total number of professional indoor tanning facility businesses of 14,000 in thousands of towns throughout USA, employing 84,000 persons, and each year about 10% of the American public visits an indoor tanning facility [32]. However, it should be noted that a 10% tax on tanning services introduced in 2009 shuttered 9600 tanning business and killed 80,900 jobs [3]. A survey of commercial indoor tanning facilities in 116 large cities in the United States, conducted before the enforcement of the tan tax, found that the average number of tanning salons exceeded the average number of Starbucks cafés or McDonald's restaurants [29]. In Europe, the European Sunlight Association represents some 20,000 indoor tanning facilities with approximately 120,000 sunbeds in use, employing nearly 100,000 people in Europe. The annual turnover of the indoor tanning industry (manufacturers, dealers and studios) is about 2.1 billion Euros [20]. This is really a huge business, and it is not surprising that the tanning industry practices an intense lobbying to promote the many benefits of exposure to sunbeds while denying any link between indoor ultraviolet (UV) tanning and skin cancer [6].

28.2 UV Tanning

Development of a tan is a response of the skin (human skin, most animals do not tan) to the aggression by UV rays. It results from the increase of melanin pigment synthesis by epidermal melanocytes and transfer to keratinocytes. This is triggered by UVB, and takes a few days to appear, while UVA (mainly involved in skin ageing) only induces an immediate pigment darkening, i.e. a redistribution of preformed pigment without increased pigment synthesis. Modern canopy-like UV-tanning units as used by professional tanning salons are equipped with low-pressure fluorescent lamps with a spectrum mainly emitting in the UVA range plus some UVB (necessary for the induction of a deep long-lasting tan). In Europe, sunbeds are currently regulated by a standard which prescribes that their maximum erythemally-weighted irradiance should not exceed 0.3 W/m^2, equivalent to an UV index of 12, i.e. to a tropical midday sun, which WHO terms extreme. Equivalent, but not identical to natural sunlight because of a higher proportion of UVA, and there are large variations in the UV spectrum emitted by different tanning appliances. Recent surveys showed poor compliance with the standard; in England, Tierney et al. [45] found only 10% of sunbeds surveyed within the recommended limit, in Greece, approximately 60% of the measured sunbeds exceeded the 300 mW/m^2 limit [40], and in Norway, although compliance had increased in recent years, the UV spectrum emitted by sunbeds had evolved towards higher UVA irradiance and solariums UV had become even less similar to natural sun [39].

28.3 Prevalence of Sunbed Exposure

We are currently obtaining more and more information on who uses sunbeds and the frequency of use. The prevalence of sunbed exposure varies greatly from one country to another and according to time period, sex, age and areas of residency. Commercial indoor tanning facilities are mainly located in urban areas, and this may help explain the higher usage of indoor tanning by urban populations.

Prevalence of indoor tanning can be approached by considering exposure of the controls in case-control studies of sunbed exposure and occurrence of melanoma and other types of skin cancers, or through specific surveys in the general population or in selected populations: younger women, university students, teenagers etc.

Numerous surveys have been conducted in Europe, USA and Australia, to specifically address the characteristics of sunbed users, their motivation and their perception of the risks of tanning. Twenty six of these surveys have been summarized in a recent review [19], and 25 additional recent surveys, 8 among adults and 17 among adolescents and children, have been summarized in recent opinion from a European Commission scientific committee [41].

Few recent surveys have addressed the prevalence of sunbed use among adults in general populations (Table 28.1). This prevalence of ever or recent use of sunbeds appears rather low: 5.6 to 10–14 or even 23%, but concentrates in specific sub-populations. In USA, 5.6% of adults reported indoor tanning in the past 12 months, but a higher prevalence of indoor tanning was found among whites, women, persons aged 18–25 years (12.3%), (CDC [14]); the highest prevalence of indoor tanning was found among white women aged 18–21 years (31.8%) and aged 22–25 years (29.6%). Among white adults who reported indoor tanning, the frequency of use was higher among women (average of 20.3 sessions per year; 57.7% reported tanning ≥10 times in the past 12 months), and more especially among white women aged 18–21 years (average of 27.6

Table 28.1 Prevalence of sunbed use among adults in general populations

Country	Period	Age (years)	Sample size	Sample source	% sunbed use	References
France	September 28–October 20, 2011	≥18	1502 (787 female, 715 male)	Nationwide telephone survey (quota method). 9209 contacted (participation rate: 16,3%)	*Current or past users:* Total: 10 Female: 14.5 Male: 5.0 (mean age at 1st use: 27.6 y) <50 yrs.: female: 18.9; male: 5.1 Skin phototype 1 and 2: 15.6	[24]
France	April 3 – August 7, 2010	15–75	3359	National telephone survey (fixed line and mobile) "Baromètre cancer 2010" (acceptation rate 60%)	*Ever use*: 13.4 Women: 19.4 Men: 7.1 *Use in the last 12 months*: 3.5 Women: 5.0 Men: 2.0 women 20–25 y.o.: 13.7 men 20–25 y.o.: 6.1	[9]
Denmark	2007–2009	15–59	13,229 6049 M 7180 F 15–19: 1359 20–29: 1958 30–39: 3049 40–49: 3552 50–59: 3301	Population based annual web and telephone surveys (following a campaign in March 2007)	*Recent users* (past 12 mo.): Mar 2007: 29.9 (M 21.8, F 35.9) Aug. 2007: 27.8 (17.2, 35.3) Aug. 2008: 26.7 (17.5, 35.4) Aug. 2009: 23.3 (16.7, 30.1) *Age:* (Mar 2007; Aug 2007; 2008; 2009) 15–19: 50.3; 47.4; 44.2; 32.9 20–29: 46.7; 45.4; 37.6; 31.5 30–39: 30.6; 30.8; 27.9; 22.0 40–49: 25.7; 22.3; 22.6; 22.5 50–59: 17.8; 15.8; 14.6; 13.8	[34]

(continued)

Table 28.1 (continued)

Country	Period	Age (years)	Sample size	Sample source	% sunbed use	References
USA	2011	≥18	315	Data from 2011 national Youth Risk Behaviour Survey (YRBS) of high school students	*non-Hispanic white female high school students:*	[25]
					use in the previous 12 months:	
					43.8% [95%CI: 36.0–52.0] ()	
					frequent use (≥ 10 times in the previous 12 months):	
					29.97% [95%CI: 23.0–37.8]	
	2010	18–34	1857	Data from 2010 National Health Interview Survey (NHIS) for adults aged 18 to 34 years.	*non-Hispanic white women:*	
					24.9% (use in the previous 12 months)	
					15.1% (frequent use ≥10 times in the previous 12 months).	
					Highest use among 18–21 year (31.8%), lowest among 30–34 year (17.4%).	
USA	2010	≥18	25,233	Data from National Health Interview Survey (NHIS) Cancer control supplement (response rate: 60.8%)	*Ever use in the past 12 mo.*:	[14]
					Total: 5.6	
					Men: 2.2	
					Women: 8.9	
					18–21 years: 21.2	
					22–25 years: 20.4	
					White women: 12.9	
					18–21 years: 31.8	
					22–25 years: 29.6	
					Frequency of use (among white adults reporting indoor tanning):	
					Men: 14.6 sessions/year	
					% ≥ 10 times: 40.0	
					Women: 20.3 sessions/y.	
					% ≥ 10 times: 57.7	
					Women aged 18–21 y.:	
					27.6 sessions/year	
					% ≥ 10 times: 67.7	

Selected reports from Europe, USA

sessions per year; 67.6% reported tanning ≥10 times in the past 12 months). The same trend was observed among younger women in France and Denmark [24, 34]. Some surveys in Europe have further shown that indoor tanning is not infrequent among sun-sensitive individuals, e.g. individuals with phototypes I or II (Fitzpatrick scale) [24], or individuals with fair skin (19% prevalence) or freckles (25%) [44].

Many studies have specifically addressed the prevalence of sunbed use by children and adolescents in Northern Europe, USA and Australia (See SCHEER [41] for a detailed presentation of 16 recent surveys). These surveys show that if prevalence of sunbed use is usually low before the age of 15, the age of first use may be very young e.g. <13 years. The highest figures were observed among US high school students [25] and in Scandinavia among girls and teenagers 15–18 years old (in 2008, 43% of them had used a sunbed in the previous 12 months [35]. However, in Denmark, annual surveys conducted by the Danish Cancer Society to evaluate campaign initiatives in the Danish population has shown that the proportion of sunbed users in the age group 15–19 years who first used a sunbed before the age of 13 fell from 13% to 8%, and first use at the age of 13–15 years decreased from 75% to 65% between 2007 and 2009. During the same period, the proportion of sunbed users in the age group 15–19 years having used a sunbed in the previous 12 months decreased from 50% to 33% [34]. A more recent Danish Cancer Society survey confirmed that the prevalence of sunbed use in Denmark in the age group 15–19 years has declined substantially [8].

A recent review and meta-analysis of 76 records published between 1966 and 2013, reporting a prevalence of indoor tanning in selected (and frequently of high risk) populations of 16 Western countries and including 406,696 total participants confirms the above mentioned figures, and more especially as far as adolescents and US university students are concerned [47] (Table 28.2).

Surveys among adults consistently show that prevalence of sunbed use is highest in white-skinned populations from Northern Europe, and in young or middle-aged women, and allow drawing a robot portrait of sunbed users. Typically, the sunbed user is is female, between 17 and 30 years old, and tends to smoke cigarettes and drink alcohol more frequently and eat less healthy food than non-users [42]. Sunbed users lack knowledge about health risks of sun and UV radiation exposure. This lack of information about safety of solariums is especially true for young people who believe that, as repeatedly put forward by the tanning industry; sunbeds are not as harmful as sun exposure. In this respect, a recent Italian survey noted that 83% of 191 students fully understood the risk of developing cancer through sun exposure, but only 65% of students believed that sunbeds could be dangerous [21].

Motivation for indoor tanning is mainly the desire to be more attractive. The perceived cosmetic attractiveness of a tanned skin has been reported by sunbed users [13]. Additionally, sunbed users may be prompted by the use of sunbeds by friends or family members or by the experience of positive emotions and relaxation by indoor tanning [22]. There is some recent evidence that, in a small proportion of sunbed users, frequent/excessive tanning could be considered as a dependence/addictive behaviour. However, further studies are required to determine the validity of an addiction diagnosis and to improve our understanding of tanning dependence (SCHEER [41]).

28.4 Health Effects of Exposure to UV Radiation Emitted by Sunbeds

UV radiations exert numerous health effects both beneficial, such as adaptation of skin to protect from the damaging effects of UV exposure and vitamin D synthesis, and adverse, such as immunosuppression, skin ageing, mutagenicity and carcinogenicity. Although different in their spectral composition and intensity, UV radiations emitted by sunbeds are not different in nature from natural solar UV radiations. Advocates of "responsible", moderate, sunbed exposure insist on the many

Table 28.2 International prevalence of indoor tanning [47]

Overall			Female participants		Male participants	
Exposure by group	Summary prevalence (95% CI)	No. of records	Summary prevalence (95% CI)	No. of records	Summary prevalence (95% CI)	No. of records
Adults						
Ever exposure	35.7 (27.5–44.0)	22	39.8 (30.0–49.7)	9	20.4 (12.4–28.3)	7
Past-year exposure	14.0 (11.5–16.5)	21	19.0 (14.7–23.4)	15	9.0 (6.6–11.5)	13
US University students						
Ever exposure	55.0 (33.0–77.1)	11	69.3 (45.4–93.2)	5	40.0 (14.1–66.0)	3
Past-year exposure	43.1 (21.7–64.5)	7	64.9 (41.2–88.5)	4	26.8 (15.6–37.9)	4
Adolescents						
Ever exposure	19.3 (14.7–24.0)	23	31.5 (22.3–40.8)	16	14.1 (10.5–17.7)	17
Past-year exposure	18.3 (12.6–24.0)	23	21.3 (8.5–34.1)	14	7.5 (4.1–11.0)	14

Source: Ref. [41]

benefits of controlled UV exposure, and more especially on "optimal" vitamin D synthesis.

28.4.1 Vitamin D and Human Health

Vitamin D (actually a steroid hormone) plays an important role in human health. Its main function is the regulation of the phosphocalcic metabolism and the control of bone growth and mineralization. But, it has also many more functions: it plays a role in cell growth, and a great number of genes are regulated by vitamin D (or its metabolites), and many cells have vitamin D receptors. The association of vitamin D status and diseases including cancer is still a matter of debate. Recent reviews confirm the association of low vitamin D status and colon cancer but not with other cancers, and suggest that low vitamin D status could rather be a consequence than a cause of poor health [4].

Several studies have shown serum levels of 25-OH vitamin D are raised in sunbed users, following exposure to the small fraction of UVB in the radiation emitted by sunbeds [18]. However, there is a general consensus among health authorities that sunbeds should not be used to raise or maintain vitamin D status: a few minutes of exposure of face and hands to midday sun in summer is sufficient, and dietary sources or supplements should be considered.

A number of studies (mainly ecological) have claimed that sun exposure is associated with a decrease in cancer incidence or mortality, a decrease in various diseases and in all-cause mortality, and that sun avoidance may be a risk factor for major causes of death. An effect generally attributed to vitamin D [7, 38]. However, in a large cohort study, these benefits of sun exposure have not been confirmed for sunbed exposure [50]. The risk of death from all causes and from cardiovascular diseases was reduced in women who took sunbathing vacations more than once a year over three decades of life, but the risk of death was not reduced for women using sunbeds. Actually, solarium use (during two or three decades between 10 and 39 years of age) was associated with an almost doubled all-cause mortality compared to women with no solarium use (HR = 1.9, 95% CI 1.3–2.7).

28.4.2 Sunbed Exposure Is a Risk Factor for Melanoma and Non-melanoma Skin Cancer

Undoubtedly, sunbed exposure is associated with increased risk of cutaneous melanoma and, to a lesser extent, of squamous and basal cell carcinomas of the skin. In 2009, a working group of the International Agency for Research on Cancer (IARC, an agency of the World Health Organization) classified the whole spectrum of UV including UV radiations emitted by sunbeds as a Group I carcinogen: carcinogenic to humans (IARC [30]). This classification was based upon a meta-analysis of 19 informative studies which concluded that "there is convincing evidence to support a causal relationship, particularly with exposure before the age of 35 years" (IARC [31]), and upon a comprehensive review of biological effects of UV radiation, showing the plausibility of a biological mechanisms, i.e. the mutagenicity of UVA that was shown to induce C-T transition DNA mutations formerly considered as UVB signature mutations (IARC [30]).

Since IARC's evaluation, several major epidemiological studies have confirmed the association of sunbed use with increased risk of melanoma and other skin cancers, especially when first exposure took place in younger age. Among these, a large population-based case-control study in the USA [36], a population-based case-control study including patients younger than 40 years old in Australia [17], and two cohort studies, the prospective US Nurse's Health Study [26] and the Norwegian–Swedish cohort study [46], all pertaining to the risk of melanoma, were previously reviewed [19]. Other recent epidemiological studies pertaining to the risk of non-melanoma skin cancers have been recently reviewed [41].

All these new epidemiological data have enabled the conduct of four new meta-analyses, including up to 27 studies (Table 28.3). These recent meta-analyses confirmed the association of sunbed exposure with risk of cutaneous melanoma, especially when first exposure takes place at an early age, and strengthened the previously observed association of sunbed exposure with risk of basal cell carcinoma, again more especially when first exposure takes place at a younger age.

Interestingly, as pointed out by Alberg [2], the value of the analysis of a melanoma epidemic in Iceland [27] is beyond that of a simple ecologic study. By showing a sharp and rapid increase in incidence of melanoma of the trunk in women younger than 50 years of age, closely following the development of tanning salons in Iceland, it suggests a promoter effect of UV exposure. UV exposure in adulthood is likely to enhance melanoma development in the following months or years and may thus be associated with seasonal variation in melanoma incidence [11] or with increased melanoma risk, more especially among young sunbed users [17, 27].

In this respect the recent Ghiasvand et al. study [23] not only confirms previous evidence of association of sunbed exposure and melanoma risk, further supporting a dose-response association and a more pronounced effect when initiation of sunbed exposure took place at an earlier age, but also contradicts the view put forward by the tanning industry that new UVA-emitting sunbeds are safer than the older ones, and add a clear demonstration of an earlier onset of melanoma (albeit likely to be underestimated) among sunbed users. In a cohort study of 141,045 Norwegian women followed for a mean of 13.7 years and subsequent diagnosis of 861 melanoma (and the young age at diagnosis), Ghiasvand et al. present compelling evidence that early exposure to tanning beds brings forward the date of diagnosis of melanoma by at least 2 years: those who started tanning under the age of 30 were on average 2.2 years younger at diagnosis than those who had never tanned. Although this latter finding is not statistically significant, it is on line with recent findings of Lazovich et al. [37] who found that women younger than 30 years were 6 times more likely to be in the case than the control group if they tanned indoors (crude OR, 6.0; 95% CI, 1.3–28.5), and of Cust et al. [17] showing that sunbed exposure is linked to earlier onset of melanoma.

Table 28.3 Recent meta-analyses of sunbed exposure and risk of skin cancer

	Summary relative risk (95% confidence interval)	No. studies	No. cases	No. controls	References
Melanoma					
Ever exposure *vs.* never	1.22 (1.07–1.39)	21	7885	24,209	[28]
Ever exposure *vs.* never	1.20 (1.08–1.34)	27	11,428	222,053	[12]
Age at first exposure <35 years old	1.59 (1.36–1.85)	13	–	–	
Ever exposure *vs.* never	1.16 (1.05–1.28)	31	14,956	233,106	[15]
Age at first exposure <25 years old	1.35 (0.99–184)	6	–	–	
Non-melanoma skin cancer					
Ever exposure *vs.* never	1.34 (1.05–170)	6	1812	2493	[28]
Ever exposure *vs.* never (SCC)	2.23 (1.39–3.57)	5	1242	75,415	[12]
Ever exposure *vs.* never (BCC)	1.09 (1.01–1.18)	6	6995	75,810	
Ever exposure *vs.* never (SCC)	1.67 (1.29–2.17)	7	1683	75,972	[48]
Ever exposure *vs.* never (BCC)	1.29 (1.08–1.53)	8	7407	76,211	
Age at first exposure <25 years old (SCC)	2.02 (0.70–5.86)	3	–	–	
Age at first exposure <25 years old (BCC)	1.40 (1.29–1.52)	3	–	–	

Finally, it should be noted that, ruining an argument frequently put forward an according to which "responsible" sunbed exposure prevents sunburns (which, everyone knows, are the main risk factor for melanoma!), a new analysis of the Lazovich et al. [36] data set restricted to non-burning sunbed exposure (i.e excluding those who had reported burns from indoor tanning use), showed significantly increased melanoma risks across all sunburn (from outdoor sun exposure) categories among participants who had tanned indoors compared with those who never tanned indoors, the highest risk being for those who reported zero lifetime sunburns (OR = 3.87; 95% CI 1.68, 8.91) (Table 28.4).

28.5 Conclusion

From the long list of studies that have addressed the risks of sunbed exposure, it is now clear that exposure to sunbeds is carcinogenic. There may be beneficial health effects of sunbed use, e.g. adaptation of the skin and protection from subsequent UV exposure, contribution to vitamin D status, but these are out weighted by risks. UV

Table 28.4 Risk of melanoma by ever use of indoor tanning among individuals who tanned indoors without burning and never users stratified by lifetime burns from sun as estimated using logistic regression (n = 1852)

No sunburns	Cases (n = 906) n (%)	Controls (n = 946) n (%)	OR crude (CI 95%)	OR ajust (CI 95%)
0	32 (78.1)	67 (40.3)	5.29 (2.01–13.96)	3.87 (1.68–8.91)
1–2	142 (56.3)	199 (45.2)	1.56 (1.01–2.41)	1.78 (1.28–2.46)
3–5	172 (54.7)	188 (47.9)	1.31 (0.87–1.99)	1.49 (1.10–2.01)
>5	560 (48.9)	492 (41.1)	1.38 (1.08–1.76)	1.42 (1.19–1.69)

From Ref. [33]

radiation is a complete carcinogen, both an initiator and a promoter. UV exposure early in childhood positively influences (initiates) melanoma risk in adulthood [5, 16], while UV exposure in adulthood is likely to enhance (promote) melanoma development, as shown by the association between sunbed exposure and early onset melanoma, and by the quasi experiment represented by the rapid increase in incidence of melanoma on the trunk in young women following massive development of artificial tanning facilities in Iceland [27].

Although the increase in melanoma risk may appear modest in the general population (+16 to 20%, according to the more recent meta-analyses), this risk concentrates among younger women: +59% for first exposure before the age of 30 years [12], and even +200% for frequent use in the 10–39 years period [46]. The fraction of risk attributable to sunbed use in patients diagnosed with a melanoma before the age of 30 may be very high: 76% in Australia among those who had ever used a sunbed and were diagnosed between 18 and 29 years of age, [17]. It has been estimated that in 2008, in Europe (15 countries of the European Community and 3 countries from the European Free Trade Association), of 63,942 new cases of melanoma diagnosed each year, an estimated 3438 (5.4%) may be related to sunbed use, women representing most of this burden with 2341 cases (6.9% of all melanomas in women). And, based on incidence/mortality ratio, about 498 women and 296 men may die each year from a melanoma as a result of being exposed to indoor tanning [12].

Even if there are currently discrete signs of a decrease in prevalence of sunbed use in some countries and specific populations, and even if some data tend to show that UV exposure may be associated with a better prognosis of melanoma [10, 11], there is no safety limit in irradiance or dose to ensure protection of the sunbeds users [41]. And the impossibility to define a safety limit was the base of the total ban on sunbeds in Brazil (ANVS [1]). Melanoma remains a potentially lethal tumor. The 2014 European Code against Cancer simply says "do not use sunbeds" [43], and World Health Organization [49] strongly advise against use of sunbeds. Public Health authorities should strongly discourage use of sunbeds, particularly by young women and, since unlike tobacco, tanning is not an addiction but a cosmetic fashion for which alternative solutions are available, even consider a total ban as enforced in Brazil in 2009 and in all Australian states in 2015.

References

1. Agência Nacional de Vigilância Sanitária (ANVS) (2009) Resolução n°59 de 9 de novembro 2009. Proibe em todo território nacional o uso dos equipamentos para bronzeamento artificial, com finalidade estética, baseada na emissão da radiação ultraviolet (UV). Diário Oficial da União – Seção 1, no. 215, quarta-feira, 11 de novembro 2009
2. Alberg AJ (2011) Re: A melanoma epidemic in Iceland: possible influence of sunbed use. Am J Epidemiol 173(7):845
3. American Suntanning Association (2015) http://americansuntanning.org. Accessed Nov 2016
4. Autier BM, Pizot C, Mullie P (2014) Vitamin D status and ill health: a systematic review. Lancet Diabetes Endocrinol 2(1):76–89
5. Autier P, Doré JF (1998) Influence of sun exposures during childhood and during adulthood on melanoma risk. EPIMEL and EORTC Melanoma Cooperative Group. Int J Cancer 77(4):533–537

6. Autier P, Doré JF, Breitbart E, Greinert R, Pasterk M, Boniol M (2011) The indoor tanning industry's double game. Lancet 377(9774):1299–1301
7. Baggerly CA, Cuomo RE, French CB, Garland CF, Gorham ED, Grant WB, Heaney RP, Holick MF, Hollis BW, McDonnell SL, Pittaway M, Seaton P, Wagner CL, Wunsch A (2015) Sunlight and vitamin D: necessary for public health. J Am Coll Nutr 34(4):359–365
8. Behrens CL, Schiøth C, Christensen AS (2016) Danskernes solarievaner 2015 – en kortlægning Kræftens Bekæmpelse og TrygFonden smba (TryghedsGruppen smba) 2016 (in Danish). Danish Cancer Society
9. Benmarhnia T, Léon C, Beck F (2013) Exposure to indoor tanning in France: a population based study. BMC Dermatol 13:6
10. Berwick M, Armstrong BK, Ben-Porath L, Fine J, Kricker A, Eberle C, Barnhill R (2005) Sun exposure and mortality from melanoma. J Natl Cancer Inst 97:195–199
11. Boniol M, Armstrong BK, Doré JF (2006) Variation in incidence and fatality of melanoma by season of diagnosis in New South Wales, Australia. Cancer Epidemiol Biomark Prev 15(3):524–526
12. Boniol M, Autier P, Boyle P, Gandini S (2012) Cutaneous melanoma attributable to sunbed use: systematic review and meta-analysis. BMJ 345:e4757. PMID: 22833605. (corrections in BMJ 2012;345: http://www.bmj.com/content/345/bmj.e8503)
13. Brandberg Y, Ullén H, Sjöberg L, Holm LE (1998) Sunbathing and sunbed use related to self-image in a randomized sample of Swedish adolescents. Eur J Cancer Prev 7:321–329
14. CDC (Centers for Disease Control and Prevention) (2012) Use of indoor tanning devices by adults – United States, 2010. Morb Mortal Wkly Rep 61(18):323–325
15. Colantonio S, Bracken MB, Beecker J (2014) The association of indoor tanning and melanoma in adults: systematic review and meta-analysis. J Am Acad Dermatol 70(5):847–857
16. Crump C, Sundquist K, Sieh W, Winkleby MA, Sundquist J (2014) Season of birth and other perinatal risk factors for melanoma. Int J Epidemiol 43(3):793–801
17. Cust AE, Armstrong BK, Goumas C, Jenkins MA, Schmid H, Hopper JL, Kefford RF, Giles GG, Aitken JF, Mann GJ (2011) Sunbed use during adolescence and early adulthood is associated with increased risk of early-onset melanoma. Int J Cancer 128(10):2425–2435
18. de Gruijl FR, Pavel S (2012) The effects of a midwinter 8-week course of sub-sunburn sunbed exposures on tanning, vitamin D status and colds. Photochem Photobiol Sci 11(12):1848–1854
19. Doré JF, Chignol MC (2012) Tanning salons and skin cancer. Photochem Photobiol Sci 11:30–37
20. European Sunlight Association (2016) http://europeansunlight.eu/. Accessed Nov 2016
21. Fabbrocini G, Mazzella C, Marasca C, De Vita V, Savastano R, Monfrecola G (2012) Sunbathing and sunlamp exposure: awareness and risk among Italian teenagers. Photodermatol Photoimmunol Photomed 28:224–225
22. Feldman SR, Liguori A, Kucenic M, Rapp SR, Fleischer AB Jr, Lang W, Kaur M (2004) Ultraviolet exposure is a reinforcing stimulus in frequent indoor tanners. J Am Acad Dermatol 51:45–51
23. Ghiasvand R, Rueegg CS, Weiderpass E, Green AC, Lund E, Veierød MB (2017) Indoor tanning and melanoma risk: long-term evidence from a prospective population based cohort study. Am J Epidemiol 185:147–156
24. Grange F, Mortier L, Crine A, Robert C, Sassolas B, Lebbe C, Lhomel C, Saiag P (2015) Prevalence of sunbed use, and characteristics and knowledge of sunbed users: results from the French population-based Edifice Melanoma survey. J Eur Acad Dermatol Venereol 29(Suppl. 2):23–30
25. Guy GP Jr, Berkowitz Z, Watson M, Holman DM, Richardson LC (2013) Indoor tanning among young non-Hispanic white females. JAMA Intern Med 173:1920–1922
26. Han J, Colditz GA, Hunter DJ (2006) Risk factors for skin cancers: a nested case-control study within the Nurses' Health Study. Int J Epidemiol 35(6):1514–1521
27. Héry C, Tryggvadóttir L, Sigurdsson T, Olafsdóttir E, Sigurgeirsson B, Jonasson JG, Olafsson JH, Boniol M, Byrnes GB, Doré JF, Autier P (2010) A Melanoma epidemic in Iceland: possible influence of sunbed use. Am J Epidemiol 172:762–767
28. Hirst N, Gordon L, Gies P, Green AC (2009) Estimation of avoidable skin cancers and cost-savings to government associated with regulation of the solarium industry in Australia. Health Policy 89:303–311
29. Hoerster KD, Garrow RL, Mayer JA, Clapp EJ, Weeks JR, Woodruff SI, Sallis JF, Slymen DJ, Patel MR, Sybert SA (2009) Density of indoor tanning facilities in 116 large U.S. cities. Am J Prev Med 36(3):243–246
30. IARC Working Group on the Evaluation of Carcinogenic Risks to Humans Radiation (2012) IARC Monogr Eval Carcinog Risks Hum 100(Pt D):7–303
31. International Agency for Research on Cancer Working Group on artificial ultraviolet (UV) light and skin cancer (2007) Int. J. Cancer 120:1116–1122
32. Indoor Tanning Association (2016) www.theita.com. Accessed Nov 2016
33. Isaksson Vogel R, Ahmed RL, Nelson HH, Berwick M, Weinstock MA, Lazovich D (2014) Exposure to indoor tanning without burning and melanoma risk by sunburn history. J Natl Cancer Inst 106(7). pii: dju219. doi:10.1093/jnci/dju219.Print 2014 July
34. Køster B, Thorgaard C, Philip A, Clemmensen H (2011) Sunbed use and campaign initiatives in the Danish population, 2007–2009: a cross-sectional study. J Eur Acad Dermatol Venereol 25:1351–1355

35. Krarup AF, Koster B, Thorgaard C, Philip A, Clemmensen IH (2011) Sunbed use by children aged 8–18 years in Denmark in 2008: a cross-sectional study. Br J Dermatol 165:214–216
36. Lazovich D, Isaksson Vogel R, Berwick M, Weinstock MA, Anderson KE, Warshaw EM (2010) Indoor tanning and risk of melanoma: a case-control study in a highly exposed population. Cancer Epidemiol Biomark Prev 19(6):1557–1568
37. Lazovich D, Isaksson Vogel R, Weinstock MA, Nelson HH, Ahmed RL, Berwick M (2016) Association between Indoor tanning and melanoma in younger men and women. JAMA Dermatol 152(3):268–275
38. Lindqvist PG, Epstein E, Nielsen K, Landin-Olsson M, Ingvar C, Olsson H (2016) Avoidance of sun exposure as a risk factor for major causes of death: a competing risk analysis of the Melanoma in Southern Sweden cohort. J Intern Med 280:375–387
39. Nilsen LT, Aalerud TN, Hannevik M, Veierød MB (2011) UVB and UVA irradiances from indoor tanning devices. Photochem Photobiol Sci 10:1129–1136
40. Petri A, Karabetsos E (2015) Effective ultraviolet irradiance measurements from artificial tanning devices in Greece. Radiat Prot Dosim 167(4):490–501
41. SCHEER (Scientific Committee on Health, Environment and Emerging Risks) (2016) Opinion on biological effects of ultraviolet radiation relevant to health with particular reference to sunbeds for cosmetic purposes. http://ec.europa.eu/health/scientific_committees/scheer/docs/scheer_o_03.pdf. Accessed Aug 2017
42. Schneider S, Krämer H (2010) Who uses sunbeds? A systematic literature review of risk groups in developed countries. J Eur Acad Dermatol Venereol 24:639–648
43. Schüz J, Espina C, Villain P, Herrero R, Leon ME, Minozzi S, Romieu I, Segnan N, Wardle J, Wiseman M, Belardelli F, Bettcher D, Cavalli F, Galea G, Lenoir G, Martin-Moreno JM, Nicula FA, Olsen JH, Patnick J, Primic-Zakelj M, Puska P, van Leeuwen FE, Wiestler O, Zatonski W, Working Groups of Scientific Experts (2015) European code against cancer 4th Edition: 12 ways to reduce your cancer risk. Cancer Epidemiol 39(Suppl 1):S1–10. doi:10.1016/j.canep.2015.05.009
44. Stanganelli I, Gandini S, Magi S, Mazzoni L, Medri M, Agnoletti V, Lombi L, Falcini F (2013) Sunbed use among subjects at high risk of melanoma: an Italian survey after the ban. Br J Dermatol 169:351–357
45. Tierney P, Ferguson J, Ibbotson S, Dawe R, Eadie E, Moseley H (2013) Nine out of 10 sunbeds in England emit ultraviolet radiation levels that exceed current safety limits. Br J Dermatol 168:602–608
46. Veierod MB, Adami HO, Lund E, Armstrong BK, Weiderpass E (2010) Sun and solarium exposure and melanoma risk: effects of age, pigmentary characteristics, and nevi. Cancer Epidemiol Biomark Prev 19:111–120
47. Wehner MR, Chren M-M, Nameth D, Choudhry A, Gaskins M, Nead KT, Boscardin WJ, Linos E (2014) International prevalence of indoor tanning. A systematic review and meta-analysis. JAMA Dermatol 150:390–400
48. Wehner MR, Shive ML, Chren MM, Han J, Qureshi AA, Linos E (2012) Indoor tanning and non-melanoma skin cancer: systematic review and meta-analysis. Br Med J 345:e5909
49. World Health Organization (2017) Artificial tanning devices. Public Health interventions to manage sunbeds. World Health Organization, Geneva
50. Yang L, Lof M, Veierød MB, Sandin S, Adami HO, Weiderpass E (2011) Ultraviolet exposure and mortality among women in Sweden. Cancer Epidemiol Biomark Prev 20(4):683–690

Part VIII

UV Dosimeters

Dose Quantification in UV Phototherapy

29

David Robert Grimes

Abstract

Ultraviolet light has long been used to alleviate a number of skin conditions, and its efficacy is well known. However, over-exposure to ultraviolet radiation has a number of detrimental effects and thus it is vital to maintain a dose to skin within the therapeutic window. To maximise treatment gain whilst circumventing potential side-effects of over-exposure requires accurate determination of irradiance and skin-dose. This is complicated by the fact that ultraviolet radiation is essentially absorbed at the skin surface, which means that changing orientation of the patient and source can modulate dose received. In addition, irregular patient shapes mean dose must be carefully calibrated. This chapter focuses on methods of determination of dose, clinical protocols for quantifying radiation dose received and mathematical models for estimating these quantities.

Keywords

Dose models • Ultraviolet radiation • Quantification • Irradiance measurement

29.1 Introduction

Ultraviolet radiation (UVR) therapy has long been known to have a measurable effect on numerous skin conditions. Informally at least the benefits of sunlight for human skin have been recognised since at least Greek antiquity, and indeed this knowledge was widely acknowledged throughout many ancient cultures [1–3]. More formal scientific study began in the early nineteenth century, with the discovery of the ultraviolet portion of the solar spectrum by Johann Ritter

D.R. Grimes (✉)
Cancer Research UK/MRC Oxford Institute for Radiation Oncology, Gray Laboratory, University of Oxford, Old Road Campus Research Building, Off Roosevelt Drive, Oxford OX3 7DQ, UK

Centre for Advanced and Interdisciplinary Radiation Research (CAIRR), School of Mathematics and Physics, Queen's University Belfast, Belfast, UK
e-mail: davidrobert.grimes@oncology.ox.ac.uk; d.r.grimes@qub.ac.uk

S. Ahmad (ed.), *Ultraviolet Light in Human Health, Diseases and Environment*, Advances in Experimental Medicine and Biology 996, https://doi.org/10.1007/978-3-319-56017-5_29

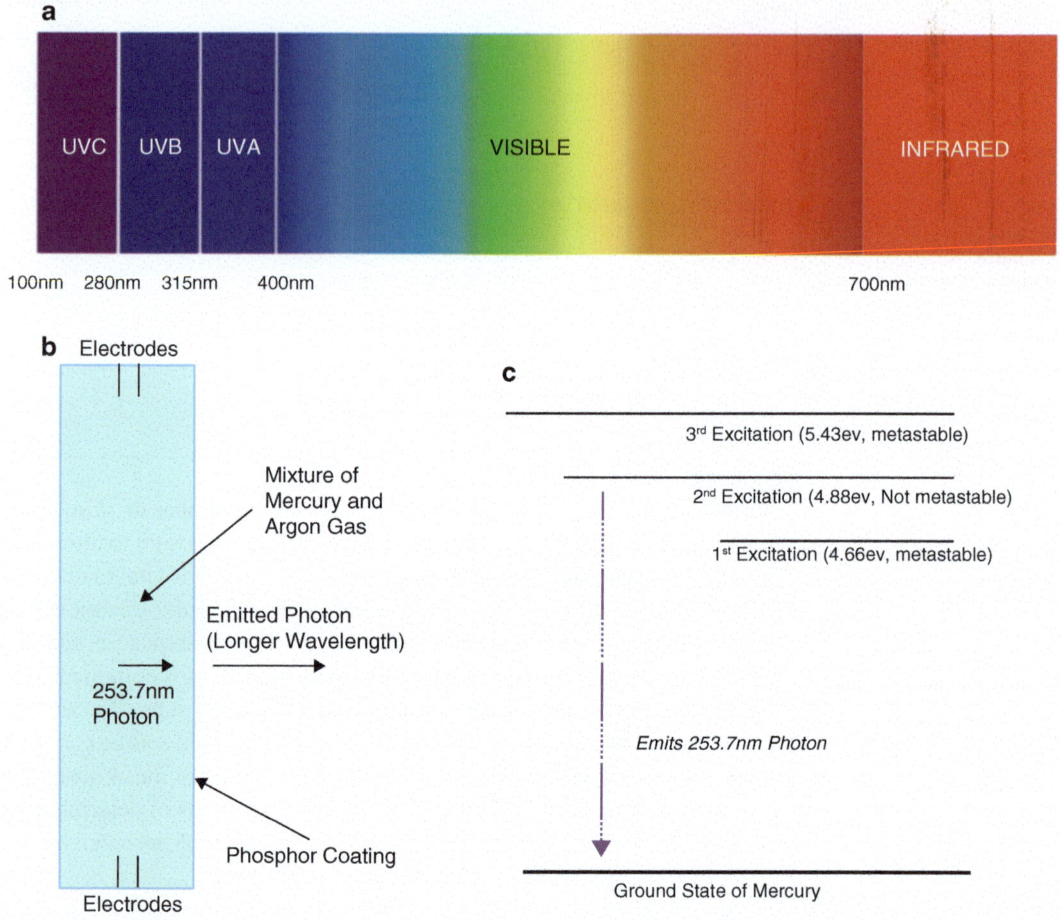

Fig. 29.1 (**a**) UVR portion of the Electromagnetic spectrum (**b**) UVR tube (**c**) Energy states of mercury gas (Figure taken from Grimes 2015 [1])

in 1801 [3]. Just over a hundred years later, Niels Finsen was able to experimentally prove that sunburn (erythema) was caused by ultraviolet radiation (UVR) and not radiant heat as such a name might suggest. His efforts remain a cornerstone of modern ultraviolet phototherapy, and for his dedicated work into the application of UVR to medicine and biological processes, he was awarded the 1903 Nobel prize in Medicine and Physiology for his research in 'the treatment of diseases, especially lupus vulgaris, with concentrated light radiation, whereby he has opened a new avenue for medical science.'

Since Finsen's initial discoveries at the beginning of the twentieth century, UVR phototherapy has become a vital treatment for a great many skin diseases. It's worth initially defining what we mean by UVR. This refers to the portion of the electromagnetic spectrum of wavelength 100 nm–400 nm, lying between the visible and X-ray part bands as illustrated in Fig. 29.1a. The UV band is usually divided into three further subdivision; UVA, UVB and UVC based on their respective biological effects. The most commonly encountered classification is that defined by the International Commission on Radiation (CIE) [4] given in Table 29.1. Other classifications exist

Table 29.1 CIE ultraviolet classification

Classification	Wavelength band
Ultraviolet A (UVA)	400 nm–315 nm
Ultraviolet B (UVB)	315 nm–280 nm
Ultraviolet C (UVC)	280 nm–100 nm

also; – some authors take 320 nm as the boundary between UVA and UVB, and define 290 nm as the boundary between UVB and UVC [5].

On Earth, the sun is the primary source of UVR incident upon the planet surface. Earth's atmosphere is remarkably efficient at attenuating the more biologically harmful bands of UVR by absorption, with wavelengths less than 290 nm effectively removed by the ozone layer in the atmosphere. Of the sun's radiation that reaches Earth, only 5% is in the UVR range. Rayleigh scattering by particles of oxygen and nitrogen has a significant effect on reducing UVR with wavelengths longer than 310 nm [6]. Of the UVR that reaches the Earth's surface, 96.65% is UVA and 3.35% is UVB [7].

In ultraviolet phototherapy, UVR is produced most commonly with fluorescent tubes, which operate on the same principle as a gas discharge lamp. The tube consists of a low pressure gas or gas mixture which is ionised by running a current through it {(Fig. 29.1b}. Excited atoms fall back to their ground state, emitting a photon whose wavelength depends on the gas mixture used. The inner walls of the tube are coated with a phosphor, so when emitted photons are incident upon the tube walls they stimulate the emission of a photon of a different wavelength through the mechanism of fluorescence. UVR lamps use a mixture of mercury vapour and inert argon gas. The energy levels for mercury are shown in Fig. 29.1c. Excitation at 4.88 eV is not metastable, and excited atoms revert rapidly to ground state. As a result they radiate a UVC photon of wavelength 253.7 nm [8]. These photons then impinge on the phosphor coat of the tube and fluoresce, emitting a photon with a wavelength dependent on the phosphor used.

29.1.1 Ultraviolet Radiation and the Skin

There is a wide-range of skin disorders for which ultraviolet phototherapy is the primary means of treatment. Chief amongst these ailments is Psoriasis, a common chronic non-infectious disease of the skin which presents as raised patches. This condition afflicts millions world-wide, and UVR phototherapy is an exceptionally effective way of alleviating the condition. Other conditions treated with UVR include chronic and stubborn eczema [9, 10], Vitiligo, [11], Polymorphic light eruption (PMLE) [12], acquired perforating dermatosis (APD) [13], Lichen Planus [14, 15], and Mycosis fungoides [16].

Ultraviolet radiation is non-ionising, but despite this it is nevertheless damaging to the molecular integrity of DNA through both direct and indirect interactions [17, 18]. Humans have adapted the defence in the form of melanin pigmentation [19] or tanning to counteract the negative repercussions of ultraviolet exposure. Negative impacts of UVR over-exposure range from the trivial to severe. Perhaps the most well-known detrimental effect is erythema, more commonly known as sunburn. This causes painful blistering effects on skin and deep reddening, ubiquitous throughout the human species as a consequence of sun overexposure. The required minimum dose to induce erythema (MED) depends on skin-type, and is given in Table 29.2. Acute ocular UVR exposure can induce photokeratitis (snow blindness) and chronic overexposure results in increased incidence of cataracts [20]. UVR can also lead to damaged collagen, with subsequent decrease in skin elasticity, promoting advanced aging and wrinkling [3, 21]. These effects of UVR exposure are considerably unpleasant, yet the primary concern with this spectrum of radiation is the potential for carcinogenesis. Exposure to high amounts of ultraviolet radiation has long been known as a major risk factor in developing skin cancers [22]. In particular, basal cell carcinoma, squamous cell carcinoma, and malignant melanoma [5] are associated with over-exposure to UVR.

Crucially, acquiring the correct dose in UVR phototherapy is vital. Dose should be sufficient to achieve maximum treatment efficacy whilst not excessive enough to induce the detrimental effects of over-exposure. Conditions treated with UVR tend to be chronic and require several exposures over a life time, exacerbating the issue further. To achieve maximum therapy effect whilst avoiding damaging side-effects, quantification of dose is of vital importance in UVR treatment.

Table 29.2 Fitzpatrick Phototype scale

Type	UVR response	Skin colour	UVA MED (mJ/cm^2)	UVB MED (mJ/cm^2)
I	Burns easily/Never Tans	Ivory white	20–35	15–30
II	Burns easily/Tans little	White	30–45	25–40
III	Burns moderately/Often Tans	White	40–55	30–50
IV	Burns minimally/Tans easily	Olive	50–80	40–60
V	Burns rarely/Tans profusely	Brown	70–100	60–90
VI	Never burns/Tans profusely	Black	100	90–150

Adapted from Fitzpatrick [19, 23]

29.1.2 Radiometric Units

There is considerable ambiguity in radiometric and photometric terms between different fields. Unlike ionizing radiation, UVR is absorbed extremely superficially, typically only a few cells deep at skin depth [20] with penetration depths of <1 mm. For this reason, surface dose is a more appropriate concept than the volume doses typically calculated in radiotherapy. At this juncture, it's worth specifically defining the quantities and units we are most interested in from a phototherapy perspective to avoid any confusion.

- **Radiant Energy** (***Qe***) – Energy of the electromagnetic radiation, measured in Joules.
- **Radiant Flux** (***φe***) – Radiant Energy received or transmitted per unit time. Measured in Watts or Joules per second.
- **Irradiance** (***E***) – Radiant flux received by a surface per unit area. Measured in Watts per metre squared or mW per cm^2.
- **Radiant Exposure** (***He***) – The radiant energy received by a surface per unit area. Measured in Joules per square metre.

Of these quantities, Irradiance is the most widely measured and typically the value that radiometers record [24]. Radiant exposure can be readily calculated from this by integrating over the time exposed, so that

$$H_e = \int_0^t E dt = Et \quad (29.1)$$

For most applications, radiant exposure is equivalent to dose received. In practice, this means any desired dose can be obtained by exposing the patient to a known irradiance for the requisite time as per Eq. 29.1. It is also important to note that irradiance and surface dose depend upon angle the surface normal makes with the incoming photons, so that surface angle can substantially modulate total dose received. This aspect will be considered in the dose model section of this chapter.

29.2 Clinical Dose Estimation Technique

For UVB treatments, a clinician typically ascertains the minimum erythemal dose (MED) for a patient. For treatments such as PUVA which involve a phototoxic agent (8-methoxypsoralen), the minimum phototoxic dose (MPD) must be found. These values are usually found by using a thin plastic template with eight small windows. This template is positioned over an area of skin relatively unaccustomed to UV light, such as buttocks or back. The slits are exposed to ultraviolet light at the treatment wavelength, with each successive slit being $\sqrt{2}$ above the dose at the previous exposure [25].

In UVB treatments, erythema peaks between 8 and 24 h after initial exposure, and the template sites are then employed to find the lowest exposure dose that yields an erythemal effect. UVA effects by contrast peak between 48 and 72 h after exposure, and MPD can be determined by similar visual inspection. Treatment is then begun at a fraction of the MED/MPD dose, typically 50–70%. This is incremented along the course of treatment to account for skin photoadaptation until a marked improvement of the condition is observed by the physician or clinician

[25]. Related methods for ascertaining the starting dose include photo-testing templates with foil apertures of differing sizes, which attenuate the incident UVR by varying amounts leading to differing irradiance at different hole sites. From this, MED/MPD is then inspected visually again. Crucially however, these starting dose methods rely on visual inspection and can be somewhat subjective. Once starting dose is estimated, cabin exposure time can be calculated. While clinically useful, skin testing methods only give information about the skins response to that particular test source and so the problem remains of comparing two sources with an objective method.

29.2.1 Radiometers

UVR radiometers usually consist of a photodiode mounted behind a filter which limits incoming radiation to the wavelengths the radiometer is designed to measure; for example UVB radiometers for narrowband therapy will have input optics designed to filter the UVB portion of the spectrum. The head photodiode is placed behind filters and often embedded deep in the probe head, and consequently diffusers are a vital part of the radiometer set-up [26]. UVR meters can have varying responses to UV light at different incident angles. A perfect detector would have a cosine response [27] but in practice when photodiodes are nested inside the receiving head then cosine responses may not be obtained. If this is the case, a diffuser head is employed [26, 28]. Pye and Martin [26] examined a number of different detector set-ups and found that detectors with polytetrafluorethylene (PTFE) diffusers perform admirably with less than 5% deviation from perfect co-sine response. For phototherapy, it is recommended that detectors have less than 10% deviation from ideal cosine response [29]. Typically, UVR radiometers measure Irradiance.

29.3 UVR Measurement in Practice

29.3.1 ScUVido Protocol

Maintaining dose homogeneity and ascertaining patient dose is important, and complicated by the wide array of treatment cabin designs available. One approach is common usage is that of the Scottish photochemotherapy audit board, who specified a protocol so that UVR therapy sources could be correctly compared and contrasted over the lifetime of a unit and even between units and phototherapy centres. These guidelines originally laid down improved PUVA treatment doses, and were updated in 2001 [30] to account for NB-UVB sources. The fundamental premise of the Scottish ultraviolet dosimetry (ScUViDo) protocol is to provide a standard for UV irradiance in treatment centres for quality assurance and cross-comparison. Initially, UVR meters used are calibrated against the source which it is designed to work with. The cosine response error of the meter should be low with an f2 error (deviation from perfect cosine response) of less than 10%. The calibration of all meters used must be traceable to the National Physical Laboratory, and the accuracy should be ±10%. Meter calibration should be performed annually and any anomalies corrected.

Cross-comparison in ScUViDo pivots on the concept of designated patient irradiance (DPI). This is the average irradiance on a patient of average height and builds standing in a phototherapy cabin at chest, waist and knee height. To determine this, an investigating physicist in appropriate UVR protection gear stands in the cabin and adopting the position of a patient in treatment and makes a series of measurements at various positions (Fig. 29.2). Tubes in the cabin are warmed up 5 min prior to measurements being made, and a hand held UV meter appropriate for the wavelength band of the lamps is employed for measurements at the twelve body sites. This gives mean DPI at each body site without requiring recourse to a known body correction factor. It is important to ensure that clothing does not obstruct any emitting sources, as this

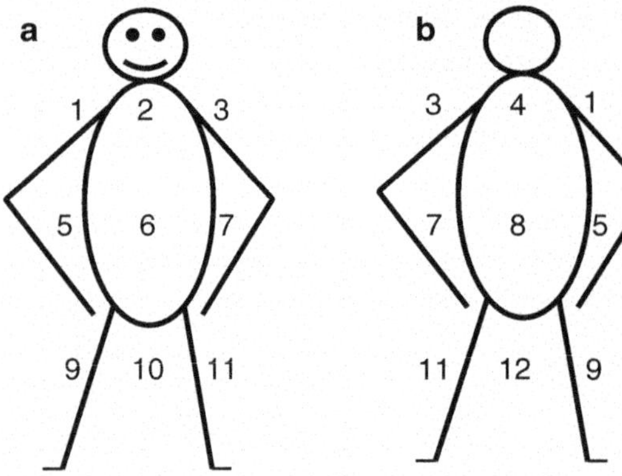

Fig. 29.2 DPI measurement sites (**a**) Anterior (**b**) Posterior (Adapted from Moseley (2001) [30])

can lead to self-shielding problems. The colour of the clinicians clothing can lead to variations of approximately 5% [30]. There is also an indirect method for obtaining the DPI by placing a retort stand with a clamped meter in place and multiplying by a correction factor for the particular cabin in question.

The major advantage of ScUViDo is that is allows comparison of treatments between different centres and units. It also indicates when lamps need to be replaced, typically when DPI has changed by 10%. Despite the practicality of this approach, clinicians and patients can have wildly differing body types and will shield different regions, resulting in large differences. However, the protocol is immensely useful in determining if tubes have failed or reduced in output – something that typically happens beyond a 1000 h of active life [5]. This is exceptionally important, as there is a wide array of functional cabin designs [31, 32].

29.3.2 Automated Detectors

As the ScUVido protocol is useful for providing localized calibration and comparison, and can indicate when irradiance has dropped due to lamp failure, aging or some other degradation. It is inexpensive to implement and as a consequence is used not only in Scotland but across many European phototherapy centres. Another less common method to examine and calibrate UVR cabins involves the use of an automated detection system. This are not widely used, but a well-known example is the system developed by Currie et al. [33], which comprises of two detectors facing opposite directions from each other; one is a wide angle UVR photodiode detector with a raised polytetrafluorethylene (PTFE) diffuser and the other is another UVR sensitive photodiode housed at the end of a 200 mm tube with slots at either end measuring 10 mm by 1 mm to provide collimation. The entire mount rotates on a stepper motor, recording irradiance at 800 points in a full rotation for both the wide-angle and collimated detector. The data is sent to a laptop computer, and the resulting data can be displayed as either a linear or polar plot. There are major advantages to such a system; firstly, it does not require an operator so self-shielding by the investigator is not a problem. Secondly, the collimated detector allows the user to see specifically which individual tubes are failing or have diminished output. Finally, it offers greater repeatability than the ScUViDo method and less uncertainly as readings are automated and human error is a less of a factor. As many hundreds of readings are made in a full rotation, specific dose incident upon the detector at various heights can be ascertained. The downside is that the system is quite costly and so far it has not been widely adopted despite its advantages. It also does not factor self-shielding into the analysis, meaning results would need to be considered and interpreted with this in consideration. The setup is shown in Fig. 29.3.

29.3.3 Cabin Detectors

Several modern phototherapy cabins also include in-built irradiance meters. In general, these are of limited use as self-shielding by the patient or even cabin geometry casts doubt on their readings. Despite this failing, they can indicate if a substantial change in cabin homogeneity has occurred.

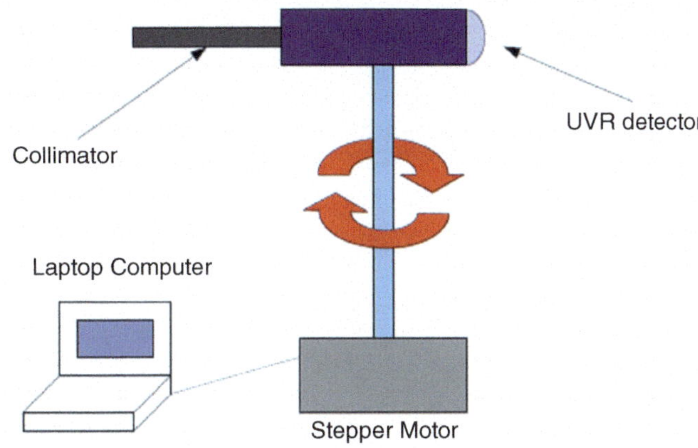

Fig. 29.3 Automated detector system. *Arrows* denote rotation direction (Adapted from Grimes (2015) [1])

29.4 Quantification of Ultraviolet Dose

Given the difficulties of measuring UVR dose in practice, modelling of dose is important in determining dose received. The most basic approximation to any light source is a point source approach, where the intrinsic geometry of the source is disregarded and is treated as a point in space. Under this approximation, the familiar inverse square law is yielded, and irradiance falls with the square of distance. This simple model is exceptionally useful in many branches of physics, but the vital caveat is that one must be sufficiently far away so that source geometry can be disregarded. However, if one is close to an extended source such as the tubes used in UVR therapy, such an approximation breaks down – a general rule of thumb is that for inverse square approximations to hold, one must be at least 5 times further from the source than the magnitude of the greatest source extent. In the case of UVR therapy, source lamps are typically 1.75 m along their vertical axis, suggesting that one would have to be at a distance of greater than 8.75 m before such an approximation would hold. As patients tend to be less than a metre from the source lamps, we require more involved methods to estimate dose from these sources.

Given the distinctive tube geometry associated with UVR sources for phototherapy, it might be tempting to model these sources as cylindrical emitters, modelling emitted light as being perpendicular to the tube surface. This approximation can give accuracy to within 10% if lamp to detector distances are small, but fails at predicting irradiance from small surface elements. There is a further issue which compounds this failure – UVR dose is absorbed primarily at skin depth, and the angle which photons strike the detector or skin surface is important. As a consequence of this, cylindrical approximations are in general not adequate for UVR dosimetry. Accordingly, more involved models have to be considered if ultraviolet dose incident upon a patient or detector is to be fully quantified.

29.4.1 Line Source Models for UVR Phototherapy

UVR lamps can be treated as a line sources to a high degree of accuracy by modelling the tube source as a linear array of point-sources and integrating over the physical extent of the tube. For simplicity, we place the tube at the origin of our co-ordinate system as depicted in Fig. 29.4. A detector, or patient skin-site, is located at the point (d, h, z), as illustrated. If we define the surface normal of the detection site as $\vec{n} = A\vec{x} + B\vec{y} + C\vec{z}$, then for a tube of length L and with a constant SR related to the power per unit length of the source, irradiance is given by

$$E(d,,h,,z) = \frac{-S_R}{|n|v} \left(\frac{(Ad + Cz)(L-h) - Bv}{\sqrt{v + (L-h)^2}} + \frac{(Ad + Cz)(h) + Bv}{\sqrt{v + h^2}} \right) \quad (29.2)$$

where $v = d^2 + z^2$ and $|n| = \sqrt{A^2 + B^2 + C^2}$. In the case of a detector or skin site directly facing the tube, $\vec{n} = -\vec{x}$, this identity reduces to

$$E(d,h) = \frac{S_R}{d} \left(\frac{L-h}{\sqrt{d^2 + (L-h)^2}} + \frac{h}{\sqrt{d^2 + h^2}} \right) \quad (29.3)$$

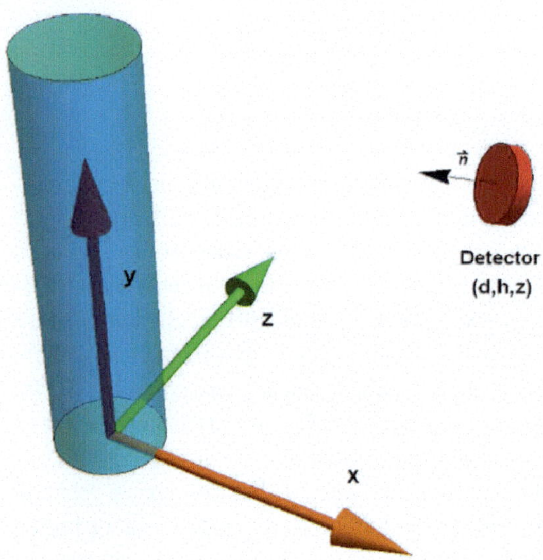

Fig. 29.4 Vector convention for a tube standing at the origin (The detector has surface normal $\vec{n}$ and stands at co-ordinates (*d*, *h*, *z*) in the x, y and z directions respectively)

The derivation of Eq. 29.2 has been left out of this chapter for brevity, but a full treatment can be found in Grimes et al. [34]. The line source model captures irradiance dynamics exceptionally well, typically with errors of less than 1% between simulated and measured values. It is important to note that the angle of incidence modulates the recorded irradiance, and an example of this is shown in Fig. 29.5 for a rotating detector a distance of 248.5 mm from the tube centre.

The line source model predicts that irradiance falls with distance from the tube source, as illustrated in Fig. 29.6a for a detector directly facing the tube. In this instance the fall-off in irradiance obeys an approximately inverse relationship with distance from the source, but it changes when the surface normal between tube and detector is not constant. Figure 29.6b depicts the variation in irradiance along the entire length of a 1.72 m tube.

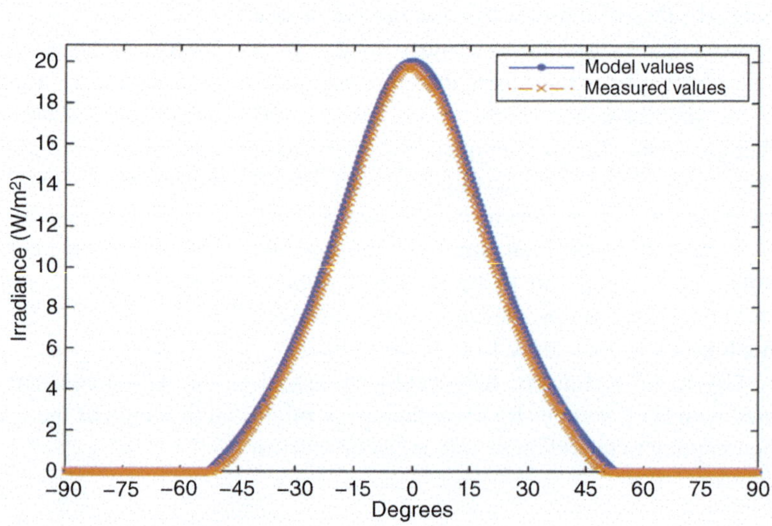

Fig. 29.5 Model/measurement data for rotating detector 248.5 mm from tube (Taken from Grimes et al. 2010 [34])

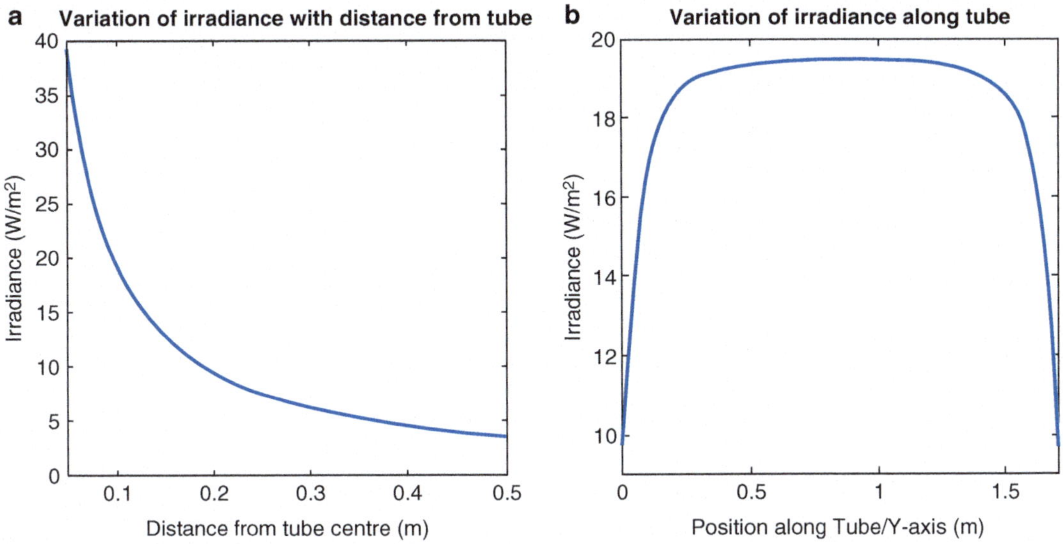

Fig. 29.6 Irradiance profiles for a tube of length $L = 1.72$ m and $S_R = 0.98$ W/m with (**a**) increasing distance from a source at a height of $h = \frac{L}{2}$ (**b**) Irradiance along tube length at a distance of 10 cm from tube centre

29.4.2 Reflection Modelling and Full Cabin Dosimetry

In practice, UVR phototherapy tubes are typically surrounded by shaped and aluminium reflectors. Evaporated aluminium has a reflectivity of up to 0.92 [35] at 311 nm but to protect the reflectors they are typically anodized. This reduces reflectivity markedly, depending on the choice of anodization material. Anodization with Coilzak reduces reflectivity to a low of 0.3, whilst Alanod gives a reflectivity of 0.85. Experimentally, modern cabin reflectors have been determined to have reflectivity $R_f \sim 0.79$ [36]. The positions of the tube images in the reflective planes can be readily determined and treated as a tube reduced by the reflectivity of the mirror. If the tube image at is some new point, we may write $v_o = d_o^2 + z_o^2$ and then for mirrors with a reflectivity of R_f, Eq. 29.2 can be written as

$$E = \frac{S_R R_f}{|n| v_o} \left(\frac{(Ad + Cz)(L-h) - Bv_o}{\sqrt{v_o + (L-h)^2}} + \frac{(Ad + Cz)(h) + Bv_o}{\sqrt{v_o + h^2}} \right) \quad (29.4)$$

This broadly describes the technique for extending the model to cover reflections, but there are a number of complicating factors. Firstly, reflections 'originate' from various mirror points which have to be calculated prior to simulation. This is beyond the scope of this chapter, but a full mathematical treatment is handled in Grimes et al. 2011 [36]. Most importantly, there are zones in the plane where reflections from the aluminium panels cannot reach, and reflections are effectively clipped. This can also be

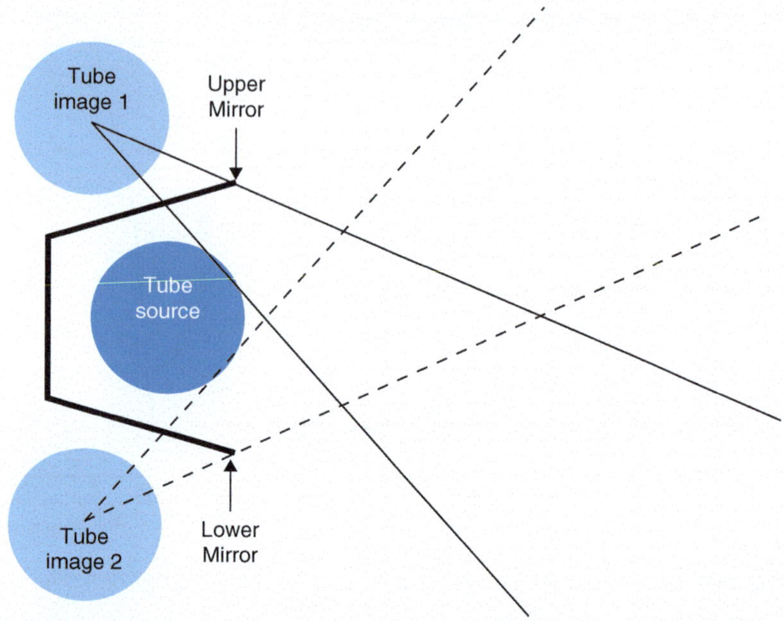

Fig. 29.7 Zones of reflection due to the shaped reflector shown in *black* around the tube source (In the region between the *dotted lines*, reflections from the *upper mirror* contribute. In the region between the *dashed lines*, reflections from the *lower mirror* contribute. There are zones where these overlap and both contribute to total irradiance, and others where mirror clipping reduces reflected irradiance to *zero*)

calculated, and care must be taken to ensure it is taken into account to avoid over-estimates of the reflection contribution. Figure 29.7 depicts a conventional shaped mirror arrangement, depicting zones of reflection from the two tube images formed in the shaped reflector.

The reflection properties of the shaped mirrors found in most UVR cabins has distinct implications for the patient dose received. In particular, the angle which the reflectors sides make relative to the back panel modulates the reflection zones, with smaller angles resulting in less reflected dose [1, 32, 37]. One of the quirks of this is that across cabins with multiple tubes and reflectors, there can be markedly zones discontinuity in irradiance profile due to these mirror properties [37]. While there has been some previous suggestion that parabolic mirrors might alleviate this, but simulation results [38] suggest that this exacerbates cabin heterogeneity and reduces reflection contribution due to the strong absorption of UVR by glass [39]. Cabin dosimetry is also complicated by variations in patient shape and size, and currently there is no set method for readily quantifying all these factors. There is also huge variation in cabin design and reflector options, which mean that in practice methods like ScUVido are currently the best we have at ascertaining dose received by a patient undergoing UVR dosimetry, though future combinations of mathematical modelling and improved measurement would be of substantial benefit in dose quantification.

29.5 Conclusions

Quantification of UVR is essential if adequate dose control is to be achieved, especially when the ailments treated tend to be chronic. However, there are a number of sizeable obstacles in the way of this goal. UV dose is absorbed at skin depth, meaning that angle of incidence is an important factor in determining surface irradiance. Yet both, treatment cabins and patients have complex geometries which render estimating this a non-trivial problem. Protocols like ScUVido are useful in ascertaining an approximate value for a particular build and for ensuring that UV output is approximately constant, and can be readily implemented. If exact tube geometry and reflector arrangement is known, it is possible to accurately predict the radiation incident on a given surface by applying the line source model, but this does not natively take into account the complex shapes and sizes of human patients, nor the potential for patient self-shielding and must be applied to clinical settings with caution. The methods and modalities outlined in this chapter are useful for various applications, but it should be noted that there is still considerable scope for improving how we determine and measure UVR dose to patients.

References

1. Grimes DR (2015) Ultraviolet radiation therapy and UVR dose models. Med Phys 42(1):440
2. Ellinger F (1958) Medical radiation biology. Cancer 11(4):872–872
3. Diffey BL (1980) Ultraviolet radiation physics and the skin. Phys Med Biol 25(3):405
4. Commission de l'Eclairage (1970) International lighting vocabulary
5. Diffey B, Hart G (1997) Ultraviolet and blue-light phototherapy: principles, sources, dosimetry and safety. Institute of Physics and Engineering in Medicine, York
6. Moseley H, Association HP et al (1988) Non-ionising radiation: microwaves, ultraviolet and laser radiation. Adam Hilger Bristol, Philadelphia
7. Diffey BL (2002) What is light? Photodermatol Photoimmunol Photomed 18(2):68–74
8. Murdoch JB (1985) Illumination engineering: from Edison's lamp to the laser. Macmillan, New York
9. Grundmann-Kollmann M, Behrens S, Podda M, Peter RU, Kaufmann R, Kerscher M (1999) Phototherapy for atopic eczema with narrow-band UVB. J Am Acad Dermatol 40(6):995–997
10. Reynolds NJ, Franklin V, Gray JC, Diffey BL, Farr PM (2001) Narrow-band ultraviolet B and broad-band ultraviolet A phototherapy in adult atopic eczema: a randomised controlled trial. Lancet 357(9273):2012–2016
11. Bhatnagar A, Kanwar A, Parsad D, De D (2007) Comparison of systemic PUVA and NB-UVB in the treatment of vitiligo: an open prospective study. J Eur Acad Dermatol Venereol 21(5):638–642
12. Honigsmann H (2008) Polymorphous light eruption. Photodermatol Photoimmunol Photomed 24(3):155–161
13. Ohe S, Danno K, Sasaki H, Isei T, Okamoto H, Horio T (2004) Treatment of acquired perforating dermatosis with narrowband ultraviolet B. J Am Acad Dermatol 50(6):892–894
14. Pavlotsky F, Nathansohn N, Kriger G, Shpiro D, Trau H (2008) Ultraviolet-B treatment for cutaneous lichen planus: our experience with 50 patients. Photodermatol Photoimmunol Photomed 24(2):83–86
15. Wackernagel A, Legat FJ, Hofer A, Quehenberger F, Kerl H, Wolf P (2007) Psoralen plus UVA vs. UVB-311 nm for the treatment of lichen planus. Photodermatol Photoimmunol Photomed 23(1):15–19
16. Diederen PV, van Weelden H, Sanders CJ, Toonstra J, van Vloten WA (2003) Narrowband UVB and psoralen-UVA in the treatment of early-stage mycosis fungoides: a retrospective study. J Am Acad Dermatol 48(2):215–219
17. Parrish JA, Jaenicke KF, Anderson RR (1982) Erythema and melanogenesis action spectra of normal human skin. Photochem Photobiol 36(2):187–191
18. Ribeiro DT, Madzak C, Sarasin A, Mascio PD, Sies H, Menck CFM (1992) Singlet oxygen induced DNA damage and mutagenicity in a singlestranded SV40-based shuttle vector. Photochem Photobiol 55(1):39–45
19. TB F. The validity and practicality of sun-reactive skin types i through vi. Arch Dermatol 1988; 124(6):869–871
20. Sliney DH (1997) Ultraviolet radiation effects upon the eye: problems of dosimetry. Radiat Prot Dosim 72(3–4):197–206
21. Fisher GJ, Wang Z, Datta SC, Varani J, Kang S, Voorhees JJ (1997) Pathophysiology of premature skin aging induced by ultraviolet light. N Engl J Med 337(20):1419–1429
22. de Gruijl FR (1999) Skin cancer and solar {UV} radiation. Eur J Cancer 35(14):2003–2009
23. Fitzpatrick TB (1975) Soleil et peau. J Med Esthet 2(7):33–34
24. Grimes DR (2010) Development of a computational dose model for use in ultraviolet phototherapy, PhD thesis
25. Taylor D, Anstey A, Coleman A, Diffey B, Farr P, Ferguson J et al (2002) Guidelines for dosimetry and calibration in ultraviolet radiation therapy: a report of a British Photodermatology Group workshop. Br J Dermatol 146(5):755–763
26. Pye SD, Martin CJ (2000) A study of the directional response of ultraviolet radiometers: I. Practical evaluation and implications for ultraviolet measurement standards. Phys Med Biol 45(9):2701
27. Coleman AJ, Collins M, Saunders JE (2000) Traceable calibration of ultraviolet meters used with broadband, extended sources. Phys Med Biol 45(1):185
28. Martin C, Pye S (2000) A study of the directional response of ultraviolet radiometers: II. Implications for ultraviolet phototherapy derived from computer simulations. Phys Med Biol 45(9):2713
29. Moseley H (2005) Ultraviolet A dosimetry in photopatch test centres in Europe. J Eur Acad Dermatol Venereol 19(2):187–190
30. Moseley H (2001) Scottish UV dosimetry guidelines, ScUViDo. Photodermatol Photoimmunol Photomed 17(5):230–233
31. Amatiello H, Martin CJ (2006) Ultraviolet phototherapy: review of options for cabin dosimetry and operation. Phys Med Biol 51(2):299
32. Grimes DR, Martin CJ, Phanco G (2012) Investigations of cabin design in UV phototherapy. Med Phys 39(6):3019–3025

33. Currie G, Evans A, Smith D, Martin C, McCalman S, Bilsland D (2001) An automated dosimetry system for testing whole-body ultraviolet phototherapy cabinets. Phys Med Biol 46(2):333
34. Grimes DR, Robbins C, OHare NJ (2010) Dose modeling in ultraviolet phototherapy. Med Phys 37(10):5251–5257
35. Jackson C, Thomas R (1979) The specular reflectivity of bright anodized aluminium. Trans Inst Met Finish 57(3):105–109
36. Grimes DR, Robbins C, Martin CJ, Phanco G, OHare NJ (2011) Reflection modeling in ultraviolet phototherapy. Med Phys 38(7):4312–4320
37. Grimes DR (2012) A computational simulation of reflector and tube effects in ultraviolet phototherapy. Phys Med Biol 57(20):6661
38. Grimes DR (2016) Simulation of parabolic reflectors for ultraviolet phototherapy. Phys Med Biol 61(16):N394
39. Duarte I, Rotter A, Malvestiti A, Silva M (2009) The role of glass as a barrier against the transmission of ultraviolet radiation: an experimental study. Photodermatol Photoimmunol Photomed 25(4):181–184

Index

S. Ahmad (ed.), *Ultraviolet Light in Human Health, Diseases and Environment*, Advances in Experimental Medicine and Biology 996, https://doi.org/10.1007/978-3-319-56017-5

The manufacturer's authorised representative in the EU is Springer Nature Customer Service Centre GmbH, Europaplatz 3, 69115 Heidelberg, Germany. If you have any concerns regarding our products, please contact ProductSafety@springernature.com

Printed and bound by CPI Group (UK) Ltd, Croydon, CR0 4YY
05/10/2026
02228133-0013